D1234773

Nurturing Science-based Ventures

Ralf W. Seifert • Benoît F. Leleux
Christopher L. Tucci

Nurturing Science-based Ventures

An International Case Perspective

 Springer

Ralf W. Seifert, PhD
IMD
Chemin de Bellerive 23
P.O. Box 915
CH-1001 Lausanne
Switzerland

and

EPFL-CDM-TOM
Odyssea
Station 5
CH-1015 Lausanne
Switzerland

Benoît F. Leleux, PhD
IMD
Chemin de Bellerive 23
P.O. Box 915
CH-1001 Lausanne
Switzerland

Christopher L. Tucci, PhD
EPFL – CSI
Odyssea 1.04
Station 5
CH-1015 Lausanne
Switzerland

ISBN 978-1-84628-873-9   e-ISBN 978-1-84628-874-6

DOI 10.1007/978-1-84628-874-6

British Library Cataloguing in Publication Data
A catalogue record for this book is available from the British Library

Library of Congress Control Number: 2007943244

© 2008 Springer-Verlag London Limited

Chatpen™ is a trademark of Sony Ericsson Mobile Communications AB., Nya Vattentornet, SE-22188 Lund, Sweden, http://www.sonyericsson.com/

EASYBAND®, FloWatch® and LAP-BAND® are registered trademarks of Allergan, Inc., P.O. Box 19534, Irvine, CA 92623, USA, http://www.allergan.com/

Expressionist®, Phylosopher® and Screener® are registered trademarks of Genedata AG, Maulbeerstrasse 46, CH-4016 Basel, Switzerland, http://www.genedata.com/

Friskies® is a registered trademark of Société des Produits Nestlé S.A., Avenue Nestlé 55, 1800 Vevey, Switzerland, http://www.nestle.com/

IMC-Hall® is a registered trademark of NV Melexis SA, Rozendaalstraat 12, 8900 Ieper, Belgium, http://www.melexis.com/

io™ is a trademark of Logitech, Inc., 6505 Kaiser Drive, Fremont, CA 94555, USA, http://www.logitech.com/

Post-it® is a registered trademark of 3M, 3M Center, St. Paul, MN 55144-1000, USA, http://www.3m.com/

ThinkPad® and TransNote® are registered trademarks of Lenovo, Raleigh, 1009 Think Place, Morrisville, NC 27560, USA, http://www.lenovo.com/

Venformis™ is a trademark of Venture Valuation AG, Kasernenstrasse 11, CH-8004 Zurich, Switzerland, http://www.venturevaluation.com/

VideoFinish™ (now known as SimulCam™) is a trademark of Dartfish, Rte de la Fonderie 6, CP 53 - 1705 Fribourg 5, Switzerland, http://www.dartfish.com/

VPen™ is a trademark of OTM Technologies Ltd., P.O. Box 4004, Herzliya 46766, Israel, http://www.otmtech.com/

Apart from any fair dealing for the purposes of research or private study, or criticism or review, as permitted under the Copyright, Designs and Patents Act 1988, this publication may only be reproduced, stored or transmitted, in any form or by any means, with the prior permission in writing of the publishers, or in the case of reprographic reproduction in accordance with the terms of licences issued by the Copyright Licensing Agency. Enquiries concerning reproduction outside those terms should be sent to the publishers.

The use of registered names, trademarks, etc. in this publication does not imply, even in the absence of a specific statement, that such names are exempt from the relevant laws and regulations and therefore free for general use.

The publisher makes no representation, express or implied, with regard to the accuracy of the information contained in this book and cannot accept any legal responsibility or liability for any errors or omissions that may be made.

Printed on acid-free paper

9 8 7 6 5 4 3 2 1

springer.com

Praise for Nurturing Science-based Ventures

While timing and luck are important factors for entrepreneurs, they are certainly no substitute for perseverance, hard work, and the ability to spot opportunities before others. Opportunity springs from many sources, not the least of which are breakthroughs in science and engineering. To close this gap is a critical issue, and that is why I am enthusiastic about this book, which I think will help entrepreneurs in their journey. In fact, this book will be a key resource for scientists and engineers looking to take research into development by creating exciting growth ventures.

Dietmar Hopp, Cofounder SAP

Major R&D organizations in Europe and around the world increasingly adopt a venture mindset to manage internal and external technologies within a common innovation pipeline. Therefore, this is a timely and practical book. The work provides case studies of both market-based and corporate-venture initiatives, appropriately bridging the two sources of financing into a common theme. *Nurturing Science-based Ventures* will surely provide an effective teaching guide to the subject of modern science and technology management.

Werner Bauer, Chief Technology Officer, Nestlé SA

While there is no blueprint for success, this book should be compulsory reading for entrepreneurs, ready to embark in the thrilling world of high-tech ventures! With its unique European focus, readers will learn through an impressive set of real world examples. If "the journey is the reward," this book will give readers a host of useful hints on how to "travel" best!

Daniel Borel, Founder and President of Logitech International

University Libraries
Carnegie Mellon University
Pittsburgh, PA 15213-3890

This book will be a practical resource – and an inspiration – to a new breed of global entrepreneurs interested in starting science- and engineering-based businesses. It takes an easy to follow step-by-step approach to starting a company and has a host of stimulating, well-researched international cases that illustrate the key concepts in a clear, succinct fashion.

Mary Tripsas, Harvard Business School

Moving science out of the labs and into the markets has always been one the tallest challenges; managers often disregard the basic science under the products or services, while scientists dismiss the contribution of the business people. Success requires the best of both worlds and the ability to work together over time. This book offers invaluable insights in the difficult Art of transforming science into successful businesses, building on the experiences of various European technology companies. Must read for anyone involved in or considering technology ventures.

Martin Velasco, Business Angel, Chairman & CEO, Speedlingua SA

There is a growing consensus in Europe that its future lies in knowledge, technology and innovation. That is why politicians call so much on scientists and engineers to be entrepreneurial and to transform the results of their research work into commercial success. This book is not only showing that Europe has its success stories and very inspiring ones, it gives above all a methodical view on how to go about setting up a successful science- or technology-based venture. It is recommended reading for all those who want to contribute to overcoming Europe's (in)famous innovation paradox.

Janez Potocnik, Commissioner for Science & Research, European Commission

To all past and future entrepreneurs.
To Laura, Milan and Aiyana – Ralf
To Véronique, Julie and Marine – Benoît
To Carolyn, Enrico, Romano and Matteo – Chris

Preface

Few would deny that small entrepreneurial firms play an important economic and social role. Not only do they generate a significant number of jobs but they also contribute a large proportion of gross national product (GNP). Not all small firms qualify as entrepreneurial entities, however. While "small" refers mostly to size, "entrepreneurial" refers to growth and a value-creation orientation. The vast majority of small firms have no growth aspirations, nor do they have the means and skills to grow. As such, they may still provide employment and local value but would not embrace the high-potential aspirations of entrepreneurial ventures. This book clearly addresses those entrepreneurs who are interested in leading high-growth-potential companies (*Table 1*).

Table 1 Growth Typology of Small Firms [1]

Type of venture	Desired sales range	Future employees
Lifestyle	0 to $1 million	0 to 4
Smaller high potential	$1 million to $20 million	5 to 50
High potential	over $20 million	Over 50

High-innovation technology-based startups assume a very special role in high-growth entrepreneurship. Although these startups constitute a comparatively low number of small businesses, they produce proportionately far more jobs than their low- and medium-innovation counterparts. The aim of achieving rapid growth is typically referred to as *high-expectation entrepreneurship*. An area of major concern to us is a fact revealed in the latest GEM[1] report: The rate of European high-expectation entrepreneurial activity is among the lowest in the world. In light of this, we pursue several goals with this book. First, we wish to demonstrate that there are, in fact, highly promising high-growth-potential ventures in Europe.

---

[1] GEM – the Global Entrepreneurship Monitor – was created in 1997 to study the economic impact and determinants of national-level entrepreneurial activity across 44 nations. GEM operates as a not-for-profit international academic research consortium.

Second, we feel that business and management schools need to do a better job of sharing the many hard-earned lessons across such ventures to help guide their development. Finally, we hope that our examples may help inspire future entrepreneurs to take on the challenge and to follow suit – equaling or even exceeding the achievements presented in our cases.

Why This Book on Science-based Entrepreneurship?

There has been a burgeoning interest in high-tech, high-potential entrepreneurship, much of it directed toward Internet-type businesses. A considerable share of high-technology entrepreneurship is, however, contributed by startups in the realm of science. This can, of course, include specific Internet businesses, but there is much more emphasis on commercializing research from disciplines in engineering and the natural sciences.

This book focuses mostly on ventures stemming from technological breakthroughs rather than from ongoing incremental product innovations or modifications. The companies we feature have established their business around ideas and findings from fields such as biotechnology, biomedicine, micro- and nanotechnology, and information science. Many of the companies are located in science parks close to universities or research centers or have capitalized on research projects supported by the resources of these institutes.

This interface between the generators of fundamental research-based knowledge (universities and research centers) and the marketplace is a complex and poorly understood one, in which Europe has indeed suffered from a much delayed reaction time. With the majority of research centers financed with public money, and the academic community's disregard for business applications, there was little incentive to try to market what was created in the labs. Researchers were mostly state employees, with no interest in (or sometimes even prohibited from) considering potential business applications and pursuing them. When public financing started to dry up, research institutions were forced to consider secondary sources of funding, in particular finding ways to capitalize on their knowledge and intellectual property. Many leading-edge institutions initiated "commercialization" efforts, actively promoting the dissemination of their research and the generation of streams of royalties. What was begun out of dire necessity proved to offer them a new level of autonomy vis-à-vis their public backers to pursue research efforts that otherwise might not have received official blessing. In other words, earlier financial motives have led the way to a warmer and deeper embracing of the need to collaborate effectively with the business world.

We wanted to create this book because science- and technology-based startups account for a disproportionate share of major, radical innovations. Frequently, they lead their industry in adapting to changing circumstances, often far ahead of larger, more established firms. These high-tech ventures assume a critical role in the "creative destruction" process underlying a flexible, dynamic economy [2].

However, their position is usually precarious, carrying high levels of risk that are inherent in these disruptive conditions. Managing the extremely complex and uncertain process of discovering and developing new technologies and markets at the same time is the focal point of our cases. They show that a great many factors need to come together to create a successful technology venture. A first-rate idea and superior technology do not suffice. Engaged and experienced management teams, a solid network of partners, proper support infrastructures, qualified employees, and the like are just as critical. Professional management, in particular, can raise the odds of success considerably, as many of the entrepreneurs we talked with readily confirmed:

> It is hard to tell which exactly sets you apart, whether technology or management. But I have seen so many companies run into difficulties along the way – and it was always good management that got the company out of danger.
>
> *Dr Andrea Pfeifer, CEO, AC Immune, Swiss biotech startup*

In addition, science-based ventures typically consume plenty of resources, and it is common knowledge that resources are one of the biggest constraints, if not the single biggest constraint, in the entrepreneurial process. We cannot guarantee that by reading this book the would-be entrepreneur will overcome these constraints and suddenly draw level with the Yahoo!s of this world, but our cases and the related material illustrate the challenges ahead and help pave the way for mastering them properly.

Supporting European Technology Entrepreneurs

It is precisely this link between technology and management that is the focus of this book. It covers in depth the challenges and rewards of new venture creation based on scientific or engineering products. Our collective experience in researching, teaching, and consulting in the field of technology-based ventures has resulted in material designed specifically, but not exclusively, for technically trained students and managers. Guided by our teaching practice in many different graduate and executive teaching programs at renowned business schools in the USA and Europe, we have put together a set of current and effective case material that highlights what technology entrepreneurs face when building their ventures.

The book primarily targets engineering and science students who wish to learn about and actively engage in entrepreneurial activities by creating their own company, by working for a startup, or by spurring innovation within a larger organization. The number of educational programs targeting these students has increased significantly over the last few years. But we believe that the majority of these programs fail to address the particularities of creating viable businesses based on science and engineering. Motivating engineers and scientists to become entrepreneurial has particular appeal in view of the abundance of technological knowledge hidden in today's research institutes and technology-based enterprises waiting for commercialization. Furthermore, with the case studies in this book, we hope to foster a general appreciation of technological venturing among current and future business leaders.

We do all this with an intentionally European perspective. While we have all marveled at the many quasiproverbial and highly publicized "garage to riches" Silicon Valley stories, the reality is that the majority of startups take very different routes to success, and success itself is often much less spectacular, rarely making front-page news. Nonetheless, they can be exciting journeys in their own way and contain invaluable learning regardless of whether the venture is ultimately successful. The ups and downs, the failures and mistakes, the rollercoaster ride from exhilaration at early market successes to depression at the viciousness of competitive reactions, although less publicized, might at times be closer to most entrepreneurial realities. To capture and preserve such learning – rather than it being lost during the termination or acquisition of a startup company – is an important objective of this book. Finally, while the examples and principles in the book are applicable globally, we wanted to bring forward examples from the science and engineering space in the specific context of Europe. Even though it was slow in making a concerted effort in this field, Europe is now fighting back to create a more entrepreneurial culture and exploit its research infrastructure.

In 2007, seven years after the European Council put forward its Lisbon Agenda to transform the EU into "the most competitive and dynamic knowledge-based economy in the world," Europe still does not fully exploit its entrepreneurial potential. One might debate whether perpetual comparison with other economies helps, but in fact the natural benchmark – the USA – is without a doubt the most successful country when it comes to nurturing startups and entrepreneurial spirit. Using this as a point of reference, we find that Europe indeed seems to lag behind. Let us discuss some statistics on the gap that Europe is attempting to fill.

When it comes to general entrepreneurial intent, only 45% of Europeans would ever like to be their own boss, as opposed to 61% of Americans, who are keen to try entrepreneurship [3]. Likewise, the effective total entrepreneurial activity, *i.e.,* the number of people actually involved in nascent startups and companies younger than 3.5 years, is double the rate in the USA (12%) than it is in Europe (6%).[2] In addition, a survey on attitudes toward entrepreneurship estimated that 26% of Americans would refrain from starting a business because of the risk to income and career. Double that figure and you will get closer to the European view, where 46% hesitate to start a business because of the risk factor [4]. This is also mirrored in the motivation factor to start a new venture: according to the Eurobarometer on Entrepreneurship, a regular income is the main motivation for those Europeans who prefer employee status – 30% of respondents cited this reason compared with only 16% of Americans [5]. However, what gives particular cause for concern is the

---

[2] The European figure is a mean of all European countries as reported in: Minniti, Maria, William D. Bygrave, and Erkko Autio. *Global Entrepreneurship Monitor: 2005 Executive Report.* GEM, 2005.

lack of entrepreneurial talent with high ambition. As the 2005 GEM report on high-expectation entrepreneurship reveals, in the USA 1.6% of the adult population has the intent to employ more than 20 people within a time horizon of five years; this is three times as high as in Europe [6].

The perennial question is, Why do these differences exist? While it is unlikely that we will be able to provide sufficient answers in this book, we hope to provide ample evidence that the European startup scene is very much alive. By making this known, we hope to contribute to a rising awareness and confidence that successes are reproducible – even in "the Old World." A lack of examples is what many sources state as a distinguishing factor between Europe and the USA. Thus, it boils down to educating people, which is one of the primary means of influencing positive attitudes.

For a long time, the educational process has been identified as a crucial ingredient in encouraging entrepreneurship and shaping attitudes, skills, and culture. While the USA offered its first entrepreneurship classes as early as the 1930s at Harvard University, Europe's research institutes and universities only started doing so in the 1990s, with most research centers only founded after 2000 [7]. Ongoing efforts are therefore under way to spur entrepreneurial education on the European side of the Atlantic. Yet, better integration and sharing – and in particular more public-private cooperation to foster the exchange of knowledge and examples – is needed, as a recent conference on entrepreneurship education in Oslo in the fall of 2006 revealed [8]. Europe still needs to catch up but is attempting to do so with full vigor, which will surely bear fruit in the years to come.

Entrepreneurship Clusters: Providing Infrastructure, Critical Mass, and Example

Startups tend to locate in clusters [9]. This is logical when one considers the power that examples can exert on potential entrepreneurs, the infrastructure (including human and financial resource availability) that can be exploited, and the atmosphere that propels entrepreneurial intent. To try to re-create the dynamism of Silicon Valley and the greater Boston area, Europe is actively promoting the development of entrepreneurial "clusters." So where are these clusters? Several regions come to mind, for example London and Cambridge (UK), Dublin (Ireland), Munich (Germany), Stockholm (Sweden), Zurich (Switzerland), and Sophia Antipolis (southern France). However, depending on which expert one asks, they will likely think of even more clusters, presumably in their home country. There is no such thing as *the* commonly agreed-on European startup cluster. So far, no place has reached the critical mass of ventures to be the single magnet for entrepreneurs. One of the problems with these European clusters is that they seem to be without either internationally known pillar companies – like HP, Intel, or Xerox in Silicon Valley – that would serve as huge success stories or internationally known research institutes, such as MIT in Boston. While we cannot easily conjure up these institutions, active policy and cross-country coordination in Europe might allow for the development of a few regions as potential clusters in this vein.

Legal and Administrative Systems: Harmonizing for a Larger Market

Europe is still highly fragmented compared with the USA, which has about 300 million people living and working in a relatively cohesive administrative framework, speaking one language, and sharing roughly the same culture. Yes, Europe has managed to harmonize many regulations and to introduce the euro – which has become a well-accepted common currency – and many borders have effectively been abolished. However, the euro is used only in parts of the EU and we still face the challenges of having different languages and many different national legal systems, which certainly hamper the much needed mobility of resources. There is no easy way around this if Europeans want to preserve what they are so proud of – diversity. It is important to acknowledge this difference, try to harmonize where possible, but also see it as a promoter of a unique entrepreneurial culture. The examples in this book show that many startups face the challenge of internationalization very early in their development – something their American peers can wait much longer to do. This does not need to be a disadvantage for the Europeans. Indeed, quite the opposite – thinking internationally from the outset can make for much bigger opportunities.

In addition, recent research shows that, at least for venture-capital-backed companies, the differences in the legal framework are less influential than commonly assumed. The bigger difference seems to be made by the level of sophistication of financiers, in particular venture capitalists (VCs) [10].

Access to Venture Capital: Improving Europe's Biggest Infrastructure Challenge

All startups need capital, and many of the startups we talk about in this book need vast amounts of money, which can rarely come from the usual friends and family or out of early revenues. They need venture capital or private equity money, which has historically been less available in Europe (*Figure 1*).

For a long time, Europe's financiers have been less active and also less effective than their US counterparts [11]. Studies show that US VCs generate significantly higher returns and tend to be more active in their investments, insofar as they invest on average twice as much in their portfolio companies, tend to apply stronger control and contingency financing, and organize themselves in larger syndicates [12]. However, on both sides of the Atlantic VC activities suffered from a severe downturn at the start of the new millennium and only showed signs of revival in 2005. Totaling US$25.5 billion in 2006, US VC investments have, for example, reached a five-year high. Europe, by contrast, still offers a mixed

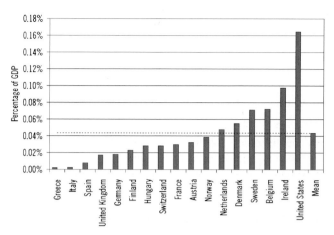

Figure 1 Classic High-Tech Venture Capital Investment in Europe and the USA as a Percentage of GDP in 2004
Source: Minniti, Maria, William D. Bygrave, and Erkko Autio. *Global Entrepreneurship Monitor: 2005 Executive Report.* GEM, 2005.

perspective: while investments in eastern Europe are predicted to make a splash of 59% growth in 2007, western Europe is expected to slow down a bit [13].

Still, we must keep in mind that VC investments are suited to the big market opportunities for startups with a defendable competitive advantage that grow rapidly and thus demand external capital. Historically, European startup founders have been less concerned with growth. They tend to build their companies more slowly and to less ambitious sizes, thus requiring smaller amounts of money. Yet Europe has also lacked a culture of smaller investments beyond those from family and friends. Awareness of microloans, for example, has only recently started growing. Furthermore, Europe has long suffered from a significant lack of a business angel culture because it did not have that many people who had made their wealth through business rather than inheriting it. The former would typically want to give something back, but their numbers are expected to grow relatively slowly in Europe.

Key Learning Points

Often we are asked how to realize greater value from the investment and effort going into nurturing science and technology entrepreneurs, in particular in terms of generating more growth for the economy. This interest arises from many different stakeholders, including policymakers, educational institutions, and the entrepreneurs themselves. While no two ventures are the same, many share common challenges that need to be mastered. These challenges lie at the heart of this book. In our fieldwork, we have encountered and documented many ventures, some clearly successful and others less so. A common denominator from the successful cases, though, is the ability to exploit technology to create unique and high-value solutions for customers. Even though the stars are likely to operate in highly attractive markets and industries, we also provide ample evidence that there are niches in less attractive environments, which creative and innovative entrepreneurs take advantage of. But how do they do this? Let us share with you some of our findings:

- It is all about markets and customers. Identify them properly and listen to your customers; they know about the problems *and* the solutions. Get them involved early and frequently during the development of the new technology.
- Provide early proof of concept. Prototypes talk better than long stories and also help you pinpoint major flaws rapidly.
- Get exposure quickly and repeatedly. This will not only help you refine the product or service, but the feedback may likewise reveal deadly flaws in your project. Early warnings are better than later ones.
- Adoption is a slow process, even more so in fields based on technical sales. Many parties, including layer after layer of engineers, will be involved, and each will take its time. You are in a hurry; they are not. Your rush is *not* their concern. Actually, making you wait is just a test of your staying power, *i.e.*, your ultimate business credibility. It makes sense for them to have you wait. Your sales cycles will be *very* long, and you will have to finance them.
- There is no such thing as overfocusing. You really need to fine-tune your product/market mix. You do not have the time or the money to pursue broadly defined markets, even though your product could probably serve all of them. Unfortunately, startups often die of opportunity overload, not lack of opportunity. Strategy is about choice, and making choices when you have been bootstrapping for years, as is often the case with startups, is even more difficult. It is nevertheless a *sine qua non* condition for success.
- Greed kills with great certainty. To succeed, you need the help of many parties. Never underestimate their contributions and make sure they are aggressively co-opted into your project. Management and technologists are equally important: make sure to recognize both. You will leave a lot of money on the table, but that is indeed the cost of building a sizable company.
- As a startup, even with the best idea or technology, you have zero credibility with potential customers. They *do not* need your solution. They *do not* want to deal with startups. They actually *hate* startups in general. Remember, you are a threat to them, and your ways and means are totally foreign to them. You will have to convince them that you can add value to them and that you are a credible partner.
- Building credibility often involves your choice of financial partners. Convincing a top-notch VC, a specialist in early-stage, high-potential, technology-based ventures, for example, to invest in your firm can make the difference between obscurity and a seat at the table.

- Financing in growth-oriented companies is *not* a one-shot process. You will be back at it repeatedly, exploiting each milestone reached in the project to raise the next tranche (round) of money for your venture. Who you bring in to finance the project says more about you and the potential of the venture than the amount of money involved. Financing is the ultimate signaling device. As the saying goes, "Money talks."
- Never underestimate the value of a great management team. Technology does not sell itself – it never did and never will. You will need the best managers with deep, industry-specific know-how on board.
- The plan will only be valid for as long as it takes for the ink to dry on the document. What you need to develop and build into the venture is the ability to respond instantly and permanently to evolving market conditions. You will learn every day, and every day will bring its new challenges. Nothing is set in stone, and you can expect your competitors to wreck the best laid plans rapidly.
- You do not win because of the best plans or because of the best management team. You win because your competitors let you win. Don't expect this to happen too often. So luck is indeed important. But luck tends to favor the prepared mind.

Organization and Features of This Book

This book has primarily been designed for teaching purposes, *i.e.*, for class discussion in the field of entrepreneurial studies, and is thus geared toward lecturers and students. However, it will also be helpful reading for self-learners and prospective entrepreneurs, private investors, or VCs interested in the subject.

Each case in the collection has been written to stimulate discussion and to allow the student to develop an understanding of the processes behind creating a new company. For those who specifically plan to launch their own business or are in the midst of the process already, the case collection will present a valuable source of ideas and opportunities to evaluate their own situation. Those who, by contrast, are not actively seeking to become involved in a startup might still learn to value the entrepreneurial mindset and apply certain concepts to their own mode of working. For example, as large corporations increasingly adopt some of the concepts of open innovation – for example, sourcing innovation from outside and forging alliances with small companies – we hope this book may also help readers to more fully understand and appreciate the environment and constraints of startup companies so that such collaborations can become more successful. To reflect that entrepreneurial activity indeed does not only occur in independent new ventures but also in large corporations, the book also includes a set of cases dedicated to corporate venturing. They might be of particular interest to managers in the R&D function of larger organizations who are increasingly bringing on board concepts and valuation models that VCs apply when assessing pure play startups.

Target Market

In essence, our intended audience is:

1. Technically trained managers from technology- or knowledge-based industries, some of whom may actively be considering starting their own company. A manager in such a situation, perceiving an opportunity but without much business management experience in a small company, will find the text helpful in laying out the steps necessary to take the opportunity to the next level.
2. Graduate students in engineering and sciences with almost no prior practical training in business or management. These students fall into two groups. The first group is doctoral student researchers who have developed some technology they would like to commercialize or spin off from their university. The second group is master's level students in science or engineering who would like to take elective classes in entrepreneurship or are motivated to learn more about the topic by themselves.
3. Graduate business school students thinking of starting a high-tech company. There are several different contexts in which this book may be considered within business schools. The first is as a standalone elective in technology entrepreneurship as part of a larger entrepreneurship program. The second is as a module in a "traditional" entrepreneurship course where the instructor would like to pay attention to science- and technology-based ventures. The third is as a module in a course on technology commercialization from universities. The fourth is as a module in a course on corporate venturing, technology strategy, or corporate renewal.
4. Graduate students in public policy or studying university-industry relations. They will find useful examples of how firms in regions that were traditionally less viable in entrepreneurial activity have lately begun to show great promise.
5. Managers inside large technology companies who are responsible for inculcating a more entrepreneurial spirit or who may need to evaluate external technology ventures.

The Case Study Approach

The book features close to *30 individual real-world ventures*, many of which we will accompany for a period of time, through a series of case studies (which brings the total of cases to more than 40). All but two cases are based on original field research supported by the respective companies and entrepreneurs – reporting actual events and management dilemmas. They give examples of successful and less successful startups and thus provide a complete and authentic learning experience. The majority of the cases have been successfully used in undergraduate, graduate, or executive education programs at IMD and EPFL. Some have already won case awards, for example, *The "mi adidas" Mass Customization Initiative*

(2006 ECCH Case Award, 2004 POMS International Case Writing Award); *Innovation and Renovation: The Nespresso Story* (2003 efmd Case Writing Competition); *Optima Environnement S.A.: Turning a Wonder Tree into an Eco-business* (2006 efmd Case Writing Competition, Special category); and *Boblbee (D): The Urban Backpack* and *Boblbee (E): Inventing the Urban Nomad* (2006 efmd Case Writing Competition, Entrepreneurship category).

Case studies are an indispensable tool for modern education as they permit interactive learning based on rich, complex, and authentic situations. From our long-standing experience, using cases allows for lasting impact as the students actively participate in the analysis and solution of practical problems. The exchange with both teachers and class participants can compound the group's knowledge and create a highly dynamic learning environment. All our cases are written in a language and style that not only academics but also managers appreciate and understand. Whenever possible, they include reflections and direct quotes from the entrepreneurs themselves. And while none of the presented cases individually aspires to holistically span all topics prevalent in entrepreneurship, taken together we hope they paint a true picture!

Book Outline

To create a comprehensive picture of entrepreneurial activities, the book is arranged in five chapters tracking the typical life cycle of creating and growing a new venture. This includes recognizing and evaluating opportunities, creating viable business plans, securing financial resources, managing growth, and, eventually, harvesting the created value. In the sixth and final chapter, we also account for the view of large firms by including studies on corporate entrepreneurship and the integration of internal and external knowledge to successfully seize new business opportunities. Each chapter consists of a general topic raiser introducing the key learning points and giving a brief rundown of the case studies included in that chapter to highlight particular aspects. The rest of the chapter consists of the case studies themselves. The case mix includes both brief "issue raisers," focusing on one specific problem, and more substantial cases, which can easily expand over a teaching block or two for their richness and complexity.

A schematic overview of the book outline and the corresponding cases is given in *Exhibit 1*.

We begin where each venture starts – with a good idea that shows the firm's business potential, *i.e.*, an actual opportunity. In fact, it is imperative for the aspiring entrepreneur to understand the difference between a good idea and a real business opportunity. This is the focus of *Chapter 1: Opportunity Recognition*. Here we discuss what it is that defines a promising business opportunity and how to check for it before sinking resources into it. Quickly separating the outstanding opportunities from the rest is crucial, and this ability usually distinguishes more experienced entrepreneurs from less experienced ones. Furthermore, for the inexperienced or would-be entrepreneur, business opportunities often emerge like a miracle. We show that this is typically a misunderstanding: ideas might appear to pop up like mushrooms after a rain storm, but there is indeed a process to shaping business concepts that will have a chance to thrive. Also, we discuss some prototypical arbitrage situations from which promising concepts can emerge, for example, when industries converge or macroeconomic trends shift, and the like. Four cases to illustrate possible processes when shaping an entrepreneurial business opportunity complete the chapter.

Quickly extracting those ideas with star potential is just one step in the long journey to a new venture. The next phase directs the entrepreneurial activity toward more detail and focused planning activity. Of course, there is some ambiguity in planning and there is not a "one-size-fits-all" approach. Despite all the planning, successful venturing remains a stochastic and iterative process, requiring dedication, perseverance, and the ability to learn. An essential ingredient in any business is a rock-solid business plan, and this is the focal subject of *Chapter 2: Planning the New Venture*. The chapter covers the contents and the "how-to" of a business plan in detail. We highlight particular aspects that are key for success, such as the new venture team and the marketing strategy, including competitive advantage and positioning of the venture. The nine cases included here deal with various aspects of the business plan, from drawing it up for the first time to revising it repeatedly – an essential process, as cases like the VistaPapers series show.

The aspiring entrepreneur will realize that a fundamental characteristic of many entrepreneurial endeavors is the imbalance between the resources currently in their possession and those needed to capitalize on opportunities. *Chapter 3: Venture Financing* therefore focuses on how to develop a proper financing strategy for early-stage, higher-risk ventures, which investors to target, and how to develop a compelling investment case. Notably, few of these ventures truly face pure financing problems but more complex "resourcing" issues, *i.e.*, how to get access to the extensive collection of resources needed to succeed, such as management skills, distribution channels, networks, technology, and the like. In many instances, money will indeed provide access to those resources. However, attracting investors is purportedly the most challenging task for small organizations, in particular if the entrepreneur is inexperienced and without a track record demonstrating his or her ability to generate the returns commensurate with the risks involved in new ventures. In situations of funding constraints, where access to finance is limited or associated with huge costs, it becomes critical to use the fundraising exercise in a more creative manner, as a holistic approach to gaining resources for the firm. In addition to the case studies that portray the creativity necessary in entrepreneurial finance, we include a comprehensive technical note focusing on the venture capital contracting process. It provides an overview of the main contractual features of venture capital contracts, explaining their *raison d'être* and their theoretical grounding and rationale.

Assembling a skilled team, writing a compelling business plan, and securing the initial financing are just the first steps in the long journey to a viable venture. As *Chapter 4: Growing the Entrepreneurial Firm: Building Lasting Success* shows, new challenges and requirements await the aspiring entrepreneur along the way: in fact, only one in ten firms can measure up to the challenges of growth and manage to employ more than ten people. Many firms fade away in their first two years, without ever having experienced "good times." Most of them perish due to a lack of resources, legitimacy, and coordination (*liabilities of newness* and *smallness*). So, how do the more successful companies actually manage to acquire legitimacy, status, and reputation? What distinguishes them from the less successful ones? There are a myriad of factors making a firm successful – from leadership and management qualities to well-thought-out growth strategies – as portrayed in the case collection of Chapter 4.

For all the time, money, and personal energy the entrepreneur and other stakeholders invest in building a venture, they certainly expect adequate rewards at some point in the future. The process of turning some of the value created into hard cash for the investors is referred to as "harvesting," and various strategies for this are the focus of *Chapter 5: Harvesting*. Often, harvesting is not at all important in the entrepreneur's considerations, especially if the business is more a lifestyle affair, only required to generate a living for the founder. By contrast, the types of ventures we showcase in this book are unlikely to be run "as is" forever. They have a significant number of outside investors, who may be looking for a return on their money and value they helped create. Defining harvesting goals and crafting a strategy to achieve them, however, goes well beyond satisfying stakeholder interests. It also keeps up motivation and helps stay focused beyond the next quarterly results. In fact, harvesting should not be conceived of as a terminal activity or the end of the venturing process. Rather, it is the necessary step toward recycling the entrepreneurial talent and capital, offering the opportunity to both external investors and the entrepreneur to pursue other plans and ventures. Having a clear, ambitious exit target, such as an initial public offering (IPO), can keep all parties focused and driven to a commonly satisfactory outcome. Thus the cases we have chosen for Chapter 5 demonstrate that harvesting should be seen as another of the fundamental steps in the entrepreneurial process, requiring full preparation and management.

While Chapters 1 to 5 talk about the creation of new businesses as independent startups, *Chapter 6: Corporate Entrepreneurship* throws light on another facet of entrepreneurial spirit. Today, many well-established and large companies engage in entrepreneurial activity, *i.e.*, corporate entrepreneurship (CE). CE may be broadly viewed as (1) the creation of new businesses within existing organizations, either through internal innovation or joint ventures and alliances, or (2) the transformation of organizations through strategic renewal [14]. Large firms embark on it for various reasons and with various approaches. Mostly, the aim is to rejuvenate a firm, create value, and sustain competitive advantage. We will explore the different strategies pursued in this regard in the cases in this chapter, which include examples from five prominent companies.

Exhibit 1
Building Blocks of this Book

1	2	3	4	5	6
Opportunity Recognition	**Planning the New Venture**	**Venture Financing**	**Growing the Entrepreneurial Firm: Building Lasting Success**	**Harvesting**	**Corporate Entrepreneurship**
Topic lead-in and case introduction	Topic lead-in and case introduction	Topic lead-in and case introduction	Topic lead-in and case introduction	Topic lead-in and case introduction	Topic lead-in and case introduction
Optima	Shockfish (B)	VC Primer	IR Microsystems (B)	EndoArt (B)	Ducati (A)
Cognosense	IR Micro-systems (A)	AVIQ Systems	IR Microsystems (C)	Sentron (A)	Ducati (B)
Shockfish (A)	InMotion Technologies Ltd	Tatis Limited	AC Immune	Sentron (B)	Logitech – Culture
Redigo	Boblebee (D)	Genedata	Covalys (A)	4M Technologies	Logitech – Digital Pen
	VistaPapers (A)	Venture Valuation	Dartfish	Generics	mi adidas
	VistaPapers (B)	Novartis Venture Fund	Netcetera (A)	GigaTera	Tetra Recart
	VistaPapers (B-2)		Netcetera (B)		Nespresso
	VistaPapers (C)		Netcetera (C)		
	Lyncée Tec		Boblebee (E)		
			EndoArt (A)		
			Google		

Exhibit 1 (continued)
Case List

Page	Case Name (abbreviated)	Business Focus	Topics
1. Opportunity Identification and Evaluation			
11	Optima Environnement	Commercialization of raw material (plants)	Opportunity assessment; finding a business model
27	Cognosense	Applications for mobile content services	Opportunity assessment, business model, and positioning
47	Shockfish (A)	Conference networking systems	Evaluation of a business opportunity
53	Redigo	Environmentally friendly vehicle engineering	Evaluating and defining the relaunch of a product
2. Business Planning: Defining Entry Strategies, Protecting Competitive Advantage and Dealing with Uncertainty			
107	Shockfish (B)	Conference networking systems	Shaping the business
111	IR Microsystems (A)	Infrared spectrometer: devices and applications	Market validation
131	InMotion Technologies	Sports training and entertainment applications	Definition of business model, marketing mix
143	Boblbee (D)	High-tech backpack design and manufacture	Defining the target clientele, integrated marketing strategy
167	VistaPapers (A)	E-printing	Developing a business plan
179	VistaPapers (B)	E-printing	Developing a business plan
187	VistaPapers (B-2)	E-printing	Revising the business plan
189	VistaPapers (C) Vistaprint	E-printing	Revamping the business model – reacting to external change
199	Lyncée Tec SA	Holographic microscopy: devices and applications	Drawing up a viable business plan
3. Financing New Ventures			
241	VC Primer	Overview of venture capital market/structures	Roles and key clauses in investment contracts
295	AVIQ Systems	Integrated audio and visual entertainment devices	Assessing startup financing alternatives/financing growth
311	Tatis Limited	Business-to-government IT solution	Finding the right (financial) partners
323	Genedata	Bioinformatics, analytical software	Preparing for the valuation
333	Venture Valuation AG	Venture valuation	Early-phase venture evaluations
355	Novartis Venture Fund	Pharma/biotechnology	Evaluating an investment

Exhibit 1 (continued)

Page	Case Name (abbreviated)	Business Focus	Topics
4. Managing Growth: Redefining Business Strategies			
383	IR Microsystems (B)	IR spectrometers	Back to the drawing board – revising business goals
395	IR Microsystems (C)	IR spectrometers	Moving forward – refocusing resources and extending development partnerships
397	AC Immune SA	Biotech – research into Alzheimer's cures	Going from research to development and managing options
413	AC Immune SA Handout	Biotech – research into Alzheimer's cures	Strategic alliance
415	Covalys (A)	Biotech – components	Revving up the business
429	Covalys Handout	Biotech – components	Expansion plans
431	Dartfish	Sports training and entertainment applications	Redefining the business model/product portfolio
443	Netcetera (A)	Enterprise information applications	Growth issues: hire an external CEO?
453	Netcetera (B)	Enterprise information applications	Restructuring the organization for sustained growth
463	Netcetera (C)	Enterprise information applications	Strategic growth options
465	Boblbee (E)	High-tech backpack design and manufacture	Building up a brand name
491	EndoArt (A)	Medical devices	Defining the right application to make the firm profitable
505	Google	Internet information search	Revising strategy and business model
5. Opportunities to Harvest the Value and Exit Strategies			
535	EndoArt (B)	Medical devices	Defining the right application for making the firm profitable
537	EndoArt Handout	Medical devices	The acquisition
541	Sentron (A)	Magnetic sensor electronics	Evaluating the future of the firm and potential exits
555	Sentron (B)	Magnetic sensor electronics	Evaluating exit decision
559	4M Technologies	Optical disk integrated manufacturing systems	Evaluating IPO options
579	Generics	Technological advisory services, incubator	Startup incubation; startup investment; hybrid business model
599	GigaTera	Manufacturing of optical network component	Evaluating continuation of company

Exhibit 1 (continued)

Page	Case Name (abbreviated)	Business Focus	Topics
6. Corporate Entrepreneurship: Rejuvenating Business			
619	Ducati (A)	Motorcycle manufacturer	Rebuilding and strategically repositioning brand
635	Ducati (B)	Motorcycle manufacturer	Rebuilding and strategically repositioning brand
645	Logitech – Creating Culture	IT, computer accessories, and devices	Spurring internal innovative capability
669	Logitech – Digital Pen	IT, computer accessories, and devices	Launching a new product on the market
697	mi adidas	Sporting goods	Developing and launching a new product internally
723	Tetra Recart	Carton solutions for food and liquid storage	Developing and launching a new product in a newly created venture
753	Nespresso	Lifestyle coffee concept	Building up a new market and creating a brand

References

1   Baird, Michael L. *Starting a High-Tech Company.* New York: Institute of Electrical and Electronics Engineer, 1995, p. 11.
2   Spencer, Aron S. and Bruce A. Kirchhoff. "Schumpeter and New Technology Based Firms: Towards a Framework for How NTBFs Cause Creative Destruction." *The International Entrepreneurship and Management Journal,* 2006, **2** (2), pp. 145–156.
3   "Are Entrepreneurs Born, Made, or Just Encouraged?" European Commission, 2005. http://ec.europa.eu/enterprise/enterprise_policy/survey/eurobarometer83.htm (accessed April 17, 2007).
4   "Giving Ideas Wings: Business Angels." *The Economist,* September 14, 2006. [US Edition.]
5   "Are Entrepreneurs Born, Made, or Just Encouraged?"
6   Autio, Ekko. *Global Entrepreneurship Monitor: 2005 Report on High-Expectation Entrepreneurship.* GEM, 2005.
7   Wilson, Karen. *Entrepreneurship Education at European Universities and Business Schools.* European Foundation for Entrepreneurship Research (EFER), 2006.
8   *Entrepreneurship Education in Europe: Fostering Entrepreneurial Mindsets through Education and Learning.* European Commission, 2006.
9   Graham, Paul. "Why Startups Condense in America." 2006. http://www.paulgraham.com/america.html (accessed April, 16, 2007).
10 Hege, Ulrich, Frédéric Palomino, and Armin Schwienbacher. "Venture Capital Performance: The Disparity between Europe and the United States." HEC Paris, Amsterdam Business School, Working Paper, 2006.
11 Leleux, Benoît. "Venture Capital and Private Equity: A Status Report." Workshop on Financing Science-based Start-ups, Lausanne, Ecole Polytechnique Fédérale de Lausanne. June 13, 2006.
12 Hege *et al.* "Venture Capital Performance."
13 National Venture Capital Association (NVCA), "Venture Capitalists Bullish on 2007." 2006. www.nvca.org/pdf/07nvcapredictions.pdf (accessed April 16, 2007).
14 Dess, Gregory G., G. T. Lumpkin, and Jeffrey E. McGee. "Linking Corporate Entrepreneurship to Strategy, Structure, and Process: Suggested Research Directions." *Entrepreneurship: Theory & Practice,* 1999, *23* (3) pp. 85–102.

Acknowledgments

Coordinating the work of three busy academics across two dynamic institutions (IMD and EPFL) is a difficult enough task. While these three individuals share the same passion for startups and new ventures in general, they also juggle multiple claims on their time, and a long-term book project is clearly an easy victim when the pressure gets too intense. Keeping them focused and organized required an exceptionally dedicated individual, combining the diplomatic skills of a professional negotiator, the rigor of a researcher, and the dogged determination of a PhD student... We found this in Jana Thiel, research assistant at the College of Management of Technology at EPFL. Without her constant help and supervision, this project would never have been completed. We are extremely grateful to her.

This book would never have come to fruition without the help of numerous other individuals and companies. Their contributions were critical to our learning, and hopefully will be to yours as well. They shared without counting their time and effort, making this a true collective effort to enhance our understanding of the issues facing science-based startups and how to tackle them.

First and foremost, a very special thank you goes to all the *entrepreneurs* and *startup managers* for their invaluable time and willingness to share with us and with future generations of entrepreneurs their passion for this challenging career path. We are particularly grateful for their willingness to let us participate in *all* moments of their journey — highs and lows. Some of these extraordinary people will make their appearance in the cases, others will remain in the background — we would like to salute them all here!

We would also like to thank all of our *students* and *participants in executive education programs* who enrolled in our courses for their engagement and questioning. This helped motivate us to push further and greatly enlarged our approach to the intricate topics covered in these cases.

Likewise, we would like to thank all the speakers who volunteered to share their experience in a series of special events that we organized to initiate or validate our work on nurturing science-based ventures. Specifically, we would like to thank: Yves Emery (Lyncée Tec), Dominique Foray, Jan-Anders Månson, Francis-Luc Perret and Martin Vetterli (EPFL), Mike Franz (Netcetera), Andrea Pfeifer and Armin Mäder (AC Immune), Christian Schott (Sentron), Henry Baltes and

Peter Seeberger (ETH Zurich), Andreas Seifert and Bert Willing (IR Microsystems), Nikos Stergiopulos (EndoArt), and Yves Paternot and Martin Velasco (Angel Investors and Entrepreneurs) for their contributions to the Mastering Technology Enterprise (MTE) program (a joint program between IMD – International Institute for Management Development, The Ecole Polytechnique Fédérale de Lausanne (EPFL), and ETH Zurich), the MTE alumni event, as well as the first Entrepreneurship Day hosted at the College of Management of Technology (CDM) at EPFL.

We thank our colleagues at CDM and IMD for many fruitful discussions and ideas, in particular, Marc Gruber and Anu Wadhwa at EPFL and Bettina Büchel, Jean-Philippe Deschamps, and Georges Haour at IMD. In the same spirit, we would like to thank Hervé Lebret, Industrial Relations Office (SRI) at EPFL and James Pulcrano of IMD for their willingness and hands-on support in the identification of suitable company leads for this book.

We explicitly would like to highlight the contributions of case authors who gave us permission to include their cases in this book to help complement our own works. In particular, Professors Jean-Philippe Deschamps, John Walsh, Andy Boynton, Joachim Schwass, Georges Haour, Dominique Turpin, and Kamran Kashani of IMD deserve special mention for the fantastic case development work they have done.

Likewise, we owe special recognition to all the research associates who have provided most valuable help in researching and preparing the case material: Jana Thiel, Atul Pahwa, Alastair Brown, Michèle Barnett Berg, Anna Lindblom, Lisa (Mwezi) Schüpbach, Victoria Kemanian, Katrin Siebenbürger, Inna Francis, Rebecca Meadows, Rebecca Chung, and Joyce Miller on the IMD side, as well as Laurent Piguet (Event Knowledge Services), Frederic Martel (Constellation Partners) and Olivier Courvoisier (EPFL).

A special thank you goes to select student teams at EPFL and IMD for helping us prepare summaries or early versions of some cases such as AC Immune (Olivier Fortin, Harjeet Grewal, Anders Jepsen, Francesco Lissandrin, Defne Saral Antypas, and Roman Tarnovsky); Google (Guillaume Basset, Guillaume Gay, Sébastien Grisot, Bertrand Rey, and Aurélien Schmitt); Lyncée Tec (Satoshi Kuroda, Sean Blackburn, Laurenz Kirchner, Lynn McClelland, Chukwuemeka Oragwu, and Francesco Selandari); IR Microsystems A (John Buckingham, Jr., Kazushige Kinoshita, Derk Jan Kwik, Jan-Alexander Nagy, Michael Staudinger, and Jussi Vanhanen); IR Microsystems B and C (Alberto Andolina, Andrey Barkin, Maxim Kuzyuk, Paul Ploumhans, and Jodok Reinhardt); and Redigo (Nicolas Bruno, Adrian Ferrero, Sandra Forster, Marc Gouyon-Rety, Andrés Moreno, Vitalie Robu, and Jie Yin). Their project work with these companies laid the foundations for superb case material.

Often less visible but of critical importance was the dedicated support of Lindsay McTeague and Michelle Perrinjaquet (IMD editors) as well as Persita Egeli-Farmanfarma (IMD case services manager) for endless hours of proofreading, editing, and formatting our chapter drafts. They have been exceptionally professional, understanding, and cheerful throughout the entire project.

Administrative support was further provided by Ilse Evertse (freelance editor), Thierry Gachet (IMD graphics specialist), Chiara Maggia, Joelle Flach, Séverine Milon, Alexandra von Schack, and Jennifer Rae Willenshofer for general administrative support and permission requests.

We also would like to thank Gordon Adler (IMD), and Anthony Doyle and Simon Rees (Springer) for their coordination efforts. Getting busy academics to work together is indeed a lot like trying to herd cats.

Finally, we are grateful for the financial support we received for this project from the Alliance for Technology-Based Enterprise (a joint endeavor between EPFL, ETH Zurich, and IMD), the Chair of Technology and Operations Management (TOM) at EPFL, and IMD Research and Development.

Contents

Chapter 1
Opportunity Recognition

All ventures start out as an idea. However, a good *idea* is not necessarily a good *opportunity* and thus does not necessarily lead to a good venture. Hence, the successful entrepreneur-to-be should be on the lookout for good opportunities rather than good ideas. But what turns a good idea into a good opportunity? How exactly is a good opportunity defined, *i.e.*, how do we find out that it is a good opportunity, and where are the potential sources of promising ventures?

In this chapter we look at where opportunities come from and whether you should sit around waiting for "lightning to strike." We also describe what a window of opportunity is and how to verify quickly whether an idea has promise, before considering whether you should tell anyone else about your idea.

1.1 Starting from the Idea

Often nonentrepreneurs regard the ability to have a great idea as a gift of ingenuity bestowed on the entrepreneur. However, the image of the lone genius awaking at night with a brilliant idea is usually far from the truth. A good idea is generally the result of an iterative process that includes carefully weighing the pros and cons of an idea and discussing them with friends, colleagues, potential customers, and so on, and then developing the idea further.

Another oft-held belief is that ideas must be unique, new-to-the-world thoughts. However, we would propose that they do not necessarily need to be unique. Almost any good idea will already have occurred to someone else [1]. What counts is developing it, changing it, and crafting it so that it addresses a market and attracts paying customers. Then, and only then, does an idea start to become an opportunity.

1.2 From Idea to Opportunity: A Deliberate Process

The myth of the entrepreneur in his or her garage pursuing a singular idea that came in a flash of inspiration persists, closely connected to the notion of endless work and bootstrapping of resources in order to chase the entrepreneurial dream. Indeed, most entrepreneurs do work relentlessly; yet again, it is usually not genius that sets them apart from the average person. Instead, they tend to build heavily on long-standing industry experience and social connections. Their expertise, often acquired from previous (nonentrepreneurial dependent) work, allows them to see opportunities that others do not recognize. That does not mean that there are no inexperienced successful entrepreneurs, but most of the cases in this book clearly illustrate the need for expertise and strong networks.

Indeed, there is no magic at work. Opportunities are deliberately created by combining knowledge with hard work and focus [2]. They are built by individuals with ideas, creativity, and openness to what is happening around them. Successful opportunity finders look at figures, work analytically, and study expectations, values, and needs. Most importantly, however, they talk to people, trying to verify ideas [3] and incorporate all the feedback they can get.

1.3 Where Do Opportunities Occur?

One thing successful entrepreneurs are good at is having an optimistic belief that opportunities are everywhere, waiting to be uncovered. Opportunities may, however, be more abundant in some areas than others. For example, they tend to emerge where change occurs [4]. There are seven typical sources of change and opportunity to look out for, as Drucker illustrated in his seminal work on innovation and entrepreneurship [5]. Four of them come from inside a company or industry, whereas the remaining three emerge from the social and intellectual environment of firms. These sources often overlap, and many new businesses draw on a mix of different types of opportunity.

1.3.1 Unexpected Occurrences

Opportunities can emerge from unexpected failures or successes in existing businesses. Failures, such as failed experiments, however, can often be punished or played down in bureaucratic companies instead of being exploited. Conversely, unexpected successes, such as surprisingly high sales of a noncore product, often go unnoticed in such companies because people tend to see what they want to see. These unexpected occurrences are powerful sources of opportunity because most companies dismiss them.

1.3.2 Incongruities

Incongruities are discrepancies between what is and what *ought* to be or what everybody assumes it to be. A prominent example is growing industries that continue to be unprofitable, such as, for example, the numerous Web 2.0 startups that have sprung up like mushrooms since 2005. In these cases, something about their economics might not have been discovered yet, which opens up superb opportunities for those who can respond first. However, the nature of incongruities makes them typically only visible to industry insiders.

1.3.3 Process Needs

Opportunities that arise from a process need usually emerge out of process improvements, *i.e.*, eliminating a weak link in an existing process (manufacturing, managerial, *etc.*) or redesigning the process around newly developed knowledge. These opportunities are task rather than situation focused. They are small and sharply focused and require a clear understanding of exactly what is required. Process needs are likely to be recognized only by process users.

1.3.4 Industry and Market Changes

Managers sometimes believe that market structures are fixed, but in fact industries can change quickly. Fast-growing industries are always a good indicator for emerging opportunities because often the entire industry is not organized to fit the new requirements. This type of opportunity is more often seen by industry outsiders than by insiders, who often perceive the changes as a threat. As a consequence, new firms that identify the trend early on are often able to get ahead of established players. Industry change can, for example, be driven by converging technologies, *i.e.*, technologies that were hitherto regarded as distinct entities. A recent example is the telecommunications market, where Internet applications have increasingly become mobile and thus opened up a wide range of new business opportunities.

1.3.5 Demographic Changes

The most reliable source of new opportunities lies in demographic changes. Shifts in population – its size, age structure, employment, educational status, and income – can have a major impact on what will be bought by whom. Demographic events are a powerful source because they have a known lead time and are relatively "easily" identified by those who can interpret figures or spot trends.

1.3.6 Changes in Perception

Much less obvious than demographics are changes in perception. Examples are developments in the attitude to food and exercise in recent years. With obesity reaching epidemic proportions in many parts of the Western world, there is a lot of emphasis in the media and even by some governments on personal health, thus raising awareness and changing perceptions of it. However, it is usually difficult to identify such long-term trends and distinguish them from short-lived fads that will not create viable opportunities.

1.3.7 New Knowledge

Opportunities based on new knowledge are the "superstars" of entrepreneurship. Indeed, these opportunities will be the focus of the cases developed in this book (although not the exclusive focus). They differ from all other opportunities in the time they take, in their casualty rates, and in their predictability, as well as in the challenges they pose to entrepreneurs. This is because usually there is a long time span between the emergence of new knowledge and its manifestation in new products. In addition, large opportunities based on new knowledge often require a combination of different knowledge sources, making exploitation more difficult and uncertain. Think of biomedicine: A myriad of opportunities are looming, but they all require years of disciplined and systematic work to result in viable businesses.

1.4 Validating Opportunities

We said earlier that good ideas may not turn out to be good business opportunities. An opportunity has the quality of being attractive, durable, and timely and is anchored in a product or service that creates or adds value for its buyer or end user [6]. This comes back to the question of whether the need is real and the potential market big enough. It also requires assessing the feasibility of the product or service, of the industry and market, and of the organizational and financial implementation [7]. In addition, each entrepreneur should also gauge the match between the opportunity and his or her personal traits [8].

1.4.1 Product/Service Feasibility

Would-be entrepreneurs must make sure that the product or service is actually what customers want and what they are willing to pay for. If an entrepreneur cannot name the customers, he or she still has only an idea, not an opportunity, and certainly not a product [9].

1.4.2 Industry/Market Feasibility

Assessing the industry and market feasibility includes evaluating the market structure for entry barriers and attainable market share. Many investors would hesitate to invest in a company with less than 5% of attainable market share, especially considering that potential market leaders may capture as much as 20% of the market or more [10]. To be able to do so, the product or service must provide a clear, quantifiable competitive advantage over existing or soon-to-be-competing products (or services) in the market. The selling positions, including features, pricing, distribution, and timing, need to differentiate the new venture sufficiently from its competitors. In terms of industry structure, the degree of control over access to customers and traditional bargaining power along the supply chain are important criteria. Good opportunities are usually found in growing markets, well before any signs of maturity are evident. Of course, opportunities do exist in mature markets, but then the new product or service has to be highly novel and disruptive, for example addressing a major process need not previously met.

One must assume that competitors and copycats will enter quickly once the new product or service is on the market, and this is likely to make the market more competitive and less profitable eventually [11]. The obvious questions therefore are: When are copycats likely to enter the market? Do they have the resources to pursue the opportunity? Could they readily acquire those resources? How will the new venture defend its position or move on quickly to something else [12]?

1.4.3 Organizational Feasibility

Whether or not the right management capabilities are available usually determines organizational feasibility. Ideally, the new venture team should include members with strong industry background in the same technology or market and with proven profit-and-loss experience. The team should have complementary skills as well as being philosopically and ideologically compatible. In addition, the advisory board of the company typically assumes a critical role in the nascent venture. Here, the entrepreneur wants to have highly committed and experienced members with high integrity and personal honesty.

Another organization issue is the business model (*Chapter 2*). The would-be entrepreneur needs to question whether or not he or she can use the available resources, relationships, and interfaces with customers to sustain the venture based on the profit it makes [13].

1.4.4 Financial Feasibility

An opportunity is viable if it is able to sustain the business built around it. That means gross margins need to be durable and high enough to translate into sustainable profits of around 10% to 15% or more. Moreover, two crucially important financial milestones (*Chapter 3*) need to be estimated: the time to positive cash flow (when the business starts earning a gross profit) and the time to breakeven (when the original investment has been recouped). Approximately two years to reach positive cash flow is usually considered a high-growth opportunity. In addition, other figures to examine are the potential return on investment (ROI), free cash flow, and additional capital requirements over time [14].

The businesses we are discussing here are typically sold later on – either wholly or partially – to private investors, strategic investors (companies), or the public. Even at the outset, some reflection about the best way to arrange a financial exit for the owners and initial investors is necessary. If, for example, an opportunity provides strategic benefit to a potential acquirer, the purchase price or value will be much higher than the profit-and-loss statement might imply.

1.4.5 Personal Fit

Another important criterion when gauging an opportunity is of a rather personal nature. The entrepreneur should evaluate the fit of the opportunity to his or her personal traits in terms of risk acceptance, opportunity cost of pursuing other ventures instead, and the (financial) downside should the venture fail. In general, before starting a venture, the would-be entrepreneur is strongly advised to be clear about his or her goals in pursuing an entrepreneurial career, what to expect in terms of stress and responsibility. Many of the opportunities discussed in this book are what might be termed *high-growth* or *high-potential* opportunities, but the entrepreneur may also decide to purse a lower-growth *lifestyle* opportunity.

1.5 The Window of Opportunity [15]

"I was seldom able to see an opportunity until it had ceased to be one," said Mark Twain, referring to the fact that each opportunity has a particular time frame

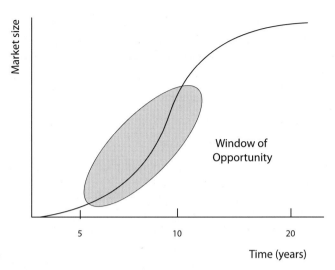

Figure 1.1 Window of Opportunity

within which it can be seized. Otherwise, it may be lost. Hence, having not only the ability to recognize a potential opportunity but also the sense of timing to seize it becomes critical. With the correct timing, a good opportunity can become a great one. Conversely, if the business is launched too early or too late, all efforts could be in vain. The difficulty is that it is much easier to see an opportunity that has been there for a while than to see it just as the window is opening (*Figure 1.1*). If the entrepreneur does indeed intend to pursue a high-growth opportunity (as opposed to a lifestyle opportunity), timing becomes crucial.

There is no best way to get the timing right, but it does help to focus on three aspects: customers, competitors, and context [16]. Successful entrepreneurs – especially in the technology and science domains – pitch their ideas to potential customers to test market receptiveness. In addition, a careful look at the competition may reveal a lot about the state of the market. If especially large competitors are heavily entrenched in the domain, the new venture launch might already be too late. Trying to identify the window of opportunity and perhaps to quantify it becomes a worthwhile exercise, since it sets the frame for the speed at which the new venture can or must grow and likewise helps define possible exit options.

1.6 The Business Concept Proposal

Once they have an idea, entrepreneurs should prepare a business concept proposal (BCP) as soon as possible. The purpose of a BCP is to see whether there is any merit in pursuing an opportunity. It is intended to be a quick, back-of-the-envelope exercise that entrepreneurs can use to screen ideas and see which ones are worth pursuing further. Unlike a business plan (*Chapter 2*), which is for further development of opportunities, the output of the BCP should be a simple go/no-go decision. The BCP is short and is intended to be performed *early* and *often*. It should take up no more than two pages and include the following points:

- Six-line (maximum) description of the opportunity or idea
- Benefit to paying customers
- Benefit to end users, if different
- Sources of the opportunity, as outlined above
- Market potential, including assumptions about the size of the market and potential market share
- Five-year projection of revenues and gross profits, including positive cash flow and breakeven
- Critical assumptions, which, if wrong, could result in a no-go decision for the business venture.

After reviewing the above points, the entrepreneur can decide whether the idea needs to be changed, investigated further, or scrapped.

1.7 The Big Question: Should You Tell Anyone Else Your Idea?

It should be obvious by now that we advocate telling people about your idea. It is difficult to obtain feedback if you keep your idea secret. Many entrepreneurs tend to be overly secretive and therefore do not get the feedback, brainstorming, feature changes, concept modifications, and the like that their more open colleagues receive. If you share your idea initially with some trusted colleagues, and then with potential customers, complementors, and/or users, your idea has the chance to go through a much more rigorous review and modification process, which in most cases will save you time and money and produce a more valuable opportunity.

1.8 Cases in This Chapter

The cases in this chapter deal, naturally, with opportunity assessment and recognition. The venture described in the first case, *Optima Environnement SA: Turning a Wonder Tree into an Eco-business,* is a biotech startup that has developed an environmentally safe alternative to chemicals in the water treatment industry using a plant-based extract. Subsequent research led to the discovery of proteins with medicinal properties and an oil extract with several cosmetic applications. With limited resources and promising opportunities in multiple industries, the company must attempt to break into markets where it has no prior experience and generate a revenue stream to keep it afloat through the complex sales cycle in introducing (potentially) disruptive technologies in existing industries. The case study provides a good basis for discussing elements of entrepreneurship, business strategy, and new product development within mature industries as well as the importance of developing strategic alliances to enter new markets. It allows for a complex assessment of potential business opportunities spawned by the cutting-edge knowledge created in the venture domain. The student is left to decide which opportunity might provide the most viable business in the future.

The case *Cognosense SA: Validating an Opportunity* traces the development of a very early-stage company that is in the process of clarifying its business model and pursuing the best possible opportunities that its founders' research has opened up. Like the previous case, the Cognosense story touches on opportunity recognition, as well as on the "organizational feasibility" issues necessary to fully exploit the opportunity. In addition, it covers the related issue of how external forces, such as the power of supply-chain partners and complementors, help shape the founders' decisions on which opportunity to pursue. As in the other cases in this chapter, the Cognosense case revolves around several possible routes for the venture and raises the question of which of them, if any, is actually a highly promising opportunity.

The *Shockfish* case likewise confronts us with a typical problem for technology-based startups. The technology available to the venture allows for many uses, but the question is, which will be the most fruitful? Is there a clear need for what Shockfish can offer, and are the opportunities big enough to fund the business long term? The (A) case of this two-part series introduces CEO Jim Pulcrano, who has to decide about the long-term focus of the business against the pressure of generating immediate revenues. The nascent venture has the opportunity to receive its first revenues from a project that is beyond its originally planned scope. However, the project would significantly delay work on the core technology. In many science- and technology-based ventures, the underlying invention often generates a myriad of opportunities, among which the business founders have to choose wisely.

The venture team in the last case in this chapter, *Redigo: Putting a New Electric Vehicle on the Road*, has already successfully launched its product – a new compact electric vehicle called the Mobi3. Yet since the car's introduction at the Swiss Expo 2002, it remained in the prototype stage until 2005. when a new generation of lithium-ion batteries became available that had the potential to provide more autonomy and require half the recharging time. Major improvements in the performance of electric vehicles were expected as well as new market opportunities. Inspired by the success of the early prototypes, the venturers planned to bring an enhanced version of the Mobi3, now renamed Redigo, to market. The case examines the various mobility vehicle target markets and users of electric vehicles and questions if there really is an opportunity and if a new launch is feasible. Just how big is the potential market for Redigo?

References

1   Bygrave, William D. "The Entrepreneurial Process." *The Portable MBA in Entrepreneurship*. Ed. W. D. Bygrave. New York: John Wiley & Sons, 1997, p.13.
2   Timmons, Jeffry A. *New Venture Creation: Entrepreneurship for the 21st Century*. Boston: Irwin, 1994, p. 87.
3   Jolly, Vijay K. *Commercializing New Technologies: Getting from Mind to Market*. Boston: Harvard Business School Press, 1997, p. 62.
4   Dollinger, Marc J. *Entrepreneurship: Strategies and Resources*. Upper Saddle River, NJ: Prentice Hall, 1999, p. 69.
5   Summarized from Drucker, Peter F. *Innovation and Entrepreneurship*. New York: Harper Collins, 1985.
6   Timmons, Jeffry A. "Opportunity Recognition." *The Portable MBA in Entrepreneurship*. Ed. W. D. Bygrave. New York: John Wiley & Sons, 1997, p. 27.
7   Barringer, Bruce R. and R. Duane Ireland. *Entrepreneurship: Successfully Launching New Ventures*. Upper Saddle River, NJ: Pearson Education, 2006, p. 54.
8   Timmons, *New Venture Creation*, pp. 102–103.
9   Bygrave, "The Entrepreneurial Process," p. 14.
10 Timmons, "Opportunity Recognition," p. 41.
11 Oster, Sharon M. *Modern Competitive Analysis*. Oxford: Oxford University Press, 1999.
12 Sull, Donald N. and Yong Wang. "The Three Windows of Opportunity." *Working Knowledge HBS*, 2005.
    http://workingknowledge.hbs.edu/archive/4835.html (accessed July 24, 2006).
13 Barringer and Ireland, *Entrepreneurship*, p. 100. See also Afuah, Allan and Christopher L. Tucci. *Internet Business Models and Strategies*. New York: McGraw-Hill, 2003.
14 Timmons, *New Venture Creation*, p. 98.
15 Bygrave, "The Entrepreneurial Process."
16 Sull and Wang, "The Three Windows of Opportunity."

CASE 1-1
Optima Environnement SA: Turning a Wonder Tree into an Eco-business

Jean-Philippe Deschamps, Benoît Leleux, and Atul Pahwa

NYON, SWITZERLAND, DECEMBER 2003: As he sat in his office preparing a presentation for the board, Yann Poirier, CEO and cofounder of the agro-startup Optima Environnement SA, looked back over the past seven years. The company had spent this time developing a market for a natural alternative to the chemical compounds used in the water treatment industry. Poirier believed that an extract from the seeds of the moringa tree could offer an economical and environmentally friendly replacement for the flocculants currently in use. Yet after significant investments from its investors, various government agencies, and nonprofit organizations – and many accolades from both technology leaders and industry visionaries – the company was still far from reaching its initial sales targets.

Company-sponsored research had yielded two more business opportunities in addition to flocculants. First, protein extracts from moringa seeds seemed to counter several strains of bacteria that were traditionally immune to standard antibiotic treatments, suggesting enormous applications within health and skin care. Second, moringa oil, initially considered a byproduct of the flocculant preparation process, turned out to have several potential applications in cosmetics.

Armed with patent applications from the existing research and promising new opportunities in multiple industries, Optima clearly had to revisit its vision and business model. Poirier was all too aware of the challenges of introducing potentially disruptive new compounds into entrenched mature industries with complex sales cycles. To make matters worse, these industries were controlled by companies with very deep pockets, companies that could easily bypass Optima.

A recent shareholders' meeting had made it clear that Poirier could not count on a continuous supply of funds from the investors unless there were significant successes to report – and rapidly. Poirier wondered how best to tackle the challenges that lay ahead.

Copyright © 2002 by IMD – International Institute for Management Development, Lausanne, Switzerland. Not to be used or reproduced without written permission directly from IMD.

1 Company Background

After a career as a New York art dealer, Paris stock trader, and consultant to a French bank, Frenchman Yann Poirier became something of a serial entrepreneur. During the 1980s and early 1990s he created and built over ten companies, some highly successful, others less so. In July 1996, inspired by the vision of developing a natural plant-based filtering product to replace flocculants and coagulants in the water purification process, he cofounded Optima Environnement SA.

Earlier in the year Poirier had read an article about research carried out at the University of Leicester, UK, on the seeds of the *Moringa oleifera*[1] tree, which had shown the potential to clean water. He was convinced that solving the world's water problems was more than just a technical or financial undertaking. The development of water treatment systems worldwide was not keeping pace with the explosive demographic growth. And here was an opportunity to participate in an ambitious and altruistic agenda of providing a potential solution to the sustainable management of water services worldwide. Poirier was also convinced that future generations would be against using chemical compounds in water treatment when natural and environmentally friendly alternatives existed.

After making inquiries into the initial research, Poirier cofounded Optima with three other independent investors and a buy-in[2] from the University of Leicester (*see Exhibit 1 for company organization chart*). Researchers at the Swiss Federal Institute of Technology in Lausanne (EPFL) were contracted to investigate the properties of moringa seed extracts. Shortly thereafter, Optima filed patents to protect its intellectual property (IP) related to the process and applications of moringa seed extracts in water treatment. In EPFL, Optima had found a strong partner to carry out further research into moringa seed applications (*see Exhibit 2 for details of the moringa tree*).[3]

Since there were no large-scale plantations or seed-processing capabilities for moringa, a subsidiary in Africa – Optima of Africa – was set up. It purchased a 99-year concession on a 2,000-hectare farm (Pajaroya Estates) in northern Tanzania. In 1999, 120 hectares of moringa were planted, with the goal of increasing

---

[1] *Moringa oleifera* is one of 13 species of moringa, which are native to India and grown in several parts of Africa. A drought-tolerant tree, it can grow on a variety of different soils. The leaves are extremely rich in vitamins A, B, and C and are the best vegetable sources of minerals and protein. The pods are eaten like green beans; the flowers can be eaten, or drunk in tea; the wood makes good firewood; the bark is used to make mats and rope; and its roots are known to have antibiotic and fungicidal properties.

[2] Optima bought the rights to the research findings and patent applications associated with the use of *Moringa oleifera* in water treatment applications from researchers at the University of Leicester, in exchange for equity in the company.

[3] Although the IP rights belonged to Optima, the IP ownership of any research findings that the company did not commercialize would revert to EPFL. This meant that Optima had to follow through on all potential finds or possibly lose ownership of the IP. EPFL was also to receive royalties on future business revenues.

the plantation size once business solidified. A demonstration site, training center, and experimental farm were set up along with the plantations.[4]

Optima outsourced the production of moringa seeds by signing up hundreds of contract farmers. The company provided them with seeds and contracted to buy back 100% of the seed production. Not only would this keep costs under control, but it would also – Optima hoped – allow it to maintain good relations with the local authorities by creating local employment and providing new job skills.

In 2001, as new business opportunities in moringa oil emerged, Poirier hired the first dedicated business development person, Lars Moeller, a seasoned executive from the specialty chemicals industry, to develop this market. Poirier concentrated on developing the water treatment market and simultaneously managed new activities in the protein market. After devoting five years to R&D, Optima now focused its efforts on commercializing its IP and technology.

2 Phytofloc™ and the Water Treatment Industry

The water treatment industry is commonly divided into two major segments: (1) wastewater treatment, which cleans wastewater from commercial, industrial, and residential sources; and (2) clean water treatment, which disinfects clean water for human consumption and for industrial and commercial purposes. Players in this market include chemical manufacturers, equipment suppliers, and service companies.

In 2002, water treatment chemicals accounted for approximately $6.2 billion in revenues annually, and the worldwide market for services and equipment was significantly more. Veolia Environnement (formerly part of Vivendi Universal) and Suez, both based in France, dominated the global water treatment industry. They controlled over 70% of the world's municipal water service market,[5] and their combined water revenues in 2002 exceeded $23 billion.

Mergers and acquisitions within the industry had resulted in a concentration of players, and, as with other utilities (gas, electricity, and telecommunications), further privatization and consolidation was expected over the next few decades. Incumbents, with their vertically integrated business models[6] and active participation in water policy think-tanks and lobby groups, made it difficult, if not

---

[4] One hectare could support 960 trees. The trees produced seeds for up to 25 years and yielded 0.5 kg of seeds in the first year, 1 to 1.5 kg in the second year, 2.5 kg in the third year, and 4 kg in the fourth year and beyond.

[5] Such companies traditionally bid to manage municipal water treatment infrastructure and personnel. Long-term contracts of up to 30 years are not uncommon, to allow these companies to make infrastructure investments. Generally, operations and implementation are outsourced to private operators.

[6] The companies providing services also provided water treatment equipment and chemicals.

impossible, for an outsider to penetrate the industry, even with revolutionary or superior technologies.

Yet the untapped market was huge. Over a billion people worldwide lacked access to basic water supply and twice as many lacked access to basic sanitation facilities. Water treatment was expensive in developing nations, which often lacked the precious foreign exchange to import chemicals and equipment.

2.1 Coagulants and Flocculants in Water Treatment

Among the chemicals used in the water treatment process, coagulants and flocculants are used to cleanse the water of impurities. Generally aluminum based, they help to separate out suspended particles in water. Coagulants neutralize the repulsive electrical charges of particles, allowing them to stick together in clumps, or flocs. Flocculants make it easier for the coagulated particles to group together in larger floccules, thereby hastening gravitational settling. Some coagulants serve a dual purpose, creating large flocs that readily settle.

In 2002 organic coagulants and flocculants had a worldwide market of about 290,000 tons with a value of $1.2 billion. Annual sales were growing at 10% in developing countries *vs.* 4% in developed ones.

2.2 Phytofloc™: Optima's Natural Organic Coagulant

Through a proprietary process, developed with technical assistance from Campden and Chorleywood Food Research Association (UK), proteins extracted from moringa seed press cakes[7] were used to produce Phytofloc™, Optima's organic coagulant. Phytofloc™ was pH neutral,[8] biodegradable,[9] and nontoxic. It also appeared to be more effective than alum,[10] the chemical coagulant most widely used currently (preliminary tests had shown it to be five times more effective.) With economical storage and transportation possibilities and an environmentally friendly image, Phytofloc™ was a revolutionary product with which Optima hoped to break into the existing commoditized market.

---

[7] The press-cakes were obtained by cold-pressing the seeds (the oil extracted in the process was initially discarded).

[8] This minimized the use of pH softeners and adjusters – a worldwide market of $450 million – which could lead to cost savings.

[9] Much of what was put into the treatment process ended up in the resulting sludge. A biodegradable compound therefore reduced the environmental problems associated with sludge.

[10] There were growing concerns over recent research suggesting that aluminum-based compounds were carcinogenic (*e. g.*, http://www.waterquality.crc.org.au/hsarch/HS10h.htm). Another study pointed to an increased incidence of Alzheimer's disease in people exposed to aluminum-treated water over a 20-year period.

2.3 Market Opportunities for Phytofloc™

Optima was not in the business of providing water treatment solutions, nor was it interested in entering this market. It would have to work with the service companies responsible for supplying solutions to end users, such as municipalities or firms that generated industrial sewage. This did not prove to be an easy task. First, the product would need to go through several levels of certification and testing before being authorized for use in water treatment. As Optima had realized, different types of water required different types and concentrations of additives in the cleaning process due to the differing levels of contamination. Unless the product was thoroughly tested against several types of effluents and the cleansed water then tested for impurities, it would not pass the minimum standards set by the authorities. Second, several of the larger players in the industry were vertically integrated and had their own chemical manufacturing facilities. Thus, they were reluctant to cannibalize these upstream investments.

Costs associated with testing procedures and the general industry aversion to new products explained Optima's initial failure to make inroads into the mainstream water treatment market. However, this led Optima to consider alternative niche applications for Phytofloc™.

The sugar industry was a potential market for such a product. Water used in sugar processing has to be treated before being discarded. However, access to this market required approval from Food Safety (European Union) or the Food and Drug Administration (USA), which could cost $250,000 and take several years.

Sekurit, a global leader in car glass manufacturing, was looking for a way to clean the water used in its production process without relying on acidic chemicals that tended to corrode the metal water pipes. A natural organic solution of Phytofloc™ provided the required results, using just one third of the quantity of chemicals normally used. Sekurit's annual requirements of Phytofloc™ could amount to 250 tons, and it was interested in testing the product.

Optima also began to target municipalities to show them the comparative advantages of using Phytofloc™ over traditional chemicals. The city of Geneva had shown an interest in learning more about the product and its stated benefits. Optima hoped that they would ultimately steer vendors providing water treatment services toward using natural organic ingredients wherever possible. Such opportunities, however, tended to be rare, as environmental issues rarely superseded financial considerations at municipal and local government levels.

Another market opportunity Poirier evaluated was the production of mobile water treatment units. They could be used to clean water in rural areas in developing countries where there were no such facilities. Optima had a mobile water treatment unit built for display and testing purposes but had yet to find any customers.

In the course of his investigations, Poirier discovered that Phytofloc™ would have to compete against another natural product billed as a "natural" flocculent: chitosan.[11] Several Japanese companies had already begun commercial production of chitosan, and some US and European firms had also shown an interest in commercializing it. In February 2003 a Russian research project had concluded that chitosan was a stronger purifying agent for water treatment than any other commercially available product.[12]

2.4 Market Penetration Strategy

2.4.1 Short Term

Given the novel concept of Phytofloc™, Optima decided to use a direct sales approach, at least initially, to demonstrate its advantages first-hand. The company targeted Swiss end users of flocculants in the cantons of Vaud and Geneva, which focused on specific industrial and commercial effluents (wastewater from the watchmaking, wine-growing, and paper and pulp industries). The market size in Switzerland was expected to be between 10,000 and 20,000 tons of Phytofloc™ per year, and the sales objectives for 2003/2004 were set at 1% of this market.

The pricing strategy was based on four parameters:

- Concentration of protein in the product
- Effectiveness compared to existing chemical alternatives
- Average selling price of chemical alternatives
- Carefully monitoring production costs to reach a target gross margin

2.4.2 Long Term

In the long term, Optima envisaged selling Phytofloc™ through a specialized distribution network. The potential market for organic coagulants and flocculants was estimated at 290,000 tons/year, about 10% of the overall market in Europe. Given the market opportunity, sales targets for 2005 were set at 500 tons/year.

Optima targeted major distributors for partnerships including Rhodia, Ciba Specialty Chemicals, and Suez. Going after this market would mean building industrial production capacity for press cakes and protein extraction beyond the

---

[11] Nontoxic and biodegradable, chitosan is made of chitin, which is found in the shells of crustaceans like shrimp, crab, and lobster. Shells are ground and processed to remove calcium and protein. Utilizing a chemical process, the refined mixture is either combined with lactic acid and then dried or combined with acetic acid and water and used as a liquid product.

[12] "Far East Scientists Discovered a New Method of Drinking Water Purification." Financial Information Service Regional Internet Company. February 21, 2003.

initial pilot installations in Tanzania and Switzerland. This would require an investment of several million Swiss francs, which the company could not afford given that it was due to run out of cash at the end of 2004. Subcontracting the protein extraction process was also a possibility. Companies that processed oil-cakes included Monsanto, Cargill, and Provimi. However, if Optima were to go through these giants, Poirier wondered what the company would be bringing to the table besides ownership of the application rights.

Although the large water treatment market was clearly the most lucrative, Optima had hit a wall in its efforts to persuade the big players to consider Phyto-floc™'s attractiveness. Now several niche applications were mushrooming and Optima needed to prioritize them in order to target those that would not only yield long-term sales contracts but also deliver much needed cash flow to the company. It also needed to build production capacity somehow.

3 Moringa Proteins

Further research into the moringa seed extracts led to the discovery of proteins with strong antibacterial properties. Three large groups of bacteria – staphylococcus, streptococcus, and legionella – that were known to show resistance to standard antibiotic treatments seemed to respond to moringa protein extracts in laboratory tests.

The opportunities in this market seemed tremendous. Recent publications from Harvard Medical School concluded that eczema and other skin irritations were caused by missing proteins in the skin and that a potential cure was to apply skin creams with active ingredients. Such treatments, with their antibacterial and microbiological effects, were supposedly far superior to the steroids and cortisone generally prescribed for extreme cases.[13]

Potential customers for such applications included pharmaceutical companies. They could utilize the moringa seed extract for prescription and over-the-counter drugs, as well as specific applications in women's health and sanitary products. Cosmetics companies could also be suitable prospects. One of the major trends in the industry over the coming decade looked set to be antiaging products. Soy proteins had been successfully marketed as organic ingredients and had found their way into several cosmetics applications. Opportunities clearly existed for moringa proteins to play a similar role.[14]

---

[13] The market potential for eczema alone was estimated at over $2 billion worldwide. Recognizing this, Optima rushed to file a patent to protect its discoveries, without necessarily having a targeted strategy to approach the opportunities that lay ahead.

[14] Pharmaceutical firms traditionally used petroleum- or mineral-oil-based products, technology that had hardly changed since the 1950s. A move to plant-based extracts could be revolutionary. Cosmetics technologies, however, were more advanced and utilized modern emulsions that could more easily take advantage of new ingredients such as moringa protein extracts.

However, once again, Optima found itself in unfamiliar territory. Testing the protein for various applications and obtaining certification was a notoriously rigorous, expensive, and time-consuming process for both the ingredient manufacturer and the pharmaceutical or cosmetics company utilizing the ingredient. It became increasingly evident that cosmetics firms would not accept the initial laboratory research and claims made by Optima. Cosmetics industry regulations demanded at least two years of testing. Intercosmetica, a small but well-respected Swiss cosmetics manufacturer, was aware of moringa's potential in cosmetics applications, unequivocally interested in Optima's research findings, sympathetic to the company's predicament of lack of adequate testing, yet unwilling to take on any of the financial risks associated with the process. Optima therefore needed to find an ally in a global cosmetics ingredients company.

Cognis[15] was already researching skin care products based on moringa proteins. The company supplied ingredients to most of the global cosmetics companies and it had also applied for patents on moringa use for several applications in the cosmetics industry.

After a year of R&D, in the summer of 2002, Cognis began to test market a line of products based on moringa protein extracts. Known as PuriCare™, this line stressed the antipollution effects of the active ingredients in the moringa peptide extract used in a shampoo. Results showed a 35% reduction in pollution levels in hair caused by city pollution and dust particles in the air. It also seemed to reduce hair damage from ultraviolet rays by up to 64%. Tests also indicated good results in strengthening, conditioning, and repairing hair in general. Cognis was convinced that future cosmetics applications could include shampoos, hair care products, conditioners, leave-on gels, and tonics.

Initial contacts and discussions with Cognis were positive. Optima's know-how in protein extraction and long-term plans for large-scale production of moringa seeds matched well with Cognis's industry expertise and commitment to developing a commercial line of moringa products. In December 2003 Optima and Cognis made an agreement whereby Optima would begin to supply Cognis with moringa proteins. Shortly thereafter, Clarins, Chanel and Lancôme began to market moringa-based products.

Optima's efforts in this market were finally beginning to bear fruit, but the way ahead was not totally clear. If several major Cognis customers decided on a global launch of hair care products with moringa proteins, Optima's current protein manufacturing capacity would not satisfy these requirements. Optima needed to carefully balance its sales development activities with its raw material production.

---

[15] Cognis, a Henkel spin-off and a global specialty chemicals company, had sales of over $3 billion in 2002.

4 Moringa Oil

When moringa seeds were pressed to obtain the press cake from which Phytofloc™ and proteins with antibacterial effects were extracted, oil was the byproduct – initially seen as waste.[16] Depending on the pressing process, between 20% and 40% oil by weight could be extracted.

The oil proved to be highly resistant to oxidation, which suggested applications in the food and cosmetics industries. Here, Moeller saw an opportunity to bring in much needed revenue.

Studies had shown that not only was moringa oil fit for human consumption but, like olive oil, it also contained antioxidants that could help in the prevention of heart disease. The approval process for a food ingredient was, however, time consuming and expensive, requiring toxicology results and safety data from animal and human testing.

Industrial applications for the oil were also emerging. Optima was beginning to evaluate moringa oil as a mineral oil replacement in the watchmaking and fine mechanics industries. But, once again, it was evident that Optima had no understanding of the market or the distribution channel, and entering yet another industry would mean taking the focus and time away from existing activities. Nor could it command necessary premiums to justify investments in these areas.

Cosmetics companies usually buy their ingredients from specialist manufacturers. These suppliers are responsible for detailed testing – at their own expense – for allergies related to the ingredient by itself or in combination with thousands of other ingredients. Optima was not in a position to perform these tests itself and therefore would have to sell to an existing ingredient manufacturer.

A specialized cosmetics ingredient supplier based in the US had shown interest in Optima and its moringa-based extracts. The US company believed that moringa oil had label appeal because of its oxidative stability. In early 2001 it applied for patents for several cosmetics applications of the oil in lotions, creams, and other products. The company had pioneered the use of jojoba oil in the cosmetics industry over the past few decades, and if it bought into Optima's moringa technologies, this would legitimize Optima's efforts in the eyes of key cosmetics ingredient manufacturers.[17]

---

[16] Optima had an oil press in Tanzania, where the oil was extracted from the seeds. The press-cake was then sent to Switzerland, where a second extraction process yielded the proteins and Phytofloc™.

[17] At that time, world production of jojoba oil was 1,100 tons per year, most of it used in the cosmetics and hair care industry. A small proportion was used as a specialist lubricant. In 1994 a study identifying new applications for jojoba oil pegged the worldwide market at 200,000 tons. (<www.dpi.vic.gov.au/notes/> Search: AG0678 [NRE: Information Series: Jojoba, January 2000]).

In late 2002 Optima received a purchase order from this US company for $1 million worth of oil – the Swiss company had just made its first major sale. By December 2003 the US company's patent applications had been successful, and it was eager to start commercial production.

However, storm clouds loomed on the horizon. Harvests from Optima's plantations in Tanzania were extremely poor in 2002 and 2003. It had been very difficult to manage the plantations from Switzerland. Internal audits in 2002 had revealed mismanagement of cash and inaccurate seed collection statistics. Contract farmers who had received seeds from Optima had not planted all of them. There were reports that some had been consumed as staple food instead. When the seeds had indeed been planted, this had not been done correctly, resulting in lower-than-expected yields. Instead of the 10 million trees reportedly planted, Optima's audit counted less than 1 million. Optima was forced to consider outsourcing the management of its plantations to a specialist in the field but had yet to find one that both found the economics of a moringa plantation appealing and believed in the future potential of moringa.

5 Licensing Moringa Technology

Optima realized that it would be unable to meet the future potential demand for moringa seeds and anticipated that others would soon get into moringa seed production. To control the supply of seeds, Optima decided to license its know-how on moringa cultivation and press-cake production. Outside of Europe, it was estimated that the most promising markets were in West Africa, the Middle East, Australia, and Brazil.

The licenses would give Optima upfront fees and would require that all seeds produced be sold back to the company at prevailing market rates. Although the deal could be potentially lucrative for licensees, there was always the concern that they had just one customer – Optima. Furthermore, given the company's shaky financial and market position, Optima's sales pitches involved a fair amount of persuasion and salesmanship.

Potential licensees included:

- Local manufacturers of coagulants and flocculants
- Industrial companies with a presence in developing countries that wished to capitalize on a new cash crop by selling the seeds to Optima
- Oilcake processors that wished to increase their value added from byproducts

In 2002 Optima signed its first license deals in Senegal and Ghana. By late 2003 it was also negotiating deals in Argentina and Uganda.

6 Financial Investments

In 1996 the Optima founders had initially invested CHF 200,000 in the venture. But it soon became evident that the R&D expenditure to get products to a commercial stage would require several years of significant investment before any financial returns were seen.

Thanks to his missionary selling, Poirier had secured external sources of public financing: about CHF 500,000 from the European Union and an equal amount from two Swiss institutions – the Federal Office for the Protection of Water and Forests and the Swiss Commission for Technology and Innovation. The latter's R&D grant was exclusively for carrying out moringa research at EPFL. Local authorities in the canton of Vaud had also granted Optima tax-exempt status for a ten-year period and the ability to carry forward losses for up to a maximum of five years.

By 2003 the investors had put in close to CHF 10 million without seeing much in the way of returns or, for that matter, revenue flows into the company. Poirier's shareholding had already been reduced to a minority stake. The other investors were increasingly worried about the lack of progress on the business front and were threatening to stop the inflow of further funds. One investor had already pulled out, and this had caused Poirier to focus his energies on alleviating the remaining shareholders' concerns.

One of Poirier's strategies was to highlight the positive media publicity the company had received on different occasions. At a trade fair in Dar es Salaam, Tanzania, in 2001 Optima won first prize in the agricultural products category, but the publicity generated no interest from the local investment community. Poirier had looked into the possibility of venture financing in both Africa and Europe, but the sense was that the capitalization was too small for venture capital firms, yet too big for further angel investments.

In 2002 revenues[18] reached CHF 370,000, while operating expenses were CHF 700,000. In 2003 revenues climbed to CHF 492,000, with expenses of CHF 690,000. Unless Optima was able to secure major contracts with upfront cash payments, Poirier expected a negative operating cash flow in 2004 as well.

7 Optima's Challenges and Options

As he pondered his options, Poirier knew well that seven years and CHF 10 million was a lot of time and money for angel investors to spend without recouping their investments. He had spent several board meetings describing Optima's opportunities and obstacles. He knew, however, that it was a matter of time

---

[18] All revenue figures stated include product sales, licensing fees, and grants.

before one or more investors refused to increase their funding, unless Optima signed big contracts for one or more of its businesses.

A quick review of the company's limited marketable assets convinced Poirier that an exit strategy of selling its assets outright would be rather unattractive for its investors.

Optima had neither specific industry knowledge nor the financial means to develop and validate specific industry applications. It was also hoping that all its patent applications would be approved. Valuation of patents was an expensive proposition, and Optima did not currently assign much value to these applications. Besides, a quick search through the patent database revealed 29 applications relating to moringa extracts filed since 2001 by major pharmaceutical and cosmetics companies, which potentially restricted Optima's IP.

Optima had no large-scale manufacturing facilities for the extraction of Phytofloc™, protein, or oil. A pilot plant in Tanzania for extracting oil had been upgraded to a small industrial capacity processing line in 2003, and a small pilot plant in the EPFL laboratories produced protein and Phytofloc™ samples for potential customers. The company's physical assets were limited to its plantations in Tanzania and its water treatment facilities, amounting to about CHF 2 million.

To make the promising opportunities with Cognis or the American cosmetics ingredient company materialize, Optima would need to secure a consistent supply of raw material. Poirier had a number of options for solving his current seed production problem. One possibility was to outsource the management of Optima's moringa plantations. This would free the company from the headaches of remote management and operational issues associated with the farming business. Another alternative was to use Optima of Africa's facilities purely to showcase the company's capabilities. The third option was to become a plantation specialist itself, removing the uncertainty of depending on a third party for raw material. But this would require significant investments of time and money, and the patience of Optima's board members was running thin.

Phytofloc™, moringa oil, and proteins were all huge potential markets, but chasing them without the necessary resources proved an impossible task. So Poirier wondered if it made sense to spin off one or more of these opportunities and build alliances with stronger partners to allow a more focused approach. But if partnership was a solution, whom should Optima partner with, and what would it bring to the party?

8 2004 and Beyond

8.1 Rays of Hope

As he put the finishing touches to yet another presentation to his board, Poirier felt encouraged by a number of recent positive developments.

In late 2003 a fruitful meeting with the city of Geneva's agency SIG, traditionally responsible for managing water, gas, electricity, and thermal energy supplies, resulted in an agreement to develop a pilot project involving the use of Phytofloc™ for its water treatment services. Such a project could generate tremendous interest in other towns in Switzerland as well.

The Wall Street Journal had just announced the winners of the 2003 European Innovation Awards, and one of the recipients was Aqua+Tech Specialties, a Swiss company promoting its proprietary chemical and natural products in water treatment processes. Optima knew of this company and believed Phytofloc™ to be a superior solution. Given that Aqua+Tech had created press coverage on novel water treatment approaches, Optima was hopeful that it could ride this wave to generate even more press interest for itself.

A recent documentary on public television in Switzerland had featured Optima as an up-and-coming entrepreneurial effort and had gained significant press coverage. However, this had not yet led to any concrete business opportunities.

A multibillion-dollar global leader in food supply worldwide had expressed interest in working with Optima to introduce moringa as the next big plant-based protein source for human consumption. Bird flu and mad cow scares in the past decade had increased the need for alternative protein sources to be assessed for use in the event of larger-scale epidemics. This company would be an ideal partner for Optima, also given its oil production capabilities.

In December 2003 a world leader in the water treatment industry, which had been working on next-generation alternatives to chemical coagulants, had been urged by the city of Buenos Aires, Argentina (which itself was under pressure from environmentalists), to consider natural alternatives in its water treatment services. This resulted in dialogue between Optima and this company on the possibility of large-scale production of moringa in Argentina, as well as the use of Phytofloc™ for its municipal water treatment. A contract with one of the world's largest companies in water treatment services would undoubtedly propel Optima into the limelight.

Poirier was bent on making the most of each of these developments. He believed that people were becoming more sensitive to the use of natural products, and this was not a passing fashion. And he truly believed in the opportunities that lay ahead. He constantly had to juggle between pacifying investors, selling the concept to potential customers, and remotely managing the plantations in Tanzania. With all this activity, he was not about to entertain the option of selling out in the short term just to obtain a positive return on the investment. Instead he was adamant about unlocking all the potential value that existed in Optima.

To turn his vision into a commercially viable enterprise, however, he realized that he had to overcome several obstacles and provide specific recommendations to his board on the following questions:

- Which of the exciting applications and product lines should Optima focus on for 2004, given that it would run out of cash by the end of the year?
- How should Optima develop and structure partnerships with allies in the water treatment, oil, and protein businesses?
- What should be done about the business opportunities that could not be pursued immediately?
- What other financing sources could Optima consider to bridge the gap until it could develop significant commercial applications?
- If none of these opportunities materialized in an acceptable timeframe, should Optima finally pull the plug and sell? What was an acceptable timeframe? To whom should it consider selling? How should it best position its physical and intellectual assets?

Exhibit 1
Organization Chart November 2003

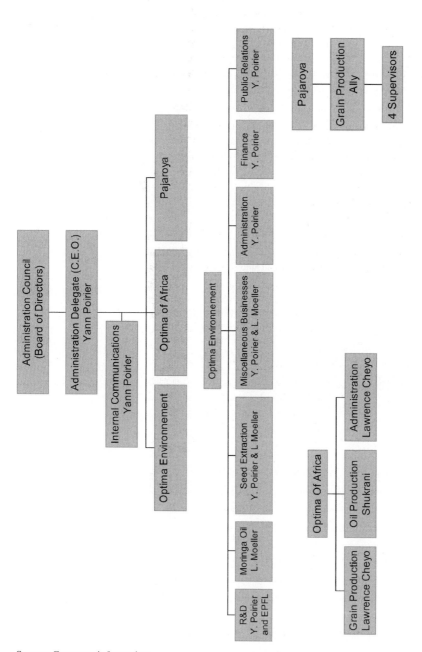

Source: Company information

Exhibit 2
The Moringa Tree

Ounce for ounce, Moringa Oleifera leaves contain:

4 times as much vitamin A as carrots, to shield against disease.

4 times as much calcium as milk, for strong bones and teeth.

3 times as much potassium as bananas, for healthy brain and nerves.

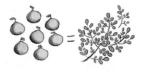

7 times as much vitamin C as oranges, to help fight illness.

Nearly as much protein as eggs, to build muscles.

Source: www.treesforlife.org

CASE 1-2
Cognosense SA: Validating an Opportunity

Christopher L. Tucci and Jana Thiel

Sylvain Paillard, chief technology officer (CTO) of Cognosense SA, entered his unusually quiet office. Only a few days earlier, on April 1, 2006, two of his colleagues and cofounders had moved to China to build the software development team there. Paillard would follow in a few months. He had stayed for now to continue managing the business in Switzerland, where the majority of his company's customers still resided.

Cognosense had been conceived two years earlier and since then had developed intelligent content search solutions for websites. The solutions were based on a ground-breaking technology, which soon led the management team to aspire to bigger opportunities. Trying to seize the global trend of converging Internet and mobile communications, the team embarked on turning Cognosense into a mobile solutions provider. By January 2006, the firm had already signed a first contract with a Chinese partner, thus gaining a foothold in the communications business at the 2008 Olympic Games in Beijing. Paillard explained:

> Our partner aims to develop an intelligent information system for the Olympic Games in Beijing. They got us on board to provide the intelligence to their virtual characters. We also promote the project for them and hope to sell it to the Olympic Committee.

Until now, the venture had lived from seed funding and initial sales. Paillard and CEO Sébastian Meng, however, aimed to accelerate the company's growth, knowing full well that they otherwise ran the risk of missing the window of opportunity. Additional funding was needed to boost the business.

Paillard occupied himself in those days preparing and giving presentations in front of potential investors. That particular afternoon, he was supposed to meet a renowned venture capitalist (VC), who would be an ideal partner for Cognosense. Paillard took his notes from his briefcase and started a final run-through of the presentation to make sure he had done everything possible to make this pitch a success.

Copyright © 2006 by Ecole Polytechnique Fédérale de Lausanne (EPFL). All rights reserved.

1 The Company

1.1 Origins

Cognosense had its roots in a term project of seven master's students of the computer science program at the Ecole Polytechnique Fédérale de Lausanne (EPFL). Building on concepts of artificial intelligence (AI), they had developed the application Eye.KIDZ, which aimed at guiding kids in their use of the Internet and controlling their Web access [1]. The students' technical approach, however, held huge potential for other application areas, in particular, Web-based content services.

Team member Sylvain Paillard was keen on the idea of commercializing their developments. Together with Sébastian Meng, whom he had met earlier at a conference, Paillard set about founding a new venture. Knowing that they needed additional people, Paillard invited six of his fellow students to become cofounders. In June 2004 Cognosense SA saw the light of day.

The role of CEO was assumed by Meng, who had previously worked for an American company in China and had then run his own Swiss-based business with main operations in Japan. He was well connected and knew people with influential relations to the Chinese government, thus opening business possibilities for Cognosense. Quickly, research collaborations were established with Chinese institutes and eventually, by April 2006, all development activities were installed in Beijing to increase efficiency. The company's headquarters remained in Switzerland.

New employees were hired in China, bringing the number of people in the firm to 12. However, only four of the cofounders remained, including Meng and Paillard, who commented on the developments:

> We have gone through a strenuous process of increasing efficiency and productivity – bringing the firm up to speed, so to speak. Moreover, goals needed to be aligned, all of which had of course an impact on the composition of the core team. Along that process it boiled down to four out of the original eight people [2].

Paillard had taken on the CTO position. He shared the management tasks with Meng.

1.2 Current Sources of Revenue

Building on its Eye.KIDZ experience, for the first two years Cognosense focused on selling content-browsing solutions to website and portal operators. A case in point was the "Centre d'Information jeunesse Assisté par Ordinateur (CIAO)" – an information portal for adolescents. Since February 2006, visitors to *www.ciao.ch* could navigate through the content of the website using an animated character (*Figure 1*). This guide asked and answered questions formed in natural language. It was able to adapt to the user profile, *i.e.*, suggesting actions based on the user's previous questions and activities. Cognosense had developed and customized the agent technology and designed the character. The project had brought high visibility for the startup in the French-speaking part of Switzerland.

More business opportunities of similar scope and requiring similar capabilities emerged: Cognosense also worked for The Guide Company – a Swiss-based provider of city guide information – and with the Swiss mobile operator "sunrise." Over time, however, the startup team realized that it needed to expand in both the scope of the application and geographically if the company was to grow significantly. The natural size of the Swiss market seemed too small and in general customers were perceived as less receptive to new technologies than in other regions of the world.

Cognosense's hopes for a larger business were in particular nurtured by the booming mobile communications market. Increasingly, Internet and mobile content was converging, calling for content accessibility across different mobile devices. Cognosense hence turned its attention to building a mobile applications portfolio. It focused on preference- and location-based content delivery.

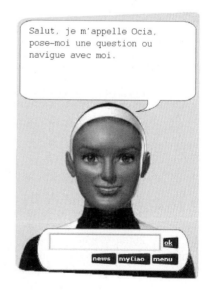

Figure 1 The Virtual Character Ocia
Source: www.ciao.ch

2 Product Portfolio

Cognosense's approach merged several aspects of AI, including technologies from fields such as natural language processing, classification, user profiling, Web content management, voice recognition, and text to speech. The portfolio comprised two main modules – ICP and UMP – that were complemented by special solutions. (*See Exhibit 1 for additional details.*)

Cognosense's *ICP (intelligent customer profiling)* application was able to create customer profiles in any given content and product delivery situation (e-commerce, mobile content, *etc.*). It made it possible to recommend content to customers depending on their current context (*i.e.*, current activity) and could moreover connect customers in communities. ICP added value to existing content and product delivery platforms such as e-commerce, Web portals, mobile delivery platforms, and the like. It accomplished this by allowing personalized content and interfaces for the customer. In addition, ICP provided full analysis of both customer behavior and the various products' level of success in the market.

With *UMP (ubiquitous multiplayer programming)* Cognosense served the needs of the increasingly ubiquitous computing for Web and mobile services. UMP offered an environment to develop those services while reducing development costs and enhancing service features. The platform allowed the programming to be separated from the user interface/look and feel. It also allowed the application to be adapted for execution on any kind of fixed or mobile device. The programming logic stayed the same, even though the layout and possible interactions of the user interface could vary greatly between devices (screen size, keyboards, *etc.*).

In addition, Cognosense provided *solutions*, such as intelligent dialogue agents, for information retrieval. The solutions were suitable for all websites and appeared as an animated character that interacted with the visitor in a dialogue as if he or she were communicating with a human. The dialogue could occur by writing sentences, clicking on buttons, *etc.*

These modules formed a unique set of technologies, as Paillard explained:

> Cognosense provides a new way to create applications and services, addressing the ubiquity and personalization demands of end-customers while, at the same time, fulfilling the needs of providers for cost reduction and advanced customer understanding [3].

However, the venture still had to figure out how to bundle its technologies and market them as a saleable product that would generate the desired profits.

3 Cognosense's Mobile Business Opportunity

By early 2006, Cognosense had laid the first stepping stone for entering the mobile business. Together with a Chinese partner firm, the Swiss venture aimed to provide content services for the 2008 Summer Olympic Games in Beijing.

Cognosense had agreed to provide intelligence for the "avatars" that the Chinese partner planned to create from the official mascots of the Games (*Exhibit 2*). The "intelligent" avatars were to guide onsite visitors around the city and sports facilities, pushing customized information and lifestyle services to the users of cell phones, Internet sites, PDAs (personal digital assistants), or other communication devices. The application could, for example, provide event schedules and last-minute updates, referring a user to the nearest taxi and show the target address in the taxi driver's home language, *etc*. Additionally, users could receive news, weather forecasts, TV schedules, movies, and restaurant recommendations. Finally, customized information could be offered not only to individuals onsite but also to Internet users in their homes anywhere in the world. (*For specific-use case scenarios see Exhibit 2.*)

However, Cognosense's ability to actually seize this opportunity depended on a profound understanding of the market, the business drivers, the risk factors, and the competitive landscape.

4 Understanding the Dynamics of the Mobile Market

4.1 Business Drivers

By 2006, the long-predicted mobile economy had finally started taking shape. Huge opportunities loomed, driven by a number of complex factors, such as converging content and improved technical standards:

- *Improvements in wireless technologies and standards*
  Compared to the early days, the latest generation of mobile communication devices had significantly improved access to the Internet. By default, cell phones were equipped with microbrowsers and other applications to bring Web content to users' pockets. Infrastructurewise, the ongoing move toward third-generation (3G) networks caused a leap in bandwidth availability that subsequently enabled the use of multimedia applications such as streaming audio and video and large file transfers on mobile devices [4].
  Users increasingly adopted these offerings. For example, in the first quarter of 2005, PDAs with integrated WLAN (wireless LAN) and cellphone capabilities accounted for about 55% of all PDAs shipped [5].
- *Converging Internet and mobile content*
  The technical advances ultimately made it possible to sell more complex content, which had previously been reserved for the fixed Internet (see also Figure 2) [6]. In October 2005 mobile content downloads were forecast to triple over the next 12 months, creating a global market of €7.6 billion by mid-2006 [7]. European consumers alone were expected to spend €3.3 billion in 2006 on content for their mobile phones, compared to €1.7 billion on their PCs. Finally, mobile content and entertainment services were expected to contribute 11% of

the total mobile service revenue and 34% of nonvoice revenue by 2008 [8]. New products emerged to exploit this unique potential. In particular, demand for personalized services had significantly increased, and such services were starting to attract myriads of game developers, broadcasters, handset manufacturers, and content owners [9].

- *Location awareness of mobile applications*
  The next-generation mobile phones were likely to act as "personal tracking devices." They would be able to receive services that relied on pinpointing a user's location. This would profoundly affect the way people interacted with their environment. On-demand services and intelligent cyber friends combined with location-based services were regarded as the next killer application [10].
  The total available market for mobile location-based services (MLBS) was projected to rise from $1 billion at the end of 2005 to $8.5 billion at the end of 2010. The largest geographic market for these services would be Asia Pacific, followed by Europe and North America, second and third, respectively [11].

4.2 Risk and Success Factors

Since the early 2000s, industry experts had been predicting that m-business would succeed the hype of e-business [12]. However, it was only during the course of 2005 that the market had picked up, and mobile business for consumers as well as enterprises had markedly increased. Whether or not the market would continue to thrive as forecast depended on some key risk and success factors:

- *Network infrastructure*
  The timely deployment of 3G networks was crucial for the success of multimedia services [13]. Furthermore, positioning technologies needed to mature: Information on user presence, preferences, and enabling authentication was imperative to improve the user experience of mobile business.
  Another infrastructure issue was the lack of international standards. The general rise in worldwide physical mobility made international roaming highly important. Collaboration was needed to further facilitate the interoperability in existing network standards such as GSM in Europe and TDMA or CDMA in the USA and some Asian countries like South Korea [14].
- *Device technology*
  Device functionality still needed to improve. The availability and affordability of Internet-enabled handsets were prerequisites for mass penetration [15]. In particular, their ability to communicate with other devices and to report on location and status were catalysts for increased business. Handsets featuring hybrid functionality, *i.e.*, mobile telephony and Internet access, became the centerpiece of the convergence trend.
- *Content and applications*
  The development of unrestricted and nonproprietary mobile content was

essential to drive revenues. It required finding local providers to supply up-to-date, personalized, and credible content [16].

Subsequently, applications that could quickly extract information relevant for the user were to become a distinguishing feature. Their success depended on personalizing and automating this process.

Furthermore, the rapid conversion of all communication channels required improved cross-platform and device functionality. Users increasingly demanded content to be consistent over the various platforms such as Palm Pilot, cell phones, Internet hotspots, *etc.*

- *Payment and billing*
  Investment in the development of innovative mobile applications and services would only occur if companies could charge for them appropriately. Hence, content and application providers were naturally concerned about standardized and accepted mobile payment procedures for end users. New methods such as micropayments and mobile wallets had emerged but still had to stand the test of time [17].

- *Security and privacy*
  Improvements in user localization and profiling had resulted in increasing privacy concerns. User education and mechanisms to protect privacy were still widely lacking.

  From the providers' point of view, content protection was critical. Standards to handle proprietary data needed to be further advanced and adapted to the latest device capabilities.

Once regulators, technology providers, operators, and developers were able to overcome the existing issues, huge market opportunities loomed – and numerous startups and established companies hoped to reap their share.

4.3 The Competitive Landscape

The question arose as to which technologies and services would become dominant, who would be first to deploy them, and how Cognosense could ensure it was among the pioneers. Paillard noted:

> Today, it is all about content and intelligent access across all devices – in particular and increasingly mobile devices. Cognosense follows a very different approach to enhancing and enabling new kinds of services in these emerging markets [18].

However, the idea behind the Cognosense offering was not entirely new: By April 2005, the South Korean telecom giant SK Telecom had launched its 1 mm service [19]. Like Cognosense's technology, the 1 mm offering featured an animated figure that connected a handset user to various context-, presence- and preference-based services. Unlike the Swiss venture, however, SK Telecom was able to spend a reported $200 million per year on research into advanced mobile services [20]. At the time, SK Telecom had already successfully exported its wireless Internet solution (NATE platform) to other countries in the Asian region and it aspired to expand further in order to provide additional global market opportunities [21].

On the other hand, device manufacturers also demonstrated genuine interest in intelligent content services. In April 2006 Nokia and MIT announced a research collaboration in artificial intelligence services that would allow people to use their phones to make queries over the Web [22].

Other potential competitors were business services of large companies (*e.g.*, Siemens Business Services) that had already started to manage and edit content to fit the format of the delivery mechanism [23]. By the same token, those companies could also be regarded as potential customers to buy or license Cognosense's technology.

Unlike all its competitors, however, Cognosense, with its artificial intelligence engine and the ubiquitous programming environment, had the potential to revolutionize the design of e-commerce and mobile applications. Yet, to turn this into profit, the venture would need to specify its target markets, target customers, and appropriate revenue models in detail.

5 Developing the Appropriate Business Model

Individualized services, well adapted to the user's habits and use patterns, would become the centerpiece of future market development in mobile communications. Yet using the Internet on mobile devices required entirely new business models. Firms competed on who was fastest and best able to educate and convince end customers about the value of mobile Internet applications. After all, the m-business market was still an early adopter market, only slowly transitioning toward addressing a broad audience. To benefit from this development, small companies such as Cognosense had to carefully position themselves, find profitable revenue models, and define appropriate value propositions.

5.1 Defining the Value Proposition

The advantages of mobility were undisputed. They included [24]:

- *Freedom of movement:* Services used while on the move
- *Ubiquity:* Services used anywhere, independent of the user's location
- *Localization:* Services exploiting information on the user's location
- *Convenience:* Always at hand
- *Instant connectivity:* Always on
- *Personalization:* Personal device, designed to store personal information

However, these were not yet propositions that would convince the immediate customers of the young Swiss venture's viability.

5.1.1 Understanding Who the Customer Is

Cognosense's technology platforms targeted different parts of the value network. The UMP platform, on the one hand, was built first and foremost as an OEM product for mobile device manufacturers. Cognosense would thus assume the role of an applications provider. It addressed developers – a very different clientele from consumers with different requirements, as Paillard explained:

> The actual problem for a developer is not the process as such but the adaptation to all platforms. This needs to be done manually and costs a lot of time! Our approach frees them from this work by providing a new platform that is content-centric. Programmers will spend much less time with our system, because they don't need to worry about the format of the stored data. Our platform fully masters the way the application is executed by taking generic decisions depending on the device. This has never been used before in the software application domain [25].

The ICP application, on the other hand, was clearly an e-commerce application. It addressed mobile content delivery platforms, hosted by either network operators or content providers. The operators and content providers were more concerned with driving their average revenue per user (ARPU). By offering a personalized content environment, accessible from all platforms and devices, customer loyalty and activity could be increased, which would eventually allow providers to generate higher income from targeted advertising.

In addressing different customers with its two main applications, Cognosense had to place itself at different spots in the m-business value network.

5.2 Positioning the Firm in the Value Chain

The ICP application was clearly placed as a gateway to the network operator's platform or at the content provider level. UMP, by contrast, addressed application developers working for device manufacturers, network operators, and dedicated application development companies. At the time, Cognosense continued working on a strategy to approach these potential UMP customers.

One of the challenges in positioning the firm was that m-business, and in particular the wireless Internet value chain, involved numerous players forming a complex system of value-adding service providers (*Figure 2*). The complexity of the system was further increased as a result of the various technologies and standards involved [26].

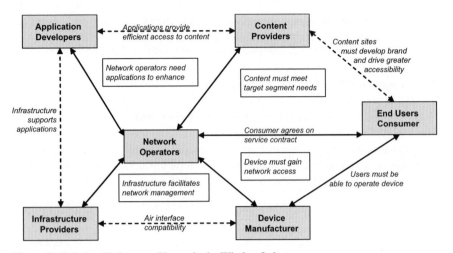

Figure 2   Relationship between Players in the Wireless Industry
Source: Esther Teo. *Market Entry Strategies of Wireless Startups*. Fisher Institute for the
Strategic Use of Informations Technology, Haas School of Business, University of California,
Berkeley. 2002.

5.2.1 Blurred Boundaries in the Value Network

In elaborating a successful strategy, the venture team had to account for increasingly blurred boundaries between the market players. Whereas three or four
years earlier wireless software vendors had created their own market by providing
their technology as a hosted service, the market in 2006 was drifting toward
vertical integration: Increasingly, fixed-line and wireless operators sought to
leverage data services as their own value-added offering and differentiator from
the competition. They hoped to increase customer loyalty and boost the ARPU.
Consequently, they set about transforming themselves into a new class of full-
service providers by offering everything – the network, the applications, and
integrated services [27].

Nonetheless, these big carriers, and likewise the big content portals such as
Yahoo! and MSN, were no longer able to keep track of and deal with the scattered
landscape of all the small content and application providers. Similarly, small
development companies faced difficulties dealing with the big players. This
setting opened up opportunities for aggregators, who collected offerings from
developers and media companies and distributed them to operators around the
world [28].

5.2.2 Software Development Company *vs.* Aggregator

At the outset, Cognosense was clearly a software development company. However, with the opportunity at the Olympics in 2008, the firm was in fact moving toward the aggregator realm. Cognosense's technology was allowing the company to become a central connection point between the principal (the Olympic Committee in this case), network operators, and content providers. Essentially it hoped to operate as a broker. Paillard explained:

> From our previous guide experience, we know how it works. Besides the Olympic Committee, we would basically have three other types of partners here: First, the content providers, then the LBS [location-based service] company to provide real time positioning, maps and directions, and third the mobile operator. Our technology allows us to be the hub in this, *i.e.*, negotiate a contract that grants us the rights to develop and sell the platform *and* make all the contracts necessary with the content providers. However, we are aware that this might raise an issue for operators as they usually assume this role. Therefore, at the moment, we focus less on where the application will finally run but on how we can leverage our technology to build a product that puts us in the center of the different types of partners [29].

5.3 Identifying a Viable Revenue Model

At the time, the company was typically paid per development milestone and for the final delivery of the working application. Moving into m-business, however, the startup hoped to generate revenues not only by selling technology but also by sharing in the content traffic, as Paillard made clear:

> Ideally, we would receive a fixed-price contract, by which the Olympic Committee pays for what we create, *i.e.*, the application on the web and on cell phones, with their content. They would then sell this product to the end user at a certain price and we, in turn, would receive x percent of the content and of any additional revenue that is created by the platform. The basic idea is to have an initial payment for our development – probably at no win [profit] – and then earn money at the later stage by people using this application [30].

Revenue sharing was the predominant pattern in the mobile market (*Figure 3*). Typically the operator had control over the end customer's payments and shared revenues with the content and application providers. However, finding the right revenue strategy became more and more difficult as the network complexity increased. Different approaches competed against each other. Media and broadcast firms, for example, demanded payments for content in advance, whereas operators tended to charge and pay for content on a per-click basis with returns depending on demand and delivery [31]. Application developers and providers were caught in the middle. Their revenue models typically included software license fees, software maintenance fees, consulting services, and, increasingly, software hosting and operation services [32].

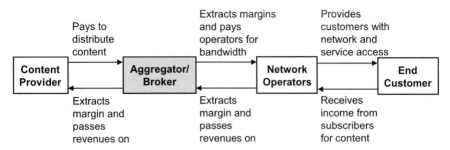

Figure 3   Standard Revenue Sharing Model
Source: Teo. *Market Entry Strategies of Wireless Startups.*

As a software hosting company, the options for revenue generation were many, ranging from per-click transactions or charging for development effort in time and material to subscription or licensing fees. Different approaches were typically chosen when addressing different customer segments, such as enterprises (B2B) or consumers (B2C). As for the B2B approach, hosting companies generally extracted their margins from selling or licensing applications and content to operators. For the B2C domain, they could follow a content portal approach, *i.e.*, offering services directly to the end customer against subscription fees (premium content) or advertising. The advertising model held particular promise for location-based services as they allowed for very targeted promotions [33].

However, any approach that Cognosense finally followed needed to prove its viability over and above the opportunity at the Olympic Games.

6 Setting the Course for Growth

Receiving additional investment was crucial in order to grow the company as envisioned and to help seize the opportunities ahead. In the next few weeks, the executive team planned to approach the Olympic Committee to get a written contract to proceed further with the development and implementation of their mobile information platform. Meanwhile, they also had to continue managing their existing clients in Switzerland and further establish operations in China.

At this point, Paillard finished checking his presentation. He was positive he could attract the VCs' interest. So far, Cognosense's management team had gathered significant expertise in the Internet domain, and Paillard was sure that their thorough market analysis and relationship with the right investors would also help bolster their case for mobile communications.

Cognosense was just at the beginning, yet it had to choose its next steps carefully.

Exhibit 1
Product Details

ICP – Intelligent Customer Profiling

Cognosense–ICP connects to your customer base and can be connected to any content source, such as products, commercial ads, information, news, *etc.*

Quality

ICP dynamically chooses the most efficient recommendation techniques and improves its logic depending on the situation. Therefore, the system is able to maximize the quality of its recommendations, integrating the differences between one platform and another.

Performance

ICP's back-end–front-end architecture provides for scalability: Its front-end architecture is based on fast clustered searches, maximizes the use of a data cache, and minimizes concurrent access; at the back end, the system's processing requirements are independent of the total number of customers. It even reduces the per-user processing time as the total number of users increases.

Integration

ICP provides a full Java and XML over HTTP communication gateway, which allows the connection (load and update) of any user or content database. Simple configuration interfaces allow the definition of any content architecture. The recommendation results and feedback are to be integrated like a simple method call.

Customer and Content Communities

ICP provides an efficient visual application to administer communities of customers and other content sources such as products. Communities are automatically suggested according to ICP's unique Communities Segmentation System, based on advanced clustering techniques in machine learning. The resulting communities can be managed and adapted to additional constraints.

Exhibit 1 (continued)

Analysis and Forecast

ICP analyzes the shifts of interest in customer and product communities over time and therefore provides multifaceted views on market trends. ICP can also predict the future success of a product given its definition. After the product launch, the system is able to adapt its predictions dynamically and to explain the potential difference between the initial prediction and reality.

UMP – Ubiquitous Multiplayer Programming

Ubiquitous Layout Communication

As the layout and possible interactions of the user interface vary greatly from one device to another (different screen size, different keyboards, *etc.*), the programming logic with UMP stays the same. The environment of UMP allows the complete separation between interface-specific layout elements and generic client-server programming.

Client-Server Architecture for Content-Centric Applications

One of the most common issues in ubiquitous application is data access and data synchronization. Cognosense's UMP environment simplifies the creation of content-centric applications, as the inner architecture of any application is based on a client architecture. This provides unified data consistency. Programmers and users no longer need to worry about accessing and saving data.

Deploy and Distribute

UMP also includes a centralized deployment repository that makes applications automatically available – as easy as uploading a file by FTP. Once the application is deployed, users of the UMP distribution platform will be directly informed of the launch of the new application and will be able to access it.

Exhibit 1 (continued)

Solutions

Information Retrieval

Unlike search engines, Cognosense's solutions are able to process any sentence (keywords and natural language) and correct common mistakes. The result is projected onto the database and a selection is drawn from the database. The system then selects a question that best separates the current results most homogeneously. After only a few questions, the system is able to narrow down the results with the highest matching probabilities.

Virtual Character

Cognosense's solutions include an advanced animated virtual character (avatar) manager that allows for constant smooth and situation-adapted animation of the character (emotional and contextual animations). The avatar is connected with a generic lip synchronization module that can be connected to any sound file, enabling the character to read displayed text. Solutions include a Web administration interface for voice recording and can be connected on demand to text-to-speech systems.

User Profiles and Statistics

Solutions include a user profile system, connected to the dialogue management scripting language. Therefore, visitors' profiles can be extracted from their navigation behavior and their answers to the system. A full range of automatic statistics is included. The statistics engine is created over a modular platform that allows for the easy addition of many additional views adapted to specific needs.

Source: Company website http://www.cognosense.com/public/index2.html

Exhibit 2
Case Scenarios for Use of Cognosense Technology
at Beijing Olympics

福娃 Friendlies

福娃贝贝 Beibei   福娃晶晶 Jingjing   福娃欢欢 Huanhuan   福娃迎迎 Yingying   福娃妮妮 Nini

Friendlies: Mascots of the Olympic Games 2008

	Scenario 1 **Follow the games, be informed and guided, and have fun from anywhere in the world!**	**Scenario 2** **I will not only inform you about the games, I will guide you in Beijing!**	**Scenario 3** **I can follow you everywhere**
User	John (American), avid fan of the Olympics	Susan (British), doesn't know anything about China and is not at all into technology	Martin (German), lived for more than 10 years in Beijing; wants to capture every day of the event by writing about what he saw and publishing it including pictures
Location During Olympics	United States	Beijing	Beijing (M. took a week of vacation)
Chosen Friendly	Huanhuan, the Olympic flame, because this represents the passion for sports	Beibei, carries the blessing of prosperity, is patient, and always has great things to say about Chinese culture – a real tour guide	Jingiing, the Panda

Exhibit 2 (continued)

	Scenario 1 **Follow the games, be informed and guided, and have fun from anywhere in the world!**	Scenario 2 **I will not only inform you about the games, I will guide you in Beijing!**	Scenario 3 **I can follow you everywhere**
The Friendly will:	• Easily understand that John loves baseball and will always remember to tell him the latest news and create a snapshot of recent results based on his interests (medals, participants, dates of each event, *etc.*). • Start to hold a baseball bat and will even suggest John play baseball with him as a small fun game. • Allow John to make research by flexible queries and will guide him in the navigation of the Olympic Games website.	• Find a hotel for Susan. • Tell Susan which bus to take; can even display a destination's name in Chinese to a taxi driver. • Be conveniently available on Susan's rental mobile phone. Susan does not need to be knowledgeable about technology. Mostly, she only has to tell her guide yes or no.	• Analyze Martin's writing and thus know what books he will like. • Connect Martin to other people who also chose Jingjiing, which introduces Martin to new friends. The cross-device functionality allows Martin to take pictures with his cell phone and comment on his PDA when he is in the stadium. Once back home, Martin will access his "virtual book" on the Internet and put all the pieces together. People from everywhere can connect to the content and contribute.

Source: "Olympic Games Info System: Beijing 2008 Olympic Games Intelligent Information System." Cognosense SA, Project Proposal, February 14, 2006.

References

1   For more information see the website http://www.eyekidz.com.
2   Interview with Sylvain Paillard on April 5, 2006.
3   Interview with Sylvain Paillard on February 23, 2006.
4   Nicholas D. Evans. *Business Agility: Strategies For Gaining Competitive Advantage Through Mobile Business Solutions.* Upper Saddle River, NJ: Prentice Hall. 2001.
5   Sumner Lemon, "First quarter was good to PDA vendors." May 4, 2005. http://www.thestandard.com/movabletype/datadigest/archives/003213.php (accessed April 2, 2006).
6   Roest, Jerry. "Are you content with your content?" Telecommagazine Online, 2004. http://www.telecommagazine.com/search/article.asp?HH_ID=0409i14&SearchWord= (accessed March 26, 2006).

7   Mahesh Pattani, "Time for a re-think." Telecommunications Online, 2005.
     http://www.telecommagazine.com/search/article.asp?HH_ID=AR_1195&SearchWord=
     (accessed March 26, 2006).
8   Roest, "Are you content with your content?"
9   Pattani, "Time for a re-think."
10   James Harkin. *Mobilisation: The growing public interest in mobile technology.* DEMOS.
     2003.
11   "Mobile Location Based Services Finally Begin to Deliver – $1 bn This Year to $8.5 bn by
     2010." *Press Release.* June 15, 2005.
     http://www.marketwire.com/mw/release_html_b1?release_id=88832 (accessed March 29,
     2006).
12   Ravi Kalakota and Marcia Robinson. *M-business: the race to mobility.* New York:
     Donnelley & Sons. 2002.
13   "Hong Kong (China) and Denmark top ITU Mobile/Internet Index." I. T. Union, *Press
     Release.* September 17, 2002.
     http://www.itu.int/newsarchive/press_releases/2002/20.html (accessed March 24, 2006).
14   Steve Kropla, "Mobile Phone Guide." http://kropla.com/mobilephones.htm (accessed May
     13, 2006).
15   "Hong Kong (China) and Denmark top ITU Mobile/Internet Index."
16   "Hong Kong (China) and Denmark top ITU Mobile/Internet Index."
17   Michaela Denk and Manfred Hackl, "Where does mobile business go?" *International
     Journal of Electronic Business*, 2004, 2 (5), 460-470.
18   Interview with Sylvain Paillard on February 23, 2006.
19   "SK Telecom Launches Personalized Wireless Internet Service Called '1mm'." *Press
     Release.* April 20, 2005.
     http://www.sktelecom.com/eng/cyberpr/press/1196759_3735.html (accessed March 27,
     2006).
20   "Niche Cell Players Strike." February 27, 2006.
     http://www.redherring.com/PrintArticle.aspx?a=15832&sector=Industries (accessed March
     26, 2006).
21   "SK Telecom - R&D Platform Center Introduction." 2006.
     http://www.sktelecom.com/eng/rnd/platform/introduction/index.html (accessed March 26,
     2006).
22   Martin LaMonica, "Nokia and MIT think artificial intelligence." April 25, 2006.
     http://www.zdnet.co.uk/print/?TYPE=story&AT=39265295-39020351t-10000010c
     (accessed May 13, 2006).
23   Ken Wieland, "Mobile content: where's the value?" Telecommunications Online, 2005.
     http://www.telecommagazine.com/International/Article.asp?Id=AR_501 (accessed May 23,
     2006).
24   Giovanni Camponovo and Yves Pigneur, "Business Model Analysis Applied to Mobile
     Business." Internat. Conference on Enterprise Information Systems, ICEIS'2003, University
     of Lausanne. 2003.
25   Interview with Sylvain Paillard on February 23, 2006
26   Evans. *Business Agility.*
27   Evans. *Business Agility.*
28   Christoffer Andersson, Daniel Freeman, Ian James, Andy Johnston and Staffan Ljung.
     *Mobile media and applications, from concept to cash: successful service creation and
     launch.* New York: John Wiley & Sons. 2006.
29   Interview with Sylvain Paillard on February 23, 2006.
30   Interview with Sylvain Paillard on February 23, 2006.
31   Pattani, "Time for a re-think."

32 Booz Allen Hamilton (Communications, Media & Technology Group), "The m-Business Opportunity: Capturing Value in the Enterprise Wireless Market," 2001, 7 (2), 1–12.

33 "Hong Kong (China) and Denmark top ITU Mobile/Internet Index."

CASE 1-3
Shockfish (A)

John Walsh, Andy Boynton, and Alastair Brown

On the night of September 10, 2000, the CEO of Shockfish, James Pulcrano, was agonizing over a crucial question. Should he sign a contract in the morning for the future delivery of a wireless voting system or not? The implications of signing were clear: his company would receive CHF2.5 million over the next six months, yet it would have to abandon its core business for four months in order to fulfill this obligation.

Shockfish was located in Lausanne, Switzerland, and it was exclusively in the business of producing and selling Spotme, a wireless networking device with many capabilities. Spotme was used as part of a system for people at an event to meet, conduct business, and interact with one another. The company was the first-mover in this area and although it sensed that other large corporations could enter this field, no direct competitors had emerged.

As a startup, Shockfish sought organic growth. To date, there had been no sales of Spotme. In fact, the company had not even chosen the market it would target from among many large opportunities. If Pulcrano signed the contract, Shockfish would realize its first revenue and these funds would give the startup leverage in arrangements currently being made with its principal investors. Voting systems did not excite the Shockfish employees though; Pulcrano had come on board thinking of other larger markets Shockfish might serve. Was it prudent to put some money in the bank now, or would it cost the company its advantage as it shot for a bigger prize?

Copyright © 2002 by IMD – International Institute for Management Development, Lausanne, Switzerland. Not to be used or reproduced without written permission directly from IMD.

1 Beginnings

Bendicht Ledin was a 27-year-old humanities student at the University of Zurich in mid-1997 when the idea for a wireless networking device (Spotme) occurred to him. Ledin had been reading about wireless devices from Japan and felt "if these things could communicate with each other, this would greatly enhance the product." Communication between devices could "bring surprise to the user," and Ledin imagined people being introduced to each other, knowing they had common interests. The idea was not entirely new, and variants were very popular at the time, but Ledin was determined to forge ahead.

Ledin knew he could not go it alone. He quickly contacted a friend at Ecole Polytechnique Fédérale de Lausanne (EPFL), who passed the information to his professor, Roland Siegwart. Within months Ledin and a master's student named Remy Blank, who chose to join the endeavor, were working in Professor Siegwart's laboratory in Lausanne. They headquartered their fledgling business in a 10-ft.-by-10-ft. room for graduate students on the EPFL campus and set to work.

In September 1998 the group was incorporated as Shockfish, and a few months later a business plan was sent to a competition at IMD – the International Institute for Management Development, also located in Lausanne. It was evaluated by Pulcrano, who quickly realized its potential. In January 1999 Pulcrano, Ledin, and Blank met to discuss Shockfish. Pulcrano commented:

> The idea was one that just hit me...there are a lot of business/technology ideas where you wonder why would anyone pay for this? And how will this work? But I remember I was sketching stuff out right there in that first meeting...you could take this and do this with it...thinking this is really good stuff.

The plan was chosen as a finalist, and as the year passed Pulcrano continued to provide free support. By August 1999, however, Pulcrano let Shockfish know that he could no longer help out – after all, he had a full-time job. Ledin commented on Shockfish's predicament: "We were three students not taken too seriously by investors or industrial partners, and we thought adding a senior manager would improve us." Thus, the three employees of Shockfish responded by offering Pulcrano 10% equity on the spot if he joined the company, with the potential for a higher percentage in the future. After deliberating whether this was too generous and discussing the opportunity with others, he accepted.

2 The Product

The Spotme device was one piece of a system of technologies that allowed wireless networking at an event. A complete system, depending on the size of the event, generally included one database server, one photo station, a few base stations, several charge suitcases, and hundreds of Spotme devices. For an event, the database server was loaded with information about all participants who would be using the Spotme device. At check-in, users had their photo taken at the photo station and were given the small wireless handheld Spotme device. Base stations were used in the area where networking would take place to facilitate the wireless communication of information. For example, participant photos were beamed to the Spotme devices. Charge suitcases were used for charging and storing Spotme devices. These technologies worked together to provide a wireless networking environment. The entire system was designed for ease of use; Shockfish expected users to be nontechnically oriented.

Once the wireless networking environment was enabled, users could do many things with their Spotme devices. First, the Spotme device could be loaded with static information that would be helpful at an event. For example, a map of the layout of the theme park you were visiting or times of various meetings at a trade show. The Spotme system also allowed for information to be sent to all participants or selected groups or even individuals during the course of an event. For example, if lunch was going to be served at a new time or place, all event participants could be notified. Or if a specific person received an urgent phone call, he/she could be alerted.

For networking purposes Spotme offered several options. If event participants did not know who they wanted to meet, they could search the database of participants in different ways to find someone. For example, at a night club you might search on having an interest in "trance music" and find someone with a common interest. Spotme's "radar" function let you know who was within 30 meters at any time and you could browse through the information to see if you would like to network with them. The "spotting" function of the device notified you when

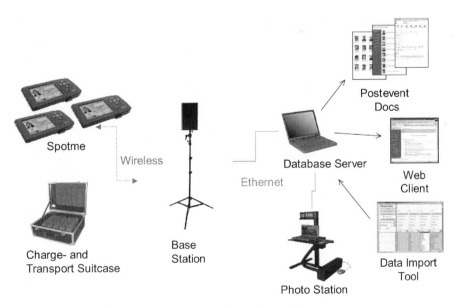

Figure 1 System of Technologies Allowing Wireless Networking
Source: Company literature

a preselected individual was within 10 meters. Spotme also allowed for device-to-device messaging. For example, at a business conference you might want to ask someone if they would like to have lunch with you. Spotme stored all information in the central database server. Therefore, statistics on the location and activities of people could be reported after an event. Additionally, activities of a single user could be sent to them after an event.

3 What Was the Best Market for Spotme?

Spotme's versatility as a networking device for any event with many people led to difficulty in targeting a market. Shockfish saw huge potential in a number of areas but had not journeyed into any specific one:

3.1 Theme Parks

The Spotme system had uses in a theme park that helped both the visitor and the proprietor. For the visitor, Spotme's spotting capability could be helpful in keeping track of children or finding information about which rides and events had long lines. This information could be dynamically updated through the base stations. Visitors could also access maps to see where the closest bathrooms or foodstands were located at any given time. If visitors wanted to meet new people, the spotting feature could also help. It was also possible to play a game on your Spotme device with someone else in the park while you waited to enter an attraction.

Park owners could use the Spotme device for location-based sales, *i.e.*, if you were passing an ice-cream stand and it was very hot outside, the Spotme device might indicate a promotion for ice cream. This type of selling had increased efficiency, as sales could be tailored to user profiles. Park owners might also send surveys to people who had the device. Lastly, visitor statistics could be tallied from the system. For example, the overall flow of people into attractions could be monitored with Spotme, helping to increase the efficiency of the park as a whole. This type of information might prove very valuable for park owners seeking to maximize revenues.

3.2 Cruise Ships

The ability to access static information such as maps and schedules on Spotme would be useful. The spotting and radar functions of the device would be highly valuable aboard pleasure cruises as people sought new acquaintances with similar interests and located those they had already met. Furthermore, ordering and billing at restaurants and bars could be automated. For example, someone at the pool could order a drink after looking at an electronic menu and have the bill auto matically charged to his/her room. Spotme devices could also be used to control access to various areas of the ship. Lastly, calculated visitor statistics could help cruise organizers.

3.3 Business Conferences

Networking was often a priority at business conferences, and using Spotme as a facilitator could greatly enhance participants' productivity at an event. Spotme users could locate people they wished to speak to easily and discreetly. Organizers could increase the efficiency of their events by sending updated information to all participants instantaneously. For example, if the location of a class were changed, this information could immediately be communicated to everyone. Shockfish had explored this path by distributing a survey at IMD, using Pulcrano's connections, in the summer of 1999. Shockfish also thought that business conference users, once impressed, were influential enough to aid in capturing other market segments.

3.4 Event Venues

Venue owners would be interested in offering Spotme to their clients as a pre-mium service, thereby attracting more bookings at their location. Shockfish had started exploring this avenue with the owners of ExCel, a new event venue at Canary Wharf in London. ExCel sought to position itself as the most technologi-cally advanced venue, complete with wireless network, and sought the exclusive rights to the Spotme product in England.

The data on people movement that Spotme tracked also provided valuable in-formation for selling advertising at event venues. For example, it was possible to say exactly how many people at a trade show passed by a specific advertising sign. In this way it might be possible for owners to earn more advertising revenue, not to mention the possibility of running advertisements directly on the Spotme device.

3.5 Social Spots

Large nightclubs or meeting places where people went to find people they did not previously know might benefit from Spotme. Imagine avoiding awkward romantic advances by approaching prospective matches with information in hand about mutual interests!

4 The Contract

On August 4, 2000, Shockfish was approached by a firm from Zurich. The firm specialized in running the services associated with shareholder meetings for large corporations. This meant that they provided all the equipment for presentations, seating, lighting, and voting, as well as support during the event. The company had previously used wireless voting devices produced by an Italian firm but now felt that Shockfish could provide a better mechanism that would be quick and easy to use and work well for large meetings. The firm believed it could attract more clients with Shockfish's help.

After a month of negotiations the final contract for signing was at Shockfish on September 10, 2000. The contract specified the delivery of an entire wireless voting system for running a shareholder meeting. Shockfish would be given CHF 100,000 upfront and then another CHF 2.5 million upon delivery of 5 base stations, 3,450 handsets, 46 charge suitcases, 1 base server PC, the voting software (database and applications), and the custom interface software. Shockfish was so confident in its abilities that it felt it had the upper hand in negotiations. It managed to include a CHF 500,000 payment in the contract simply for abandoning its core operations.

If Pulcrano signed the contract, Shockfish would be entirely devoted to the project for the next four months, making no progress on its core product, which was already three to four months behind schedule. Would four months away from its task allow competitors a chance to catch them before it reached one of the larger markets it had identified? However, Shockfish had generated no revenue thus far, and CHF 2.5 million would completely change the company. Specifically, it would be possible to leverage the success into a more favorable structure for the funds the company was to receive from angel investors. Although investors had promised CHF 2 million in May 2000, the division between equity and loans was undecided and due diligence was still being carried out. Additionally, the CHF 2.5 million would allow Shockfish to survive for a few more years without securing additional rounds of financing.

Knowing he must decide by morning, these issues pounded in Pulcrano's head. It was going to be a long night.

CASE 1-4
Redigo: Putting a New Electric Vehicle on the Road[1]

Jean-Philippe Deschamps and Michèle Barnett Berg

At a highly publicized launch during the Swiss national exhibition, Expo 2002, a small startup company named Triax AG introduced a new compact electric vehicle called the Mobi3. Swiss retailer chain KUNZLI had ordered 100 of the small electric three-wheelers to move people across exhibition sites. At the end of the exhibition, the Mobi3 vehicles were quickly resold to the general public. Three years later, the cars were still in circulation, fetching resale prices competitive with their original list price.

Hubert Grandjean, a member of the concept teams of the Swatchmobile and the smart, was one of the creators of the Mobi3. Inspired by the success of his early prototypes, he was wondering how to bring an enhanced version of the vehicle, later renamed Redigo, to market. Since 2002 Triax AG had become dormant and Grandjean had formed another innovation-oriented startup – Ingenio Ltd. – with partner Gilles Morvan.

In 2005 a new generation of lithium-ion batteries had become available. They were four times lighter than traditional lead-acid alternatives, provided more autonomy, and required half the recharging time. The Ingenio Ltd. partners believed that it would lead to a major improvement in the performance of electric vehicles and hence would enhance their attractiveness. The time appeared right to launch Redigo. Three years after the Mobi3, Grandjean and Morvan wondered how to kick-start their new concept:

- How could they attract investors to help them develop, manufacture, launch, and distribute the Redigo on their own?
- Should they partner with or license an external company in related vehicle industries to launch it together? And which type of partners should they approach?
- How should they position the vehicle, define the target user, and prioritize the geographic markets to be addressed?

Copyright © 2002 by IMD – International Institute for Management Development, Lausanne, Switzerland. Not to be used or reproduced without written permission directly from IMD.
[1] All names of individuals and some of the Swiss companies listed have been changed.

1 History

In the early 1990s the producer of Swatch watches, SMH AG (Swiss Corporation for Microelectronics and Watchmaking Industries, Ltd.) came up with the idea of an environmentally friendly but stylish supercompact car: the Swatchmobile. Hubert Grandjean, a creative design engineer and specialist in advanced plastics technology, was a key member of the project team. The Swatchmobile was conceived as a light, inexpensive, easily recyclable vehicle for city traffic that used alternative energy sources. The head of SMH, Nicolas Hayek, proposed the project to Volkswagen, who declined in 1993. He then approached Mercedes-Benz AG, who bought the idea and formed a joint venture with SMH – Micro Compact Car AG – from which came the smart. The name came from the combination of the words Swatch, Mercedes, and Art.[2]

Grandjean was sent to Daimler-Benz in Germany to continue working on the Swatchmobile, which would eventually turn into the smart. However, he became disillusioned and frustrated as the outcome of the smart strayed significantly from the original Swatchmobile concept that he had passionately promoted. The smart was no longer electric, nor was it made completely out of recyclable materials, and the cost had increased significantly since it needed to be produced like a car. Determined to remain loyal to his original concept, Grandjean left Daimler-Benz in 1996 to join Triax AG.

1.1 Creation of Triax AG

Ecology, research, creation, and engineering led to the creation of Triax AG. The company was founded in western Switzerland by a handful of entrepreneurs to develop, produce, and market an internationally patented ecological local transport vehicle. Its initial round of funding through private sources had brought them approximately € 8.5 million.

The initiators and founders of Triax AG were vehicle design and development professionals experienced in the fields of propulsion technology, environmental and sustainable technologies, transportation design, advanced plastics technology, strategic company management, corporate finance, and global sourcing. The organization consisted of four full-time employees and a steering committee of three well-known executives from the banking, investment, and retail industries.

While at Daimler-Benz, Grandjean had continued to refine the concept of a four-wheel electrical vehicle in his free time. This was the proposal he brought to Triax AG when he joined the firm. After further discussion and analysis, the Triax AG partners agreed that an electric three-wheeler would be less complicated

---

[2] Yüksel, Özlem, and Prof. Hau L. Lee. A Smart Supply Chain for a Smart Product: Micro Compact Car Smart GmbH. Stanford Case No. GS-32 SGSCMF-001-2004, 2003.

and expensive to develop and produce than a four-wheeler.[3] This decision was determinant and the Mobi3 concept began to take shape. (*See Exhibit 1 for pictures of the Redigo – assembled and unassembled.*)

1.2 Events of 2002

Triax AG managed to attract the attention of Swiss hypermarket giant KUNZLI, who was interested in the vehicle as an image builder for the 2002 Swiss World Exposition, which was about to start. KUNZLI saw this as a very attractive promotional platform as its focus was on ecologically friendly products and reusable energy and it was a highly publicized event. At that time, KUNZLI was considering selling and servicing alternative mobility vehicles in some of its larger stores. They placed an initial order for 80 of Triax's Mobi3s. However, upon further analysis, the cost to set up distribution appeared too high and they decided not to pursue the project further. Regardless, the initial vehicles were well received and were later resold to private individuals at the original target price of € 6,500. With the remaining components they had ordered, Triax then assembled 20 vehicles, which were sold to private individuals in France, Germany, and Switzerland.

Having learned about the Mobi3, a European performance car manufacturer expressed an interest in purchasing licenses of the vehicle as a way to explore the alternative energy vehicle market without allocating R&D efforts in that area. It conducted an intensive review of the Mobi3, producing a business plan outlining how to best produce and distribute these vehicles. Ultimately, it decided not to engage further. (*See Exhibit 2 for excerpts of review findings.*)

In the fall of 2002 Grandjean decided to leave Triax AG, which had begun to disband following the Expo. After his departure, the Triax AG partners went their separate ways to pursue other interests. The patent for the development and design concept of the Mobi3 (and what would be the basis of the Redigo) continued, however, to be legally owned by the Triax AG entity. Although the company was dormant, it was never liquidated and ownership was still divided among the original partners.

---

[3] In electric cars, motors are generally placed directly on the wheel. A four-wheeler requires at least two electronically synchronized power units. This leads to a higher cost in terms of electrical, electronic, and mechanical components than for a three-wheeler, which requires only one motor.

1.3 Ingenio Ltd.

Grandjean formed Ingenio Ltd. in 2003 in western Switzerland to offer consultancy services on various engineering and design-related projects and develop patents on product innovation related to the alternative mobility market. Eighteen months later Gilles Morvan, a marketing specialist, joined the firm.

While the mobility concept was intriguing and the Mobi3 had been well received and continued to resell at its initial list price, the investment community was reluctant to take a stake. Ingenio Ltd. met with over two dozen investors (venture capital firms and private angel investors) looking for a € 13 million investment, of which € 2.5 million would be used to reengineer enhancements for the Redigo. While their pitch and proposal sounded intriguing, it did not have the same appeal as other popular investment opportunities that were being offered in the biotech, pharmaceutical, and telecom industries. As no additional funding was gained from their efforts, the focus on Redigo dropped significantly. The two partners only worked on it between paying engagements.

The same environmental considerations attracted the attention of the Chinese government of the province of Guangzhou, which was interested in promoting eco-friendly transportation vehicles since traffic congestion and pollution issues were on the rise. Grandjean and Morvan were asked to develop a streamlined five-seater/four-wheeler electric vehicle that could be sold for less than € 7,000 based on a yearly production of 200,000 vehicles. The vehicle was to be assembled in China and would be distributed locally. Ingenio Ltd. had partnered with China's largest battery manufacturer, and the prototype was expected to be delivered in 2006.

Learning about new technology developments in batteries prompted the Ingenio Ltd. partners to rethink their efforts. They approached a former, influential advisor of Triax AG, Peter Stalder, one of the iconic Swatch innovators and a serial entrepreneur. When he heard about the benefits and performance of the new lithium-ion batteries, Stalder expressed an interest in being involved. His personal commitment to help financially and to serve as an active advisor with a significant and influential network looked attractive to Ingenio Ltd. Additionally, he would be pivotal in managing Triax AG's ownership of the Mobi3 patent.

2 The Vehicle

2.1 Basic Concept

Triax's aim when launching the Mobi3 was to bring a patented ecological local transport vehicle to market. It was an urban vehicle for short-distance (maximum

200 km) travel that was designed as a cabbed three-wheeler with two seats, one behind the other. It looked like a cross between a car and a scooter and required no car driving license in Europe, just the registration and authorization needed for a scooter below 125 cc.[4] The vehicle used an extremely simple concept of modular construction and assembly. The assembly kit was its guiding principle of production.

The entire cab of the vehicle consisted of only three plastic components. A single mold form was used to create the shell, made of 100% recyclable thermoplastics. The body required no chemical coating or paint as it would never rust.

The vehicle ran on batteries stored in a simple electric drawer attached to part of the chassis. The typical lifespan for batteries was up to three years. An electric motor drove the rear wheel through a Kevlar belt, as found in many up-market motorcycles.

Due to the very small number of individual components, the vehicle could be produced in an "assembly center" requiring no heavy infrastructure, which lowered production costs significantly. This allowed different partners to produce it in their own factories with an investment of less than € 10 million and to sell it through their own efficient distribution networks. Triax's initial plan was to design and build the first assembly center module of this kind and begin production in Switzerland. Following this, they had planned to sell production licenses for the vehicle and identical assembly centers could be established in other countries.

Manufacturing costs were a fraction of the costs of conventional small cars due to a low level of investment in machinery and tooling. Standard automotive off-the-shelf components could be used whenever possible. Ecological sustainability could be maintained in production through the use of recyclable materials. No compromises in terms of safety and comfort were taken and the design was focused on being original and trendsetting.

2.2 Performance/Features

At its target price of € 6,500 without batteries, the Redigo was competitively priced between scooters and light motorcycles, which ranged from € 3,250 to € 7,000, and small cars like the smart, which ranged from € 9,000 to € 14,000. Users would have to pay an extra € 1,600 to lease a set of batteries for three years, beyond which, having lost 20% of their nominal charge, they would have to be replaced.

---

[4] European driving license requirements were generally as follows: Below 50 cc displacement and from 14 years of age – no driving license required. Between 50 cc and up to 125 cc and from 16 years of age – either a (B) driver's license or A1 license (obtained after completing a highway code exam). Above 125 cc and from 18 years of age a driving license was required (obtained upon successful completion of a highway code exam and driving test).

With its new and much lighter lithium-ion batteries, the Redigo could reach a top speed of 120 km/hour on motorways, a significant improvement over the Mobi3's 85 km/hour. It had autonomy of up to 150 km on one battery charge, which took one hour, but 80% of the power could be regained within twenty minutes. Compared with the lead-acid batteries present in most electric vehicles, including the Mobi3, the new lithium-ion batteries improved vehicle autonomy and shortened battery recharge time by 50%.

Body styling could be modified simply since it was manufactured in a mold and required no welding. Body colors could also be modified easily to adapt to changing consumer tastes.

2.3 Customer Satisfaction

As of 2005, there were 80 Mobi3 vehicles on the roads in Switzerland, Germany, and France. The Maverick Garage in western Switzerland, owned and managed by one of the original members of Triax AG, continued to resell Mobi3s and other electric and hybrid vehicles, in addition to providing service and repairs.

Resale values of the vehicles remained identical to or higher than the original purchase price three years after being introduced to the market. Ingenio Ltd. continued to check in with the Maverick Garage quarterly to gather market intelligence and customer feedback on usage and repairs. Their overall findings were positive and the negative feedback was incorporated into developing a better vehicle with Redigo. However, very little data was available on customer profiles except that it was almost exclusively male. (*See Exhibit 3 for customer feedback.*)

3 Market Developments in Urban Mobility

As environmental concerns escalated globally, fuel prices rose, and traffic congestion became more of a political issue, the emphasis on improving transport and alternatives to petrol cars began to garner more attention. For example, in 2003 the traffic congestion in central London was so bad that the city began charging motorists a congestion fee. To enter the eight-square-mile zone of central London by car, a driver was required to pay a toll of about £5. Similar programs to reduce traffic began in Toronto, Singapore, Melbourne, Oslo, and, on an experimental basis, in Seattle in the USA. The authorities in Norway had equally ambitious policies for the increased use of electric vehicles, including incentives offering considerable financial and time-saving benefits. This included no taxes, no road toll, free parking, and use of the bus lane. In 2004 China had hosted the Challenge Bibendum – Michelin's effort to promote sustainable mobility. The race inspired many small and large companies to create alternative energy vehicles.[5]

3.1 The Rise in Popularity of Scooters

To combat these new tolls and beat traffic, many large cities were experiencing a rise in scooter usage. In Europe scooters had always been popular, and Italian-based Piaggio SpA held a large share of that market. For example, scooters were gaining new popularity as commuters looked for other transportation options to combat traffic. In the UK, scooter sales increased by 130% between 2000 and 2004 with approximately 155,000 scooters on the streets. Some of this increase was due to the availability of larger-capacity machines in the 100-cc to 150-cc range. To meet this demand, Piaggio launched the 139th of its legendary Vespa, the LX. Piaggio had a selection of scooters to appeal to all ages and budgets, and prices ranged from € 1,000 to € 7,000.

However, this surge was creating a new category of problems. Some large metropolitan cities like Paris were contemplating banning or limiting the use of scooters due to the noise and pollution they created.

The scooter market was also booming outside Europe, particularly in emerging markets and a large number of producers, notably from Japan, China, and India, competed fiercely for their share of the market. Piaggio entered the Chinese market and hoped to capture the top-end segment, expecting that urban residents with higher disposable income would be able to move from low-cost utility scooters to fancier models with better features. It would be a challenging market to capture since Honda and Yamaha and many other manufacturers were already well known to consumers. Piaggio distributed its scooters through a range of dealers focused on selling scooters and motorcycles. Some of the larger distributors had Piaggio Centers that offered a wide range of Piaggio products, not only scooters, but also three-wheel industrial minitrucks, such as Piaggio's popular Ape. (See Exhibit 4 for further information on various Piaggio scooters.)

3.2 BMW and the C1

Described as a small motorcycle with a safety cage, the BMW C1 had short-lived success. It targeted urban European markets as an appealing alternative to small cars and crowded mass-transit systems. Sold as the ultimate congestion beater, it was easy to maneuver and park while simultaneously making a fashion statement.

---

[5] www.challengebibendum.com

The C1's most innovative design feature was its emphasis on safety. BMW claimed that its crash tests showed that, in a head-on collision, the C1 offered a standard of accident protection comparable to a modern subcompact car. It was the first two-wheeler of its kind to promote such safety features. Because of these safety claims, the German, French, and Spanish authorities allowed an exemption to the helmet law for the C1. BMW failed to obtain similar legislation in other larger markets like the UK, which greatly impacted its sales. Moreover, the C1 was considered expensive in comparison to the performance and features it offered. Many customers found it was underpowered for a BMW cycle and overpriced at a starting price of €7,000. After selling 10,614 in 2001 and only 2,000 in 2002, BMW ceased production in October 2002. (*See Exhibit 5 for additional information and features of the BMW C1.*)

3.3 Developments in Small and Hybrid Cars

Ever-increasing urban congestion and pollution was putting renewed pressure on car manufacturers to develop eco-friendly vehicles.

3.3.1 DaimlerChrysler's smart

The smart had been designed to respond to these trends and was priced competitively at a starting price of €9,000. However, while building a very strong brand and loyal following, the smart had a difficult start and continuously missed its sales and profit targets. In early 2005 DaimlerChrysler (Daimler-Benz became DaimlerChrysler in 1998) carried out a significant restructuring plan in order to focus on profitability for the smart division by 2007. This included the discontinuation of the smart Roadster as well as the anticipated Formore model, a mini-SUV designed to serve as the brand's entry into the US market. On the positive side, however, within the first four months of 2005, the company sold 47,700 smart vehicles in Europe, which was an increase of 33% over the sales figures for the same period in 2004. Additionally, the smart car made its way into the US market through an American alternative vehicle distributor called ZAP Technologies (Zero Air Pollution). ZAP received the patent from the United States Environmental Protection Agency to convert the smart car to meet all American emissions and safety standards. There was a significant US interest in the smart as gas prices continued to rise. By mid-2005, ZAP reported that it had received over $1 billion in advance purchase orders from US car dealers who would be selling the Americanized smart cars. (*See Exhibit 6 for more information and features of the smart car.*)

3.3.2 Hybrid Vehicles

A few experimental hybrid vehicles (*i.e.*, combining petrol and electric propulsion) had been launched in the past decade, notably by Volkswagen with its hybrid Microbus taxi and Mazda with its Titan truck. However, this type of vehicle did not catch much public attention until Toyota started selling its Prius model, which gained fast acceptance in the USA, Japan, and some European markets. In 2004 Toyota sold over 50,000 hybrids and exported 23,000 Prius hybrids to North America in the first quarter of 2005 alone. This accounted for 10% of Toyota's entire shipment to that region. Despite its high starting price of € 17,000, the Prius began to gain a status quality as many celebrities bought it, and it was even used to chauffeur actors to the 2003 Academy Awards. (*See Exhibit 7 for additional features and information on the Toyota Prius.*)

3.3.3 Fuel-cell Vehicles

In the long term, most car manufacturers have been working on fuel-cell powered electrical vehicles that would convert hydrogen into electricity and were believed to be the ultimate solution to the emissions problem. The expectation was that fuel-cell technology would be widely available in 2015.

3.4 The Microcar Category

As the streets grew more crowded, smaller vehicles like the microcars – or quadricycles, as they were know in some markets – offered alternatives to combat traffic congestion. In Japan, microcars (called Keicars) were produced by a number of specialized manufacturers. They were also attractive to consumers because owners received tax and insurance credits and special parking benefits.

In a number of European countries people were allowed to drive microcars below a certain engine displacement, for example 125 cc, without a driver's license – except on motorways. Microcars allowed drivers to make short-distance trips necessary for daily life. Legislation on the driving of this type of vehicle differed from one country to another and in some countries young people could drive a microcar from as early as 14 years of age, on the condition that they passed a highway code test. Because of this, a new market had started to emerge with microcars being bought by parents as a safer alternative to a scooter or motorcycle for their under-driving-age children.

As of 2004, 30,000 new microcar vehicles were sold on average each year in Europe, and the registration rate had increased by 25% since 1993 (the 10,000 registrations threshold was crossed in France in 2004). About 300,000 microcars were on the roads in Europe, including 140,000 in France, 42,000 in Italy, and 39,000 in Spain – the three main markets for such vehicles. While Austria and Belgium had become stable markets, Germany, Great Britain, and Russia were poised to open their market in 2005. Finland and Sweden, and also, in the longer term, Romania, Turkey, and Hungary, were seen as potential markets.[6] Microcars were popular because they were affordable, compact – thereby making parking in congested areas easy – and simple to access either through rental or purchase. Their downside was a lack of performance as speed was limited. Besides, they were often noisy and created pollution.

The interest for microcars was particularly developed in France where three specialized manufacturers targeted the 80 million Europeans without a driving license. The microcar served two specific markets. As no driver's license was required, these small cars appealed to drivers who had lost their licenses due to a traffic offense. Many rented a microcar for the suspended license period. The other group was mostly senior citizens in rural locations who had never obtained their license (70% were men over age 50). In France, three major companies led the competition, all with a range of diesel-powered models:

- Aixam held first place in the microcar market with a 40% market share and 14,000 units sold in 2004. Cars cost between € 7,500 and € 13,000.
- Microcar, produced by yacht-building group Bénéteau, held second place with approximately 25% of the market. Their cars cost between € 9,500 and € 12,500.
- Ligier, a specialist of quads and minicars, created by a former racing-car builder, held third place with a 24% market share and exported 70% of its production. Ligier cars cost between € 7,500 and € 12,000.

(See Exhibit 8 for more features and information on microcars.)

Microcars were distributed through specialized dealer networks selling scooters and motorcycles, which had exclusive selling and servicing relationships with one or two microcar manufacturers. Aixam boasted 150 distributors and branches within France and other French territories in addition to a network of 900 agents throughout the world. Additionally, Aixam created partnerships with specialized garages in France to be part of their network. Bénéteau's Microcar had extensive distribution through approximately 100 dealers in France, as well as representation in other French territories and the other European countries. In 2005 Ligier had close to 100 Ligier distributing dealers in France and at least one other in every country in Europe.

---

[6] www.cyburbia.org

3.5 The Electric Vehicle (EV) Market

Electric cars are not really a novelty. The first electric car was designed in 1881 and Paris had electric taxis equipped with lead-acid batteries in 1911. However, electric technology failed to live up to initial expectations because battery capacity had not increased enough to be competitive and comparable to internal-combustion-engine cars. Additionally, sales of electric vehicles were low and had not progressed because of their high prices and low performance. However, as technologies continued to evolve, the demand for electric vehicles remained steady. Two types of consumers were interested in such vehicles: drivers that considered themselves mavericks, always having the latest inventions and cutting-edge technology, and the eco-enthusiasts, who wanted products that were environmentally friendly and ecologically sustainable. On the commercial front, fleet buyers for organizations like the energy and postal companies were showing renewed interest in electric vehicles.

3.5.1 Three-wheel Electric Vehicles

Most of Redigo's direct competition was expected to come from this category, dominated by two products:

- TWIKE: Launched in 1986 and produced in Switzerland. As of 2005, there were fewer than 1,000 in circulation worldwide. The Twike was a hybrid mobility vehicle with pedals and an electric motor. Both could be used to power the vehicle. Its performance was low and it was distributed through a handful of specialized electric vehicle dealerships, many of whom were Twike exclusive. Prices for the Twike started at € 12,500 plus € 2,100 for the batteries.
- CityEl: First manufactured in Germany in 1995, the CityEl was a small single-passenger electric vehicle. Over 5,000 had been sold, and CityEl differentiated itself by the fact that buyers could take a two-day course to learn how to assemble their own vehicle. Due to limited speed, these vehicles were not allowed on the motorway. CityEls were distributed by the manufacturer and through a specialized network of dealers.

(*See Exhibit 9 for additional information and features of three-wheel electric vehicles.*)

3.5.2 Four-wheel Electric Vehicles

Within the electric vehicle (EV) market, the four-wheel/four-passenger segment appeared to have the most potential to capture car consumers. It was a tough segment to crack due to cost, competition from large automotive makers, and the effort to bring to market a vehicle that would appeal to the budget, taste, and expectations of mainstream drivers. Two types of electric cars had attracted a fair amount of attention:

- Kewet Buddy: Originally created in Denmark but later produced by ElBil Norge AS in Norway, the Buddy aimed to be the least expensive electric four-wheel car on the market. In January 2005, the newly developed sixth generation of the Buddy was introduced to the market. It was delivered with battery packs including lithium-ion technology in May 2005. Production planning was for 80 vehicles a year and with no plans to distribute outside Europe. Kewet's starting price was € 10,500.
- BlueCar: French state-owned electric power group EDF formed a joint venture with investor Vincent Bolloré to produce an electric city car designed along the principles of a "modern Mini." The prototype was launched at the Geneva Motor Show in 2005 and sales were not expected until 2008 at a starting price of € 15,000.

(See Exhibit 10 for further information and features of four-wheel electric vehicles.)

Electric vehicles could be purchased directly from the manufacturers or were distributed through specialized dealers who carried other electric mobility vehicles.

3.6 Other Developments in Mobility

To ease traffic congestion, a number of organizations had appeared in Europe to promote car sharing. The most important of them was Mobility CarSharing in Switzerland. Created in 1987 by the combination of two cooperatives (ATG and AutoTeilet-Suisse) and ShareCom, Mobility CarSharing offered its 60,000 customers 1,750 small cars at 1,000 locations in 400 towns, making it the leading car-sharing enterprise with by far the highest customer density in the world. The success was attributed to a strong market presence, enhanced by partnerships both within and outside the domain of public transport and the continuous adaptation to customer requirements. As of 2005, the system could be accessed in Switzerland, Austria, and some parts of Spain.

Mobility developed its own fully automated car-sharing technology that allowed for all processes from reservation to invoicing to be carried out electronically: via Internet (56%), telephone voice computer (12%), or the car's onboard computer (15%). Only 17% of reservations were still made through a 24-hour service center, which served more as a place for special orders. All Mobility cars came equipped with an onboard computer including display and keyboard. As of 2004, 15% of the cars in Mobility's fleet were accessible with a Mobility card allowing the driver to open the car, turn on the ignition, and release the immobilizer – all accomplished with the pressing of a start button.

Some industry observers believed that organizations like Mobility CarSharing offered ideal conditions for launching new-generation electric vehicles for two reasons. Firstly, their cars were typically used on small distances within urban areas. Secondly, they were generally found in fixed locations, for example in the parking area of large Swiss railroad stations, which could easily be equipped with a power outlet to allow battery recharges between uses.

4 Where to Go with Redigo

As they reviewed the market and competitive environment in which the Redigo would have to operate, Grandjean and Morvan were aware that they would have to proceed in two steps. Firstly, they would have to clarify their product concept and value proposition and determine the profile of the most likely customers. Secondly and because of their limited resources, they would have to choose between several strategic options involving very different types of partners. Thirdly, they would have to work on a detailed marketing plan focusing on the most attractive geographical market. These three questions were obviously linked as the partners to be approached would be different depending on the product concept and vehicle positioning. Each type of partner would also react differently to Ingenio Ltd.'s business proposition.

4.1 Defining Redigo's Product Concept and Positioning

Although the media had talked of the Mobi3 as being a small electric car, Grandjean and Morvan were aware that their vehicle was cutting across different vehicle segments and could be marketed to widely differing customer segments and geographic areas. How should they position the Redigo and whom should they target?

For example, they asked themselves the following questions:

- Is there a market for the Mobi3 as it exists today? And in which segment of the market? For which type of customer?
- If yes, should they market Redigo as a basic and modern transportation mode for congested and polluted cities in emerging markets like China, or as an alternative urban car for eco-enthusiasts in highly developed markets?
- Was Redigo really competing with small cars like the smart? With microcars? Or should it be positioned as a safer and environmentally friendlier alternative to the scooter for commuting young people?
- Should the Redigo be offered as a vehicle people would buy and use for themselves, or should it be viewed as a vehicle for institutional, public service (*e. g.*, mail distribution), mobility, or promotional applications?

The answers to these questions would undoubtedly be influenced by the operating costs of the Redigo *vs.* competing vehicles. (*See Exhibit 11 for a comparison of Redigo's operating costs with those of some potential competitors.*)

- Should Triax AG develop another type of electric vehicle using the same technology and construction method? And which type – a larger four-wheeled vehicle to accommodate more passengers or a vehicle that seats two people side by side? Or other styles and sizes, *etc.*?

4.2 Exploring Strategic Options

Depending on the answers to the above questions, Grandjean and Morvan knew that they had at least three broad and nonmutually exclusive options:

- Option 1: Do it on their own. That meant finding investors willing to contribute the € 13 million required to complete the car design, order tooling, build an assembly plant, and set up a distribution network, probably piggy-backing on an existing dealer network. Grandjean and Morvan had prepared a business plan under that option for the assembly and sale of 5,000 vehicles over five years in Switzerland.

(*See Exhibit 12 for a copy of Grandjean and Morvan's business plan.*)

This option, however, would provide a limited coverage of markets peripheral to Switzerland – Austria, France, and Germany.

However, given the low manufacturing investment required, the company could easily design and license another body on the same platform and license it to partners in other geographical areas of the world.

- Option 2: Partner with a vehicle manufacturer to launch the Redigo. The type of partner to be approached would, of course, have to reflect the positioning of the vehicle and volumes expected. That is, should they partner with a microcar company, a scooter company, or another type of company?
- Option 3: Sell licenses and keep developing variants and design upgrades of the Redigo while the partners handle their own production and distribution. Here again, many issues remained as to the nature of potential partners and potential financial arrangements?

Grandjean and Morvan were sure that they had the potential to make a great contribution to the world's urban transportation problems. However, they still had to kick-start the adoption process for their small revolutionary vehicle, and that was a real challenge!

For further information on the EV industry, market trends, and related information, see the following websites:

4.2.1 General information on alternative vehicles

www.eere.energy.gov
www.energy.ca.gov
www.evfinder.com
www.evuk.co.uk
www.evworld.com
www.extremetech.com

4.2.2 Vehicles

www.aixam.com	www.autos.msn.com	www.batscap.com
www.bmwworld.com	www.cityel.com	www.kewet.com
www.ligier-automobiles.com	www.microcar.com	www.micro-cars.co.uk
www.mobility.ch	www.smart.com	www.twike.ch
www.zapworld.com		

Exhibit 1
The Redigo

- Starting price   € 6,500 plus € 1,600 for 3-year lease of batteries
- Length   3,162 mm
- Height   1,583 mm
- Width   1,553 mm
- Weight   600 kg
- Seats   1 + 1, cab style, one in front of another
- Maximum Speed   120 km/h
- Battery   Lithium-ion
- Battery charging time   Range 150 km and total recharge would take 1 hour
                          and 80% recharge could be done in a few minutes
- Body   Double-layer polyethylene – 100% recyclable and
                          available in many colors
- Chassis   Aluminum frame

Exhibit 1 (continued)

Assembled Redigo car

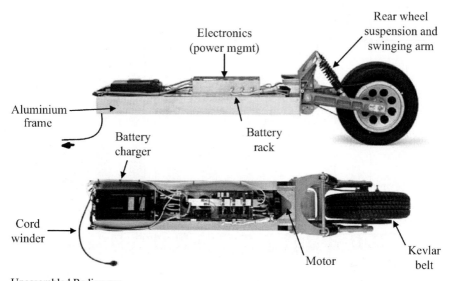

Unassembled Redigo car
Source: Company literature

Exhibit 2
Tentative Buildup of Redigo Vehicles per Month[7]

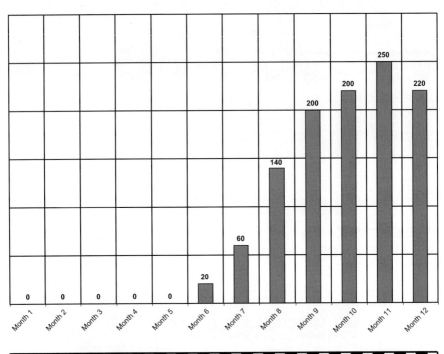

Week	19	20	21	22	23	24	25	26	27	28	29	30	31	32	33	34	35	36	37	38	39	40	41	42	43	44	45	46	47	48	49	50	51	52
Mobi3 per day	0	0	0	0	1	1	1	1	2	2	4	4	6	6	8	8	10	10	10	10	10	10	10	10	10	10	10	10	10	10	11	11	11	0
Mobi3 per week	0	0	0	0	5	5	5	5	10	10	20	20	30	30	40	40	50	50	50	50	50	50	50	50	50	50	50	50	50	50	55	55	55	0
Mobi3 per month	0	0	0	0	20				60				140				200				200				250					220				
Total	0	0	0	0	5	10	15	20	30	40	60	80	110	140	180	220	270	320	370	420	470	520	570	620	670	720	770	820	870	925	980	1035	1090	1090

Triax AG-Business-Proposal foresees production output of 1,000 vehicles in the first year of production. SOP is possible approx. 12 to 18 months after funding is obtained.

---

[7] Within 6 months after start of production, 250 vehicles could be produced per month.

Exhibit 2 (continued)

The use of supplied systems (modules) increases the efficiency of the assembly line and the value-added chain.

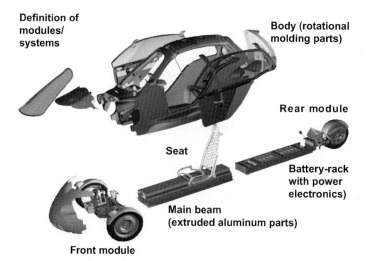

Definition of modules/ systems

Body (rotational molding parts)

Rear module

Seat

Battery-rack with power electronics)

Main beam (extruded aluminum parts)

Front module

Current costs are far higher than planned target costs.

Modules	Initial Cost (€) 100 units	Target Costs (€) 1,000 units/year	Target Costs (€) 5,000 units/year
Front module	1,555.00	767.00	647.00
Chassis – main beam without battery	3,361.00	1,475.00	971.00
Rear module	2,250.00	1,174.00	971.00
Body	2,930.00	1,726.00	1,619.00
Seats	416.00	190.00	162.00
Total	10,512.00	5,332.00	4,370.00

Source: Tentative manufacturing plan prepared by a European performance car manufacturer

Exhibit 3
Customer Feedback from Mobi3 Owners

Positive Comments	Negative Comments
• Great concept • Very convenient for short distances • Great feeling in traffic jams: while the other motorists keep polluting, you know you are not at anytime polluting the environment • Easy handling • Good acceleration • Enjoy the feeling of really contributing to reducing impact on global pollution • Problem-free vehicle seems to need less maintenance than a regular car • A breakthrough design • A smart and different way to show off	• Absolutely not designed for women: There needs to be more space in the back to put groceries and/or to accommodate 2 kids if you are going to use this to take them to school • Definitely spartan. The design of the interior needs to be improved to accommodate space for handbags and other items • Opening and closing the door takes a while to understand • Steering is a bit physically demanding, driving position awkward • Range definitely has to be improved • Ventilation has to be improved for foggy, rainy, and/or cold weather as condensation forms all over windows • In general, and on cobblestones in particular, very noisy, making radio listening impossible • Sporty concept with some disadvantages, rough suspension and strange steering position. "Mustang" seat like the WWII fighter aircraft. You need to be fit to get into the car, a bit of an acrobat

Source: Company literature

Exhibit 4
Piaggio Scooters

	PX 150 €1,878	Granturismo 200L €4,338	LX 125cc 4T €3,495	X9 Evolution 500 €6,863
Engine	Single cylinder catalyzed 2-stroke	Single cylinder, 4-stroke, 4-valve, with two-way catalytic converter and secondary air system	Single cylinder 4-stroke LEADER, two-way catalytic converter, and secondary air system SAS	Single cylinder, 4-stroke 4-valve Piaggio MASTER with counterrotating balancer shaft and two-stage catalytic exhaust
Displacement	151 cc	198 cc	125 cc	460 cc
Bore	57 mm			92 mm
Stroke	58 mm			69 mm
Bore x stroke		72 × 48.6 mm		
Fuel	Unleaded petrol	Unleaded		Unleaded
N.O.R.	min. 95			
Max. power / Max Power / Power	8.2 CV	14.7 kW/20 bhp at 8,500 rpm	7.6 Kw (10.3 bhp) at 8,000 rpm	39 bhp at 7,500 rpm
Induction	Rotating valve in crankcase	SOHC		
Induction/Distribution			SOHC, 2 valve	
Fuel supply				Electronic injection system
Start	Electric and kick starter	Electric with automatic choke	Electric and kick starter	
Lubrication		Wet sump		
Gear change	Four-speed, on handlebar			
Chassis		Pressed-steel monocoque		
Length		1,940 mm		
Length/width			1.800/740 mm	
Wheel base			1,280 mm	
Saddle height			785 mm	
Width to handlebars	Pressed steel monocoque	755 mm		
Dry weight			110 kg	
Frame	755 mm			Double-cradle, high-strength steel tube

Source: Dealership website

Exhibit 4 (continued)

PX 150 €1,878		Granturismo 200L €4,338		LX 125cc 4T €3,495		X9 Evolution 500 €6,863	
Length	1,810 mm	Wheel base	1,395 mm	Fuel tank capacity	8.6 liters	Length/width	2,140 mm/ 880 mm
Width	740 mm	Seat height	790 mm	Consumption (at 60 km/h)	39 km/l	Wheel base	1,540 mm
Wheel base	1,260 mm	Dry weight	138 kg	Top speed	91 km/h	Seat height	780 mm
Running weight	116 kg	Fuel tank	10 liters, of which 1.9 liters reserve	Acceleration to 30 m	4.6 seconds	Running weight	213 kg
Fuel tank capacity	8 liters (of which 1.6 liters reserve)	Max. speed	119 km/h	Emissions (exhaust, noise)	Euro 2	Fuel tank	15 liters
Emissions	Euro 1	Acceleration	At 30 m: 4.1 seconds –at 60 m: 5.9 seconds			Top speed	158 km/h
		Consumption (ECE40 cycle)	25 km/liter			Emissions	meet European Directive 97/24/CE chap.9 (noise) and 02/51/CE (exhaust) Euro 2
		Consumption at 60 km/h:	37 km/liter			* License	Requires motorcycle license
		Emissions (exhaust, noise)	Per Euro 2 Multidirective				

Exhibit 5
BMW C1

- Starting price   €7,000
- Length   2,075 mm
- Width   1,026 mm
- Height   1,766 mm
- Seat height   701 mm
- Fuel Capacity   9 liters
- Empty weight without fuel 185 kg
- Permitted total weight   360 kg
- Displacement   124.00 ccm (7.57 cubic inches)
- Engine type   Single cylinder
- Stroke   4
- Power   15.00 HP (10.9 kW) @ 9250 RPM
- Dry weight   185.0 kg
- Top speed   100.0 km/h
- Power/weight ratio   0.0811 HP/kg
- Fuel capacity   9.00 liters
- Economy   3 liters/100 km

Source: Dealership website

Exhibit 6
DaimlerChrysler's smart Car

smart fortwo coupe made by DaimlerChrysler

- Starting price   € 9,000 to top price: € 14,500
- Type   All-aluminum SOHC, 3 cylinders, 2 valves per cylinder, turbocharged, rear-mounted
- Displacement   698 cc
- Bore × stroke   66.5 × 67 mm
- Max power   45 kW/61 bhp at 5250 rpm
- Max. torque   95 Nm at 2,000 ~ 4,000 rpm
- Compression ratio   9.0 : 1
- Fuel type   Petrol RON95
- Driving wheels   Rear
- Transmission   Sequential 6-speed with manual/auto modes
- Final drive ratio   3.92:1

Dimensions and Weights

- Wheelbase   1,812 mm
- Front track   1,272 mm
- Rear track   1,354 mm
- Length   2,500 mm
- Width   1,515 mm
- Height   1,549 mm
- Curb weight   730 kgs
- Fuel tank   33 + 5 liters

Source: smart dealer website

Exhibit 7
Hybrids – Toyota Prius

- Starting price   € 17,000

2005 Toyota Prius Specifications
- Body Styles   4-Dr Sedan
- Engines   1.5L inline 4
- Transmissions   Continuously variable
- Drivetrains   Front wheel drive

2005 Toyota Prius Performance and Efficiency Standard Features
- 1,497 cc 1.5 liters inline 4 front engine with 75-mm bore, 84.7-mm stroke, 13 compression ratio, double overhead cam, variable valve timing/camshaft, and four valves per cylinder 1NZ-FXE
- Unleaded fuel with additional electric 87
- Fuel economy EPA highway (mpg): 51 and EPA city (mpg): 59
- Multipoint injection fuel system
- 11.9-gallon main unleaded fuel tank
- Power: 57 kw, 76 HP SAE @ 5,000 rpm; 82 ft. lb., 111 Nm @ 4,200 rpm
- Secondary power: maximum power (kw): 50, maximum power (hp): 67, maximum torque (ft lb): 295 and maximum torque (nm): 400

2005 Toyota Prius Handling, Ride and Braking Standard Features
- Four-wheel ABS
- Brake assist system
- Two disc brakes including two ventilated discs
- Electronic brake distribution
- Electronic traction control via ABS and engine management
- Immobilizer
- Spacesaver steel rim internal spare wheel
- Strut front suspension independent with stabilizer bar and coil springs, torsion beam rear suspension rigid with stabilizer bar and coil springs

Source: Dealership website

Exhibit 8
French Microcars

Aixam 500
- Starting price   € 7,500, €13,000 for larger models
- Height   2,674 mm
- Width   1,474 mm
- Wheelbase   1,737 mm
- Number of seats   2
- Drive   Front-wheel, Speed reducer-reverse gear (Vario-matic)
- Motor   Kubota Diesel, 2 cylinder, Liquid cooling system
- Cylinder   400 cc
- Power rating kW   4
- Power in HP rpm   5.4 at 3,200
- Battery   12 V, 40 Ah
- Boot capacity   600 liters
- Reservoir capacity   16 liters
- Fuel consumption   3 liters/100 km
- Weight   350 kg
- Total weight allowed   640 kg
- Maximum speed   45 km

Microcar by Bénéteau
- Starting price   € 9,500, top price: € 12,500
- Length   2,788 mm
- Width   1,493 mm
- Height   1,420 mm
- Engine   Petrol multi point fuel injection – Lombardini
- Fuel   Petrol unleaded
- Displacement   Twin cylinder four stroke in line 505 cc
- Power   21 bhp
- Transmission   CVT variable system auto via Comex geared differential 8:1
- Fuel capacity   16.1 liters
- Fuel consumption   Average rural 3.6 liters/100 km estimated
- Body   Composite formed with rigid aluminum safety cell structure
- Seating capacity   Two front (long wheel base four)
- Seatbelts   Three-point front and rear inertia type
- Maximum speed   116 km/h estimated

Exhibit 8 (continued)

Ligier Nova X-Size
- Starting price   € 7,750, top price: € 11,750
- Length   2,640 mm
- Width   1,400 mm
- Height   1,530 mm
- Number of seats   2
- Fuel consumption   3 liters/100 km
- Battery   12 V 38 Ah
- Frame   Aluminum frame and fiberglass body
- Fuel consumption   Average 3 liters/105 km estimated
- Maximum speed   120 km/h

Source: Dealership websites

Exhibit 9
Three-wheel Vehicles

Twike
- Starting price   € 12,500, up to € 15,500, additional cost of batteries € 2,100
- Length × width × height   2,650 × 1,200 × 1,200 mm
- Unloaded (curb) weight   220–250 kg, including battery (depending on equipment)
- Payload   2 people plus luggage
- Maximum speed   85 km/h
- Starting gradeability   > 20%
- Energy consumption   4–6 kwh per 100 km, starting with a fullcharge
- Chassis   3 wheels, independent suspension, suspension struts front and rear
- Brakes   rear hydraulic drum brakes, front mechanical disk brake (CH), front mechanical drum brake (D), rear parking brake, electric brake (regenerative)

Exhibit 9 (continued)

- Steering   Control tiller with adjustable damping, turning circle radius 3.5 m
- Chassis   Aluminum frame construction, (space frame with roll bar)
- Body   Thermoplastic Luran® S
- Pedal drive   5-speed gearing
- Electric drive   AC synchronous motor, rated 5 kW
- Battery   Ni-Cd, 2 kWh (2-battery blocks), or 3 kWh (3-battery blocks)
- Battery charger   2 kW (charge time approx. 2 hours)
- Range   40–80 km (depending on driving style and battery)

City-El
- Price   € 7,100
- Length   2,741 mm
- Width   1,060 mm
- Height   1,260 mm
- Unloaded (Curb) Weight   290 kg
- Payload   400 kg
- Maximum speed   50 km/h
- Starting Gradeability   > 16%
- Energy Consumption   4–10 kWh/100 km
- Chassis   3 wheels
- Brakes   3 drum, hydraulically operated, 2 in front and 2 in back
- Steering   Pulse with modulation
- Body   Plastic sandwich structure self-supporting PURELY foam with PMMA + ABS bowl
- Battery   3-piece traction battery Yuasa 90/100 Ah C5 36 V
- Battery Charger   Charging time to 75% 3 hours
                            – charging to 100% 8–9 hours
- Range   30–50 km

Source: www.evfinder.com and company websites

Exhibit 10
Four-wheel Electric Vehicles

Kewet Buddy by elbil NORGE

- Starting price — € 10,500
- Homologation — e11*2002/24*0153*00, category L7e
- Seating capacity — 3 adults
- Vehicle length/width/height — 2,440 × 1,430 × 1,440 mm
- Unladen mass (without batteries) — 400 kg
- Curb weight (without driver, Lead-acid batteries) — 785 kg
- Maximum allowed weight — 1.020 kg
- Luggage compartment capacity — 150 liter/base 920 × 470 cm
- Motor — SepEx 72V DC, 12 kW
- Transmission — Gear wheel transmission with differential rear-wheel drive and fully automatic electric gear shift system, gearing 1:7
- Top speed — 80–90 km/h depending on battery technology
- Acceleration — 0–50 km/h in 7 seconds
- Maximum hill-starting ability — 20%
- Body — Fiberglass-reinforced polyester and ABS
- Safety cabin — Welded tubular steel space frame, optional hot dip galvanized

Exhibit 10 (continued)

- Batteries   Maintenance-free lead batteries, ca. 8.4 kWh
                          or optional ca 10.5 kWh available. Safe Li-
                          ion technology, ca. 7.2 kWh or optional ca.
                          14.4 kWh available
- Charging time   0–100% in 6–8 hours, 30–100% in 3 hours
- Range   Lead-acid batteries 50–100 km, depending on
                          road conditions, tempera-ture, and driver.
                          Li-ion batteries up to 150 km

2 BlueCar by Bolloré
The car will not be available until 2008 and prices will start at € 15,000.
- Length × width × height   3,050 mm × 1,702 mm × 1,610 mm
- Seating   3 seats + 2
- Maximum speed   125 km/h
- 0–60 kph acceleration   6.3 seconds
- Average operating range   200–250 km
- Maximum power output   50 kW
- Min/max battery voltage   243 V/374 V
- BatScap battery pack   27 kWh
- Battery pack weight   < 200 kg
- Total weight with batteries   980 kg
- 100% charging time   6 hours
- Rapid recharging time   A few minutes for a range of 20 km
- Body   High-resistance steel, aluminum, recyclable
                                 composite

Source: Company website

Exhibit 11
Redigo's Operating Costs Compared
to Potential Competitors (Euros)

Assumptions

Purchase price	Amortized in 4 years
Insurance	Full coverage
Mileage	10,000 km/year
Maintenance	Based on official recommended schedule increased by 10% for disposable items (tires, oil, *etc.*)
Gas consumption	City running cycle
Gas price	€ 1.20/liter
Winter tires	Not included
Cars	Gas engine running with unleaded fuel.
Electricity cost	€ 0.17/kWh (slightly above the official rate)
Battery recharge	4 hours for 150 km

Annual operating costs	BMW C1	REDIGO	smart	Microcar	Kewet	Renault Scenic
EUR per km	0.47	0.39	0.70	0.59	0.55	1.24

Source: Company literature

Exhibit 12
Redigo Business Plan

Volumes	Y	Y + 1	Y + 2	Y + 3	Y + 4	Y + 5
Number of vehicles sold/year	300	4,450	5,000	5,000	5,000	5,000
Cumulative vehicle sales	300	4,750	9,750	14,750	19,750	24,750
Profit andLoss (euros)						
Net sales	2,478,500	22,058,650	24,785,000	24,785,000	24,785,000	24,785,000
Cost of goods sold	1,574,000	14,008,600	15,740,000	15,740,000	15,740,000	15,740,000
Commercial margin (value)	904,500	8,050,050	9,045,000	9,045,000	9,045,000	9,045,000
Commercial margin (%)	36.49%	36.49%	36.49%	36.49%	36.49%	36.49%
Costs (euros)						
Personnel	1,357,333	1,457,000	1,878,333	1,956,433	2,065,333	2,065,333
Tooling investment	7,300,000	300,000	250,000	170,000	170,000	170,000
R&D	370,667	393,433	133,333	224,533	224,533	224,533
Administrative (IT, freight out)	646,667	670,667	754,356	754,356	754,356	754,356
External services	86,667	86,667	86,667	86,667	86,667	86,667
Warranty costs	114,667	755,333	855,000	858,000	578,667	578,667
Marketing expenses						
* POS material	200,000	200,000	200,000	140,000	130,000	120,000
* Advertising	190,000	440,000	440,000	440,000	440,000	440,000
* Promotion (fairs and shows)	49,570	113,433	113,433	113,433	113,433	113,433
* PR operations	25,000	35,000	40,000	40,000	40,000	40,000
* Coop advertising	53,240	85,600	97,333	97,333	97,333	97,333
Operating profit (euros)	-9,489,311	3,512,917	4,196,545	4,164,245	4,344,678	4,354,678
Financial costs and income	215,333	231,333	238,000	40,667	39,333	104,000
Net income after tax (euros)	-9,273,978	3,744,250	4,434,545	4,204,912	4,384,011	4,458,678
Depreciation	404,000	434,000	434,000	375,333	375,333	375,333
Cash flow (euros)	**-8,869,978**	**4,178,250**	**4,868,545**	**4,580,245**	**4,759,344**	**4,834,011**

Source: Company literature

Chapter 2
Planning the New Venture

Having great ideas and singling out the great opportunity is only the first step in a long journey to a viable venture. The next steps typically need some planning, mostly packaged into what will eventually become a "business plan." The business plan is a document that evolves from the initial idea into a business concept proposal (BCP) and from there into an opportunity assessment. At each stage of this evolution, the entrepreneur can decide to proceed or move on to a more promising venture.

The exercise of putting a business plan together is a helpful one, and by the end of the process you will have a document that helps you and potential investors think about how promising the venture truly is. This chapter covers the whys and how-tos of a business plan in depth, looking at the main parts and why they are important, as well as what to do with your business plan. In addition, we will look at other common themes characteristic of the early phase of technology startups such as product development and marketing and communications strategy, as well as issues of intellectual property.

2.1 What Is a Business Plan?

A business plan is a document describing a venture's opportunity, its product or service, context, strategy, team, required resources, and potential financial returns [1]. It is guided by three basic questions [2]:

- Where are we now?
- Where do we want to be?
- How are we going to get there?

There is ample material available on how to structure and write a successful business plan. We could even go so far as to say that the art of writing business plans has been commoditized over the years. Therefore, the mechanism behind the document should not present the entrepreneur with insurmountable challenges. What makes successful business planning so difficult is the time and thinking necessary for the entire process. It requires – usually for the first time – sitting down and carving out and putting in writing every single detail about the what, whys, and how-tos of the business.

2.1.1 Why Write a Business Plan?

Business plans serve many functions . Mostly, they serve to do some or all of the following [3]:

- *Sell yourself on the business:* A carefully elaborated plan can help convince you that starting this business is the right thing for you to do.
- *Obtain financing:* A business plan is an essential prerequisite for convincing potential investors to finance the new venture.
- *Motivate and focus the management team:* Developing a business plan gets everyone thinking about the business goals and can ensure a joint understanding of the company's roadmap.
- *Obtain large contracts:* The fact of having thought about the future and put some strategy down on paper provides credibility, and presenting a sound business plan may facilitate larger contracts.
- *Attract key employees:* The business plan might help prospective employees to decide whether to join the venture.
- *Arrange strategic alliances:* As we stress in various chapters throughout this book, strategic alliances have become a dominant pattern in recent years, especially between small and large companies, and the bigger partner generally likes to see a business plan to guide the selection process.
- *Complete mergers and acquisitions:* In these cases, the business plan serves as a company résumé, helping to demonstrate that the value of the business is the highest possible.

In terms of targets for the plan, one emphasis is on raising capital. This means that while answering the three questions about where the company is today, where it should be in the future, and how to get there, the business plan also needs to convincingly demonstrate the ability of the business to generate satisfactory profits [4].

2.1.2 Who Should Write the Business Plan?

Always short on resources, entrepreneurs are sometimes taken with the idea of "outsourcing" the preparation of the business plan, using consultants or interns (often the case for startups in close proximity to research institutes). The help of such outsiders can indeed be highly beneficial, but it cannot replace the personal commitment of the business owners and management team. Only they have the right level of insight and imagination as to the direction their business should take. And it is they who need to present and defend their plan and eventually build a venture on it.

2.1.3 Who Will Read the Business Plan?

It might be surprising to hear, but most entrepreneurs will say that they have written hundreds of different business plans – all for the same business! The issue here is that a business plan needs to be adapted to the audience to which it will be presented. A business plan can, for example, be directed at internal people such as the management team or employees. In this case it is more an operational document to help align individuals within the company. Much more common, though, in the startup phase of a firm is the type of business plan that is presented to the outside world – mostly to raise money from potential investors, sometimes also to present the firm to customers. Here, the audience and their interest in the plan is highly diverse, ranging from bankers, angel investors, venture capital providers, institutional investors and investment advisers to business partners, managers of other companies, and government agencies. Each presentation of the company and its business plan needs to be tailored to the interests of these different stakeholders.

2.1.4 What Should Be in the Business Plan?

Even though each business plan might be a little different, the overall structure will include some core elements that stay the same. Only their length and focus might vary with the audience. To be on the safe side, the plan should include the following points [5]:

- Cover page and table of contents.
- Executive summary.
- Description of the current situation: Basic company information, products/services, management team, business organization, future goals, vision, and mission
- Description of opportunity and market: Who are the buyers, who are the competitors, what are the competitive advantages of the company?
- Description of the business model, the marketing and sales strategy.
- Basic facts on the financials: Cash flow projection (life line), income statement (bottom line/profit and loss), balance sheet (business health/assets, liabilities, *etc.*), funding requirements.
- Risk analysis and possible exit strategies.
- Conclusion and appendixes: Résumés, literature, technical descriptions.

2.1.4.1 Focus on the Important Parts

The preceding list makes it clear that there is a plethora of information to include in the document. So how can this be accomplished if the document should only contain between 10 and 40 pages? The advice here is: Focus, and highlight the particular strengths of your business. For example, if the team is perhaps not as seasoned as the textbooks recommend, but the venture has already gained renowned customers, then of course the team section will be less extensive than the parts on the customers and track record [6]. The business plan is a marketing instrument that requires honesty but at the same time allows you to direct attention to the positive features of the new venture [7]!

Furthermore, although this may seem self-evident, the plan should follow the basic rules of style. Information should be clearly organized, segmented, and logically integrated. Make the executive summary as strong and compelling as you can; most people will judge the venture on it. Creating a truly compelling plan takes time and many, many revisions. Before the plan is declared final, it should in fact pass several checks by outsiders. In many cases, investors will also ask you to revise your plan after the final version!

> The Business Concept Proposal (BCP)
> As soon as you think you may have a potential business based on new technology, you should put together a preliminary BCP, as discussed in Chapter 1. *Under no circumstances should the document be more than two pages.* If it is, this indicates that you have not developed the idea enough, or that you are thinking about it in too complex a manner, or that the idea itself is not very good!
>
> Doing this preliminary exercise should give a quick go/no-go decision and also give you some immediate material for the first draft of your business plan.

2.1.5 Common Pitfalls

Venture capitalists (VCs) receive hundreds of plans each week, and each week hundreds of plans are rejected. In addition, there are thousands of business plans spread all over the world, collecting dust in drawers instead of being lived and implemented. Some of the most common reasons that business plans fail are described in *Table 2.1*.

Table 2.1 Planning: Its Ailments and Symptoms [8]

Pitfall	Symptoms
No real goals	• Goals are vague, general • Goals are not specific, measurable, or time-phased • No subgoals or action steps • Activity oriented, not goal oriented
Failure to anticipate obstacles	• Excessive optimism • No alternative strategies • No conflicts recognized • Missed delivery date • Missed lead time forecasts • Didn't get support when needed • Crises prevail
Lack of milestones and progress reviews	• Don't really know how you are doing • Short-term orientation • Can't recall when you last reviewed how company was doing • No recent revisions of plan
Lack of commitment	• Procrastination • Focus on routine, daily activities • Failure to meet goals, milestones • Failure to develop specific action steps to meet goals • Lack of priorities • Missed meetings, appointments
Failure to revise goals	• Plan never changes, lacks resilience • Inflexible or stubborn in face of feedback dictating change • Goals not met or exceeded greatly • Unresponsive to changing situation • Help not sought when needed • Wasted time or unproductive tasks or activities • Activities don't match goal priorities
Failure to learn from experience	• Lose sight of goals • Mistake is repeated • Feedback is ignored or denied • Same routine – same crises as before • Unwillingness to change way of doing things • Not asking "What do we learn from this experience?"

2.2 Key Elements of the Business Plan

2.2.1 New Venture Team

Many factors need to come together to start and grow a successful new venture. However, first comes a great idea and directly after that the people who can realize it. It is generally believed that startups thrive and prosper when standing on the shoulders of more than one person – especially science-based and high-tech startups. A single entrepreneur typically can make a living out of the business, but startups that are led by teams create substantial value. The advantage of having a team is mostly in the greater network, the more diverse knowledge and skills, and the possibility to divide and specialize tasks, which eventually enables faster growth.

However, forming a successful team is sometimes compared with the process of courtship and marriage. And like some marriages, there are divorces. So choose your partners wisely. This handful of individuals is likely to stick around for a while and will have to fight some battles. The way team members find one another varies significantly, and it is hard to say which way – if any – is best. Some teams form by accidents of geography, others through common interest, still others by working together or simply through past friendships [9]. However, only two distinct patterns can be identified for team formation as such: Either there is an individual entrepreneur who will be joined over the first few years by three or four partners, or the team was already formed at the outset. Both can be successful.

2.2.1.1 Beyond the Founders

However, apart from the founder or founders, the team is much bigger – a fact that is typically overlooked. Many more individuals, such as key employees, financiers, and outside professionals, help build the company, and they should be chosen with almost as much care as the founder team (*refer to Figure 2.1*):

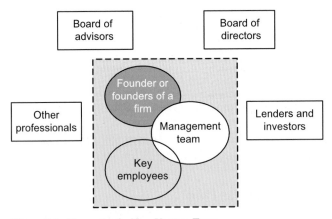

Figure 2.1 Elements of a New Venture Team
Source: Adapted from Barringer, Bruce R. and R. Duane Ireland. Entrepreneurship: Successfully Launching New Ventures. Upper Saddle River, NJ: Pearson Education, 2006, p. 128.

2.2.1.2 Team Quality

Starting a business is a demanding undertaking. The pressures can be enormous and there is no time for on-the-job training. The ideal entrepreneur has both industry and management experience, *i.e.*, he or she has worked in the same or a similar industry before and preferably has had responsibility for budgeting and profit and loss [10]. If he or she cannot provide that, there should at least be seasoned board members or other mentors to provide that sort of knowledge and insight. Having a team is another way to increase the likelihood of all necessary experience being present, but the team members must complement and balance each other.

Team quality is also demonstrated in the way the team effectively operates as a team: Is it able to establish trusting work relations with a clear definition of roles and responsibilities? Can it put effective communication processes in place and create an atmosphere that allows constructive criticism, with respect?

A word of warning at this point: For several of the above-mentioned reasons, people tend to prefer working with others of the same background and values. This can be – but is not necessarily – beneficial. In the first few years, it is usually deemed better for the team to be highly diverse to accomplish their versatile tasks.

2.2.1.3 The Role of the Board

Several of the cases in this book prove the importance of the board of directors. First, the board can provide some legitimacy for an otherwise unknown venture. But more important is the fact that well-chosen and well-matched board members can provide invaluable advice. They can and should challenge the assumptions, strategies, and actions of the entrepreneurial team. For this reason, board members should come with seasoned industry and business experience.

2.2.2 Market Analysis and Sizing

Following the evaluation of the opportunity, this part of the business plan should be quite quick to do. But a business plan does need to give credible statements on the venture's market and size, beyond the back-of-the-envelope exercise done in the BCP. Often entrepreneurs do not have a clear understanding of what a market actually is and sometimes confuse it with an industry. However, the attractiveness of each can differ. So, to clarify: A market consists of a group of current and/or potential customers with the willingness and ability to buy products – goods or services – to satisfy a particular class of wants or needs. These potential customers may be consistent in their geographical location, purchasing power, or buying attitude [11]. An industry, by contrast, consists of sellers, *i.e.*, you and your competitors [12].

Analyzing a market can and should be done on two different levels: The macro- and the microenvironment. The *macrolevel analysis* typically asks questions about things such as the number of customers, aggregate money spent, and number of units and usage occasions. Answers to these questions are often to be found in secondary data sources such as trade publications, the business press, and so on.

Also, these sources might give answers to how fast the market has been growing and the expected growth rate in the future. The analysis should also cover macro-economic trends in terms of possible political, technological, and sociocultural developments.

The *microlevel analysis* is somewhat more intricate. It is about segmenting the market and putting a name to potential customers. At the end of the day, a successful business needs to find customers who are willing to pay for that business's product or service. Successfully entering and competing in a market is frequently accomplished by solving a customer need, which does not necessarily mean selling a particular feature of a technology. It is more about delivering benefits. Convincing customers that you have the best solution to their problems, or even teaching them that they have a need you can fulfill, is actually the challenging task. Typical questions to answer in the microlevel analysis are [13]:

- Is there a target market segment where we might enter the market in which we offer customers clear and compelling benefits at a price they are willing to pay?
- Are these benefits, in customers' minds, different from and superior in some way – better, faster, or cheaper – to what other solutions currently offer? Differentiation is crucial, since the vast majority of me-too products fail.
- How large is this segment, and how fast is it growing?
- Is it likely that our entry into this segment will provide entry to other segments we may wish to target in the future?

Answers to these questions are often found in a combination of primary data, mainly gleaned from talking to prospective customers, and secondary data – collected from the Internet or in libraries or from other sources, to determine segment size and growth rate, and thus supporting the entrepreneur's learning about customer needs. A proven approach to defining a segment or portfolio of segments to begin with is to screen for market attractiveness of the potential applications and compare it with the existing skills in the company, with "ease of commercialization," uniqueness of the applications, and scope of problems addressed [14]. The results will provide a guideline as to which are the most attractive products and with which of them the company is likely to succeed.

2.2.3 Industry and Competitor Analysis

No serious investor will believe a startup that claims there is no competition. If there really were no competition, there would be no market. Furthermore, even if it were possible that there was no competition, as soon as the startup began making money, many players would enter the market seeking to gain a share of the trail that the entrepreneur had blazed. Therefore, any serious business plan needs to contain a careful analysis of the industry, its outlook, and the competitive forces inside. At a bare minimum, the plan should lay out what percentage of the market the venture could *realistically* achieve, both at the beginning and five years on once the big players and other startups enter the fray.

The typical approach here is to perform some textbook strategy analysis such as Porter's five forces or SWOT analyses, but this often reads like boilerplate material in a business plan. Instead, we would recommend tailoring the analysis to the opportunity. In no-nonsense language, explain what the market is, how quickly it is growing (backed up with references to credible market research), and who the main players are. Find a startup that was in a similar situation in a similar market and explain what its market share was and why yours would deviate (or not) from that of the other startup. Then, assuming there will be some competitive reaction, discuss how your company would respond and what your long-term market share would be. Make a table with the competitive advantages and disadvantages of all the players in the market.

2.2.3.1 Competitive Advantage

So it all comes back to whether or not you have a sustainable competitive advantage. A competitive advantage is essentially the ability to prevent others from exploiting the same opportunity, which should usually grant you the potential for higher returns than normal [15]. The critical question is how this advantage can be protected from competitors and whether it can be maintained over a long time. Two vital factors are [16]:

- *The presence of proprietary elements, i.e.*, patents, trade secrets, *etc.* (intellectual property), that can possibly prevent others from copying your business
- *The presence of an economically viable business model, i.e.*, a model that generates sufficient revenue and gross margin to cover the cost structure of the business.

We will take a closer look at both distinguishing factors in the two following sections.

2.2.4 Intellectual Property

Intellectual property (IP) is one of the most important and, at the same time, one of the most delicate assets to handle of the new technology-based venture. IP can be any product of the human intellect that has value in the marketplace, *i.e.*, products, technologies, methods, processes, new services, and new designs. Recognizing the value of the knowledge contained in these assets and identifying and legally protecting the parts that are the original property of the entrepreneur can become the heart of any commercialization strategy. Four main instruments of IP protection exist: patents, copyrights, trademarks, and trade secrets [17].

2.2.4.1 Patents

Patents are official titles to exclude others from making, selling, or using an invention for a limited time. These rights are defendable before a court. The process of obtaining a patent is usually lengthy and expensive [18]. Patenting an invention costs between $10,000 and $15,000 in most industrialized countries [19]. If patents are sought in all major countries where the invention might be practiced, the cost can easily reach around $100,000 per patent! Plus there are additional costs that may be added for renewal and for litigation, if it should arise. In general, it is advisable to consult a professional to help obtain a patent. This ensures that the claim will be airtight and – if need be – defendable, assuming that the venture's resources stretch to being able to defend its rights (*see also Table 2.2*).

2.2.4.2 Trademarks

Words, names, or symbols that identify a company, product, or service and distinguish it from others are known as trademarks. They help companies to be uniquely recognized by their customers. They need to be officially registered and are renewable every ten years, as long as they remain in use. Obtaining a trademark is typically much faster and easier than obtaining a patent.

2.2.4.3 Copyrights

Tangible outputs of a person or company, such as a book, article, software, and the like, are protected by copyright. It grants official ownership and the right of commercialization. Officially, copyright is obtained by the creation of a tangible work. It is not necessary to indicate that something is copyright protected. However, attaching a copyright note (usually in the form © [first year of publication] [author or copyright owner]) helps make it more official and explicit.

2.2.4.4 Trade Secrets

Going beyond what is written in the technical description of a patent, trade secrets include business or technical knowledge that is kept secret for the purpose of gaining an advantage in business over a competitor [20]. They are, for example, customer lists, sources of supply, faster delivery, or lower prices. The protection is established by the nature of the secret and the effort to keep it secret.

Not all forms of protection are applicable to all forms of intellectual property. However, the main controversy that still has not been finally resolved is the role and extent of patent protection of IP. There are both academics and practitioners who maintain that a small venture would not be able to defend its rights anyway or, as the famous Silicon Valley entrepreneur Guy Kawasaki, noted: "You won't

Table 2.2 Arguments for and against Patenting

For Patenting	For Keeping Technology Secret
• Patents provide a defined period of exclusivity during which others can be prohibited from commercializing the invention, even if independently developed. • Patents establish ownership distinct from teams and individuals. • Market protection is assured longer if technology is easy to reverse engineer. • Patents prevent others from preempting the technology in a world where often it is not the first to invent that counts, but the first to file forms at the patent office. • Patents give something to exchange if in-licensing is desired. • Patents facilitate and clarify research collaboration and technology marketing agreements. • Patents allow greater freedom in choosing a business formula. • Patents can impose more restrictions on a licensee than is possible with know-how licenses, including, in some countries, longer duration of contracts. • Patents motivate inventors and are a sign of achievement.	• Patents are expensive to obtain and offer weak protection. • Secrets can be kept indefinitely if well protected, while patents force early disclosure to suit market opportunities. • Sometimes it is impossible to prove that an end product infringes a patent.

have the time or money to sue anyone with a pocket deep enough to be worth suing." [21]. So in many cases keeping something as a trade secret might be a more worthwhile option. *Table 2.2* lists some of the typical arguments of the discussion on patents *vs.* trade secrets [22].

2.2.4.5 Licensing [23]

Closely linked to protecting IP are the topics of licensing or transferring IP. All IP that is protected can essentially be licensed to another company in exchange for money or access to its IP and other resources. Simply speaking, a license is a contract by which one party commits to do or pay something in return for the other party's doing or paying something. Any contingency that can be written into a contract can be written into a licensing agreement. Usually the licensee, who receives a right, pays an initial payment and ongoing royalties for permission to use the IP. This permission can be either exclusive (only the licensee is entitled to use the protected technology) or nonexclusive (others are also allowed to exploit the object of the license). A license can also be restricted to a specific purpose or geographic region.

The importance of licensing is that the owner of the property – the licensor – retains ownership [24]. He does, however, partially transfer rights to the licensee in a formal contract that binds both parties legally. This requires a careful definition of the content and scope of the contract, in particular, who has the right to exploit what, where, and when. Negotiating license contracts is frequently intricate, especially from the licensor's perspective. First of all, the entrepreneur always walks a fine line between attracting the potential licensee's interest and not revealing too much confidential information. Second, basic legal knowledge is absolutely imperative to understand the contractual obligations (of course, it is advisable to hire professional help here). Last but not least, determining the payment conditions requires some serious attention. A clear picture of the business model and interest of the licensee is needed.

Determining the payment conditions ultimately leads to the question of defining the license value on which the pricing can be based. This process should start by looking at the cost of creating the IP, *i.e.*, research and development costs, cost of commercialization, legal costs, *etc*. In a second step, the potential profitability of the business that is secured by the license should be assessed. Naturally, the licensor will want to share in this business value.

2.2.5 The Business Model

A business model essentially defines how a firm competes in the marketplace and how it earns profits from this activity. This includes, in particular, how the firm structures its relationships with customers and suppliers [25]. All else being equal, the profit that can be made from a technology depends on choosing the right business model. The tradeoff is to find a balance between quick market access and, at the same time, maximizing the returns from the investment made. It also relates to decisions on whether to make or buy, whether to sell or license products or components, and whether to sell a product or a service or a combination of both (*see also Figure 2.2*).

Internal factors that influence the choice within this continuum are the founder's long-term ambition *vs.* the immediate economic and promotional needs of the technology at the particular moment. Externally, it depends on the ability to mobilize particular market partners to deliver the technology effectively and quickly. The latter, above all, can serve as a strategic asset, preventing other companies from entering a particular market. The best product is no good if you cannot get the resources to build it or deliver it to your customer. Bearing this in mind, it is particularly important to understand that a business model goes well beyond the boundaries of the firm. It needs to be defined in relation to other market actors: Partners to deliver the product/service, customers, and competitors. As it is naturally difficult for small and unknown ventures to enter into these types of partnerships, it is advisable to seek out collateral benefits that could convince potential partners [26].

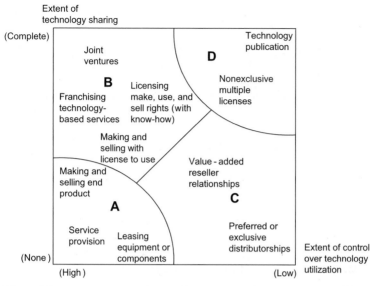

Figure 2.2 Instruments of Technology Commercialization
Source: Jolly, Vijay K. *Commercializing New Technologies: Getting from Mind to Market.*
Boston: Harvard Business School Press, 1997, p. 100.

The major questions that help determine the business model are summarized in *Table 2.3.*

Table 2.3 Elements of a Business Model [27]

Component	Questions That Help Define the Component
Customer value	Is the firm offering its customers something distinctive or at a lower cost than its competitors?
Scope	To which customers (demographic and geographic) is the firm offering this value? What is the range of products/services offered that embody this value?
Pricing	*How* does the firm price the value?
Revenue source	*Where* do the dollars come from? *Who* pays for *what* value and *when*? *What* are the margins in each market and *what* drives them? *What* drives value in each source?
Connected activities	*Which* set of activities does the firm have to perform to offer this value and *when*? How *connected* (in cross section and time) are these activities?
Implementation	What organizational *structure, systems, people,* and environment does the firm need to carry out these activities? What is the fit between them?

Table 2.3 (continued)

Component	Questions That Help Define the Component
Capabilities	What are the firm's capabilities and which *capability gaps* need to be filled? How does a firm fill these capability gaps? Is there something distinctive about these capabilities that allows the firm to offer the value better than other firms and that makes them difficult to imitate? What are the *sources* of these capabilities?
Sustainability	*What* is it about the firm that makes it difficult for other firms to imitate it? How does the firm sustain its competitive advantage?
Profit site	What is the relative (dis)advantage of a firm vis-à-vis its suppliers, customers, rivals, complementors, potential new entrants, and substitutes?
Cost structure	*What* drives costs in each component of the business model?

A word of warning needs to be issued at this point: There is no standard business model. It might even be dangerous just to copy successful business models because their success depends not only on the model as such but also on the ability to execute it. The dependence on outside partners makes each firm's situation unique. And what works for one company will not necessarily work for another. That is why spending ample time on answering the above questions and analyzing the company's capabilities and, in particular, its access to outside partners is so important.

2.2.6 The Marketing Plan

There are two distinct ways in which products emerge – either as a result of a research-and-development project (*technology push*) or by first listening to a customer need and developing a product accordingly (*market pull*). Most products in the real world are the outcome of a mix of both models [28]. It is unlikely that any successful technology company will either exclusively develop to customer needs or exclusively try to find markets for its greatly engineered products. The first rarely succeed long term because they frequently miss out on the highly innovative and high-margin products that customers did not know they would like until they had them. The second group of ventures spends too much money developing products that mostly fail to find a customer at all. Science-based startups tend to be among the second group. They often struggle to strike a balance between engineering and marketing.

Marketing in startups is indeed a challenging task, and it differs from marketing in established companies. Particularities are a result of the venture's newness and size (*see Table 2.4*).

Table 2.4 Challenges for Marketing in New Ventures [29]

Characteristics	Challenges for Marketing in New Ventures
Newness of the firm	• Unknown entity to potential customers and other parties • Lack of trust in the new firm's abilities and offerings • Reliance on social interactions among strangers • Lack of exchange relationships • Lack of internal structures, processes/routines in marketing • Lack of experience in marketing • Lack of historical data
Small size of the firm	• Very limited financial resources available for marketing • Few human resources • Lack of critical skills in marketing • Limited market presence • Limited market power, disadvantage in negotiations
Uncertainty and turbulence	• Very low predictability of market and other data • Only limited information available for marketing planning and for marketing decisions • Best practices in marketing have yet to be determined for the specific industry • Dominant design of an offering is unknown • Competitive structure of the industry is changing, relationships with suppliers, distributors, *etc.* are unstable • High risk of wrong decisions, which may have fatal consequences for a small firm with limited resources

Because of these challenges, marketing is one of the most critical operations and needs a professional approach. However, empirical evidence shows that the marketing function typically emerges together with the firm and only gradually gains in professionalism. Surprisingly, it is one of the functions that founders relinquish last. One reason for this is its known importance, accompanied by the thinking, "I am the one who understands the product best," or "If I cannot succeed in selling my product, who else can do it?" Furthermore, many ventures start with some initial success, and as long as the business continues to run, the need for change or improvement goes unnoticed. This can backfire as soon as "daily life" is part of the reality. We would like to stress here the importance of defining a solid market strategy and investing in professional marketing.

2.2.6.1 Unique Selling Position (USP)

One of the first marketing tasks, typically going together with the market and industry analysis, is to define a unique selling position (USP). This is the perceived value of the product or service compared to competitors' offerings. The main question to answer is why the customer should buy this particular product and what makes it so unique. A learning point for many technology-based entrepreneurs is that the USP is essentially about selling benefits rather than features.

Mostly, it is not the technical detail that leads to the buying decision but the ease and comprehensiveness with which a customer need is solved.

The marketing plan then needs to elaborate, in more operational terms – usually done within the classical 4-P framework.

2.2.6.2 Marketing Mix (4 Ps)

Once the more strategic marketing plans have been shaped, it is time to define how you actually wish to reach your target customers on the operational level. Here, decisions need to be made on the classic four Ps: product, price, place, and promotion (*Table 2.5*).

The *product* strategy involves several aspects, in particular the scope of the product portfolios, the approach to product development, branding, packaging, *etc.* For many technology-based startups, generating a list of potential applications that lead to products is often the easy part. There are frequently more opportunities than the startup can possibly seize. The difficulty is deciding which of these applications to pursue, when, and with how much resource commitment. Often it is a fine line between what the technology can do at a particular moment in time and what the requirements of the market are. Tradeoffs need to be made and resources juggled. Many entrepreneurs struggle to identify which of the technically feasible features the customer values and would pay for. It cannot be stressed enough that talking to customers, learning about their needs, and then deciding about the product is one of the most crucial tasks.

The *pricing* decision is an extremely important one as it determines how much money the company can earn in the end. Naturally, the decision cannot be made autonomously but will be influenced by the demand, by the cost of operations, by competitors' prices, and the like. Pricing objectives can be manifold, and the entrepreneur has to decide whether he or she prefers a strategy of survival, maximum current profit, maximum current revenue, maximum sales growth, and so on. In general, there are two approaches to pricing – cost-based or value-based.

Table 2.5 Critical Decisions for the Marketing Mix [30]

Marketing Mix Variable	Critical Decisions
Product	Quality of components or materials, style, features, options, brand names, packaging, sizes, service availability, and warranties
Price	Quality image, list price, quantity, discounts, allowances for quick payment, credit terms, and payment period
Place (distribution channels)	Use of wholesalers and/or retailers, type of wholesalers or retailers, how many, length of channel, geographic coverage, inventory, and transportation
Promotion	Media alternatives, message, media budget, role of personal selling, sales promotion (display, coupons, *etc.*), and media interest in publicity

While the first might be easier to justify, the second can yield significantly higher returns. It requires determining what customers are willing to pay and fixing the price tag accordingly. Often enough, this strategy allows for premium prices, in particular if the demand is strong, the technology is proprietary, and a compelling product (see above) has been created. Many experts warn startups against trying to compete too aggressively on price, *i.e.*, setting it deliberately low to gain market share. This strategy rarely pays off, as larger firms usually have a much stronger position in this competition and it denies the startup much-needed profits early on [31]. Certainly, if a venture is to employ this strategy, it is necessary to practice penetration pricing and not desperation pricing [32], with a concrete plan to monetize the product or service.

The *place* component of the marketing mix encompasses all activities that deal with bringing the product to the customer. In particular, it addresses the distribution channel. One of the first decisions the new venture needs to make is whether to sell direct or via intermediaries. This is not easy, as it will shape the entire sales and distribution organization – it determines the cost structure, the margins, and the dependencies of the new venture. Selling direct to a customer allows for much better learning about customer needs and for maintaining much more control over the initial product. However, it requires competences and consumes resources that might not be available in the venture. Using intermediaries can be a solution. Typically, this enables faster growth and is easier to scale. However, distributors also require a minimum commitment within the new venture: product materials need to be designed and the distributor's sales force needs to be educated. Therefore, the number of channels needs to be managed carefully. Likewise, care needs to be taken in picking the right channel partner because of the strong dependency on this relationship. A startup will inevitably lose some control over the product and the way it is sold, which requires a trust-based relationship.

Promoting a new technology-based product is best seen as a combination of market discovery (conceiving of products that satisfy a latent demand; compare with *demand pull*) and market creation (building demand where none exists: *technology push*) [33]. Most entrepreneurs think of advertising (print and broadcast media) when they think of promotion. However, such advertising can only raise awareness. It has less power to make people buy a product [34]. That is why, for most startups, spending a lot of money on ads does not pay off. Targeted customer contact, for example at trade fairs and the like, seems to be more promising for that purpose. Especially when selling technical products, direct contact with the customer can help build credibility and trust in the new venture and its product. However, some professional promotional material needs to be in place even at the early stages to demonstrate commitment and to inform the customer.

2.2.7 Financing the Venture

Financing a new venture is such a prominent function within the new venture creation process that we dedicate all of Chapter 3 to it. We will, therefore, only make a few remarks on financial planning here.

Mostly, a business plan serves to raise money. To succeed in this, it will have to lay out the investment needs and expected returns. The projections on investments and returns will usually cover three to five years. It will contain the projected income statement, a pro forma cash flow analysis (often monthly for the first year and quarterly for the following years), pro forma balance sheets, break-even analysis, and cost controls [35]. Of course, all forecasts should be as realistic as possible. However, it is unlikely that the entrepreneur can accurately anticipate how much capital and time will be required to accomplish his or her goals. Therefore, the major focus should rather be on the key drivers for business success. This includes at what level of sales the business starts making profit and when the cash flow will start turning positive [36].

By its nature, the financial plan will be based on many assumptions. To build credibility with potential investors, these assumptions need to be backed by reliable reference points or a plan for testing the assumptions. Often, market studies and comparisons with competitors will be such reference points. You can and should, for example, research material costs, competitors' prices, and the like. Still, there will be quite a few assumptions that come without quantifiable sources, where the entrepreneur relies on his or her own prior experience or sometimes just on gut feeling. What can help in these cases is trying to verify the assumptions with industry experts and other experienced people. In general, the entrepreneur should draw on different sources to cross-check the plans.

Eventually, each financial plan needs to talk about an exit strategy. It should be noted that the exit is intended to be for the investor, not (necessarily) for the founder! The end process is typically what investors are most interested in, as it is where they realize their returns. We will talk about exit again in Chapter 3 and at length in Chapter 5, but it should be noted here that investors typically prefer a wide range of possible options and that IPO may not always be the best. However, including or excluding certain exit options will attract different types of investors.

2.2.8 Talking About Risk

Because of the many assumptions about an unknown future, most people will readily agree that there is tremendous risk in any entrepreneurial venture. A serious business plan needs to account for this. It needs to include statements on the risks associated with the business and on the actions the entrepreneur envisions taking if issues occur. Risks are inherent in all parts of the business: in people, in the assumptions about industry dynamics, in the market, and, of course, in financial planning.

In some of our cases, we recognize that a proven strategy to deal with the risk is to pursue a stepwise roll-out, *i.e.*, starting the business in a certain region, sometimes with limited features of product or service. In these cases, the venture is treated as a series of experiments, which allows learning about the true economics and the ability to determine in subsequent steps how much money will be needed at what stage.

2.3 Cases in This Chapter

The *Shockfish (B)* case describes what decision the management team made in terms of business focus. However, this was only the start. Shockfish needs fresh money to continue business, and Jim Pulcrano, the CEO, is tasked with giving presentations on the company's potential. Essentially, this means a fully fledged business plan needs to be developed, *i.e.*, the Shockfish team needs to elaborate the business model and select its target customer segments. Furthermore, it still has to discover the economics behind the business and possibly experiment with the financial drivers. Hence the student is asked not only whether or not he or she would fund the venture but also what information is needed to make that decision. The case offers a superb opportunity to have the structure of a business plan crafted, which would then need to be fleshed out and filled with real market information.

Somewhat more mature than Shockfish but also at an early stage in its business model definition is the company in the next case, *IR Microsystems (A): June 2002*. The venture, founded by seasoned engineers, has developed a low-cost solution to broad-range infrared spectrometry. Up to June 2002, the company had focused on R&D and sales while all other functions were outsourced. However, the team needs to make a strategic decision about whether to move up to the next step and turn the technology into a standalone product. This would likely entail keeping a hand in manufacturing and marketing the product. In fact, the founders still need to decide on their specific business model. The options range from licensing to becoming an OEM supplier to integrating downstream. In order to decide, the team had just started to develop a better understanding of its markets. In particular, it had identified a possible target segment. The challenge, however, is that the venture team would still need to pioneer this particular market, which had significant financial and human resources implications. To obtain both, the team had to flesh out their business strategy.

The next case, *InMotion Technologies Ltd.,* presents a venture with a somewhat more developed business plan than the first two cases. However, the long-term viability is subject to analysis. InMotion Technologies is based on innovative imaging algorithms that allow two pictures to be superimposed in the same picture frame. The first application considered by the venture team is in sports broadcasting, where it can help compare the performance of two athletes over the same course. A second application was identified in sports training solutions, but the venture still needs to decide how to tap the potential here. Both markets come with distinctive requirements in terms of a business model. The first experiences in the market have shown the possibilities and limits in terms of scaling the business. Decisions need to be made and then the marketing strategy and additional investments need to be refined accordingly. At the same time, the technology is still in the process of maturing. The case invites the student to assess the strategy of the venture. Will the company take off?

This question will also be raised in the next case, *Boblbee (D): The Urban Backpack*. The new venture team is particularly challenged by marketing issues. Compared to the two preceding cases, Boblbee's product is much more clearly defined and the features developed. However, positioning it on the market is giving the company founders a headache, as is the entire marketing strategy. This includes the decision making for the classic four Ps and even adjusting the manufacturing capabilities and developing strategic partnerships to spur market penetration. The case is hence an excellent vehicle for studying the key elements of marketing in entrepreneurial ventures.

A more comprehensive overview of all parts of a business plan will be developed in the ensuing cases series, *VistaPapers A, B, B-2, and C*. It describes the efforts of the entrepreneur, Robert Keane, to develop the business plan for his project, VistaPapers, and the interaction with investors. He tries to set up Vista-Papers as a catalogue marketer of specialized stationery and related products that enable customers to use their own PC and laser or ink-jet printer to economically produce high-quality printed communications in small quantities – even one copy. We follow Keane in his learning process in adapting the business plan to investors' requirements and in his dealings with unforeseeable external factors. Whereas the (A) and (B) cases focus on the formal presentation of the business plan to potential investors, the (B-2) case highlights the risks associated with new ventures. The (C) case, by contrast, stresses that a business plan is not a static document. It is the description of business over the venture's lifetime and thus will need to be adapted if the environment changes. The (C) case is also an excellent example of ongoing opportunity recognition. Robert Keane once more shows amazing foresight as he sets about revamping his entire print business and turning his company into an e-business.

The last case in this chapter, *Lyncée Tec SA: Scaling up a Technology Venture*, covers a wide range of issues in the early phase of a startup. Lyncée Tec is a small startup commercializing holographic microscopy solutions. After two years in business, the venture started operating at breakeven, financed entirely by sales. Nevertheless, the huge potential inherent in Lyncée Tec's applications had not been fully tapped. The company needs to grow more aggressively to become a serious industry player and to stay ahead of the competition. It needs additional external funding to speed up operations. Where should the money come from? The case outlines the venture's business plan and invites the student to analyze and provide comments on the business plan proposal and on the positioning of the high-tech venture in the market. The case also provides a vehicle for reviewing and discussing the venture's financing strategy, which leads to the next chapter on financing entrepreneurial ventures.

References

1   Dorf, Richard C. and Thomas Byers. *Technology Ventures: From Idea to Enterprise*. New York: McGraw Hill Higher Education, 2005, p. 186.
2   Stutely, Richard. *The Definitive Business Plan: The Fast-Track to Intelligent Business Planning for Executives and Entrepreneurs*. London: Financial Times/Prentice Hall, 1999, p. 41.
3   Gumpert, David E. "Creating a Successful Business Plan." *The Portable MBA in Entrepreneurship*. Ed. W. D. Bygrave. New York: John Wiley & Sons, 1997, pp. 120–147.
4   Gumpert, "Creating a Successful Business Plan."
5   A more detailed list can be found in Stutely, *Definitive Business Plan*, p. 41.
6   Baird, Michael L. *Starting a High-Tech Company*. New York: Institute of Electrical and Electronics Engineers, 1995, Chapter 11.
7   Baird, *Starting a High-Tech Company*, Chapter 11.
8   Adapted from Timmons, Jeffry A. *New Venture Creation: Entrepreneurship for the 21st Century*. Boston: Irwin, 1994, p. 381.
9   Timmons, *New Venture Creation*, p. 257.
10 Bygrave, William D. 1997. "The Entrepreneurial Process." *The Portable MBA in Entrepreneurship*. Ed. W. D. Bygrave. New York: John Wiley & Sons, 1997, p.16.
11 Mullins, John. *The New Business Road Test: What Entrepreneurs and Executives Should Do Before Writing a Business Plan*. London: Financial Times/Prentice Hall, 2004, p. 7.
12 Mullins, *New Business Road Test*, p. 7.
13 Mullins, *New Business Road Test*, p. 11.
14 Jolly, Vijay K. *Commercializing New Technologies: Getting from Mind to Market*. Boston: Harvard Business School Press, 1997, p. 100.
15 Baron, Robert A. and Scott Andrew Shane. *Entrepreneurship: A Process Perspective*. Mason, OH: Thomson/South-Western, 2005, p. 241.
16 Compare also with Mullins, *New Business Road Test*, p. 15.
17 The following description is mainly based on the very comprehensive overview in Barringer and Ireland, *Entrepreneurship*, pp. 282–297.
18 For a more detailed overview see Barringer and Ireland, *Entrepreneurship*, p. 284 ff.
19 Jolly, *Commercializing New Technologies*, p. 111.
20 Iandiorio, Joseph S. "Intellectual Property." *The Portable MBA in Entrepreneurship*. Ed. W. D. Bygrave. New York: Wiley, 1997, p. 335.
21 Refer to Kawasaki's blog on the Internet: Guy Kawasaki "The Top Ten Lies of Entrepreneurs." 2006. http://blog.guykawasaki.com/2006/01/the_top_ten_lie_1.html (accessed September 22, 2006).
22 Jolly, *Commercializing New Technologies*, p. 115.
23 Iandiorio, "Intellectual Property."
24 Iandiorio, "Intellectual Property."
25 Barringer and Ireland, *Entrepreneurship*, p. 100.
26 Jolly, *Commercializing New Technologies*, pp. 249–281.
27 See Afuah, Allan and Christopher L. Tucci. *Internet Business Models and Strategies*. 2nd ed. New York: McGraw-Hill, 2003.
28 Bell, C. Gordon. *High-Tech Ventures: The Guide for Entrepreneurial Success*. New York: Perseus, 1991, p. 86.

29 A comprehensive overview of the existing literature on the challenges in new venture marketing is presented in Gruber, Marc. "Marketing in New Ventures: Theory and Empirical Evidence." *Schmalenbach Business Review*, 2004, *56* (2), pp. 164–199.

30 Hisrich, Robert D. and Michael P. Peters. *Entrepreneurship*. Boston, MA: Irwin/McGraw-Hill, 1998, p. 269.

31 Barringer and Ireland, *Entrepreneurship*, p. 265.

32 Shapiro, Carl and Hal R. Varian. *Information Rules: A Strategic Guide to the Network Economy*. Boston, MA: Harvard Business School Press, 1998.

33 Jolly, *Commercializing New Technologies*.

34 Barringer and Ireland, *Entrepreneurship*, p. 269.

35 Coulter, Mary K. *Entrepreneurship in Action*. Upper Saddle River, NJ: Prentice Hall, 2000, p. 123 ff.

36 Sahlman, William A. "How to Write a Great Business Plan." *Harvard Business Review*, July-August 1997, pp. 98–109.

CASE 2-1
Shockfish (B)

John Walsh, Andy Boynton, and Alastair Brown

On the morning of September 11, 2000, Shockfish turned down the CHF 2.5 million offer. Its potential partner for wireless voting mechanisms would eventually go bankrupt.

The company chose to work on Spotme as a business "conference navigator" product. Shockfish decided on the business conference market because it had the most leverage across other areas, as those who attended business conferences were largely important decision makers in a variety of industries. For example, when used initially at an IMD conference, 28 new leads were originated from 400 customers. Shockfish hoped that after establishing itself in the business conference market it would then be able to head into other areas it had identified.

The company now had a vision: the Spotme product would be used at business conferences to increase the networking, productivity, and efficiency of an event. Despite this focus, Pulcrano was still left with many new questions as he sought to bring the product to market. In fact, it seemed as if his task had just begun. First, who should Shockfish sell its product to in order for it to be used at business conferences? Second, should it sell or rent the device? Third, how could it spark adoption and convince customers of the product's value? Lastly, should it shoot for a service or turnkey product?

Copyright © 2002 by IMD – International Institute for Management Development, Lausanne, Switzerland. Not to be used or reproduced without written permission directly from IMD.

1 Crossing the Chasm

After deciding to focus on the conference navigator product, there were still many questions to answer:

- Who do we sell to?

There were many possible parties that could be interested in the Spotme product. First, large multinational corporations and professional organizations would have an interest because of all the large events they held. The decision makers about conferences in these companies and organizations could be CEOs, specific program managers, or a marketing department head. Second, event-management companies and professional conference organizers would have an interest in Spotme. They might use the product to differentiate their service from others, by reselling Spotme to clients. Third, venue owners might also be interested in purchasing the Spotme system in order to differentiate themselves. Lastly, technology providers that helped to service events might be interested in reselling Spotme as part of their technology package for an event. Thus, there were four distinct ways in which Spotme might successfully penetrate the business conference market. A conference often involved more than one of these parties making final decisions; therefore Shockfish needed to decide which groups were best to target.

- Do we sell the product or rent it? What price should be charged?

Shockfish initially proposed charging CHF 51,000 for renting the system for a 7-day event with 100 people. The charge would be CHF 105,000 for a 7-day event with 1,000 people. Each additional participant would cost the client roughly CHF 8 per day. Purchase of a system that had been enabled to handle 1,000 people would cost CHF 1.2 million. (These prices would end up changing drastically over time. At the time, though, Shockfish debated whether to sell or rent and whether these original prices made sense.)

Shockfish absorbed heavy upfront costs associated with developing, maintaining, and improving the Spotme system. Although the cost of producing one handset was unknown, Shockfish estimated that the cost of producing one Spotme device was similar to the cost of producing one of the most recent Palm Pilots.

- How do we convince consumers of the product's value?

Shockfish had to decide whether to "push" or "pull" the Spotme product in its marketing efforts. "Pushing" sales would mean selling a service to professional conference organizers, who would in turn sell the system to clients. "Pulling" new business would mean large-scale marketing efforts.

Additionally, the Spotme product had many different features and capabilities, so educating customers was a difficult task. Shockfish sought to identify a few attributes of the Spotme system that differentiated it from other products.

- Should Spotme be offered as a turnkey product?

If Shockfish could provide a turnkey product, then it would no longer have to bring representatives to each conference where the system was rented and would not have to spend time supporting the product after a sale. Yet, as of September 2000, a turnkey solution was at least two years away. Was it worth spending time reaching this goal so that the product could be sold or rented without support? If it chose not to shoot for a turnkey solution, should Shockfish hire new employees to specifically help users of its system?

2 Funding

Since its inception Shockfish had raised a total of CHF 130,000 through the Fondation pour l'Innovation Technologique, an organization in Lausanne geared toward helping startup companies. By early May 2000 it had largely run through this money. As the due diligence of its angel investors was concluding, Pulcrano had to give a presentation promoting the potential of Shockfish and Spotme.

Would you fund?

CASE 2-2
IR Microsystems (A): June 2002

Benoît Leleux and the 2002 MBA Startup Project Team

To the IR Microsystems (IRM) team the issues had always been very straight-forward. They had developed a unique technology to produce a detector array for infrared (IR) light that could be used – once integrated in a spectrometer – in a number of industrial sectors, most notably for composition analysis of gases and liquids in application-specific process control. The market for low-cost broad-range IR spectrometers was completely open, but the team members were slowly realizing that an empty market segment was not necessarily waiting to be filled. The resistance to new technology was high, the industry was very conservative, and industry investments in new technology after the year 2000 were down. Their positioning as a component and subsystems producer – at least two layers re-moved from the end customers – did not help to create the demand, especially in the face of entrenched OEMs.

The unique selling proposition of IRM's technology was its ability to measure a full IR spectrum at a low cost – about US$600 to $1,000. This positioned the product between the low end and the high end of existing offers on the market. The optimal positioning from IRM's point of view was thus one in which IR spec-trometry provided superior value, but low cost was also critical. The challenges facing the team were now much clearer: How could they break into that novel market space, which they would have to create from scratch? How could they transform their IR detectors into products that would fill that new market niche? And how could they transform the end-user desires for lower-cost analytical solu-tions into a serious OEM offering using IRM components?

If and when they solved those problems, they might then have to redefine the purpose of the company. All three founders were leading technology experts with PhDs: Would it be fun for them to turn the company into a sales organ-ization? What should their business model be? On the one hand, they could con-tinue to focus on technology development. Should they then just "pass on," through licensing, the rest of the process? But as suppliers to OEMs and inte-grators, it seemed important to keep a hand in the later stages, or risk becoming

Copyright © 2006 by IMD – International Institute for Management Development, Lausanne, Switzerland. Not to be used or reproduced without written permission directly from IMD.

totally disconnected from end users. Could they afford to integrate downstream? Did they want to do so?

It was also important to understand the financial implications of their decisions. Their cash burn rate was not outrageous, but someone had to finance it. And without a clear plan to market, they would have a hard time obtaining additional funding. The story could end right here! Another great technology driven into the ground.

1 InfraRed (IR) Microsystems in January 2002

1.1 The Team

The IRM team included the three founders and two employees – an electronics engineer and a lab technician. The founders shared a similar pedigree: They were all PhDs with a very strong technical background related to IR detection technologies. Together they brought 21 years of development experience to the company. Previously, they had worked together in the Ceramics Laboratory at the Federal Institute of Technology (EPFL) in Lausanne, Switzerland, from 1995 until IRM was founded in February 2000.

Bert Willing, the CEO, had some startup experience in France, as cofounder of Composite Compression Company S.A., in which he also invested about $300,000 from 1994 to 1995. The experience proved to be more valuable from a learning point of view than from a financial standpoint.

Andreas Seifert, the chief scientific officer (CSO), possessed in-depth technical knowledge of thin-film processing and characterization as a result of five years as a research and teaching assistant at the University of California from 1990 to 1995. Another five years as a materials engineer at EPFL on industry-related projects gave him further knowledge on thin-film deposition and microstructuration using clean-room facilities. He was IRM's core technical strength and had enabled the company to develop some distinctive technologies.

Markus Kohli, the chief technology officer (CTO), had gathered extensive knowledge of the design and fabrication of thin-film devices during his employment as an assistant and PhD student at the Ceramics Laboratory at EPFL (1994–1998). His position as quality control manager at SWATCH Groups CMOS foundary, EM Electronic Marin, gave him invaluable expertise in production management. He brought to the company a good understanding of sourcing issues for IR detectors and spectral engines, as well as control of the manufacturing process.

1.2 The Products

IRM's product line in 2002 consisted of two products: a microray detector unit and a compact spectral engine. A third product, a discrete multigas analyzer, had just been dropped from the lineup because of unattractive market conditions, including the strong bargaining power of suppliers (the automotive industry), low prices, low entry barriers, and not enough synergistic benefits with the other two products.

Beta versions of these products existed and were being sold to trial customers for a price of $1,000 to $3,000 per unit, a price that covered variable costs. Although the main technical problems were under control, IRM's prototypes were not generating much market demand. The limiting factor seemed to be that the products were only components (detectors) of larger systems (analyzers for composition determination of liquids or gases, used mostly in process and quality control), and IRM so far had not been able to integrate vertically, mostly for resource-related reasons. Its sole focus was on the detector and signal processing side. On the production side, IRM relied on outsourcing for standard processes like SMD mounting, ASIC production, and packaging. The key processing steps for processing the detection layer, representing the core know-how of the company, was kept in-house and handled via a clean-room walk-in service at the semi-industrial clean-room facilities at EPFL.

1.3 Funding and Burn Rate

IRM estimated that it would need some CHF 7 million to reach the market. Initial attempts to tap venture capitalists (VCs) in 2001 had proved fruitless. Their key concern was the lack of demonstrated market demand, i.e., no actual sales, no signed contracts, and no track record in the target sector. It was difficult to demonstrate the real potential of the products, and the market lacked structure.

IRM had started with a CHF 450,000 capital injection from the founders in February 2000 (personal loans, funds from a local technology innovation support group called FIT, and the DeVigier Prize, a technology startup competition). This was complemented by a CHF 200,000 line of credit from a major Swiss bank's startup initiative, putting the startup capital at around CHF 650,000. At the initial burn rate, this secured financing until around May 2001.

That month, IRM received an additional capital injection of CHF 300,000 from a local business angel. This angel investor was willing to commit the funds to secure a Swiss immigration visa on the basis of investments in local companies and the consequent creation of local jobs. That investor also proved invaluable in terms of contacts and willingness to provide advice, support, and direction.

The monthly cash burn rate was estimated at CHF 32,000 for the first quarter of 2002, including five salaries, office rent, services, and other fixed costs.

1.4 Markets

The market information was pretty patchy. IRM's market research was based on a rather diverse group of trade-show contacts with original equipment manufacturers (OEMs) and on information gathered from industrial fairs, press releases, contacts with potential customers, and hearsay. IRM targeted the OEM market as it wanted to build sales volume fast, but the initial approaches to OEMs were rather unfocused (*see Exhibit 1 for relevant OEMs*). After months of no customer contacts, the business angel persuaded IRM to focus on 12 OEMs as potential target customers.

There was no obvious competition for IRM's detectors. However, the final products in which these detectors were integrated had to compete against well-established products such as FTIR spectrometers,[1] near-IR spectrometers (high end), and various detection technologies for single gases at the low end. The market between these two products did not exist yet, so IRM would need to educate and develop the market space.

2 Infrared Spectrometry

IR spectrometry can be used to analyze the composition of solids, fluids, and gases. It is based on the principle that every molecule has a specific, distinctive absorption spectrum for IR light, which is always the same. So by measuring the IR spectrum of a mixture of gas or liquid, it is possible to determine which molecules are present and in what quantity by comparing to calibration series. IR spectrometry is, therefore, a powerful analysis and monitoring tool for a variety of applications, including process and quality control (*e.g.*, in the agricultural, food, chemical, pharmaceutical, and petrochemical industries), monitoring of toxic gases, or temperature measurements in the metal industry to observe temperature distribution in extruded or molten metals.

---

[1] FTIR stands for Fourier Transform InfraRed, a high-end method of infrared spectroscopy.

2.1 How It Works

Light from a light source – basically a bulb – is shone onto a device that splits the different wavelengths contained in the beam into separate bundles of light, each now with a specific wavelength. This wavelength-separating device can be a monochromator or simply a linear variable filter (LVF), as shown in *Figure 1*. These bundles are each aimed precisely at the surface of one pixel in an array of pixels. The pixels, deposited on a solid surface, form the basis of the detector. Each pixel measures the intensity of the bundle that is aimed at its surface. Because every pixel receives a bundle of IR rays of a certain wavelength only, its signal tells something about the intensity of that specific wavelength.

If there is nothing between the light source and the bandwidth filter, all pixels will generate the same signal. If the signals of all pixels are plotted against the wavelengths, a so-called IR spectrum is created, in the form of a straight, horizontal line.

However, if a chemical compound is present between the light source and the LVF or monochromator, this will absorb part of the energy of the IR rays from the light source. It happens that a given molecule will absorb IR radiation from one wavelength more readily than from other wavelengths; the molecule thus has a characteristic signature. Plotting the wavelength (pixels) against a relative intensity results in an absorption spectrum, where the depth of the absorption bands gives the concentration of a molecule. (*See graph in Figure 1.*)

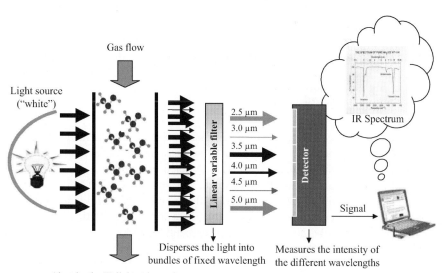

Figure 1 Schematic Diagram of an IR Gas Analyzer
Source: Company information

The detector array above has multiple pixels. Many instruments employing IR technology, however, use only one to four pixels, suitable for measuring only one or a few wavelengths, which is adequate for some applications (*e.g.*, single- or two-gas measurements). For multiple-gas/compound detection, however, a detector array with multiple pixels is superior.

2.2 Benefits of IR Spectrometry and Implications for IRM

The distinctive benefits of IR spectrometry compared to other analysis techniques are that:

- It can measure multiple gases/molecules at the same time.
- It is stable, with little need for regular recalibrations.
- It has no moving parts, making it rugged for in-line applications.
- It is based on light absorption, therefore making it possible to keep the detectors away from potentially corrosive, hot, or dirty process streams.

IRM's key product was a detector array for IR light. The company referred to the combination of detector and LVF – which together produced the spectral information of interest (readily computer processed) – as a *spectral engine*. This, in turn, was a subsystem of a *spectrometer*, or in other words a click-in, independent part (like a CD-ROM player in a PC). The spectrometer itself was the complete (OEM) measurement instrument, with buttons, a screen, and relevant software for analyzing the signals.

3 Market Analysis and Marketing Strategy

3.1 Market Size and Growth

Market data on IR technology and gas sensors in 2002 was fragmented and scarce.
The North American IR sensor market (for gases, liquids, and solids) was estimated to be $474 million in 2000. It was expected to increase by approximately 8% for the next five years. A rough estimate was that the North American market constituted about 40% of the global market, thus making the worldwide gas sensor and analyzer market worth approximately $1.2 billion. The North American IR sensor market could be divided into four segments according to technologies: IR temperature (31.6%), IR gas (35.5%), IR optical (30.6%), and IR humidity (2.2%).
The European total gas sensor and analyzer market in 2000 (including technologies other than IR) was estimated at $284 million [1]. The overall European market was mature and was expected to stagnate within the next five years (*Exhibit 2*). Because IR gas technology was replacing competing technologies, it was

the only one expected to grow in revenue terms, albeit at a modest pace of 2.4% [1]. Volume growth was expected to be higher.

3.2 Market Segmentation

Gas sensors had a wide variety of end uses for detecting and analyzing the presence and composition of different gases. The market was highly fragmented, including, for example, petrochemicals, power generation, environmental monitoring, waste incineration, and food and beverages (*see Exhibit 3 for the most important industries making use of the technology in Europe*). Similarly, there were a great number of technologies for serving the gas sensor market, each with its specific benefits and application. Few technologies competed head-on, but for many applications they overlapped. Any market analysis therefore had to be conducted on an end-use-by-end-use basis. Generic market surveys were not very useful.

Within most of these application segments, it was possible to distinguish:

- High-performance, high-cost systems, *e. g.*, laboratory equipment and the process industry, with system costs of up to $200,000, and typically used in large service contracts.
- Simple, low-cost detectors, *e. g.*, for medical and safety-related uses, with system costs of approximately $50 to $200.

There was hardly any middle segment because, until recently, there had been no technical solutions available to enable relatively high levels of performance at reasonable costs (*Exhibit 4*) for in-line and real-time analytics. Market analysts acknowledged that the development of such technologies could spur the market into stronger growth, but they had not included such developments in their projections [2]. This mid-price segment was targeted by IRM. End users, for example in the chemical industry, regularly complained about the lack of mid-priced solutions for their in-line process control. OEMs would only sell them high-priced lab instruments with big service contracts, whereas application-specific process monitoring solutions would help cut costs by indicating instantly (*i. e.*, no need to take a sample and go to the lab) when a process would go from "green" to "yellow."

3.3 IRM's Business System

IRM's business system is shown in *Figure 2*, with the companies involved in the various parts of the business system listed below.

IRM focused on R&D and marketing and sales. Almost everything else was outsourced. The client was never approached directly but via OEMs, which had the critical contacts and knowledge about end-user applications that IRM did not have.

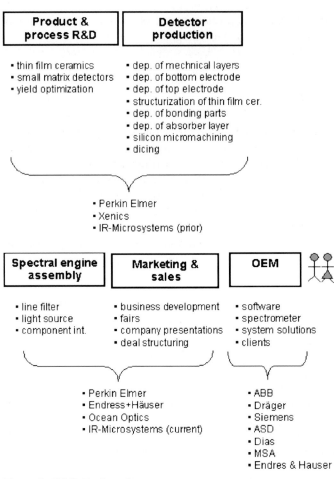

Figure 2   IRM's Business System
Source: Company literature

3.4 Market Attractiveness

3.4.1 Customers

The total market for high-end analytical systems was dominated by a few companies like ABB, Bruker, Emerson Process Management, Siemens, and Thermo, which together controlled about 60% of the worldwide market. Many other OEMs were active in a specific field – they were in general small, with a turnover of less than $10 million each.

OEMs typically procured either subsystems or components for the systems they supplied from third parties. Since know-how and understanding of the particular applications was important, customer (OEM) bargaining power was high. The number of OEMs active in any particular end-use segment was usually limited, so a subsystem provider was reliant on only a few customers.

IRM's problem was that customers at both ends of the price range were quite satisfied with the equipment they had. At the high end of the segment, customers were rarely aware of lower-cost solutions and thus were not looking for lower-priced detection systems. In some cases, as in the pharmaceutical industry, cost was somewhat irrelevant to their applications. At the low end, existing technologies met customers' needs for simple application (*e.g.*, one-gas monitoring) and they were not demanding increased. The challenge for IRM was thus to find a way to create a market for its products. Because there was such a gap between the high end and the low end, there appeared to be a void in the middle market waiting to be filled. But for that to happen, the customer had to be aware that such a novel product even existed.

3.4.2 Competition

IRM's detectors were likely to encounter a lot of competition in both the high-end and low-end segments. At the top end, it was outperformed by existing technologies, and at the low end it could not match the cost of existing technologies. For the microray detector to succeed, a new market had to be developed from scratch, where real spectroscopy analysis capabilities would add value, but costs could be contained, *e.g.*, in application specific production control, where a mid-priced system could prevent out-of-spec production and avoid frequent, time-consuming, and costly lab analysis. Competition for IRM's technology is presented in the following subsections:

Low-end: Single-pixel IR Gas Detectors

Using technology similar to IRM's, such detectors were used in inexpensive single-gas detectors, but their use could be extended to medium- to high-end applications for serious process applications (waste incineration, *etc.*).

A variation on the theme was the filter wheel instruments, in which rotating filters of different bandwidths allowed for multiple-gas detection. A disadvantage of this approach was that less information was extracted than with IRM's product and there were many moving parts that were subject to wear and tear. It was also important to know beforehand what gases were being measured. From the cost angle, the machines ranged in price from $20,000 to $50,000.

Low-end: Electrochemical Detectors

Simple and relatively cheap, these detectors dominate the gas-detection market. On the negative side, their long-term stability is questionable and they measure only single gases. Recalibration frequency is high and there is no functional safety. This compares unfavorably with IRM units, which have unlimited service time (no recalibration or maintenance required). Another issue is the "cross-sensitivity" to other gases, *i.e.*, other gases can cause interference and falsify the readings.

High End: Lab Spectrometers

FTIR and near-IR diode array spectrometers are widely used in the process control industry. For a system cost of up to $200,000 they provide excellent accuracy and resolution. The technology is proven and well accepted. Fiber-optic coupling for near-IR (up to 2.5-micron wavelength) is readily available with probes and standard telecom glass fiber. The drawbacks are lack of ruggedness and cost, which are at the same time a selling point for IRM. In many applications these instruments are "overkill" solutions where routine processes are being monitored.

3.4.3 Suppliers

IRM was at the stage of prototype production and rented clean rooms at the Swiss Federal Institute of Technology (EPFL) to manufacture the IR-sensitive microchips. Except for the deposition of the IR-sensitive film, IRM intended to outsource all production and had identified four preferred manufacturers.[2] Despite IRM's initial low volumes, there were enough alternative suppliers from which to choose if the four it had identified failed to deliver.

3.4.4 Entry Barriers

The key entry barriers to IR detection and analysis were:

- Patent protection in a few limited instances
- Specific technical knowledge about IR spectrometry
- Expertise in ceramic thin-film processing and integration (at the heart of the detector technology); few companies were active in this field at the time
- Expertise in silicon micromachining

---

[2] They were HL Planartechnik (Germany, www.hlplanar.de),
Leister (Switzerland, www.leister.com), XICS (Switzerland, www.xics.ch) and SEMEFAB (Scotland, www.semefab.com).

- Access to mixed signal *application-specific integrated circuit* technology (cus-tom-made chips)
- Access to nonstandard clean-room facility (cost $10 million)
- Spectrometer manufacturers' vertical backward integration
- Relationship (partnership) with OEM or spectrometer manufacturer.

The entry barriers for new players were relatively high because of the neces-sary technical expertise needed to succeed in this market. IRM's founders had this expertise and a first-mover advantage, estimated to be about two years. However, a big risk came from incumbents active in related technologies that could easily move into IRM's domain. IRM would need to rely on strong partnerships with OEMs or spectrometer manufacturers.

3.4.5 Substitutes

Although IR detectors seemed to have momentum, there were many substitute technologies for the detection and analysis of gases and liquids, and new tech-nologies were a possible threat. At least one company was thought to be develop-ing a competing detector technology. Since technological development in the field was relatively fast, the threat of substitutes was real. (*See Exhibit 5 for a summary of market attractiveness using Porter's five forces model.*)

3.5 Targeting

IRM management scored a number of potential end applications according to the following criteria: (1) the extent to which IRM provided unique benefits, (2) vol-ume and price potential, (3) technical/operational feasibility of entering the mar-kets, and (4) time to cash generation. The findings are summarized in *Figure 3*, with the arrows in the graph representing management's views.

Strategic importance placed equal weight on volume potential and the unique-ness of IRM's value proposition. Likewise, operational feasibility placed equal weight on technical feasibility and time-to-money.

3.6 Positioning

IRM was manufacturing and commercializing a new type of IR detector that could be turned into an IR spectrometer using a number of readily available components. The detector was not an end product in itself but was to be integrated into a final instrument produced by an OEM. Therefore, IRM sold to OEMs, not end users.

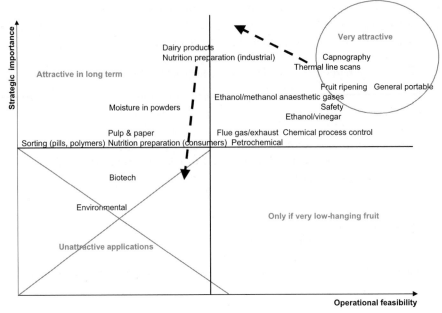

Figure 3   Strategic Importance *vs.* Operational Feasibility
Source: Company literature

The unique selling proposition of IRM's technology was its ability to measure a full IR spectrum at a low cost – about $600 to $1,000. IRM's detector could also be used in lower-end applications, for instance in simple gas detection, but had to compete there with many other established and cheaper technologies. In the high-end market, customers were less price sensitive but required capabilities that were

Table 1   Required Benefits

Benefits	Comments
Multi-compound analysis	Despite its low price and small size, IRM's detector could analyze several compounds simultaneously and so was able to deliver "finger-print spectra" of gas or liquid mixtures
Price	The detector is 1% to 10% of the total cost of present high-end spectro-meters. IRM's detector becomes interesting in applications where the detector represents more than, say, 20% of total cost, which is the case for all mid-performance applications. In high-end systems like FT/IR, the detector price doesn't matter
Small size	About as big as a cigarette box; appropriate for handheld devices
Robustness	No moving parts. Suitable for in-line applications
Stability	No need for recalibration
Digital output	Direct feed to computer
Customizability	Design can be adapted to application
No need for cooling fluids	Low hassle

beyond the reach of IRM's products. The optimal end use from IRM's point of view would be one where IR spectrometry would provide superior value, but low cost was a requirement. This meant that the application would need to have as many of the benefits in *Table 1* as possible.

Figure 4 highlights the applications in which IRM's technology was believed to add the most value ("uniqueness of product") and cross-references this with volume potential.

A number of applications seemed particularly attractive:

- Capnography (measure of CO_2 during exhalation)
  Existing instruments used single-pixel IR detection with a line filter. IRM arrays provided several advantages: They potentially offered (1) faster, more robust measurements; (2) smaller, simpler systems; (3) lower costs; and (4) little sensitivity to the high relative humidity of the breath and no direct contact with it.
- Thermal line scanning
  In steel or glass production, IR spectroscopy at low cost could potentially add value because the detector could be used to determine the heat distribution of molten steel extrudates, an important parameter for the quality of the end product.

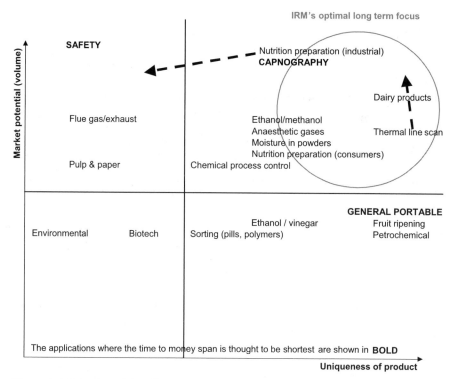

Figure 4   Market Potential *vs.* Uniqueness of Product
Source: Company literature

- Mobile gas analysis systems
  Existing systems incorporated various technologies in order to measure multiple gases such as NO_X, H_2S, CO_X, and hydrocarbons. IRM's product could measure several components using a single device.
- Other food applications
  Early detection of moisture in, for instance, grain or cacao could help prevent inventory writeoffs. IRM's product could detect the important IR peak of water at 1.8 micrometers and was economical enough to be interesting for low-margin food industries. The technology could also be used in fruit-ripening applications.

4 Strategic Plan

It was important for the company to look at the future in three distinct time horizons: short, medium, and long term.

4.1 Short-term Strategic Plan (Until New Financing)

IRM required new (equity) financing within a few months. Its attractiveness to investors had increased over the last few months due to an improved customer focus, which led to a better understanding of applications and customer needs. A number of solid letters of intent had been received, supporting the view that IRM's products and business were viable. What was still missing, though, was a formal commitment from a future customer to invest in product development through a joint development agreement (JDA) that would bring short-term revenues.

First of all, IRM should focus mainly on capnography applications where joint development initiatives were most advanced and volumes might pick up quickly. IRM already produced spectral engines, not just detectors. This would shorten product development cycles at the OEM, speeding up adoption. Reducing the focus to a few applications would enable IRM to keep any required modifications within "workable" limits.

4.2 Medium-term Plan (Up to 18 Months)

The development cycles of OEMs are quite long, so it was highly likely that IRM would need another financing round before it was able to reach positive operating cash flows. To secure additional financing, the company would need to demonstrate some volume of sales in one or two applications, as well as a solid pipeline of new applications. Actual sales would prove the viability of IRM's value proposition, and the possibility of adding new applications would enhance the appeal for investors.

Capnography held the strongest potential to generate sales. The value proposition of the microray also seemed to be strong in thermal line-scan markets, which would have to be developed. Applications in dairy products and other portable applications could also be considered. Considering the tasks at hand, it was also time for IRM to consider hiring a dedicated business development/marketing executive with OEM experience. Ramping up production volumes and managing the subcontractor relationships might create new complexities that should not be underestimated. Moving from a prototype production model to an industrial model would be a significant change

4.3 Long-term Strategy (More Than 18 Months)

The period 18 to 36 months after new financing would enable IRM to reach profitable operations. This would involve successfully ramping up operations, increasing sales volume, and developing new applications. New steps would then have to be taken.

4.3.1 Broadening the Product Range

IRM's core strength was in IR detection, with a fairly limited product range. A first step toward development was to improve the detector's functionality, for example resolution and accuracy, without compromising on cost. This would enable IRM to make further inroads into applications that required higher functionality.

4.3.2 Integrating Forward

IRM could remain a subsystem supplier of spectral engines or integrate forward and become a fully fledged spectrometer manufacturer. Remaining a subsystem supplier made sense only if IRM's strategy was to penetrate multiple applications in different fields. In such a case it would be difficult for IRM to integrate forward to serve all these markets, as each market and customer required some degree of customization in terms of detection range, resolution, and other product features. If IRM was able to penetrate only a few end-use applications, and IR detector competition in these applications emerged quickly, integrating forward was a more likely strategy, necessary to capture more of the total value creation and raise barriers to entry.

4.3.3 Make or Buy

IRM's manufacturing was appropriate for the demo prototype stage, but moving into quasi-industrial production would prove a challenge. To improve profitability,

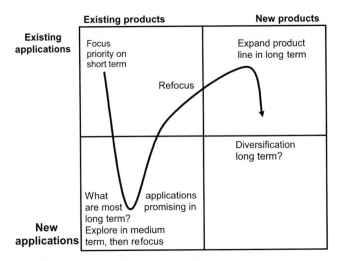

Figure 5 IRM's Staged Development Strategy
Source: Company literature

once there was a better understanding of sales volume, IRM would have to determine which steps to bring in-house. *Figure 5* gives a summary of the staged strategy rollout over the next 36 months.

5 Time to Sign off on New Plan

A sense of urgency was slowly dawning on the team. They had developed
a unique technology to produce a low-cost detector array for IR light that could be
used – once integrated in a spectrometer – in a number of sectors, most notably for
gas detection and analysis. The market for low-cost broad-range IR spectrometers
seemed completely open, but they had failed so far to generate much traction.
How could they break into that novel market space, which they would have to
create for themselves? How could they transform their IR detectors into products
that would create a new market niche? Once they had solved that issue, they
would have to revise the company structure and turn it into a viable commercial
operation. That would include not only completing the team but also developing
the supply chain to deliver and service the products on a larger scale. It was also
important to understand the financial implications of their decisions. Their cash
burn rate so far had been contained, but they would have to raise additional
financing soon. And without a clear plan to market, they would have a hard time
obtaining additional funding.

The story could indeed end right there! Unless they found a way to put all the
pieces back into a coherent strategy.

Exhibit 1
Relevant OEMs

High-end Market Segment

Company	Application field
ABB *(www.abb.com)*	Market leader in high-end segment, especially process control, pharma, and chemical engineering. Less likely IRM client due to high-end specs
Analytical Spectral Devices (www.asdi.com)	US company with strong presence in pharma, mining, food, chemical, and pulp and paper segments
Bruker (www.bruker.de)	FTIR spectrometer market leader, high end
Carl Zeiss (www.carlzeiss.com)	Present in very high-end segment of process control systems. Total turnover > €2 billion, turnover in gas sensors unknown
Perkin Elmer (www.perkinelmer.com)	Producer of laboratory equipment; also in low-end market
Oridion (www.oridion.com)	A market leader with sales of $10 million in capnography
Gottlieb Weinmann (www.weinmann.de)	Private company with $31 million turnover. OEM for capnography products
Mikron (www.mikroninst.com)	Thermal imaging technology that includes high-speed, high-accuracy, high-resolution features
Siemens Automation and Drives (www.automation.siemens.com)	Market leader in high-end segment, especially process control, automotive exhaust test benches, and chemical engineering. Less likely IRM client due to high-end specs

Low-end Market Segment

Company	Application field
Endress + Hauser (www.endress.com)	Focuses on analysis of food and water
Mine and Safety Appliances (www.msanet.com)	Large player in many segments, *e. g.*, process control, safety appliances, agriculture. Total turnover US$542 million; a low-end player
Ocean Optics (www.oceanoptics.com)	Portable and handheld spectrometers in visible, near-IR, and UV technologies; competes with PerkinElmer
Perkin Elmer (www.perkinelmer.com)	Produces portable IR spectrometers. See also in high-end market. They are only interested if there is market potential for 10000+ detectors per year per end customer
Sensitron (www.Felnet.it/sensitron/sensitron.htm)	Focuses mainly on safety applications for wide range of end uses
BCI (www.sims-bci.com)	Leading manufacturer of low-end, standalone, handheld, bedside medical monitoring devices with $29 million
Raytec (www.raytek.com)	US market leader in IR temperature-sensor market
Land-Infrared (www.landinst.com)	World leader in design, manufacture, and application of radiation thermometry, and a pioneer in IR thermometry

Exhibit 2
Growth of European IR Gas Analyzer Market

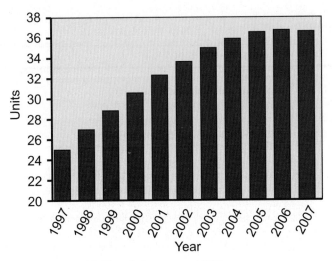

Source: Frost and Sullivan industry report, 2001

Exhibit 3
Gas Sensor Market Segmentation by Industry

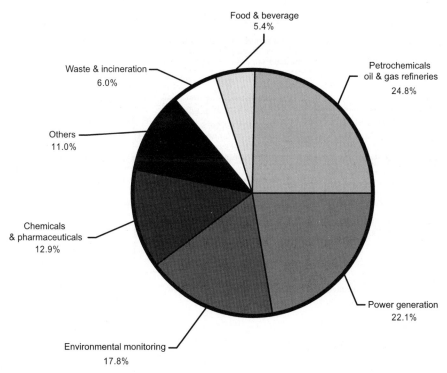

Source: Frost and Sullivan industry report, 2001

Exhibit 4
Nonexistent Middle Segment of Gas Analyzers

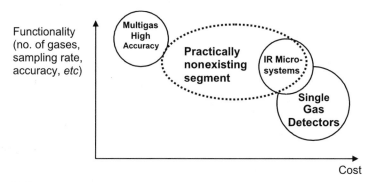

Source: IMD MBA startup project team analyses

Exhibit 5
Porter's Five-forces Model for IR Gas Analyzer Market

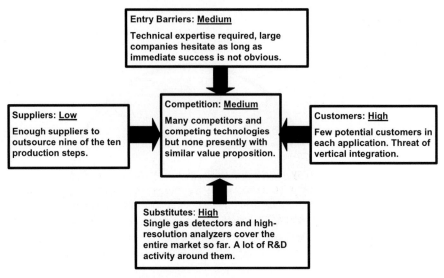

Source: IMD MBA startup project team analyses

References

1   Frost & Sullivan market report, 2001.
2   Opto & Laser Europe, November 2001; telephone interview March 15, 2002 Mr Leeward
     Bean, Ocean Optics.

CASE 2-3
InMotion Technologies Ltd.[1]

Olivier Courvoisier

WENGEN, SMALL MOUNTAIN VILLAGE OF SWITZERLAND, JANUARY 15, 2000. Amid the noisy celebrations of the Austrian downhill ski team and their supporters for their resounding victory in this year's famous Lauberhorn World Cup downhill race, another small team is having its own party behind the scenes. Indeed for Jean-Marie, CEO, and all the stakeholders of InMotion Technologies, today is a major event in their company's thus far short life, and while offering a toast to their future successes, Jean-Marie cannot help thinking about the past. Only two years ago at the same race, his brother and colleagues gave the first demonstration of the fruits of their research: a new feature for use in television coverage of sporting events that enthralled the whole sports and TV community.

Today the new version of their technology successfully underwent its first trial in a live broadcast around the world. Even so, Jean-Marie cannot totally relax. Leading the company to success will be like proceeding from one slalom gate to the next as quickly as possible with as few errors as possible. At this intermediate stage, InMotion Technologies is well ahead of the competition but, like the skiers, it has to reach the finish line and get there first.

Will InMotion Technologies be a winner?

Copyright © 2000 by UNIL-EPFL
[1] We wish to reiterate our thanks to Jean-Marie Ayer, CEO of InMotion technologies Ltd., and its team for their cooperation in developing this case.

1 Background

1.1 Technology

The innovative product that started the company is called VideoFinish™. It was developed in the laboratory for Audiovisual Communication of the Swiss Federal Institute of Technology in Lausanne by a group of researchers, including Jean-Marie's brother Serge, under the leadership of Professor Martin Vetterli.

The product uses new digital techniques to synchronize multiple sequences of video footage. The images acquired from normal television cameras are processed with VideoFinish™ identifying and calculating the various image characteristics and zoom parameters. A selected part of the video image (an athlete for example) can be isolated from the background and then reimposed on the video footage of other competitors at exactly the same spot of the race and with the correct timing. Unlike the usual display where the two competitors appear in split screens, this technology shows them both together in the same picture, apparently on a collision course racing against time. In 1998 when VideoFinish was demonstrated for the first time, it needed quite a few hours in postproduction work, so commentaries and analyses were only possible long after the race.

1.2 History

After the first demonstration in Wengen in January 1998 and the encouraging reaction it received, Prof. Vetterli's team was convinced that a market existed for their technology, but none of them really had any sound knowledge of how to start a company or had been involved in the launch of a startup company in any high-tech field, let alone the world of television.

After the success of the Wengen demonstration, Serge and Martin were first joined by Serge's brother, an experienced manager, who helped them in his spare time. At this stage, not yet fully committed to going for it, they entered a competition sponsored by McKinsey and the Swiss Federal Institute of Technology in Zürich for the creation of innovative new ventures. This meant they had to write a complete business plan. Not only did their entry finish second in the competition, which was pretty encouraging by itself, but also the whole process helped them to clearly identify the steps leading to the successful creation of a new venture.

Two years later, at the time of the January 2000 Wengen World Cup downhill race, the InMotion Technologies company employed 30 people.

2 Creation of the Company

During the summer 1998, around the time of the award from the venture competition, the whole team spent days and nights thinking about the question "Shall we start a company to sell and market our innovation? If so, how, with whom, where do we raise the money, *etc.*?" A thought bothered them all along: Would it not be easier and less risky to simply sell the innovation outright, to a TV network for example?

During these few months many contacts were made with potential customers, stakeholders, and friends. They confirmed that a market existed for their innovation, even though it needed more developments, and that the two brothers Serge and Jean-Marie, with their scientific and managerial qualifications and entrepreneurial spirit, could become the backbone of the new company. Opportunities also seemed to exist to get help from public authorities in raising the seed money needed for the first months after the official launch of the company.

By the end of summer 1998, the final decision to start a company based on VideoFinish™ was taken, and the company was officially launched in November 1998.

2.1 The Management

The main concern of the founders was to put together a well-motivated team with all the necessary capabilities for running a new company. As the idea of the company grew clearer, the need for added scientific support became evident. First Emmanuel Reusens joined the team. At the time of the company's incorporation more managerial capacities were required and Victor Bergonzoli became the fifth member.

At the time of the official creation (incorporation in the fall of 1998) this team of five was in charge of the company. Their responsibilities were: Chief Executive Officer – Jean-Marie Ayer; Chief Technical Officer – Serge Ayer; Head of Engineering – Emmanuel Reusens; Chief Scientist – Martin Vetterli; Chief Business Officer – Victor Bergonzoli (*Exhibit 1*).

To launch their new company the five founders committed a total capital of CHF 500,000 (only half of this in cash, as allowed by Swiss law).

On January 1, 1999, when InMotion Technologies started its activities in earnest, three team members opened the "headquarters" in Fribourg while the other two stayed in Lausanne at the Swiss Federal Institute of Technology (EPFL). The choice of Fribourg for the location of the offices has been greatly influenced by the support of the regional authorities, in addition to the Ayer brothers' attachment to the city they grew up in.

2.2 Product Development

At the early stage, there was not really a marketable product as such. The innovation was merely some algorithms and know-how stored on computers at the EPFL. To reach the market the VideoFinish™ system still needed a lot of technological development to reduce the delay between the video sequences and their display alongside on the same screen. For example, even though the broadcasters and commentators of sporting events were thrilled by the system's potential for analysis, they were much more interested in using it live. They wanted to show a competitor on the same screen as the leader of the race or maybe one of its chosen national athletes.

The system also had to be adapted in terms of convenience/portability so it could be moved easily from one competition site to another.

To casrry out these improvements, the new company recruited more technical/engineering people. At the same time, other possible uses of the innovative VideoFinish™ system were researched.

2.3 The Enterprise Mission and Strategy

To be leaders in creating state-of-the-art products and services in digital imagery.

This was the motto of the founders of InMotion Technologies and its implementation was not an easy task and experienced several unexpected delays. The following strategic steps were behind the successful launch of InMotion Technologies:

- From the beginning they chose to first concentrate their activities on the Video-Finish™ postproduction station, its development and future use by TV broadcasters being a perfect medium to get public attention. This public exposure should then ideally open the way to testing other markets for the new products under development.
- The second important strategic decision concerned the continued partnership with EPFL to keep abreast of the latest technology developments. This partnership was useful in particular for the development of a real-time system, with an industrial-academic collaboration grant through the Commission for Technology and Innovation of the Swiss federal government.
- At the same time an agreement was signed with EPFL solving the problems of ownership of the invention and use of the system by the newly founded InMotion Technologies company. Equally important was patenting the innovation in Switzerland (the patent was filed the day before the first Wengen demonstration); VideoFinish™ is also trademark protected. A patent filed in the USA was still pending after one year, but it nevertheless protected the InMotion Technologies system against potential competitors.
- Another important step was to partner with broadcasting companies; not only could they give InMotion access and knowledge in this particular industry, but just as important they would help to develop the products that the TV companies

would buy. The first collaboration started with Swiss-German television for the first demonstration in Wengen in 1998, and this partnership was continued for the second Wengen downhill race in January 2000.

- A delicate step was to exactly define the product they wanted to market and hopefully sell. The managing team carried out a thorough study of the different markets they wished to enter. The figure below shows the first results with indications of the possible markets for their potential products. It was quite a difficult task, as they had no competitors with a similar product, but they were able to estimate the production costs of sporting events and other media business. The figure also shows estimated market figures where they hoped to sell their innovative image processing system. (This figure was included in a preliminary version of InMotion Technologies' business plan)

- As an entry point into the market for their new technology, InMotion focused on its core competencies, deciding to become a service company at first, selling the VideoFinish™ images and not the computer systems. This meant that InMotion would have to send its technicians to every broadcasting competition site and run the system, creating images for the broadcasters for a fee.

- At the time of their first business plan, the price for this service was fixed at CHF 20,000 per competition/broadcast. This was quite a low price compared to the production budget of a ski race by a TV channel, estimated between CHF 1 and 2 million, but the strategy was to get maximum media exposure immediately.

- Lastly, as a nonconnoisseur remarked, the company had a possibly decisive advantage in that the fantastic visual effects of VideoFinish™ immediately appealed to viewers. However, the hidden complexity of the VideoFinish™ system caused problems for some of the less tech-savvy end customers, as is quite often the case with new technologies. This hampered the easy adoption by the public at large, which was needed to help convince broadcasters to buy the images produced by VideoFinish™ and also influence the different stakeholders.

Figure 1 Possible Markets for VideoFinish™
Source: Company literature

2.4 Financing

Right from the beginning, according to their calculations and predictions on the hoped-for future development of the company, the founders knew quite well that the money they were ready to invest would not be enough to cover the difficult first months of any new company that had no real product yet on the market and many technological improvements to implement. So with their winning Venture competition business plan in hands, members of the team went looking for financing aid. They had meetings with numerous development agencies, business angels, banks, and venture capitalists. At this early stage in the development of the company, their efforts had a positive reception from the regional development agency of the Canton of Fribourg.

The agency invested CHF 400,000 in stock (equity injection with a repurchase option). This, together with the CHF 500,000 invested by the founders, led to total starting capital of CHF 900,000 for the development of InMotion Technologies. In addition, the development agency also helped the launch of InMotion Technologies with a CHF 300,000 loan.

3 Development of the Company

Back in Wengen in January 2000, looking up at the slope of the Lauberhorn, Jean-Marie could name every gate of the downhill race InMotion had to negotiate during this past hectic year for the new company. Technology development delays, definitions of new products or new uses of the VideoFinish™ system, reluctance of TV producers to try out new products, difficulties in recruiting new employees, and new rounds of financing were the main gates, but InMotion seemed to be still leading the race.

3.1 Product Portfolio

To pass the new products and technical development gate, as in a downhill race, sometimes you need to slow down/brake just before the gate in order to stay on the best track. Jean-Marie similarly sometimes had to keep the young team of engineers on task. Highly motivated and imaginative, they tended to get lost in technical details or explore every possible new application for the core technology. Nevertheless, quite soon the whole team concentrated the technological developments on three product lines:

- *Broadcast*
  The VideoFinish™ postproduction and real-time system, successfully shown in Wengen in January 2000 (this is in fact two products using the same platform for the TV industry).

- *Internet*
  The adaptation of the system for the Web, making VideoFinish™ streams available on the Internet. Viewers could then choose their own simultaneous competitors on screen.
- *Trainer*
  The VideoFinish™ digital trainer: a lighter version of the postproduction system running on domestic video cameras widely available on the market, which could be used in numerous sports where one movement is repeated and can be improved. Comparing the movements of a trainee to the movements of a professional on the same screen would be much more persuasive and effective for the trainee than the verbal explanation of a trainer.

Even though the focus decision had been made early enough, delays in developments arose, and the real-time system project was completed six months later than expected in their optimistic business plan.

3.2 Portfolio Strategy

The focus on these three products was part of the commercialization strategy that was set after a complete business opportunities assessment of the images markets. Based on these results, the following target markets were chosen. Other possible use of their technology (movie industry, video editing and games, secondary sector industries, *etc.*) will be addressed at a later stage. (*Table 1* was reconstructed from a preliminary version of InMotion Technologies' business plan.)

Table 1 Portfolio Strategy

VideoFinish™ Product	Target market	Market size	Margin	Advantage for InMotion	Competition
Live production system	TV industry	Moderate	Moderate	Public exposure	None
Postproduction system	TV industry	Moderate	Moderate	Public exposure	None
Digital trainer	Sports professionals (associations and top clubs)	Moderate	Moderate	Synergies with broadcasting products	None
Digital trainer	Sports clubs	High	Moderate	Exposure	None
Digital trainer	Sports (individuals)	Very high	Moderate	Exposure	None
Internet	Sports Internet publishing	High	High	Synergies with other products	None

The TV broadcasts, even though not an infinite market and probably not an income earner, should enhance the public visibility of VideoFinish™ images, thus opening the markets for the VideoFinish™ digital trainer and Internet plug-ins. An important synergy could also develop between these last two products, as the number of sports sites grows. VideoFinish™ images could be stored in accessible data banks and displayed over the Internet for comparison and study of performance of different competitors.

Based on their business opportunities assessment, the InMotion Technologies team chose a "top-down" marketing strategy for the VideoFinish™ digital trainer. The specialized markets of professional coaches in top clubs/sports would be addressed before slowly opening up to sports clubs and, lastly, to individuals.

An unexpected problem also delayed their strategy – the reluctance of TV producers to accept new technologies, in particular in Europe. They could be classified rather as part of the "majority" than as "early adopters" in the new technology life cycle. According to a TV sports commentator, this surprising attitude to new technology is due to their "search for perfection," meaning only the real-time system would interest them, and long-term budget constraints impede the adoption of new technologies in this highly competitive TV industry. In other words "wait for the new budgets."

The way to overcome these difficulties was to convince the producers that VideoFinish™ would give them a competitive advantage rather than added production costs. But this requires a TV market completely open to multiple competitors, as is already the case in the USA.

3.3 Marketing

During its first few months, InMotion focused its marketing efforts on enhancing visibility at every level in the sports world, especially in the media, among image professionals, and in sports associations. InMotion also participated in trade fairs to open contacts with business people. The technology was used at the Ski World Championship in Vail, CO in late winter 1999, and the interest this created in the USA led the company to open a subsidiary for the US market in Portland, OR (where Nike headquarters is located).

When TV companies were not as willing as expected to buy VideoFinish™ images, the marketing team developed a new idea with the technical help of the engineers: As sponsors are present and invest large amounts of money during sporting events, why not offer them the chance to digitally add their logos on screen together with the two competing athletes?

Apart from the TV industry, which would not bring in the necessary revenue to sustain the company, the second market of individual athletes was covered; demonstrations of the potential of the VideoFinish™ digital trainer were given to sports associations and top clubs. In the near future, InMotion expects the VideoFinish™

digital trainer could become the cash cow of the company, bringing in the needed income.

For the third product, the VideoFinish™ plug-ins for the Internet, the tremendous development of the Web in itself and the public visibility of VideoFinish™ images in other media should combine easily to enter the market. For this market InMotion's sales strategy consists of selling hardware and software solutions to Web publishers together with plug-ins for their end customers at home.

At this time, following their overall strategy, the possible new markets outside of the sports world have not been researched.

When the company was formed, the quite optimistic founders had hoped that for the first year the company would already be present at 20 sporting events at CHF 20,000 each, at 50 in 2000. For the VideoFinish™ digital trainer, they counted on 10 sales in 1999 at CHF 20,000 for the software only, and more than 50 sales in 2000.

In reality during the first year, InMotion Technologies had "sales" of CHF 400,000, CHF 300,000 for broadcasts and CHF 100,000 for the first trainers.

3.4 First Round of Financing

With the rapid development of the company at the beginning of 1999, the need for new money and capital soon became obvious and quite urgent. In accordance with their plans, the importance of the amount, estimated at several million CHF, forced them to move to different capital "providers" – the venture capitalists.

Taking advantage of their public recognition and the many contacts they had made in the business community since the launch of the company, the discussions between the InMotion's managers and venture capitalists soon produced results. In less than six months, by August 1999, when the seed money was drying up, the first round of financing was completed with a venture capitalist firm providing a share of the capital of InMotion Technologies.

This first round of financing resulted in a new capitalization of CHF 2,600,000 for the company. In fact several investors brought together CHF 2,100,000 in new capital. The Regional Development Agency took back its money as allowed under the repurchase option agreement.

The appearance of investors in InMotion Technologies induced several changes in the shareholder structure as well as in the board of directors. At that time, the founders also reserved part of the capital for a generous stock option plan for the employees, board members, and advisory board members.

After this first round of financing, the shareholder structure was as follows:

- 62.3% for the founders
- 21.1% for the investors
- 16.6% for the stock option plan

The board of directors was reorganized with the immediate arrival of one member representing the investors. Over time, three external directors would be nominated, thus leaving only one founder on the board.

4 The Future

Wengen 2000 has been a key achievement for the startup: the new version of Video-Finish™ has been demonstrated successfully. The other products are almost ready to enter the markets. A well-motivated and sizeable team is working now in Fribourg, and the financing of a second round is under way. The five founders of InMotion Technologies could look back over these two last years and be proud of themselves. Nevertheless, the finish line is still way ahead and several gates remain in front of them.

How should/can InMotion Technologies manage its growth and maintain its leadership?

Exhibit 1
CVs of the Founders

Jean-Marie Ayer

Jean-Marie Ayer (1960) earned his degree in economics from the University of Fribourg in 1984 and his doctorate from the same university in 1989. He also studied at the University of Kiel in Germany (1982/83), where he was a full-time student, at the University of Texas in Austin, and the University of California at Santa Cruz, where he was a researcher from 1988 to 1989. At the University of California, he was involved in the Silicon Valley Research Group. For his doctoral dissertation he specialized in the high-tech industry and innovative environments. He is also the author of several research and educational publications in the area of economics.

In 1989, Ayer joined ABB in Zurich, where he was assistant to the group chief financial officer (CFO) for a period of three years. In 1992, he joined another unit of ABB to lead a project in the Czech Republic. The task consisted of assisting a newly acquired company (2000 employees) to integrate the ABB Group. In addition to this function, in 1993–1994 he joined ETHZ part time to coach a student research group. In 1994–1997, he joined the Power Generation Unit of ABB in Malaysia as CFO. This was the largest ABB unit in Asia with annual revenues of 1,000 MUSD. In 1997, he joined Swisscom International as head of the group in charge of monitoring international joint ventures. He led the restructuring of major projects in Malaysia, India, and the Czech Republic. He was board member of Swisscom's joint ventures in India and in France and was a member of the management board of Swisscom International.

Exhibit 1 (continued)

Serge Ayer

Serge Ayer was born in Fribourg, Switzerland. In 1985, he got his professional degree in music (guitar) from the music school of Fribourg and received a degree in computer science from the Swiss Institute of Technology, Zurich, in 1991. In that same year, he joined the Signal Processing Laboratory (LTS) of the Swiss Federal Institute of Technology (EPFL), Lausanne, where he received his Ph.D. in 1995. He was a member of a joint project between EPFL and the French company Thomson.

   During the summer of 1994, Serge Ayer was a visiting researcher at IBM Almaden Research Center, California, in the vision group, where he worked on a query–by-image-content tool for videos. In 1995, he joined IBM Switzerland, where he focused on data-mining technologies. His role was to interact with the various IBM research centers and work on data-mining solutions for selected customers. In late 1996, he joined the Laboratory for Audio-Visual Communications as first assistant. He led projects in digital photography, mosaicking techniques, and multimedia retrieval. He is the creator of the VideoFinish™ technology and one of the cofounders of InMotion Technologies.

Emmanuel Reusens

Emmanuel Reusens was born on May 14, 1968 in Brussels, Belgium. In 1991, he received an M.Sc. degree in electrical engineering and a bachelor's degree in philosophy from the Catholic University of Louvain (UCL), Belgium.

   In 1991, Reusens joined the Signal Processing Laboratory of the Swiss Federal Institute of Technology (EPEL) as a Ph.D. student. During his stay, he worked on wavelet image compression, fractal image and video compression, and dynamic coding in the framework of MPEG-4 developments. During the summer of 1993, he was visiting researcher at the Information Processing Laboratory of the University of California-Santa Barbara. He received his Ph.D. in May 1997 with a thesis entitled "Visual Data Compression using Fractal Theory and Dynamic Coding". He has authored or coauthored more than 20 scientific publications.

   In April 1996, he joined Logitech, Inc., in Fremont, CA, as senior software engineer in the scanner and video business units. He was in charge of all aspects of image processing from image acquisition to image reproduction through optical-component specification and qualification including CIS/CCD image sensors, sampling, signal datapath, host image processing algorithmic, colorimetric, automated image quality assessment. He joined InMotion Technologies in May 1998.

Exhibit 1 (continued)

Martin Vetterli

Martin Vetterli (born 1957) got his engineering degree from the Swiss Federal Institute of Technology in Zurich (1981), his MS from Stanford University (1982), and his doctorate from the Swiss Federal Institute of Technology in Lausanne (1986).

In 1986, he joined the Center for Telecommunications Research and the Department of Electrical Engineering at Columbia University in New York, where he was an associate professor of electrical engineering and codirector of the Image and Advanced Television Laboratory. In 1993, he joined the EECS faculty of the University of California at Berkeley, where he was a full professor until 1997, and is now an adjunct professor. In 1995, he joined the Communication Systems division of the Swiss Federal Institute of Technology in Lausanne, which he headed from 1996 to 1997.

He works on signal processing and communications, in particular, wavelet theory and applications, image and video compression, digital video communications, and computational complexity.

His work has won him several prizes (best paper awards from EURASIP in 1984 and from the IEEE Signal Processing Society in 1991 and 1996). He received the Swiss National Latsis Prize in 1996 and is a fellow of the IEEE. He is the coauthor, with J. Kovaoevic, of the book *Wavelets and Subband Coding* (Prentice-Hall, 1994) and of numerous research papers. He is member of the technical board of C-Cube Microsystems.

Victor Bergonzoli

Victor Bergonzoli got his business and administration degree from the HEC in Lausanne in 1990. During his studies he used to work as a software and computer salesman. After obtaining his diploma he joined Credit Suisse in Geneva to work as relationship manager in the commercial department. He left Geneva for Zurich to work in the international division at Credit Suisse's headquarters, where he was in charge of credit analysis for the North American market.

In 1996, Bergonzoli joined Swisscom International in Bern, where he was involved, as a core team member, in the acquisition of two major companies in Europe. He left for India in early 1997 to take over the position of country manager for Swisscom and later on also became chief financial officer of a Swisscom joint venture in India.

Bergonzoli oversaw a department of more than 50 people and was a member of the core management team and directed a US$250 million financing process with major international banks. Appointed president of the Swiss Business Forum in India, an association that supports large companies like Nestlé, ABB, *etc.*, he has been interacting intensively with many companies to help them in their investment decisions in India.

CASE 2-4
Boblbee (D): The Urban Backpack

Benoît Leleux, Joachim Schwass, and Anna Lindblom

The summer of 1998 would be one that Patrik Bernstein would never forget. One Friday evening, he sat in a small cottage in Torekov, southern Sweden, assembling some 200 backpacks with his partners Sam Bonnier and Jonas Blanking, their wives, and many friends. It was a world apart from his Indonesian expat lifestyle with Unilever, but he was enjoying himself. On the Monday morning they had loaded three cars with the Boblbee backpacks and set off in three different directions – to Stockholm, Göteborg, and Malmö – to deliver their first order to the Swedish sports retail chain The Stadium (*Exhibit 1*).

That first major delivery was a momentous step in the short life of their startup company. When the three partners reconvened in Torekov a few days later, and the high had had time to wear off, reality hit them: They had no clear strategic plan for the company. Their focus had been on finalizing the design of a launch product and getting the first order delivered. With that done, what next? After a few hours of discussion, they elected to postpone any decision until after the International Trade Fair for Sports and Fashion (ISPO) in Munich, a major trade event scheduled to take place three weeks later, in September.

The response they received at the fair was sensational – everyone from competitors to media loved the backpack, the urban marketing brochure, and Boblbee's unconventional stand. Marketing experts reckoned that the Boblbee backpack would be the next inline skate – it was revolutionary! The hype was totally unexpected for the company. As many as 84 distributors showed interest in distributing Boblbee. Bernstein said:

> We had rough price lists, but no real marketing plans – we were not prepared for this response. We sat in our hotel room after the first day, drinking whiskey, talking about prices and strategy. Where should we position ourselves? We knew our costs, but we had never discussed a distributor pricing strategy. We had only produced a few prototypes and made one sale. What distribution channel should we be using? How should we organize production? What should our proposition be?

Copyright © 2006 by IMD – International Institute for Management Development, Lausanne, Switzerland. Not to be used or reproduced without written permission directly from IMD.

1 Background

1.1 Global Act AB (GAAB)

Bernstein and Bonnier were old friends – they had both attended the Sigtuna boarding school in Sweden for six years. Bernstein commented:

> When you are at boarding school you develop a special bond – it is deeper than normal friendship. You go through quite a lot together, and that stays long after graduation.

In the early 1990s, Bernstein started to feel restless in his position as a "fast tracker" at Unilever. He had reached a point in his career where he needed to be "reenergized" and to face new challenges and possibly less politics after some 15 years in the corporate world. One long summer's day in 1995, Bernstein and Bonnier sat in their bathing suits on a cliff at Hallandsväderöarna on the west coast of Sweden talking about life and what to do next. Neither of them knew that the other had plans to start a business. Bernstein had only shared his plans with his wife so far, fearing not only that he would not be taken seriously but also that someone would steal his ideas. The two nevertheless opened up to each other – and were elated to discover that what they had in mind for their respective futures was quite similar and fairly compatible.

Bonnier wanted to start a concept called "toy factory"; Bernstein had dreamed up Spotlight – The Big Little Theatre – which was a way for children to learn through the arts. It took seconds to realize that they would be stronger together, so they decided to start something jointly. They invested equal amounts of money and set up Global Act AB (GAAB) in Torekov, a small, sleepy summer town with great golf courses and beaches on Sweden's west coast. For no particular reason, Bernstein took the role of CEO and Bonnier that of chairman.

They wrote business plans for both Spotlight and the toy factory but in the end decided to do Spotlight first (*see Appendix 1 for a brief description of Spotlight*).

1.2 Backpack Encounter

In 1996, during the development of the various technical components needed for Spotlight, the duo started to work with Swedish industrial designer Jonas Blanking. Bernstein remembered the first encounter with Blanking.

> It was "click" at first sight! We had so much in common, and yet were so different.

It was during the first meeting with Blanking that Bernstein noticed a prototype of a mysterious-looking backpack hanging in Blanking's studio. However, it wasn't until Bernstein was trying to launch Spotlight in the USA that he finally got a chance to experience the backpack concept. He was staying with one of

Blanking's friends – Chet Swenson – who at the time was trying to sell the backpack concept to the sports manufacturer K2 as a licensed product. The hard, funky backpack was lying beside a bed in the guestroom and Bernstein could not help picking it up and playing with it. He was fascinated, but equally surprised that it was just lying around, so he asked Blanking:

> Jonas, why is nothing happening with the backpack?

Blanking was a designer to the core and less of a business person. He had asked Swenson for help, and Swenson had made a few attempts to present the pack to some bigger sports companies but did not receive a positive reaction. It did not seem to appeal to them. Bernstein saw great potential in the concept and was also truly impressed with the work Blanking had done for Spotlight. Bernstein and Bonnier decided to suggest that Blanking become a partner in GAAB. The three of them could then develop a multibrand strategy around Spotlight and the backpack.

In 1997 the three shook hands and GAAB was split into three equal parts – each partner acquired 33% of the stock – and the backpack project was code-named Boblbee, for *Bo*nnier, *Bl*anking, and *Be*rnstein Enterprises (*see Exhibit 2 for capital structure*). The company adopted an abstract of a person in free-fall as a logo. It symbolized Boblbee's soul – free! Blanking took over responsibility for product development.

1.3 Genesis of Boblbee

The original idea for the backpack had come to Blanking when he was riding to work, usually on his rollerblades or bike. He was sick and tired of having his books, laptop, MP3 player, gym clothes, and other bag contents ruined by rain or from being knocked around. So he started developing a novel backpack that would more exactly match his needs, which he assumed would also be those of his generation.

To create a light, yet impact-resistant, casing, Blanking looked mostly at the air cargo industry for inspiration. The industry handled many small shipments for which it used special containers that had to be light yet strong and protective. It quickly became clear that the keys to a successful new backpack were (1) a sophisticated weight distribution system that would shift the effort to where it would be least felt and (2) the use of composite materials that would be both light and resistant. Blanking researched the plastic and aluminum structures used in air cargo containers to create the backpack structure and impact-protection casing. He also investigated how the containers were assembled, the types of screws used, and how the materials reacted in various stress situations and corrosive environments.

To make the backpacks comfortable, it was critical to use a soft harness, but combining the hard shell with the soft harness proved to be the most difficult part of the product development. Blanking produced several mock-ups and prototypes in 1996, which resulted in the first working prototype.

2 Getting Boblbee to Market

2.1 Technical Description

Boblbee incorporated materials that had never been used before for backpacks. It combined a monocoque hard shell with a set of soft harnesses in a unique, ergonomic, and radical design. It was highly functional, with loads of pockets inside and add-ons that could be attached to the outside of the backpack for even greater usability and flexibility.

Boblbee was built around the following features (*Exhibit 3*):

- The Lumbar Support System™, which included the S-design that separated the upper load area from the lower lumbar support area. The Lumbar Support System™ ensured excellent weight distribution and reduced stress on the back.
- The Quick Lock and unique Bellow Flex™, which allowed easy access and flexible volume adjustment.
- The monocoque hard shell, which not only provided Impact Protection™ for sensitive products such as laptops and cameras but also protected its carrier.

Height	60 cm	
Width	30 cm	
Depth compressed	15 cm	
Weight hard shell (ABS, acrylic)	700 g	
Weight harness	700 g	
Weight assembled	1,600 g	
Volume compressed	15 liters	
Volume expanded	25 liters	
Volume sweat bag	20 liters	

The monocoque hard shell was initially based on thermoplastic injection molding with materials such as ABS (acrylonitrile − butadiene − styrene terpolymer) and acrylic. This ensured high tensile strength, excellent thermal properties, and flexible surface treatment. But the hard shell could be developed further. Materials such as recycled plastic, metals such as aluminum or titanium (radically new materials in the backpack market), and crystalline materials (such as carbon or Kevlar fibers for very high-end backpacks) could also be used. The surface materials could be changed or anodized, the colors easily altered, and graphics such as logos and pictures easily added. For a true "street oriented" design, all it took was a couple of cans of spray paint. Various graffiti artists would be invited to customize the shells (*Exhibit 4*).

The harness was made of several flexible foam plates that optimally followed the movement of the body. Each plate conformed to its own area, for example pelvis/hip, waist, shoulder blades, and shoulders/arms. The inner structure of the harness was covered in DuPont 1000C Cordura – a synthetic high-performance fiber that ensured resistance even after years of wear and tear. The material was also waterproof. Countless hours were put into designing the assembly, primarily to find ways to hide all the seams to avoid fraying fibers. The external seams were sealed with Cordura-clad piping. Boblbee hoped to continue to work with agents at DuPont and in Hong Kong and Korea to make sure that the company always had the latest in fiber technology.

The nuts and bolts were essential pieces in the Boblbee system. They joined the flexible back plate to the hard shell through a watertight synthetic rubber gasket and an aluminum ledge. The nuts and bolts were also important for the add-on equipment and were designed to conform to specific attachments and strengths. The nuts and bolts were surface-treated to avoid corrosion, discoloration, and material fatigue. Each Boblbee fixture was made up of a corrosion-resistant zinc and iron blend, with a superstrong Deltacol black surface. The combination was tested in a salt chamber for 400 hours and showed the same corrosion protection as stainless steel, but it was lighter.

2.2 Patents

A provisional patent application for BOBLBEE™ was filed on January 6, 1997 by Kelly, Bauersfeld and Louwry in Los Angeles. GAAB conducted two independent patent searches, one at the US patent office and the other at its Swedish equivalent, the "Patent och Registreringsverket." The searches revealed no significant prior art for a monocoque three-wall shell and a flexible fourth wall. The patent lawyers saw the probability of getting a patent on these key features of the design as very good. During summer and fall of 1997, Awapatent (a Swedish company specializing in patent protection) prepared an International Patent Application and included ten areas of innovation.

The PCT (Patent Collaboration Treaty) application was filed on December 5, 1997. The patent was expected to come through in fall 1998, but in the meantime the hard shell of the backpacks was embossed with the inscription Pat.Pend. GAAB also sought trademark protection for the Boblbee brand in all the markets where it planned to be active.

2.3 Supply Chain

Considering the high-tech nature of the Boblbee backpack and the multiple technical competencies involved in its manufacture, it was clearly inconceivable to manufacture and assemble entirely in-house. There were a number of critical constraints. First, it was important that all suppliers of the various Boblbee components be strong, established, quality conscious, and with leading-edge technology. Second, the company was quite open to forming partnerships with its suppliers in order to make the cooperation even more successful.

One of the arrangements considered was to subcontract the entire backpack production to an equipment specialist and to focus on the design, marketing, and sales of the product themselves. This approach would require little financing up-front, but they did not have a subcontractor in mind. There were also concerns about quality control and potential conflicts when introducing further innovations, with the integrator possibly dragging its feet if faced with costly changes. Finally, with a technology product, it was deemed valuable to stay in direct contact with some of the manufacturing or assembly operations so that direct feedback could be established from marketing and sales.

A second approach would be to source the components through reputed independent producers but keep the assembly and final quality control in-house. So far, the shell and cleat were made by NOLATO Termoform, a Swedish producer of polymer materials, and the nuts and bolts were produced by Willum Nielsen, a Danish specialty producer, Nedschroef in the Netherlands, or Bufab BIX AB in Sweden. The harness was sourced in Asia, mainly China. The longest lead time was for sourcing of the harness, which could take up to three months. A potential production facility had been located in Torekov. It had a total area of $800\,m^2$, of which $160\,m^2$ were used for office space and the rest could be used for storage, assembly, and shipment. The facility would have the capacity to assemble and ship up to 8,000 Boblbee backpacks per month, as well as several hundred Spotlight kits.

Such an arrangement could possibly be further optimized by using suppliers on a global basis. They could also consider developing a number of assembly plants around the world.

2.4 Marketing

After the product launch at The Stadium, GAAB decided to commission the first marketing brochure. One of Europe's best design photographers, who had also done shoots for the trendy jeans manufacturer Diesel, contacted GAAB and offered to take the photos. Flattered, Boblbee gladly accepted. The result was a trend-setting, futuristic, and very "urban" brochure that won the company a Danish design award. It showed lots of skin and funky people and left plenty of room for the imagination (*see Figure 1*). Blanking also wrote a poem that was printed on the inside cover.

Figure 1 Image from
Boblebee Brochure
Source: GAAB Marketing
materials

2.4.1 Early Market Feedback

Without much of a marketing strategy in place, the company decided to show its products and let the potential users drive the positioning. ISPO in Munich in 1998 was Boblbee's first international appearance. The company wanted to make a lasting impression, but, with tight budgets and little understanding of its strategic positioning, it had to be innovative in its booth design. Piping from Spotlight was used to create a frame on which to hang the backpacks. Freight pallets were recycled as tables on which Boblbee displayed its "urbanistic" marketing brochures. Boblbee packing boxes were cut up to make signs on which Blanking wrote key words or slogans. (*See Exhibit 5 for early drawings of the ISPO display.*)

ISPO turned out to be a huge success not only in terms of marketing, but also product development. During the fair many mountain guides examined the harness and gave Blanking tips on how it could be improved. The company received lots of media coverage during those few days in Munich. One German journalist wrote:

> The queue to get into Boblbee's booth stretched all the way to the main train station in downtown Munich.

The German TV channel RTL was in attendance at the fair and covered Boblbee extensively. Overall the response gave the GAAB team new energy and trust in their idea – it was clear they had the new-generation backpack on their hands…if only they could figure out how to market them!

2.4.2 Customer Value Proposition

So far, GAAB had operated mostly as a product-driven organization. The product had been created to solve one of the partner's own transportation problems, and his experience was deemed to apply to a significant number of people. But several questions remained unanswered: How large was that population? What segment should GAAB treat as the target group? Where and how could they reach it? What were the key buying criteria of that segment? What price would it be willing to pay for a technologically advanced backpack? Did it even care about technology or was it mostly a fashion statement? What was GAAB's value proposition? A fashion statement based on Nordic design? A practical transportation means with unique ergonomic properties? A technology statement for techno freaks? A protection statement for safety freaks? All of the above? Without a clear understanding of what the product stood for in the market, it was difficult to develop an effective marketing strategy.

The initial marketing brochure had taken the product toward the trendy, urban group, but was that really what the founders wanted, or was it simply chance? Was this the most potent market segment or even the easiest one to reach? What implications would this positioning have for the future of the company?

The Boblbee backpack was best described as highly functional and sturdy, but also extremely trendy, and it could therefore be used by many different customer categories.

- People doing outdoor sports would find the backpack light and highly functional. It could withstand wet conditions and hence was perfect for skiing, hiking, or sailing. The Impact Protection™ not only shielded the carrier but also safeguarded the backpack's contents. The ergonomics of the pack made it suitable for carrying heavy weights for long periods, as the Lumbar Support System™ distributed the weight evenly and did not place stress on the back or shoulders.
- Another potential customer group was business people. The backpack had been developed to be able to carry sensitive equipment such as laptops, mobile phones, PDAs, and MP3 players, and the functionality was excellent. As such, it was easy for people commuting to work by train, bike, scooter/motorbike, or on foot to keep things organized.
- Students often had to use their bikes to get to school, at least in towns in Europe, and they therefore needed a functional bag for their gear, laptop, and study material.
- There were also opportunities to expand the product range to professionals such as police, ski guides, alpine rescue teams, or medical staff who needed mobility while still being able to carry valuable goods with them.
- There was also the possibility of targeting trendsetters, who wanted to be different and always first with the latest products.

2.4.3 Marketing Issues

Without a clear decision as to the positioning of the product, GAAB had postponed making precise marketing plans. One possibility was to engineer cheap ad campaigns, such as PR in various newspapers and magazines, sponsoring of key influencers, or product placement. Another option was to continue doing trade shows – which had proved successful in Munich – in company displays and event sponsoring.

Another possibility was to invest a bit more and develop a more purposeful marketing campaign (print media, TV, radio, *etc.*) that would establish the brand in the market and build brand awareness and identity.

GAAB could also agree to sell through an established consumer brand, such as Nike, Reebok, Adidas, or Oakley. This form of private label production would shield the company from having to get involved in complex marketing and distribution operations. It also provided the instant recognition of a global brand.

But to decide on a marketing mix, the partners would have to make decisions regarding the product positioning.

2.4.4 Sales Channels

Backpacks could be sold through many different distribution channels, and a product like Boblbee, with its strong functional features and revolutionary design, was suitable for most of them (*Table 1*). First and foremost, the product could be sold through traditional retail channels, such as fashion stores, travel gear specialists, sports and leisure boutiques, and even computer/IT distributors. These stores were often part of larger groups that provided regional, national, and sometimes global advertising coverage and visibility. From GAAB's standpoint, they could provide immediate credibility and sales reach. On the negative side, the multilayered distribution system, with intermediate wholesalers and distributors, ate significantly into the profit margins. There was also little control over the final sales pitch and positioning once the product was in the store.

A close variation on the first model was the use of branded channels and stores, such as Nike or Adidas stores, or some other fashionable travel brands such as Timberlake. Selling the product under their brand names provided better recognition, but Boblbee would never be visible as a brand, just as a product concept, and being at the mercy of such retail behemoths did not provide much negotiation leeway.

A more radical option was to focus directly on corporate channels, selling the backpacks strictly through VIP promotions and sophisticated corporate gift specialists. These were often more directly connected to the clientele, hence reducing the total sales costs. But most corporate gift companies were relatively small without much brand appeal, and corporate PR opportunities remained limited.

Since the backpack was originally designed with superior functional attributes (ergonomics, shock and water resistance, favorable aerodynamics, *etc.*), it was also possible to target professional users directly or companies supplying professional workwear and accessories.

The alternatives did not stop there. Large malls were appearing in suburban areas, combining the offerings of dozens of smaller stores in a single convenient location. It seemed totally feasible to rely on Blanking's design skills to create a trendy in-mall store, based on an extended version of the display used at ISPO in Munich. This would shift to the company the burden of creating hundreds of boutiques, and the partners were not sure that this was where they wanted to go. A recent trend in large

Table 1 Sales Channels

Classic Retail	Corporate	Direct to Professionals
• Fashion	• VIP promotion	• Police
• Computer/IT	• PR promotion	• Military
• Sports/leisure		• Coast guard
• Store-within-a-Store		• Rescue units

retailers was the creation of stores-within-a-store, where a company was able to sell its products under its own label within the confines (and benefiting from the infra-structure) of a larger store. This was particularly appealing because it could help build Boblbee into a distinct brand while keeping the costs under control.

2.5 Founders and Employees

2.5.1 Patrik Bernstein, Founder and Partner

Bernstein was born in 1956. He was the president and CEO of GAAB. After a short career as a steward for Deutsche Lufthansa, he acquired a degree from the European Business School (at International University Schloss Reichershausen in Germany) and started a career as an account executive at the J. Walter Thompson advertising agency in Frankfurt. After two years, he left the company to become a brand manager for Hasbro, one of the largest toy manufacturers in the world, in Nuremberg. In 1988 Bernstein was headhunted to Unilever Germany. After a few years in Indonesia, Bernstein returned to Germany in 1992 as general manager of Rimmel-Chicogo International, a division of Unilever.

2.5.2 Sam Bonnier, Founder and Partner

Bonnier, born in 1959, was the chairman of the GAAB board. He was a member of the Bonnier family in Sweden, a large publishing house (Bonnier Group) with annual turnover of € 2.5 billion.[1] Bonnier obtained a degree from the University of Denver and spent his first professional years in various companies in the Bonnier Group. In 1988 he switched careers and started selling equity funds for Alfred Berg Fond. In 1991 he started his own company – Office Complete Sweden AB.

---

[1] € 1 = US$1

2.5.3 Jonas Blanking, Partner

Blanking was the youngest of the three (born 1966). He graduated from the Art Center College of Design in Lausanne (Switzerland) before spending three years at IDEA Institute in Italy. In 1992 he was appointed senior designer at Barré design in Lyons and Paris, where he worked with Salomon (the sports equipment manufacturer). He later moved to Kontrapunkt Design in Copenhagen, where he worked with LEGO and Danish Post. In 1994 Blanking became a freelancer with his own studio in Malmö.

3 Industry Environment

Boblbee competed within the apparel industry/sporting goods equipment market. In 1996 the global apparel and sporting goods market generated annual retail revenues of €130 billion (€85 billion wholesale) and was expected to grow by about 23% by 2001. The US accounted for 50% of all sales, followed by Japan, which accounted for 16% (*Exhibit 6*).

The apparel industry/sporting goods equipment market was truly global and highly competitive, with a vast number of players. Sales growth was influenced by overall economic cycles, consumer attitudes and spending inclinations, and replacement of existing products. Other drivers for the sports apparel industry were the growing acceptance of informality (business casual) and greater health consciousness.

Analysts expected that rapidly rising costs would put pressure on profit margins and that consolidation would follow as a result.

The global backpack market (a segment within the apparel industry/sporting goods equipment market) was large and growing. It was difficult to quantify exactly how many backpacks were sold each year, but GAAB believed the number was in the millions. In Sweden alone, one of the largest sports chains sold over 100,000 backpacks per year.

3.1 Competition

GAAB had identified a few potential competitors for Boblbee depending on the segment in which it chose to position itself. (*See Exhibit 7 for a visual mapping of competitors.*) Some of the competitors are described in more detail below.

Pelican: Pelican made technically advanced bags that targeted professionals. The bags were especially developed for cameras, mobile phones, and laptops. Prices ranged from €70 to €300, with most of the products costing around €200.

MacCase: The bags were first produced for Apple Mac laptops, hence the name MacCase. The product range had been expanded to include backpacks, mail bags, and briefcases. MacCase focused mainly on the business community, but the bags were still innovative, trendy and colorful (think iMacs). They were also affordable and cost € 30 to € 80.

Nike: Over the years Nike had moved to become one of the most innovative sporting goods companies, always at the forefront of design. The company was a world leader in sports shoes, clothing, and various sporting gear. At the end of the 1990s it began integrating semirigid panels in its bags. Nike sports bags and backpacks cost between € 50 and € 200.

Oakley: Oakley was positioned as the trendsetter in sporting goods accessories. It became world famous for its sunglasses and ski goggles, but later produced a wide range of backpacks, laptop bags, duffel bags, luggage bags, and golf bags. Prices ranged from € 50 for a simple mail bag to € 250 for a wheeled travel bag. Most of Oakley's backpacks cost around € 120.

Salomon: Unexpected and innovative sporting equipment and apparel were the hallmarks of Salomon. The company had started to move toward more technical clothes and also produced a wide variety of bags from duffel bags and ski sacks to boot bags, backpacks, and various smaller bags. Prices ranged from € 50 to € 200.

Arcteryx: This company was innovative in the material and technology used to produce its technically advanced bags. Arcteryx targeted climbers and skiers in particular but also produced two bags with special laptop pockets for city dwellers. The price of an Arcteryx bag ranged from € 100 to € 500.

EastPak: Bags from this company were high-quality "street packs." Their designs were traditional, but the colors followed fashion trends. EastPak had a wide variety of bags, including backpacks, totes, duffel bags, smaller travel bags, doggy bags, and lots of accessories such as wallets, mobile phone pockets, *etc*. Prices ranged from € 20 to € 150, depending on the size and type of bag. EastPak had a no-questions-asked product guarantee of 30 years.

Hagflöfs: A Swedish manufacturer of outdoor gear, Haglöfs produced everything from backpacks – for climbing, skiing, long journeys, and city use – to sleeping bags, clothes, shoes, tents, and more. The company was famous for its high-quality, long-lasting goods. Prices ranged from € 44 to € 275.

Timbuk2: Founded in 1989 in San Francisco, California, Timbuk2 targeted students and young professionals, offering mail bags as a trendy alternative to backpacks and black briefcases. Later it extended its product line to backpacks, laptop bags, totes, duffel bags, and handbags. The bags were colorful and highly functional. It was also possible to build one's own bag at the company's Web site. The bags ranged from € 50 to € 120 in price, depending on size and functionality.

Ortovox: The company was founded in 1976 in Germany. At first Ortovox focused on producing measuring equipment and industrial automation systems. In 1985 backpacks were added to the product line, and in 1989 the company also started producing outdoor sports gear. Ortovox produced sports wear, backpacks, and safety systems for outdoor use (mainly alpine). The products were of high quality and the company claimed to be the number one choice of the pros.

Samsonite: The travel luggage segment was Samsonite's focus. In the 1990s the company started to work with designers and, as a result, became more fashion oriented, producing a wide variety of bags that were both functional and trendy. The intelligently engineered, practical bags were famous for their quality. The product portfolio also included backpacks and mail bags especially for laptops. The laptop bags were sturdy, highly functional, and very durable, but rather dull looking. A Samsonite bag cost from € 100 to € 600. The company had also started to produce clothes and shoes especially aimed at travelers.

Travelpro: Travelpro targeted business people on the go and focused on travel and computer bags. The company was innovative in its materials and functionality, but the designs were a bit dull – they usually came in black or other dark traditional colors. Also, the bags could only be purchased in the USA.

Tumi: American company Tumi focused on design excellence, functional superiority, and technical innovation. The product line ranged from travel bags, duffel bags, and laptop bags to backpacks and mail bags for urban use. Tumi produced special mail bags with a modern edge for people who wanted a functional bag but one that was a bit more casual and exciting than a traditional business bag.

Tumi's target market was professionals, but the bags varied from very traditional laptop and travel bags to more funky bags in order to accommodate the broad spectrum of business professionals. Tumi bags cost from € 100 for the simplest laptop bag to € 900 for a wheeled garment travel bag. Tumi also made stylish and lightweight bags (costing around € 150) especially for women.

Eagle Creek: The company offered a wide variety of bags from wheeled travel bags, duffel bags, laptop bags, backpacks (they even had a special line called "parent survival pack"), and shoulder bags to totes and accessories. EagleCreek targeted family-oriented professionals. The bags were very functional, but a bit boring in terms of color and design. Prices ranged from around € 20 to € 100.

Delsey: Founded in 1911 in France, Delsey initially made cases for typewriters, cameras, and record players. In 1965 the company started a department producing molded plastic items – which led to the introduction of hard-shelled suitcases

in 1970. At the end of the 1990s Delsey was one of the leading luggage producers in the world. It was present in over 105 countries on five continents, with
6,000 points of sale. The company's product range included two- and four-
wheeled hard travel bags, backpacks, duffle bags, attachés, and travel accessories.
Prices ranged from € 80 to € 300. Delsey also had a wide range of backpacks:
standard, urban, and laptop, business bags with lots of compartments, as well as
expandable and foldable bags.

Other: Many designer labels also made travel bags, laptop bags, and backpacks.
Prices started at € 500, depending on style, size, and label. Louis Vuitton, Gucci,
and Longchamp were a few of the designer labels active in the bag market. Companies such as Benetton also produced bags.

4 Working the Numbers

4.1 Next Steps

Bernstein, Bonnier, and Blanking sat in the hotel room and discussed the next
steps for Boblbee. They had been so focused on delivering their backpacks to The
Stadium and getting organized for the fair that they had not had time to think
about the future. Since the response at the fair was so enthusiastic, they needed to
come up with a credible growth strategy quickly, *i.e.*, a clear definition of the
"valued customer," a distinct value proposition, and all the elements of marketing, sales, distribution, and production strategies. A pretty tall order for an evening in the hotel.

On the hotel notepad, quick numbers jotted down seemed to imply that GAAB
would possibly sell 15,000 Boblbee backpacks during the first year of operation
(1998/1999) and that sales could grow annually by 15,000 units for the next three
years. But estimating the sales was really difficult without a clear decision as to
the key elements of the company's strategy. A great positive was that the existing
suppliers were understanding about GAAB's situation and had promised a great
deal of flexibility and responsiveness to its needs. This was important as some of
the bigger stores such as Intersport and Karstadt Sport had lead times of up to one
season, while smaller stores had greater flexibility.

With total sales so far of only 200 backpacks (in just two colors – black and
silver – even though extra colors had been tested for the ISPO fair), the founders
decided to focus only on the Megalopolis model for the rest of 1998, but to expand
the product line in 1999 to include the People's Burden and Box Jellyfish (*Exhibit 8*). These products shared many of the innovations of the original Boblbee
model, such as the harness system, which would be identical for all three – a combination of black and dark gray in high-quality Cordura. But the bags would differ

in color, material, and surface treatment of the hard shell and its various components, with Megalopolis being the most high scale. Bernstein said:

> At least we knew what our costs were and that helped a bit.

For the Megalopolis model, the cost of goods sold was € 36.74, for Box Jellyfish it was € 34.50, and for People's Burden it was € 32.89. The difficult question was where they should price the product. This would be impacted by the distribution and production structure they selected, which in turn would be chosen to match the strategic positioning of Boblbee. (*See Exhibit 9 for financial statements and projections for 1998/1999.*)

All in all, there was no way to escape the reality of the situation: They had a potential killer product in their hands, but unless they quickly defined its target clientele and an integrative marketing strategy, Boblbee would go where too many innovations had gone before, *i.e.*, the product graveyard. The partners were not about to let that happen before they had given it all they could. And it would start that night.

Exhibit 1
The Stadium

Source: Patrik Bernstein's private pictures

Exhibit 2
Capital Structure of Global Act AB

January 1998

Source	Sum in Swedish crown (SEK, 1 SEK = 0.11 €)
Founders' equity	300,000
Conditional equity	300,000
Short-term loans	1,178,000
Line of credit with Sparbanken Gripen	6,000,000
Loan from ALMI	500,000

End of 1998

Source	Sum in SEK (1 SEK = 0.11 €)
Founders' equity	500,000
Conditional equity	750,000
Short-term loans	778,000
Line of credit with Sparbanken Gripen	6,000,000
Loan from ALMI	500,000
Other loans	1,500,000

Source: GAAB internal information

Exhibit 3
Technical Information

LUMBAR SUPPORT SYSTEM™ **BELLOW FLEX™** **IMPACT PROTECTION™**

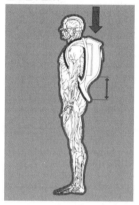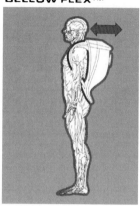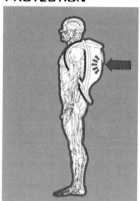

Source: GAAB internal company information

Exhibit 4
Boblbee Customized Designs

Source: GAAB internal company information

Exhibit 5
Boblbee Display System

BOBLBEE™
In-Store Display

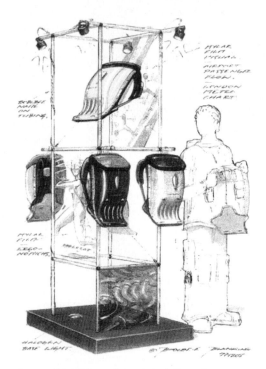

Source: GAAB internal company information

Exhibit 6
Global Sporting Goods Retail Market
Worldwide Sporting Goods Retail Market, 1996

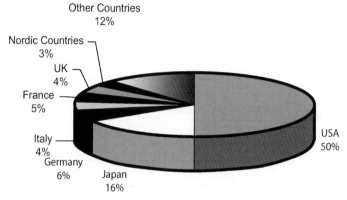

Total Market: US $130 billion

Other Countries
12%

Nordic Countries
3%

UK
4%

France
5%

Italy
4%

Germany
6%

Japan
16%

USA
50%

Source: Industry, Morgan Stanley Dean Witter Research Estimates

Exhibit 7
Strategic Mapping of Competitors

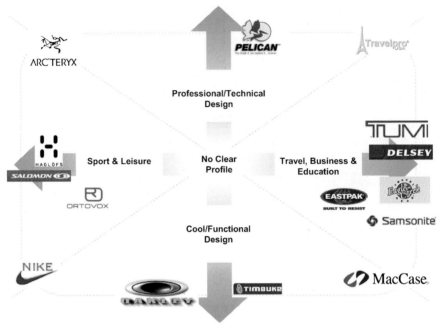

Source: GAAB internal company information

Exhibit 8
Initial Product Assortment 1998

People's burden

Box Jellyfish

Megalopolis

Source: GAAB internal company information

Exhibit 9
GAAB Financial Statements

Balance Sheet

Assets	95/96 (SEK '000) (1 SEK = 0.11 €)	96/97 (SEK '000) (1 SEK = 0.11 €)	Estimate 97/98 (SEK '000) (1 SEK = 0.11 €)
Cash	107	365	1,108
Accounts receivable	—	296	1,900
Other receivables	140	265	199
Inventory (Spotlight)	—	2,358	1,746
Spotlight in progress	321	91	96
Boblbee in progress	—	—	220
Inventory (other)	31	172	139
Activated plays	—	1,200	933
Total Assets	**599**	**4,747**	**6,341**

Liabilities and equity	95/96 (SEK '000) (1 SEK = 0.11 €)	96/97 (SEK '000) (1 SEK = 0.11 €)	Estimate 97/98 (SEK '000) (1 SEK = 0.11 €)
Share capital	300	300	300
Cond shareholders contributions	10	24	(1,402)
Shareholder loan		1,371	778
Bank loan		3,713	5,900
ALMI loan		500	500
Other			400
Short-term loans	294	98	238
Accounts payable	279	166	874
Earnings	(284)	(1,425)	(1,247)
Total liabilities and equity	**599**	**4,747**	**6,341**

Source: GAAB internal company information

Exhibit 9 (continued)

Profit and Loss Statement

	95/96	96/97	Estimate 97/98
	(SEK '000)	*(SEK '000)*	*(SEK '000)*
	(1 SEK = 0.11 €)	*(1 SEK = 0.11 €)*	*(1 SEK = 0.11 €)*
Volume in Pcs:			
Spotlight	–	192	1,170
Boblbee	–	–	–
Gross revenue	–	**1,419**	**3,601**
Net revenue	–	**1,419**	**3,601**
Cost of goods sold	–	(822)	(2,344)
Gross profit	–	**597**	**1,257**
Gross margin	–	42%	35%
Personnel	–	(249)	(558)
Rent and maintenance	–	(160)	(219)
Administration/office supply	−285	(401)	(594)
Patent/trademark	–	(147)	(207)
Marketing and sales	–	(859)	(487)
Depreciation	–	(35)	(47)
Operating income	**(285)**	**(1,254)**	**(855)**
Net margin	–	–	–
Interest	(1)	(172)	(393)
Income before tax	**(286)**	**(1,426)**	**(1,248)**
Tax	–	–	–
Net income	**(286)**	**(1,426)**	**(1,248)**

Source: GAAB internal company information

Cash Flow Statement

	Estimated for 98/99
	(SEK) 1 SEK = 0.11 €
Personnel/Administration	3,485,839
Spotlight	2,970,000
Boblbee	11,906,250
Taxes to Pay	1,824,788
GAAB Total	(20,186,877)
Spotlight sales	6,102,316
Boblbee sales	12,039,063
Total Sales	18,141,379
Tax Returns	2,462,304
Net Cash	(7,560,877)

Source: GAAB internal company information

Appendix
Spotlight

Spotlight was a mobile, easy-to-build theater stage. The stage was packed in an easy-to-carry bag and came with classroom-ready acting sets, *i. e.*, a teaching kit with scripts that had been narrated on CD (by a famous Swedish TV profile) with sound effects and music. There was also information about the play, the roles, and the backdrops, as well as a guide for the teacher on how to get started. Some backdrops were printed to suit to the plays; others were blank so the children could draw their own.

Bernstein's idea was that Spotlight would be a way for children to learn through the arts and increase their self-confidence by getting positive and constructive attention. It would also allow them to improve their speech and communication skills. But most of all, it would allow children to have fun. The idea was also that the children would have to become involved in building the scenery and coloring the backgrounds.

CASE 2-5
VistaPapers (A): December 1994[1]

Benoît Leleux and Lisa (Mwezi) Schüpbach

Robert Keane could not believe it: December 1994 had already arrived, and within a couple of weeks he would be graduating from a prestigious business school south of Paris. He should have been elated at the thought of rejoining the work force with his brand new MBA degree, but he could not get his mind off a more pressing concern.

For most of the last six months in the MBA program, he had focused his attention and energy on developing a business plan for his project, VistaPapers. Tomorrow he would present his concept formally for the first time to a panel of successful entrepreneurs and investors who had been invited to review projects for the business school's New Ventures Day. Robert would have to sell them his business opportunity.

The impact of the business plan presentation on his course grade mattered little to Robert at this point. He thought he had done a good job of presenting the opportunity in the written document and was pretty confident he could convey the full potential of the intended venture in the oral presentation. What really worried him was the possibility of having to face the realization that his business model may be materially flawed, forcing him to alter his plan even before he started his career as an entrepreneur, hopefully in a few weeks' time. So much effort had gone into the plan already and so much personal commitment generated that a sudden stop would really hurt.

Tomorrow's presentation was a chance to gain real feedback from experienced professionals. Eliciting, hearing, and acting on the panel's opinions was an opportunity not to be wasted. He checked one last time the short version of the plan distributed to the panelists.

Copyright © 2002 by IMD – International Institute for Management Development, Lausanne, Switzerland. Not to be used or reproduced without written permission directly from IMD.
[1] The case was previously registered as "Papyrus Laser (A): December 1994." The help of Robert Keane is gratefully acknowledged.

1 VistaPapers: Enabling Quality Printing for SOHO[2] Market

1.1 Business Plan (Abridged Version)

This plan describes an opportunity to invest in VistaPapers, a catalogue marketer of specialized stationery and related products that enables customers to use their own PC and laser or ink-jet printer to economically produce high-quality printed communications in small quantities – even one copy. VistaPapers focuses on the French market, with future expansion to other European markets anticipated.

Two proprietary and unique product categories form the core of the Vista-Papers business opportunity. The first is a Microsoft Windows-compatible software package, which makes the layout and production of printed literature simple and enjoyable. This software also serves as a marketing mechanism to acquire and retain customers for the second product offering: a direct-mail order catalogue of stationery, which makes it possible to produce high-quality color-printed communications using a black-and-white PC printer.

VistaPapers will capitalize on its business opportunity with a highly capable team. Entrepreneur Robert Keane previously managed the development of a similar business and has extensive management experience in a small-company environment. VistaPapers will also benefit from the expertise of specialist subcontractors in areas such as catalogue production, software publishing, software translation, and printing.

VistaPapers is seeking equity financing to meet marketing and working capital cash flow needs. Investors are projected to receive a compounded annual return on investment of 63% over five to seven years, which is expected to be realized via the acquisition of VistaPapers by another firm, as with recent acquisitions of similar companies in the industry. The current round of financing requires an investment of approximately FFr 3.5 million (French francs[3]) in return for 750 shares of equity at a price per share of FFr 4,583. One outside investor has precommitted to buy between 170 and 220 of the 750 shares.

The financial forecasts for the startup are presented in *Tables 1 and 2*:

Table 1   Summary of Financial Forecasts, December 1994 (FFr 000)

Fiscal Year (ending June)	1995	1996	1997	1998	1999	2000
Revenue	478	12,514	24,514	43,279	68,479	107,377
Net income	(1,430)	(2,743)	(654)	3,512	9,332	16,854
Cash flow from operations	(1,950)	(741)	124	2,368	9,446	19,399

Source: Company literature

---

[2] SOHO = small office/home office
[3] FFr 6 = US$1 at the time of the business plan in December 1994

The equity offer to potential investors, and the corresponding shareholding structure, is as follows:

Table 2 Proposed Financing Structure

Source of funds	Investment (FFr)	Ownership (shares)	Ownership (%)
Robert Keane and other initial investors	62,750	1,500	60
Future key employees (stock options)	0	250	10
Equity investors per this plan's offer	3,437,250	750	30
	3,500,000	2,500	100

Source: Company literature

1.2 The Opportunity

Printed materials are a primary tool used in effective communication. Aesthetically pleasing printed support materials stand on their own; they convince the targeted reader and they advance a cause.

The market today does not provide a solution to meet everyone's needs. Local printers will not consider small-volume print runs, and graphic design freelancers charge very high rates that cannot be justified for small volumes. In the recent past, some people have begun using their existing computer equipment to produce simple brochures and trifold flyers. However, layout is time consuming and complex. Furthermore, such documents are usually printed on plain paper, without graphic designs, and thus without the visual impact of a well-designed brochure. It is not feasible to coordinate brochures with other stationery products unless a large amount of money is spent.

A similar story can be told for many types of printed communications: restaurant menus, company announcements, mailers, business cards, letterheads, newsletters. Everyone knows that the terms "high quality," "color," and "professional design" simply do not belong in the same sentence as "inexpensive," "low volume," "easy," or "quick." This incompatibility represents a gap in the market that is not being met by current printing options. Simply put, there is a clear market need.

VistaPapers removes the limitations previously inherent in low-volume printed communications by providing printing that is low volume, flexible, high quality, aesthetic, quick, colorful, and affordable.

VistaPapers is a catalogue. It is sent to French businesses with small-volume printing needs (and that have PCs and laser or ink-jet printers). The catalogue contains scores of communication products, all computer printer compatible: high-quality papers with or without professional graphic designs, available individually or in coordinated sets; page after page of brochures, letterhead sheets, postcard mailers, business cards, envelopes, and newsletters. All items are delivered within 48 hours of placing a toll-free phone call.

Attractive and professional four-color graphic designs are preprinted on luxurious papers, many of which are prescored so they are easy to fold. This allows end users to customize the products with their own black-and-white text, using their own computer printer, their own photocopier, a copy shop, or a local printer. Vista-Papers products are fully compatible with all these printing methods.

As well as being able to manipulate text, customers can add clip art, scanned photos, or company logos. They can make last-minute changes and print, say, ten new brochures just before an important meeting. A restaurant owner, for example, can print weekly or even daily menus.

VistaPapers also understands the importance of a user-friendly computer product. It will offer a third-party software package called Flyers, and anyone with basic Microsoft Windows skills can produce great looking communications within 60 minutes of opening the Flyers box.[4]

To enhance the professionalism of customers' printed communications, Vista-Papers also provides a wide range of innovative complementary products. These include tabletop display stands, matching folders, and designer mailing labels that coordinate with the image of other printed items. VistaPapers acts as a communication partner, enabling customers to:

- Achieve a sense of professionalism, creativity, and satisfaction
- Utilize a new and exciting application for their existing computer equipment
- Create an image, a visual identity, for their small businesses
- Gain power and control over their printing requirements
- Find one-stop shopping for high-quality laser and copier compatible stationery.

VistaPapers' products are unique and immediately add value for customers. They are easy to purchase and use. The high-quality printing results and the flexibility in terms of quantities create customer satisfaction. Furthermore, the products are available at a cost well below that of existing alternatives. The price borders on impulse purchase levels, thus ensuring a stream of customers who are willing to try the product.

1.3 Proven Business Model and Investment Payoff

The concept originated in the United States, where approximately ten firms have built successful and growing businesses using catalogue marketing of preprinted laser-compatible papers. Most of the US firms began as independent startup businesses, indicating the feasibility of VistaPapers' approach to the business as an independent entity. PaperDirect, Inc., for example, started operating in the USA in 1989 with sales of less than $1 million. By 1993 sales had grown to approximately $70 million, an annual growth rate of 289%. In the fall of 1993 Deluxe Corporation

---

[4] Flyers is not essential to being able to use VistaPapers papers – customers can use existing word processing or desktop publishing software. However, Flyers greatly simplifies the process.

purchased PaperDirect for $90 million in cash plus additional fees to be paid based on 1993 to 1996 fiscal year earnings. Queblo Corporation investors recently realized their profit by selling out to Regency Thermographers, Inc., a division of Taylor Corporation.[5] These examples, and those of approximately eight other American competitors, illustrate the fact that the business model of catalogue sales of preprinted laser papers is highly attractive as an equity investment opportunity.

Two key drivers of the VistaPapers business model are exhibiting strong growth in France: laser printers and direct marketing. Both are following penetration curves similar to those seen in the US five years ago. In other words, the timing of VistaPapers's entry into the French market is at the same relative point as the American firms who entered the US market four to six years ago. Investors can expect to see similar market expansion in Europe over the next several years.

The belief that the window of opportunity is now open in France is supported by the fact that, in the past six months, two competitors have entered the French market. VistaPapers considers the existence of competitors to be a healthy situation, and since there is low awareness of this product category, competition reduces the cost of educating customers.

VistaPapers is confident that it has the management and technical capabilities required to capitalize on its market opportunity. These capabilities come from the quality and experience of both internal and external managers, combined with unique technical know-how. With this team, the company has the following key attributes:

- Experienced and proven leadership with startup company management skills
- French direct marketing and catalogue development expertise
- Skills in programming and publishing of personal computer software
- French market adaptation of personal computer software
- Printing and graphic design excellence
- Attractive salary and equity provisions to attract future key managers.

VistaPapers will subcontract a significant amount of work in the initial stages. First, this will enable it to enter the market more quickly and more effectively. Second, outsourcing will minimize the burden of human resource management issues at a time when the company needs to focus on business creation and opportunity exploitation. Third, by utilizing subcontractors VistaPapers will operate with a high percentage of costs as variable or discretionary, thus allowing for rapid adjustments to varying rates of market penetration should this be necessary. Some of the selected subcontractors are presented below.

---

[5] The value of the acquisition was not disclosed, but industry sources indicate it was very profitable for investors.

1.3.1 French Direct Marketing and Catalogue Production Experience

VistaPapers will work closely with Seibel Direct Marketing (Paris, France) to develop and execute its direct marketing strategy. Seibel is a leading European direct marketing firm that has taken an active part in the creation of marketing plans and budgets. The senior manager at Seibel France, with over 15 years' experience in French catalogues and direct marketing, will lead a VistaPapers catalogue team of graphic artists, direct mail specialists, copywriters, and product photographers.

1.3.2 Personal Computer Software Development

PerSoft Corp. (California, USA) is a leading publisher of small-business PC software. PerSoft will license its highly successful layout software (Flyers) to Vista-Papers for exclusive distribution in the French market and will provide project management for software modification services. Flyers will be modified to include graphic images of VistaPapers papers during onscreen layout, its product names and ordering codes, and toll-free phone numbers. As a leader in the US market for small-business software, PerSoft invests in regular upgrades to keep the product at the leading edge of the fast-moving software industry. These upgrades provide ongoing value to VistaPapers by moving it further ahead of its competitors.

1.3.3 Software Translation and Localization for the French Market

Leaders (Senlis, France) specializes in the translation and localization of PC software for introduction into the French marketplace. Its clients range from small firms such as VistaPapers to well-established software publishers such as Microsoft and Aldus. Leaders will translate the PerSoft layout software, provide disk duplication and software packaging services, and advise on adapting the software to maximize penetration into the French market.

1.3.4 Printing and Graphic Design

Manhattan (Rennes, France) is a specialized graphic layout, prepress, and printing consulting firm with extensive experience in printing, paper source selection, and management of printers for third-party firms. Manhattan will work closely with the VistaPapers team to (1) identify and manage high-quality printers for the Vista-Papers catalogue; (2) perform desktop publishing services, prepress, and printer selection for the production of VistaPapers's products; and (3) assist in the identification of competitive suppliers of high-quality blank stationery products that will be carried in the VistaPapers catalogue.

Before retiring for the night, Robert double-checked the details of his product portfolio, his analysis of the market environment and competitive positioning, his operating plan, and, finally, his financial assumptions. (*See Exhibits 1 to 4 for the financial assumptions used in pro forma cash flow and income statements.*)

He was ready for the big day.

Exhibit 1
Pro Forma Cash Flow Statements FY 1995

	Jan.	Feb.	Mar.	Apr.	May	June	FY 1995
Operating cash inflows							
Revenues	0	0	0	0	187,519	290,069	477,588
+ VAT collections	0	0	0	0	1,956	10,336	12,292
+ Interest income	0	0	0	0	0	0	0
– Δ Accounts receivable	0	0	0	0	0	0	0
Net operating cash inflows	0	0	0	0	189,475	300,405	489,880
Operating cash outflows							
+ Variable costs	0	0	0	0	60,549	112,974	173,523
+ Fixed costs depreciation	57,573	58,599	58,625	58,651	106,595	106,650	446,393
+ Marketing	117,000	271,000	340,000	366,000	54,000	94,000	1,242,000
+ Taxes	0	0	0	0	0	0	0
+ VAT payments	0	0	0	0	3,546	9,286	12,832
+ Δ Inventory	0	0	0	15,215	18,500	21,000	54,715
+ Capital expenditures	0	0	0	0	0	0	0
+ Interest expenses	0	0	0	0	0	0	0
– Δ Accounts payable	(0)	(0)	(0)	(15,215)	(18,500)	(5,000)	(38,715)
Net operating cash outflows	174,573	329,599	345,865	424,651	224,690	338,910	1,890,748
Net operating cash flows	(174,573)	(329,599)	(345,865)	(424,651)	(35,215)	(38,505)	(1,400,868)
Cumulative cash flows	(174,573)	(504,172)	(850,037)	(1,274,688)	(1,309,903)	(1,348,408)	(1,348,408)

The company plans to operate with a fiscal year ending on June 30 each year.
Source: Company literature

Exhibit 2
Pro Forma Cash Flow Statements FY 1996 (FFr 000)

	July	Aug.	Sept.	Oct.	Nov.	Dec.	Jan.	Feb.	Mar.	Apr.	May	June	FY 1996
Operating cash inflows													
Revenues	347	368	560	664	891	972	1,055	1,273	1,305	1,350	1,643	1,593	12,020
+VAT collections	16	20	34	40	69	67	74	112	107	113	165	153	973
+Interest income	0	1	0	2	0	1	1	1	1	2	1	2	14
−Δ Accounts receivable	(69)	(35)	0	(46)	(69)	(92)	(115)	(103)	(70)	(36)	(38)	(40)	(713)
Net cash inflows	294	355	594	660	892	948	1,016	1,283	1,344	1,429	1,771	1,708	12,293
Operating cash outflows													
+Variable costs	130	140	214	255	358	388	431	540	545	567	714	689	4,970
+Fixed costs depreciation	138	142	146	148	175	176	178	182	212	213	255	254	2,220
+Marketing	182	255	402	454	190	112	346	386	312	540	430	232	3,841
+Taxes	0	0	0	0	0	0	0	0	0	0	0	0	0
+VAT payments	84	100	168	159	134	126	178	206	199	246	260	219	2,078
+Δ Inventory	25	89	40	188	(14)	44	247	(33)	39	335	(77)	(9)	875
+Capital expenditures	0	0	140	0	0	0	0	0	0	0	0	0	140
+Interest expenses	0	0	0	0	0	0	0	0	0	0	0	0	0
−Δ Accounts payable	(118)	(36)	(311)	(29)	(35)	(26)	(187)	(204)	(57)	(183)	(76)	(63)	(1,089)
Net cash outflows	440	691	798	1,234	808	873	1,192	1,077	1,249	1,718	1,507	1,449	13,035
Net operating cash flows	(145)	(336)	(205)	(573)	84	76	(177)	205	96	(289)	264	259	(741)
Cumulative cash flow	(1,493)	(1,829)	(2,034)	(2,607)	(2,523)	(2,447)	(2,624)	(2,419)	(2,323)	(2,612)	(2,348)	(2,089)	(2,089)

The company plans to operate with a fiscal year ending on June 30 each year.
Source: Company literature

Exhibit 3
Pro Forma Cash Flow Statements FY 1996 to 2000 (FFr 000)

	1996	1997	1998	1999	2000
Operating cash inflows					
Revenues	12,020	27,716	47,078	75,600	109,071
+ VAT collections	973	3,168	5,998	10,409	16,064
+ Interest income	14	21	22	10	37
− Δ Accounts receivable	(713)	(593)	(791)	(1,154)	(460)
Net cash inflows	12,293	30,313	52,307	84,865	124,713
Operating cash outflows					
+ Variable costs	4,970	12,856	20,634	33,666	49,002
+ Fixed-cost depreciation	2,220	5,241	7,863	11,456	14,636
+ Marketing	3,841	6,823	10,665	14,794	19,554
+ Taxes	0	0	2,539	5,096	8,426
+ VAT payments	2,078	4,747	7,284	11,144	15,474
+ Δ Inventory	875	1,442	2,143	1,320	1,049
+ Capital expenditures	140	0	0	0	0
+ Interest expenses	0	0	0	0	0
− Δ Accounts payable	(1,089)	(1,521)	(2,390)	(3,657)	(5,077)
Net cash outflows	13,035	30,189	49,939	75,419	105,313
Net operating cash flows	(741)	124	2,368	9,446	19,399

The company plans to operate with a fiscal year ending on June 30 each year.
Source: Company literature

Exhibit 4
Pro Forma Income Statements FY 1996 to 2000 (FFr 000)

	1996	1997	1998	1999	2000
Revenues					
Catalogue sales	9,546	19,471	35,865	58,399	92,346
Desktop publishing software	2,494	3,925	6,106	8,677	13,533
Sales through distribution channel	372	1,118	1,308	1,403	1,498
Other revenues	127				
Total revenues	*12,539*	*24,514*	*43,279*	*68,479*	*107,377*
20.6% VAT collected	2,578	5,050	8,915	14,107	22,119
Less... direct costs					
Catalogue COGS [*]	3,924	6,815	12,194	19,856	29,735
Desktop publishing software COGS	1,507	2,373	3,691	5,245	6,959
Other sales COGS	287	863	1,009	1,082	1,155
Credit card charges (bad debt)	151	296	522	826	1,212
Fulfillment costs	692	1168	2,152	3,504	5,247
Total direct costs	*6,561*	*11,514*	*19,568*	*30,512*	*44,307*
Contribution	5,979	13,000	23,711	37,966	63,070
Less... fixed costs					
Salaries and wages	394	1,144	1,976	3,069	4,431
Social charges on salaries	193	572	988	1,534	2,215
Direct mail costs	200	257	316	372	424
Direct marketing consulting	242	180	191	200	207
Rent, utilities, leases, *etc.*	158	203	150	193	334
Depreciation	21	84	160	310	610
Other services (acct, bank, *etc.*)	155	200	245	288	329
Office supplies	173	223	274	322	367
Other costs	251	490	866	1,370	2,009
Total fixed costs	*1,787*	*3,353*	*5,266*	*7,758*	*10,926*
Margin before discretionary costs	**4,192**	**9,647**	**18,445**	**30,208**	**52,144**
Less... marketing costs					
Marketing expenses	5,931	9,082	13,369	18,864	25,506
Catalogue development	710	650	850	950	1,250
Product development, market research	214	450	550	850	1,050
Travel	80	119	164	212	262
Total marketing costs	6,935	10,301	14,933	20,876	28,068
Income before taxes	(2,743)	(654)	3,512	9,332	24,076
Taxes	0	0	597	2,333	7,222
Net income	(2,743)	(654)	2,915	6,999	16,854

* Cost of goods sold
The company plans to operate with a fiscal year ending on June 30 each year.
Source: Company literature

CASE 2-6
VistaPapers (B): September 1995[1]

Benoît Leleux and Lisa (Mwezi) Schüpbach

The initial presentation of the business plan to a panel of professional investors and entrepreneurs at the business school's New Ventures Day in 1994 had not generated the kind of enthusiasm Robert Keane had been hoping for. In fact, the plan had received a big blow from the critical audience. Robert had known beforehand he could expect some serious feedback from such a select panel, but he definitely got more than he bargained for.

Robert took it all in stride. He carefully took note of the major areas that needed to be addressed and decided to work on fixing these deficiencies. He was not going to abandon his entrepreneurial dream but would instead use the feedback to ensure that the dream came true.

Copyright © 2002 by IMD – International Institute for Management Development, Lausanne, Switzerland. Not to be used or reproduced without written permission directly from IMD.
[1] The case was previously registered as "Papyrus Laser (B): September 1995." The help of Robert Keane is gratefully acknowledged.

1 Feedback Received on the Initial Business Plan

Initial criticisms focused on a number of areas:

- Robert had not thoroughly studied the market for mail-order specialty papers in France. In spite of anecdotal evidence that many professionals and small businesses were interested in the product and its delivery mode, there was insufficient data to gauge the size and characteristics of the market.
- Assuming there was a market, there was no compelling and unique customer proposition. Why would target customers use VistaPapers for their printing needs? Was mail order the appropriate delivery mechanism?
- The business offered little in terms of entry barriers. Since most of the business activities were subcontracted, anybody with some money and connections could offer a similar service. The mechanisms to raise some form of entry barrier (software development, first-mover advantage, *etc.*) seemed weak and inadequate to discourage a willing and able competitor.

Existing competitors in France and similar businesses in the USA seemed to have enjoyed high levels of growth for the past five years, but there was no evidence yet of the ability of the business to generate profits.

- A number of critical components of the business model were still not locked in. In particular, even though there was an agreement in principle with a company to cross-sell its Flyers layout software in France, the final contract had not been signed. Since the business model relied heavily on the software to sell the paper products, this was perceived as a significant shortfall. Other potential software partners, some of them with large market shares, were equally uncommitted.
- Why begin the business in France? Robert was not French, spoke little of the language, and, other than his year at the international business school, had never lived in France. Furthermore, France had a well-deserved reputation as one of the most "entrepreneur-hostile" places in Europe.
- There had been no samples to look at to get a feel for the product. Even though it was not too difficult to visualize the final catalogue, not having one available at the time of the initial business plan reduced the credibility of the concept. This credibility gap was compounded by the fact that Robert had no personal experience in the direct-mail business.
- The assumptions used in the financial forecasts were optimistic, and there was no real attempt to contain the level of initial investments in the marketing of the product. The amount of money to be raised at the startup stage, some FFr 3.5 million[2], seemed too high for many potential investors. There were also concerns about the valuation of the company at harvest time.
- The pro formas were vague. In particular, some potential investors wanted to know more about the costs of the various development steps.

---

[2] FFr 6 = \$1 at the time of the December 1994 business plan.

- The financial offer proposed for potential investors (*see Table 1*) was perceived as being "greedy." It asked external investors to put in 98.2% of the capital in return for only 30% of the company's equity, a level of "sweat" equity many saw as unacceptable. There was also insufficient financial commitment from the lead entrepreneur, Robert himself.

Table 1   Proposed Financing Structure in December 1994 Business Plan

Source of funds	Investment (FFr)	Ownership (shares)	Ownership (%)
Robert Keane and other initial investors	62,750	1,500	60
Future key employees (stock options)	0	250	10
Equity investors per this plan's offer	3,437,250	750	30
	3,500,000	2,500	100

Source: Company information

2 Back to the Drawing Board: January to September 1995

Addressing all these issues would not be simple, but Robert was committed to this project and would do all he could to see it through, in one shape or another. After the initial anticlimax, Robert had to acknowledge that the points the panel raised were valid and that he would be in a much better position to capitalize on the opportunity if he addressed them first.

In fact, Robert went ahead and started the business immediately after his graduation from business school, but on more modest terms than those proposed in the first business plan. With just one office assistant, VistaPapers came into existence in the second bedroom of Robert's Paris apartment.

Early in 1995 Robert was hired as a freelance consultant to analyze the Small Office/Home Office (SOHO) market in France for Microsoft, the American software giant. The company was interested in learning more about the commercial printing requirements of the European SOHO market as it developed new products for it, such as the Office suite of applications and Publisher, an easy-to-use page layout software. In particular, Microsoft wanted to know more about the dynamics of the market, its sociological attributes, its product desires, and the best channels to reach it. The questions posed were complementary to those Robert had to address for his own products. The information gathered allowed him to get a better feel for what it would take to be successful in the competitive SOHO marketplace.

As he developed a better understanding of the market, he also built some crucial connections within Microsoft, leading to an exclusive and extensive cross-marketing agreement by July 1995. Furthermore, for VistaPapers' direct marketing campaigns, Microsoft would allow the startup to use Microsoft's proprietary customer database free of charge in return for selling a "bundle" of VistaPapers papers and Microsoft Publisher software. The agreement with PerSoft for distribution in France of the Flyers software, envisioned in the initial business plan, had yet to materialize, but the potential of the Microsoft partnership far outweighed that disappointment.

Robert grew more realistic about the financing side of things. Using personal savings, credit card debt, and investments from friends and family, Robert invested everything he had in VistaPapers – more than FFr 1 million. This money, plus consulting revenues from the Microsoft work, was used to cover the main costs of the new business: legal fees for the complexities of French incorporation, catalogue and product design, inventory buildup, travel expenses, and the salary of an office assistant. Contrary to the 1994 business plan assumption, Robert did not draw any salary for himself and did not intend to draw one any time soon.

It was clear to Robert that more money was needed to get the business off the ground. By late summer 1995, the revised business plan was ready. Robert's goal was to raise an additional FFr 2.25 million by early 1996.

3 The Revised VistaPapers Business Plan: September 1995

3.1 Executive Summary

This business plan describes an opportunity to invest in VistaPapers. Through its direct marketing catalogue, VistaPapers sells innovative paper products that facilitate low-volume commercial printing. Small businesses use the products sold in the catalogue in conjunction with their office printers to create printed materials that are higher in quality, lower in cost, and more convenient than those produced using alternative printing methods. The primary target market for VistaPapers consists of European businesses – equipped with PCs and printers – with fewer than ten employees.

VistaPapers is currently focused on the French market, with expansion into other geographic markets planned for the future. The business is a European adaptation of an American business model that has been very lucrative for investors. The first edition of the catalogue includes approximately 240 types of high-quality, graphically appealing papers uniquely suited to low-volume commercial printing – documents such as brochures, press kits, flyers, newsletters, reports, announcements, letterheads, and business cards. As a complement to these paper products, VistaPapers also markets high-quality software for the layout of printed communication.

In particular, VistaPapers has an exclusive agreement with Microsoft Corporation in which its catalogue – showing preprinted papers and examples of page layout using its Flyers proprietary templates – is included in the Microsoft Publisher software box. Microsoft Publisher holds the dominant market share in this software category, both in France and worldwide. It is an easy-to-use page layout program targeted at the small-office market. Innovative PageWizards guide the customer and allow a first-time user to create pleasing page layouts quickly and easily.

The more significant aspects of VistaPapers' market environment are as follows:

- French small offices are expected to buy 574,000 laser and ink-jet printers in 1995. This has grown from fewer than 250,000 in 1993 and is expected to increase to 672,000 by 1998.
- The black print quality of laser printers is approaching that of traditional offset printing. Ink-jet printing is quickly improving its cost and quality positions.
- Print runs of under 500 copies are now convenient and economically feasible. But there are many other as yet unfulfilled requirements that prevent the widespread use of office printers for commercial printing. These include easy page layout, secondary processing operations, a wide choice of quality papers, full-color graphics, and relevant user information.

VistaPapers fulfills these requirements and thus makes low-volume commercial printing easy and affordable. Its assets include:

- A team of managers and partners with extensive experience in managing a growing company, direct marketing, mail-order logistics, software development, and paper design and processing.
- A high-quality direct-mail catalogue.
- A strong and growing marketing partnership with Microsoft, with the following benefits:
  - Excellent market knowledge: VistaPapers was hired by Microsoft to play a leading role in a 1995 market research of the European small-office market for computer and commercial printing. Research included over 1,600 interviews with European small offices, 900 of which were conducted in France.
  - Highly responsive customer target lists: VistaPapers has obtained exclusive rights to distribute its catalogue in every box of Microsoft Publisher sold in France and is negotiating similar arrangements for other Microsoft products aimed at the small-office market. Through joint mailings, VistaPapers has access to Microsoft's registered-user database for the purposes of customer acquisition mailings. These agreements provide VistaPapers with lists of extremely responsive potential customers.
  - Low-cost direct marketing: The cost of reaching potential customers via Microsoft box inserts and mailings is approximately 50% of the cost of traditional catalogue marketing.
  - Product development: VistaPapers has access to planning and prerelease versions of Microsoft products that are complementary to VistaPapers. It is working with Microsoft on numerous product and service developments for future market introduction.
- Marketing, logistics, and product development plans: These plans are ready for market, ensuring a position as an early market leader.

- Marketing method: VistaPapers forms partnerships with suppliers of products that complement those in its catalogue. Marketing partnerships provide high-response customer targeting at low cost.

VistaPapers has been financed to date by Mr Robert Keane, its founder, and by other private investors. Capital invested to date is FFr 986,155.[3] In addition, Robert Keane is not taking a salary for the first 18 months of VistaPapers's activities and has forgone costs he has incurred in creating the company. These items represent a saving to the company of approximately FFr 535,000.

The opportunity for new investors is as follows:

- 7,750 shares are available at FFr 290 each (FFr 2,247,500 total), equal to 22.3% of PL's equity.
- Projected return on investment is 55% compounded annually, over a five-year period.
- Investment return is expected to be realized through the sale of VistaPapers to another firm. As discussed in the original business plan, precedents for such a sale exist in the US with similar firms.

3.2 Financial Aspects

Table 2 Summary of Financial Forecasts, September 1995 (FFr 000)

Fiscal year (ending June)	1996	1997	1998	1999	2000
Revenues	12,540	24,514	43,279	68,479	100,464
Net income	(2,761)	(655)	2,915	6,999	11,328

Source: Company literature

3.2.1 Dilution to be Caused by Employee Stock Options

VistaPapers believes that stock options are a highly effective tool to recruit top-quality managers and to provide management incentives that correspond to the interests of the shareholders. A stock option plan will be drawn up in the fall of 1995; 3,850 shares will be made available for future stock option awards. If and when these options are exercised, they will dilute the ownership percentages of all shareholders. Assuming 100% of the options are exercised, the percentage of VistaPapers owned by investors who invest per this business plan will drop from 22.3% to 20.08%. Accordingly, all calculations of potential investor return assume the smaller ownership percentage.

---

[3] FFr 5 = $1 as of Q3, 1995.

Table 3 Proposed Financing and Ownership Structure

Source of funds	Investment (FFr)	Ownership (shares)	Ownership (%)
Investments per this business plan	2,247,500	7,750	22.3
Of which:			
Newly issued shares		6,033	
Shares sold by owner		1,717	
Investments to date in VistaPapers	1,520,372	26,992	77.7
Of which:			
Invested capital, existing shareholders	985,372		
Foregone salaries and expenses	535,000		
	3,767,872	34,742	100

Source: Company literature

Table 4 Expected Investor Returns

- Fiscal year 2000, projected after-tax income — FFr 11,328,015
- Value of 100% of VistaPapers in June 2000 — FFr 90,626,086 (8 times net income)
- Value of 20.08% of VistaPapers in June 2000 — FFr 18,197,718 (7,750 shares)
- Price of 20.08% of VistaPapers per this offering — FFr 2,247,500 (7,750 × FFr 290 each)
- Time between investment and profit taking — 4.75 years (Sept. 1995 to June 2000)
- Compounded annual growth rate of investment — 55.32% (using above info)
- Year 2000 value as % of original investment — 810% (8.1 × investment)

Nine months later, it was time to try the new pitch for real – the business plan had been fixed, but could it fly with real investors now?

CASE 2-7
VistaPapers (B-2): November 1995[1]

Benoît Leleux and Lisa (Mwezi) Schüpbach

It was almost midnight, and in the chill of the early winter in Paris, Robert was contemplating the extent of his troubles. Riding high on the wave of an exclusive marketing agreement with Microsoft, he had planned to aggressively enter the market for specialty papers for low-volume commercial printing in the fall of 1995 with a large direct-mail campaign. This was supposed to mark the beginning of a glorious odyssey that would take him and his potential investors from France to Germany and other European countries.

The latest business plan had been well received by private "angel investors." Money had been committed by a senior manager at Microsoft who believed in the project and whose investment proved key in convincing other investors less familiar with the market potential for this type of business.

On November 18, 1995 the first VistaPapers mailing was sent out. Papyrus was about to get its first orders.

On November 20, 1995 the French postal services went on indefinite strike in support of salary increases. With all the money raised to date invested in preparations for this first mailing campaign, and the peak holiday season just around the corner, this was devastating news for Robert. Given the uncertain duration of the strike, Robert, as a first move, immediately halted all mailings that had not yet left the shipping facilities. That did his cash flows no good, though: with massive inventories and no sales for the foreseeable future, the approaching bills to be paid loomed large in his mind.

The strike lasted through the holiday season, essentially killing the year for businesses relying on mail orders and catalogues. It was back to the drawing board again for Robert, except now the priority was to salvage the business, not to get it started. Almost all the money he had put into the business had been spent, much of it having gone down the drain in the end-of-year calamity. Could he rebuild the business with what was left? Would his (mainly American) investors stick by him, or would they see this strike as proof that France was no place to make investments? He was certainly going to give it his best shot.

Copyright © 2002 by IMD – International Institute for Management Development, Lausanne, Switzerland. Not to be used or reproduced without written permission directly from IMD.
[1] The case was previously registered as "Papyrus Laser (B-2): November 1995." The help of Robert Keane is gratefully acknowledged.

CASE 2-8
VistaPapers (C): Going VistaPrint.com[1]

Benoît Leleux and Lisa (Mwezi) Schüpbach

Robert Keane was not resting on his laurels. In the four years since he founded VistaPapers, in 1995, the company survived a number of close calls – such as when its first round of startup funding was totally lost because of the longest running postal workers' strike in France. The company was now a significant player in the catalogue marketing of specialized stationery for small print runs in the European small-office market. Its innovative partnership marketing approaches, in particular with powerhouses such as Microsoft, had caught the attention of a leading French venture capital firm, Sofinnova, which led a Series A round of professional financing in November 1997.

The company had broken even operationally in 1998 and had started to show a net profit in early 1999 on revenue forecast to top $5 million. What started to bother Robert was how to deal with the explosion of the Internet as a communication and retailing channel. Sure, it was an easy thing to post VistaPapers' catalogue on the Web and offer the same level of service as with mail order. They had done so already, with both catalogues coexisting peacefully. But Robert saw this as simply "Web-enabling" the company's current business model. It barely touched the potential he could see in the Internet revolution. He now wanted to revisit his whole operational model and make VistaPapers a truly "Web-driven" business, using the Web's full potential to serve his customers' printing needs. It was time for a complete strategic overhaul and possibly a change of name as well.

Welcome to e-printing and the new, improved VistaPapers, now called VistaPrint.com SA.

Copyright © 2002 by IMD – International Institute for Management Development, Lausanne, Switzerland. Not to be used or reproduced without written permission directly from IMD.
[1] The case was previously registered as "Papyrus Laser (C): Going VistaPrint.com." The help of Robert Keane and Patrick Mataix is gratefully acknowledged.

VistaPapers had a new name: VistaPrint, which incorporated both the old (paper) and the new (printing) lines of activity. The latter focused on Internet-related e-printing activities, reinforcing the strategic shift from paper marketing to servicing the printing needs of its customers. E-printing had the potential to reduce the total costs of production significantly while speeding up the whole process.

To fund its move into e-printing, in June 1999 VistaPrint had closed a Series B financing round. This FFr 10 million[2] investment was led by SPEF, the VC arm of the French Banques Populaires. Sofinnova also reinvested in the Series B, as did multiple private investors who knew Robert Keane personally. VistaPrint deployed these newly raised funds to recruit Internet and software engineers and managers, to license a crude but functional e-printing technology from an overseas inventor, and to launch the first versions of the VistaPrint Web sites in the French, UK, and German markets.

But e-printing also implied massive new capital needs, so Robert was back on the financing circuit, looking this time for a $10.8 million investment to support his aggressive vision of what the printing world would look like next.

1 VistaPrint.com Business Plan: March 2000 [1]

This plan describes an opportunity to take part in a FFr 65 million[3] Series C financing, the total of which will purchase 25% of the postfinancing equity of Vista-Print.com SA ("VistaPrint"), the Internet printing company. Investment exit is anticipated within two years via an IPO. VistaPrint's existing VC investors, Sofinnova and SPEF, have indicated strong interest in investing again as part of this round, leaving 50% remaining for a new VC.

VistaPrint is one of a small number of Internet printing companies worldwide, and the only significant player in the European market. VistaPrint is active in Germany, the UK, and France, with expansion into four additional European markets under way. The company has spent several months planning its January 2000 US market entry. In October 1999 it opened marketing and development offices in Boston and is in the process of testing and bringing online a $7 million specialized production line in upstate New York. VistaPrint has approximately 50 employees, based in offices in France, the UK, and the US. In addition, an exclusive strategic partnership with a leading US printing firm adds 130 "virtual" employees. Vista-Print forecasts Internet revenues approaching $40 million within three years, with pretax profits of 10 to 15% of sales.

---

[2] FFr 6.07 = $1 at the time of June 1999 Series B financing round.
[3] FFr 6.03 = $1 at the time of the March 2000 business plan.

1.1 The E-printing Market

In the words of Keith Benjamin, a leading Internet analyst at BankBoston Robertson Stephens, the commercial printing market is "one of the world's biggest," at $100 billion. However, the market is highly fragmented and served by small regional firms that are often undercapitalized and technologically unsophisticated. The traditional supplier base relies on conventional printing processes that involve long delivery delays, high expenses, significant rework-due-to-error rates, and the need for long print runs, which has led to significant customer dissatisfaction.

E-printing is an Internet business model that has the potential to address all of these problems. E-printing refers to Internet technologies and systems that automate the process of defining, modifying, and ordering commercial print items such as stationery, business cards, brochures, and invitations. Specific benefits of e-printing include a streamlined ordering process, customer convenience, reduced lead times, and the ability of widely dispersed customers to control graphic templates. Furthermore, as shown in *Figure 1*, e-printing reduces total costs in the same way that bank ATM machines revolutionized the cost relationship between a bank and its customers by making the most popular transactions self-service.

Worldwide, VistaPrint is one of several e-printing pioneers. Like VistaPrint, most of these firms are VC-backed startups, have annual revenues of zero to $10 million, and have been started in the past four years. These firms have been exceptionally well received from both a market and a financial standpoint, with high growth rates and market valuations of $200 million and higher for post-IPO firms. All competitors use the Web as a front-end design and ordering mechanism and transmit data digitally to the print production floor. However, as described below, two VistaPrint technologies differentiate the company from all known competitors, and VistaPrint has a clear plan to leverage these technology assets to make it the leader in the emerging e-printing market.

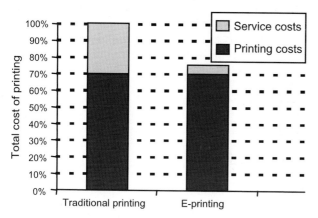

Figure 1 E-printing's self service ATM nature eliminates approximately 30% of the total cost of commercial printing
Source: Company literature

1.2 Proprietary Technologies

1.2.1 *VistaStudio*: Fast Web Design Studio with Massively Scalable Architecture

VistaStudio is VistaPrint's wizard-driven, user-friendly, Web-based desktop publishing (DTP) application. The next version (3.0) is due to be launched in January 2000. *VistaStudio* 3.0 employs dynamic HTML, or DHTML, to perform page layout functions in the browser (*i.e.*, client-side). All of VistaPrint's competitors, as well as its current version of *VistaStudio*, use either server-side manipulation of page layout data or unwieldy 100% Java technologies. *VistaStudio* 3.0 technology has an industry-leading browsing speed, enabling the user to edit much more quickly, and the ability to massively scale the VistaPrint system without degrading system performance or requiring enormous server capacities.

1.2.2 *VistaBridge*: Enabling an Order of Magnitude Cost Advantage

VistaBridge is another piece of proprietary technology developed by VistaPrint; it deals with manufacturing know-how, standardization, specialized production equipment, and process information systems. *VistaBridge* gives the company an order of magnitude cost advantage over both traditional and e-printing competitors by bridging the gap between low-volume commercial printing and low-cost industrial printing facilities (*Figure 2*). VistaPrint can, for example, produce 250 full-color custom business cards for under $3. This is less than 4% of the retail price and under 10% of the cost of competitors, both on- and offline. VistaPrint believes that no competitor has developed a technology comparable to *VistaBridge*.

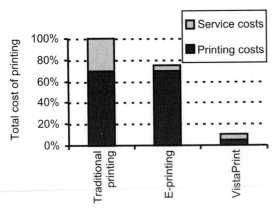

Figure 2   *VistaBridge's* cost advantage
Source: Company information

1.3 Marketing Strategy

Both *VistaStudio* and *VistaBridge* can be used for a broad range of commercial printing applications, such as brochures, letterheads, presentation folders, envelopes, and invitations. In order to place the *VistaStudio* and *VistaBridge* technologies in the leading market share position in the e-printing sector, VistaPrint has developed the multistep marketing process described below.

1.3.1 Customer Acquisition via Viral Marketing

History has shown that Internet firms that employ free product offers linked to viral marketing methods can capture unassailable first-mover market mindshare and dramatically lead in market share gains. Freeserve, e-Greetings, ICQ, and Hotmail are examples of this successful strategy, and VistaPrint intends to take a similar leadership position in the e-printing market.

To do this, VistaPrint plans to introduce, in early 2000, a radical market-breaking product offer: custom full-color, top-quality business cards FREE FOREVER.

The dramatic FREE FOREVER VistaPrint offer will be backed by a multimillion-dollar public relations and advertising campaign. Furthermore, to ensure the maximum viral spread of this free offer, the VistaPrint FREE FOREVER offer and URL will be discreetly printed on the back of each card, generating 250 advertising hits for every order. *Figure 3* shows a conceptual banner advertisement for an advertising campaign to supercharge VistaPrint's viral customer acquisition offer.

Importantly, the FREE FOREVER business card offer will incorporate a series of up-selling and margin-generating shipping charges, which will give VistaPrint a positive margin from "free" sales. The average order, inclusive of these charges, is expected to be $7 to $9, with a positive contribution margin of $1 to $2. *Figure 4* illustrates that, even with these up-selling charges included, VistaPrint pricing will remain an order of magnitude lower than comparable products from leading competitors.

Figure 3   Conceptual banner advertisement

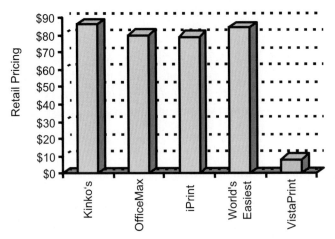

Figure 4 Retail pricing for 250 business cards, full color, including delivery in continental USA
Source: Company literature

The proprietary VistaPrint technologies are crucial to the ability to make the dramatic FREE FOREVER offer. First, *VistaStudio* permits rapid system scaling to cope with exponential increases in customer load levels typical of virally marketed free offers. Second, *VistaBridge* slashes production costs, thereby allowing positive margins despite not charging for the business cards themselves. Using competitive e-printing approaches (*i.e.*, without *VistaBridge*), the negative margin would be approximately $30 per customer, or tens of millions of dollars per year.

1.3.2 Cross-selling of a Full Product Line

Once customers are familiarized with VistaPrint through the FREE FOREVER business cards offer, they will be converted into repeat customers with the offer of popular relationship-building products such as letterheads. Like VistaPrint's business cards, these relationship-building products will be full-color market-leading quality. Prices of relationship-building products will permanently be 50% or more below those of leading competitors and will be heavily promoted on the site to reinforce VistaPrint's image of top quality at unbeatable prices. In addition, VistaPrint will offer a complete product line of other printed materials priced just below (10%) those of leading competitors. Due to the cost advantages of the *VistaBridge* system, VistaPrint expects to enjoy gross margins of up to 60% for cross-selling products.

1.3.3 Marketing Alliances

VistaPrint's management has a long and successful history of using marketing alliances to grow the business. For instance, VistaPapers catalogues are currently marketed via exclusive cross-marketing agreements with Microsoft, GSP, Corel, and other leading DTP software firms. In a similar fashion, VistaPrint will implement two types of alliance programs to maximize the rate of market penetration. First, VistaPrint is recruiting a team of marketing managers to develop alliances with major sites such as portals, office-supply retailers, and software firms. Second, an instant-sign-up affiliate program will allow any Web site in the world to become, in just a few minutes, a VistaPrint reseller. Under this program, Vista-Print will pay resellers $1.00 for every FREE FOREVER order that comes through them.

1.3.4 Corporate Sites

Business stationery and promotional material represent enormous market potential for VistaPrint. To address this market, VistaPrint is developing a full series of specially targeted Internet printing services. These capabilities include multilevel password-authorized access, centralized control of graphic layout, centralized price negotiation, decentralized purchasing and modification of variable data, preestablished delivery addresses, electronic purchase orders, and onsite display of individual company purchasing guidelines.

Corporate VistaPrint sites will be designed so customers can set them up following an easy, wizard-driven process. Given the proper authorization passwords, graphic artists, advertising firms, or other organizations working for customers will be able to create and update graphic templates in industry-standard DTP software (such as Adobe or Quark) and then upload them for immediate publication on the site. VistaPrint believes that a large number of corporate accounts will be gained via the viral advertising associated with its FREE FOREVER offer. In addition, for major potential accounts VistaPrint will deploy an internal sales force to sign up these customers more quickly.

1.4 Management and Infrastructure

VistaPrint's management team benefits from strong experience in entrepreneurial ventures and in fields closely related to VistaPrint's market focus, such as DTP, direct marketing, alliance marketing, low volume printing and development of graphics software, databases and Internet sites. All top managers have MBAs from prestigious business schools. The technical team includes the former CTO of one of Europe's most successful Internet sites plus a rapidly growing number of engineers with software and Internet development experience. These functional management skills are complemented by strong international business abilities: there are five nationalities in the top five management positions, most employees speak multiple languages fluently, and all senior executives have extensive management experience in countries other than their country of birth.

VistaPrint further distinguishes itself from early stage e-commerce startups because it already has the infrastructure in place to ensure the on-time delivery of the highest-quality products to a large number of customers. This logistical capability is the result of several years of development and over $9 million in capital investments worldwide. Key highlights include:

- Established mid-volume European print suppliers, now fulfilling orders
- North American capacity of 4 million business card orders per year, now being brought online for January 2000
- Multilingual call center with telephone and e-mail customer support in English, German, and French
- Multicurrency, multilanguage MIS systems recently upgraded for year 2000 and Euro compliance
- Next-day and standard delivery of commercial stationery packages across Europe
- Worldwide intranet and management reporting
- Offices and staff in Paris (France), Birmingham (UK), Buffalo (USA), and Boston (USA)
- Monthly financial reporting and well-established internal controls.

VistaPrint's e-printing service fulfills the same customer needs as its catalogue business, VistaPapers. Both businesses address commercial printing needs through direct marketing: e-printing uses the Internet as the technology catalyst; the other uses PCs and printers. VistaPrint is currently shipping thousands of orders per month to customers across Europe and has built a proprietary database of over 180,000 customers. The highly specific characteristics of this database provide VistaPrint with a springboard of existing customers unmatched in the European marketplace: 100% of customers have computers and have expressed an interest in using them to accomplish commercial printing tasks.

1.5 Financial Overview

VistaPrint forecasts rapid revenue growth to $40 million by the fiscal year ending June 2003 (*Figure 5*). Of this, $30 million is expected to come from e-printing and the remaining $10 million from the traditional VistaPapers business. This is a continuation of VistaPrint's four-year track record of increasing success. Revenues have grown from zero to FFr 27 million[4] for the fiscal years ending June 1995 through June 1999. The company's traditional VistaPapers business is profitable, generating 15% operating income in the most recent fiscal year (1999). VistaPrint is strongly financed, with approximately $1 million cash and no debt other than a small amount of convertible shareholder debt.

---

[4] FFr 6 = $1

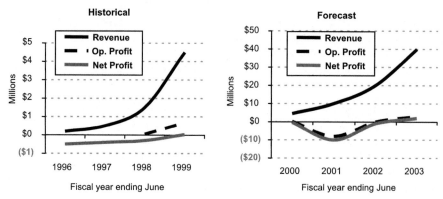

Figure 5 Financial Overview
Source: Company literature

Seriously interested institutional investors should contact Robert Keane, Vista-Print's President.

References

1   www.vistaprint.com

CASE 2-9
Lyncée Tec SA: Scaling up a Technology Venture

Ralf W. Seifert and Jana Thiel

Yves Emery, CEO of Lausanne-based Lyncée Tec, a provider of innovative holo-graphic microscopy systems, listened to the voice message he had received while he was out at lunch. The venture capitalist (VC) to whom Emery had presented the Lyncée Tec investment opportunity last week had called. He was interested in investing, provided that some additional amendments could be agreed upon. To learn about these new terms and then hopefully settle the deal, Emery had to call back by 5:00 pm at the latest that day – Tuesday, May 10, 2005 – as the VC would be out of town for the rest of the week.

Emery leaned back, pondering how to respond. He knew from experience that the amendments to expect at this stage were most likely an increase in the VC's share, an entire revamp of the board, and a definite exit after five years. It was not his first VC contact – and he had turned down other offers before.

> At the end of the day it comes back to trust. Do we trust the investor to pursue the same goals as we are? We are looking for someone with whom we can build a house together, so to speak.

Since its inception, Lyncée Tec had been in the comfortable position of being able to finance its operations entirely from sales. Nevertheless, funds were still limited, and this restricted the venture's operational freedom. Hence, Emery was looking for an additional CHF 1.3 million to grow Lyncée Tec more aggressively to get a lead over the competition and seize the window of opportunity.

On the other hand, Emery was in the middle of promising negotiations with a long-term strategic investor, although nothing was fixed yet. Financing, how-ever, was only one issue on Emery's jam-packed agenda. He would certainly appreciate having it taken off his shoulders, as it was somewhat of a gating issue. For example, the slow-moving bargaining with a potential distribution partner might speed up considerably if Lyncée Tec could boost its financial credibility.

Emery gazed at the phone, not sure how to respond to the VC's offer.

Copyright © 2005 by the alliance for technology-based enterprise (IMD, EPFL and ETH Zurich). Copyright permissions are handled by IMD, Lausanne, Switzerland.

1 Company Background

The technical foundation that led to Lyncée Tec's inception stretched back to 1948, when scientist Dennis Gabor invented holography.[1] Digitalization and the advance in computing technology and speed led to new industrial applications that benefited from this principle. In 1994 a project was launched at the Ecole Polytechnique Fédérale de Lausanne (EPFL) to investigate digital holography and its practical application, specifically in biotechnology. In 1998 the team was joined by researchers from the University of Lausanne (UNIL). Four years later the inventions were mature enough to contemplate creating a marketable product, and a new venture – Lyncée Tec SA[2] – was incorporated in May 2003.

The startup aspired to develop, manufacture, and market imaging systems based on "Digital Holographic Microscopy" (DHM). It was founded by a multidisciplinary team of engineers, physicists, physicians, and biologists. Four of the five cofounders had fathered the base invention and coauthored the patents that protected the core of the business: intelligible processing of 3D holographic images from microscopes with nanometer resolution.

By May 2005, Lyncée Tec employed seven people working full time in production and R&D (partly paid by public grants). In addition, frequent collaboration with PhD students helped to further advance the technology base.

Throughout 2004, Lyncée Tec had operated on a breakeven basis – a staggering outcome achieved solely from income from customers. For 2005, sales were expected to be CHF 1 million, strengthening the previous results.

2 The Product

Lyncée Tec leveraged its DHM technology by offering not only a product but also lab services. The services included the production of movies of a sample's temporal evolution, defect characterization, and optical measurement consulting, which frequently led to subsequent orders of the product – the image processing device.

Lyncée Tec's product comprised two components (*Figure 1*):

1. A *holographic microscope*, which recorded the information from the sample as a hologram by measuring both the amplitude and phase of the light and thus providing more detailed and complete information than other microscopes.
   The instrument consisted of a rigid frame, a translation table for the sample, a microscope objective, various lenses and mirrors fixed to the frame, a laser type source, and a digital camera.

---

[1] Holography is the recording and play back of three-dimensional images of an object which allows the viewer to move back and forth and up and down to see different perspectives as if the object were actually there. In 1971 Gabor was awarded the Nobel Prize for the invention.
[2] Initially Lyncée Tec was called New Holography but was soon renamed to ensure uniqueness and branding opportunities. The name is derived from Lynceus, a figure of Greek mythology, who had preternaturally keen sight and could even see things that were underground.

2. The DHM software, which enabled the "image reconstruction," *i.e.*, the digital processing of the holograms in real time and representation of a 3D quantitative image of the observed object on a PC screen.

The software, *i.e.*, the numerical aspect, played a strategic role. Whereas competitive techniques applied digital treatment of the image independently of the instrument and only at the end of the process, DHM integrated the digitalization at the heart of the system. The interlocking of software and hardware and the use of advanced numerical methods permitted both unrivaled *imaging performance* (resolution and speed) and *ease of use* (automation and robustness). At the same time, the *production cost* of the instrument was kept low. In general, Lyncée Tec's 3D high-resolution microscopy was perceived to be unique because it simultaneously allowed:

- *Real-time imaging,* enabling people to analyze samples very quickly or to study object movements, vibrations, or deformations of very small amplitude
- *Strictly noninvasive observation*, *i.e.*, the measurements did not disturb or even destroy the sample
- *Easy-to-perform measurement,* which did not require any controlled environment or sample preparation.

Holographic microscope

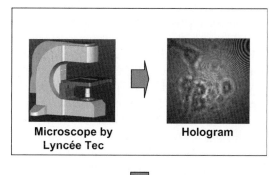

Microscope by Lyncée Tec   Hologram

Image reconstruction

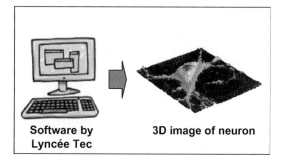

Software by Lyncée Tec   3D image of neuron

Figure 1 DHM imaging system, comprising a microscope to acquire holograms, and software to digitally process the hologram and build up a 3D image of the observed object

Besides constituting a unique selling proposition, this unprecedented performance allowed Lyncée Tec to pioneer new markets by fulfilling requirements that existing systems with nanometer resolution could not meet.[3]

The base technology could also find applications in fields other than microscopy. Although Lyncée Tec focused explicitly on commercializing the holographic microscopy, it did not rule out developing other products or services at some stage (e.g., nanobarcode readers). In any case, the key to success was the software, i.e., the numerical approach, which was protected by three patents in the USA, Europe, and Japan. Ownership of two of these was shared with EPFL, which had granted Lyncée Tec an exclusive worldwide license for the development and commercialization of measuring and imaging devices. A third patent, drawn up by intellectual property (IP) professionals, had been filed in the sole name of Lyncée Tec SA in 2005.

Both the outstanding performance and airtight protection of Lyncée Tec's technology base were key to entering and competing in the microscopy market.

3 Competing Firms and Technologies

The microscopy market in 2005 included six major imaging techniques apart from DHM: atomic force microscopy (AFM), scanning electron microscopy (SEM), optical microscopy (OM), inferometers, contact probe systems (CPS), and specific measurement sensors (SMS).

All were mature and well proven but nevertheless inferior to Lyncée Tec's DHM because of their invasiveness (AFM, SEM, and CPS), which meant they could not be applied in life sciences, or because of their high cost (high-end OM), or because they lacked speed, which hampered industrial application (SMS, inferometers). DHM took a clear lead when compared in terms of speed and measurement range (*Figure 2*); it even had the potential to replace existing systems, specifically inferometers. (*See Exhibit 1 for additional information on these technologies and how they compare to DHM.*)

Although Lyncée Tec was obviously a serious competitor on the technological level, a lingering threat was that quite a few providers of current imaging systems were large and well-known companies (*Exhibit 1*), with the corresponding resources. The startup was not able to match their track record, or their marketing, sales, and service infrastructures. Lyncée Tec's management team still had some way to go to gain market recognition for their venture and earn customers' trust.

---

[3] For additional information on the innovativeness and specific performance data of the product, visit Lyncée Tec's Web site at http://www.lynceetec.com.

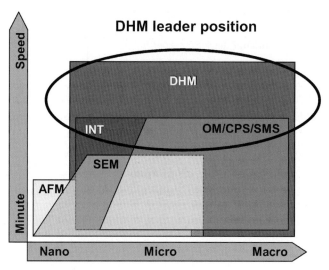

Figure 2 Lyncée Tec's Competitive Position (DHM)

4 Management Resources and Networking

4.1 Management Team and Board

Lyncée Tec was managed by CEO Yves Emery and CTO Etienne Cuche, both cofounders of the venture. Emery's prior work experience gave them access to a broad network of potential customers. This, together with his scientific background, was a key factor in bringing not only scientific but also economic success. However, Lyncée Tec was founded on science, driven forward by CTO Cuche, who had defined the major concepts of the DHM technology – specifically its numerical features – and had coauthored the corresponding patents. In 2001 his work was recognized by the Swiss Physical Society.

Emery and Cuche relied on the support of Lyncée Tec's two boards (*Figure 3*). The members of the scientific board, all still active researchers at EPFL and UNIL, fostered recognition and acceptance of the DHM technology through scientific publications and conference contributions. Their efforts were complemented by organizational support from the board of directors. Chairman Schneider directed an office of lawyers, accountants, and fiscal experts, and so could be called upon for advice on financial issues, while Bernard Rueger, who headed a company that sold measurement instruments through a worldwide network of distributors, contributed valuable knowledge in international trade. (*See Exhibit 2 for additional information about the team members.*)

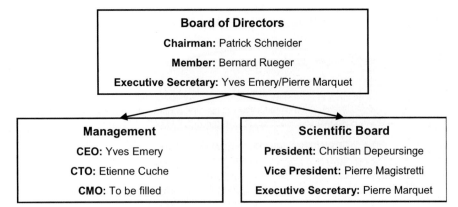

Figure 3 Lyncée Tec SA – Organization Chart

4.2 Outside Resources

In addition to the accumulated internal expertise, Lyncée Tec also relied on some external coaching and professional services: CTI[4], CimArk SA[5], and Vendbridge AG[6] helped to validate and develop the venture's strategic approach, and Pricewaterhouse Coopers (PwC) refined its administrative and financial structures. The startup further benefited from various governmental initiatives, such as TOP NANO 21 and Nano-Micro (driven by CTI), both of which granted financial support and access to the internal resources of EPFL and UNIL. The CTI funds, in particular, secured the regular employment of three R&D collaborators until June 2006. CTI had further helped by conferring its startup label on Lyncée Tec in November 2003, which created additional visibility, boosted credibility, and provided access to a broad network of industry experts and venture capitalists. Also in 2003, Lyncée Tec was ranked second in the "Lausanne Région Entreprendre" competition and was selected to be supported in the annual IMD startup competition. Additional prizes such as Venture 2004 and the W. A. de Vigier award in 2004 increased the venture's public recognition.

Integrated in an excellent network, led by a committed management team, and equipped with an innovative, proven technology, Lyncée Tec now needed to develop a viable business plan as the next part of its journey to success.

---

[4] Commission of Technology and Innovation: the Swiss innovation promotion agency with a specific focus on knowledge transfer between science and industry.

[5] CimArk SA is the cantonal agency of the CCSO network in Valais, Switzerland. CCSO is a private organization set up within the six French-speaking cantons of Switzerland to support small and medium-sized enterprises in growth and competitive strength.

[6] A service/consulting company specialized in resolving sales issues.

5 The Business Plan

5.1 Selecting Target Markets

> We expect our technology not only to replace existing approaches but also to open new
> markets. Hence, we had to see potential customers first and talk to them in order to better
> understand their concerns and subsequently tap the full potential of our business.

Lyncée Tec initially visited over 100 potential customer sites in Switzerland and
adjoining regions and became aware of a broad range of possible applications,
centered around two main industries: material and life sciences. Both industries
contained several promising subdomains, which, given the usual resource constraints of a startup, could not be addressed at the same time. A thorough assessment of technological fit, competitive intensity, market potential, and entry barriers
for each domain facilitated the selection process (*Figure 4*).

> We have deliberately chosen not to select one single segment, for we believe a small port-
> folio will reduce risk in both technology and market. In future, though, we might focus on
> whichever segment receives the greater response from the market.

Finally, four target markets were identified from the analysis: Micro-optics
(2B), MEMS and MOEMS[7] (3), Microtechnique (1), and surface analysis and
biochips (which had the same instrument specifications) (4, 7). (*See Exhibit 3 for
more detailed information on these four and other identified applications.*)

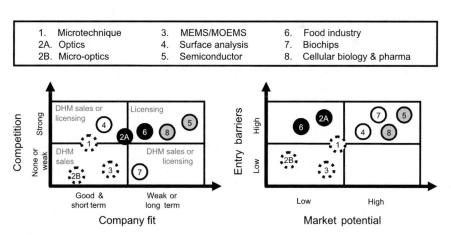

1.	Microtechnique	3.	MEMS/MOEMS	6.	Food industry
2A.	Optics	4.	Surface analysis	7.	Biochips
2B.	Micro-optics	5.	Semiconductor	8.	Cellular biology & pharma

Figure 4 Market Segmentation

---

[7] MEMS: micro-electro-mechanical systems; MOEMS: micro-opto-electro-mechanical systems:
these are printed circuits that incorporate a moving mechanical part. Examples are accelero-
meters, pressure sensors, and micropumps.

5.2 Selecting Target Customers

The segmentation into application domains was only a first step toward making money out of the instruments. Customers, naturally, wanted to know the value added of their investments. To answer them, Lyncée Tec needed to gather specific knowledge about the activities performed by the imaging systems. The findings indicated that the venture should further focus development efforts on satisfying two particular groups – *R&D laboratories* and *quality control* in manufacturing (*Table 1*).

> Our DHM offers unprecedented possibilities in the quality control domain, unlocking a market not yet served by current products. It is a unique opportunity we aim to seize.

Although R&D laboratories already used 3D high-resolution microscopes for various purposes, production quality control had so far not been able to employ these devices regularly since they did not capture images quickly enough. Hence, Lyncée Tec targeted customers that were already using competitors' devices and had to be convinced to change their existing system; at the same time it targeted new customers that had yet to discover the virtues of DHM. This was no easy task for a startup, which – like many others before it – did not have a track record or substantial funds. Entering the production environment, with its long lead and decision cycles, required high credibility in terms of robustness and financial backing.

However, Lyncée Tec's strategy of going proactively to customers, treating their concerns seriously, and assigning top priority to meeting market demands began to bear fruit. The majority of clients visited in various R&D and quality control settings had expressed a genuine interest in the product. More than 50% had provided Lyncée Tec with samples to analyze – at the clients' expense – and more than 25% of these had requested a quote after seeing the first images of their

Table 1 Value Propositions of Application of DHM in R&D and Quality Control

R&D	• DHM offers unrivaled possibilities for investigating micro- and nanomovements, deformations, and vibrations of samples. Examples are the development of MEMS and MOEMS and *in vivo* observations of morphological changes in biological cells. • DHM offers new possibilities for innovative research and product development in industrial academic laboratories. Examples are the development of capacitive membranes in microelectronics and process control in the food industry, such as observing the formation of ice cream structures.
Quality control in production	• Fast and accurate quality control in manufacturing processes, which increases product quality and reduces production and warranty costs. Examples are the quality control of microlenses and of antigen spots on biochips. • Increase in productivity and yield. Examples are the control of watch parts at an early stage of production and high-throughput screening.

samples. By May 2005, Lyncée Tec reported total sales of CHF 1.5 million and 20 more orders in its pipeline – a big success for the small venture, which until now had been self-financing.

5.3 Sizing the Market

> There is huge untapped potential in the quality control domain. By invading this realm we hope to get round the current stagnating tendency in the overall microscopy market.

By 2004, the annual sales volume of traditional 3D high-resolution microscopes amounted to US$1.5 billion worldwide. Yet this market lacked growth prospects. Therefore, even though its first sales and orders confirmed its hopes of being able to capture a reasonable share of the conventional market, Lyncée Tec aimed higher. Its technological edge allowed it to look to develop an entirely new clientele. Looking beyond the known, however, made it difficult to estimate the true sales potential.

Cautious market research, drawing on various sources, was required. The insights from extensive field research at more than 100 customer sites and from well over 400 phone calls were compared with the findings of contracted analysts and with information from market reports. The results depicted a reliable and comprehensive market portrait, conservatively estimating an annual demand for 150 of Lyncée Tec's microscopes worldwide after five years (*Table 2*), which represented less than 5% of the overall high-resolution-system market and 25% of inferometer sales.

Table 2   Forecast of Annual Sales Units for Each of Lyncée Tec's Value Activities

	Year 1	Year 2	Year 3	Year 4	Year 5
Feasibility studies	14	20	30	40	50
Customization	7	12	17	20	23
Final assembly and delivery/sale of manufactured instrument	4	20	40	75	150
After-sales services	0	3	19	54	114

5.4 Shaping the Business Model

To convince customers of the technology and gain their trust, Lyncée Tec adopted a business model with a step-by-step approach, starting with feasibility studies, followed by customization, final assembly and delivery, and after-sales services (*Figure 5*). These value activities were carried out on a go/no go basis, which

Figure 5   Step-by-step Go/No Go Decision Process Offered to Customer

meant that each step resulted in a defined delivery and was invoiced separately. This gave customers the freedom to terminate business after each milestone, and at the same time Lyncée Tec secured income.

Step 1, the feasibility studies, served two purposes: Besides demonstrating how appropriate DHM was for meeting particular needs, they also offered a separate service to clients who did not need their own complete microscopy system. For those clients that did, however, Lyncée Tec would – in Step 2 – adjust the field of view, resolution, and speed of the microscope. This customization was usually concluded with a live demonstration on laboratory equipment. Once the final configuration was agreed upon, Lyncée Tec would start with Step 3 – the assembly. For that it first had to order highly specialized mechanical parts, which typically had lead times of up to three months. The actual assembly, final test, and shipping, by contrast, took a mere two or three days. In Lyncée Tec's current after-sales process – Step 4 – services such as maintenance and calibration were offered. To streamline all steps, the venture aimed to fully modularize its components over the next two years. This would allow customers to add functionality more easily later on by purchasing additional mechanical units or software.

Needless to say, however, the venture's resources were not infinite. Soon, Lyncée Tec would have to concentrate on those activities where it added most value – advancing and innovating in digital holography. All other operations, such as manufacturing the instruments and distribution, would sooner or later require partners – most likely larger or specialized organizations with the appropriate distribution capacities (*Figure 6*).

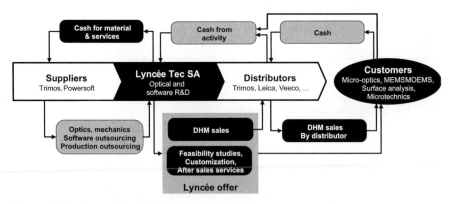

Figure 6   The Targeted Business Model

It might well happen that one of these partners eventually acquires Lyncée Tec. Therefore we have to watch out and ask ourselves whether our largest competitors, for instance, are the right partners. They might just want to shut down Lyncée's operations which would be a shame in light of the efforts we have made and the funds we have gathered. Ideally we are looking for companies distributing products complementary to our DHM.

5.5 Setting Prices

Each of Lyncée Tec's four value steps was designed as an independent source of revenue. While after-sales services and customization were priced at the market average, the price of the imaging device was set lower than competitors' prices. Customer surveys had confirmed the acceptability of Lyncée Tec's price ranges (*Table 3*).

Table 3 Target Price Ranges for Lyncée Tec's Value Activities

Value activity	Target price
Feasibility studies/engineering services (typically one day to one week):	CHF 1,500 to CHF 2,000 per day
Customization (depending on sample and degree of adaptation necessary)	CHF 50,000 to CHF 150,000
Final assembly and delivery/sale of the manufactured instrument (depending on options chosen)	CHF 150,000 for low-end products to CHF 250,000 for high-end products
After-sales services	CHF 1,500 to CHF 2,000 per day

Lyncée Tec's approach of concentrating on numerical processing rather than on complex and expensive optical systems made comfortable margins possible. The venture forecast an average net margin of 30% over the next five years, provided it received the desired financial support to scale up operations.

5.6 Establishing Distribution Structures

The extensive market research had revealed that Switzerland had a convenient number of potential customers for Lyncée Tec to begin with. For the first 12 to 18 months, sales activities had hence been confined to the Swiss region, which minimized travel, installation, marketing, and after-sales costs, and thus allowed Lyncée Tec to conduct the first sales entirely on its own, without any partner.

> Covering the entire value chain from prestudies to after sales on our own will improve our understanding of the processes and their dependencies. It will subsequently enable us to manage the links more effectively once we collaborate with external partners. Moreover, we hope to build a track record and obtain revenues in short time.

As its product matured technically, Lyncée Tec had started to approach nearby Germany and France through agents and distributors. Gradual expansion to other European countries, Japan, Asia, and the USA would follow as soon as marketing and training materials were further developed. Distribution would then be taken over by standard sales channels like specialized agents or production line integrators.

Since the software and optical modules were basically the same for all Lyncée Tec products, the road to success depended on finding the appropriate distribution channel for each application. Lyncée Tec had started to establish appropriate contacts early: It had even signed a first agreement through which it licensed its software to Trimos SA[8], a provider of industrial measurement solutions. Negotiations in other domains had also been entered into, *e. g.*, delivering standardized microscopes for a major manufacturer of microscopic devices.

5.7 *Developing Promotion Activities*

Lyncée Tec's promotion strategy focused on two activities:

1. Publishing technical advances in scientific papers and attending conferences, which would add to the recognition of the technology and hence serve Lyncée Tec in the long run.
2. Traditional advertising in specialized public magazines, on the Internet, in demonstration CDs, in marketing brochures, *etc*. This aimed not only to make Lyncée Tec's DHM as well known as the acronyms AFM or SEM but also to tie the name Lyncée Tec inextricably to the label DHM.

So far, most customer contacts had either come through Lyncée Tec's broad network or were remnants from its initial field research. To put customer acquisition on a firmer footing, the startup would soon have to add marketing capacity. Lyncée Tec planned to recruit a senior marketing director in the near future and later on to hire account managers responsible for the individual application domains.

6 Funding the Venture

Lyncée Tec had been created as a limited company with a share capital of CHF 125,000, split into equal shares between the five cofounders. Options for an undisclosed percent of shares had been granted to EPFL in exchange for licenses. The options had not yet been exercised at this stage.

---

[8] Trimos SA is headquartered in Renens (VD), Switzerland

Lyncée Tec's current operations were entirely self-financed, meaning its largest cost elements like labor, intellectual property, and material costs were fully covered by sales. The breakeven had proved stable and was likely to be boosted by the forecast income of CHF 1 million in 2005.

However, if Lyncée Tec wanted to attain the estimated profit margins of up to 30% within the next five years, it had to boost its sales volume. This would require investments in new facilities outside EPFL, new equipment, and marketing capacity. The additional funds would also allow the startup to commit to industrial projects that would subsequently enforce its credibility and enhance its market visibility.

A careful cash flow analysis that considered the investments necessary to pursue Lyncée Tec's operational plans (*Exhibit 4*) revealed the need for CHF 1.3 million over the next 24 months.

7 Outlook

Lyncée Tec had shown staggering progress over the last 18 months. It was operating at breakeven, achieved solely from customers' money, and would likely survive without external investments at the moment. Yet the question was, for how long? Emery was aiming for long-term growth, and even though a future trade sale had not been ruled out as a potential exit option, it was not a near-term scenario.

On the other hand, Emery knew that time was against Lyncée Tec. So far, competitors had not reacted, but sooner rather than later one of them would be attracted by the potential of the new market. And when that happened, any one of Lyncée Tec's competitors had the financial strength to outplay it. Lyncée Tec had to grow more aggressively to capitalize on the opportunity.

Emery picked up the phone and dialed the VC's number.

Exhibit 1
Major Competing Technologies and Firms

Competitive technique	Imaging description	Major competitors	DHM in comparison to competitor (+)/(−) advantage/disadvantage of DHM
SEM (scanning electron microscope)	Electron beam fired at sample and X-rays reflected back	Veeco (USA), JEOL (USA), LEO (USA), Hitachi (USA), KLA Tencor (USA)	(+) Noninvasive, higher image acquisition rate, better axial resolution, larger measurement range, adapted to manufacturing processes, lower utilization costs and price (−) Lateral resolution limited to 300 nm
AFM (atomic force)	Similar to SEM but use of micro-fabricated probe	Veeco (50% of market), JEOL, NT-MDT (Russia), Omicron (Germany), Nanosurf (Switzerland)	(+) Strictly noninvasive, faster, larger measurement range, adapted to manufacturing processes, lower utilization costs and price (−) Lateral resolution limited to 300 nm
OM (optical microscopes)	Glass lens used for simple magnification	Leica, Nikon, Olympus, Zeiss	(+) Strictly noninvasive, higher image acquisition rate, much better axial resolution, lower price than an LSCM (−) No morphological sectioning yet
Inferometers	Use light source and measure reflected phase change of source	Zygo (USA), Veeco, Toray Engineering (Japan), Fogale (France), Atos (Germany), Nanofocus (USA)	(+) Higher image acquisition rate, easier to use, adapted to manufacturing processes, lower utilization costs and price (−) DHM not at same technological maturity level
CPS (contact probe systems)	Mechanical probe touches surface and records movements	Veeco, KLA Tencor, Zeiss	(+) Noninvasive, better axial resolution, higher image acquisition rate, lighter and more compact, 3D surface measurement (−) DHM not ready for macroscopic measurement
SMS (specific measurement sensors)	Lasers scan a surface	Zeiss, Werth Messtechnik, KLA Tencor, Cotec, Nanofocus, Infinitesima	(+) Less invasive, better axial resolution, higher image acquisition rate (−) DHM not at same technological maturity level as most sensors
Other holographic systems		Holo3, Tropel, nLine, Extreme diagnostic, lambda–x	(+) Better resolution than usual holographic systems, higher image acquisition rate, adapted to manufacturing processes (−) DHM not at same technological maturity level

Source: Company literature

Exhibit 2
Management Team, Administrative and Scientific Board

Management

Yves Emery, CEO, cofounder, holds a PhD in physics and a business administration postgraduate certificate from HEC-UNIL. He has worked for several years as director of R&D and production in two startups in the medical industry.

Etienne Cuche, CTO, cofounder, holds a PhD in physics (development of DHM). He has long-term experience in the practical application of DHM and is the original author of the DHM software. Cuche has coached more than 20 diploma graduates and 4 PhD students in the development of the DHM technology.

Scientific Board

Professor Christian Depeursinge, cofounder and president of scientific board, holds a PhD in physics and is the leader of the microvision and microdiagnostic group at the Institute of Applied Optics (IOA) at EPFL, where he also teaches courses in optical design. He has initiated DHM technology development at EPFL and performs research in coherence imaging techniques applied in particular in nanotechniques, nanodiagnostics, and nanoassembly in biology, living-tissue analysis, and spectroscopy and *in vivo* biopsy techniques.

Professor Pierre Magistretti, cofounder and vice president of scientific board, holds an MD from the University of Geneva and a PhD from the University of California-San Diego. He is director of the Psychiatric Neurosciences Center and codirector of the Institute of Physiology at UNIL. He is former president of the Federation of European Neuroscience Societies (2002–2004) and on the editorial board of six scientific journals. He initiated the application of the DHM technology to the domains of biology and neuroscience.

Pierre Marquet, cofounder and executive secretary of scientific board, received his master's in physics from EPFL in 1991 and his MD-PhD from UNIL. Among other awards, he received the "Prix jeune chercheur européen 2001 de la société française des Laser médicaux." Currently he is group leader of Prof. Magistretti's research team at UNIL.

Board of Directors

Patrick Schneider, chairman, heads Ofisa, a law, tax, and accounting company of 100 people, and hence brings with him broad experience in negotiations, finance, and law.

Bernhard Rueger, owner and executive director of Rueger SA, a company of more than 170 people, with subsidiaries in four countries and distributors worldwide. He is a member of an institutional investment committee and is experienced in international business development.

Source: Company literature

Exhibit 3
Potential Application Segments of Lyncée Tec's DHM Technology

Applications in Materials Science

- Microtechniques and Micromechanics
  Potential: Quality control in manufacturing processes for microparts, *e.g.*, in the watch industry (spiral springs in watches).
  Advantages: DHM systems can provide objective and quantitative 3D information within a tenth of a second.
  Current situation: Quality controls are done manually (subjectivity), by slower systems (LSCM and specific measurement sensors), or in some cases by instruments providing 2D images (optical microscopes).
- Optics and Micro-optics
  Potential: R&D and quality control of transparent objects. Lyncée Tec is in contact with six companies in this domain, two in optical glasses and four in plastic microinjections. All are interested in the systems for their quality control.
  Advantages: DHM products can detect nano- or microdefects on the surface of and inside a sample in a tenth of a second, even with high aspect ratio samples.
  Current situation: Micro-optic manufacturers do not have satisfactory systems for controlling entire wafers of thousands of micro-optical elements. Optical glasses are often controlled manually with simple magnifying glasses.
- Micro-electro-mechanical Systems (MEMS) and Micro-opto-electro-mechanical Systems (MOEMS)
  Potential: R&D possibilities and quality control of moving mechanical parts. There are advanced contacts with four companies interested in DHM systems for their R&D, and at a later stage, for their quality control.
  Advantages: Thanks to their high acquisition rate and 3D resolution, DHM products are able to visualize micromovements or microvibrations.
  Current situation: None of the current systems simultaneously has both the required resolution and speed to enable 3D measurement of micromovements. Stroboscopic inferometers are the main competitors.
- Surface Analysis, Coating, Thin Film
  Potential: Alternative solution for R&D and an interesting tool for high-resolution quality control in the manufacture of thin film and coating. DHM systems can, for instance, measure layers of SiO_2 on a silicon substrate, which is not possible with inferometers.
  Advantages: High axial resolution, speed, 3D, and ease of use are key elements for users in R&D and the industrial environment.
  Current situation: Systems are too slow and not accurate enough to be used in a manufacturing process (inferometers, AFM, and SEM).

Exhibit 3 (continued)

- Semiconductors, Front End, and Back End
  Potential: This is a large market with well-established technologies, big players, and various applications such as wafer quality control, lithography quality control, packaging, and data storage. Lyncée Tec has identified applications where DHM products can add value.
  Advantages: Speed and resolution.
  Current situation: Various types of systems, most of which are slow (SEM and inferometers). A more detailed analysis of this market still has to be done to determine if and when Lyncée Tec should try to enter this market.

Other segments in materials science to be served long term are industrial metrology (macroscopic measurements), measuring instruments for multicantilever AFM, and cryptology applications. Strategic partners would be needed to develop applications for these domains.

Applications in Life Science

- Food Industry
  Potential: R&D for big food companies. A contract for specific images and movies could be signed with one of the big players.
  Advantages: Tests have shown that DHM products are suited to investigating the evolution of food substances and their stability at nano- and microlevels thanks to their noninvasive and very fast measurements.
  Current situation: Depending on the sample and the information to be retrieved, companies use various types of microscopes (confocal, fluorescence, *etc.*), which are of limited use due to their invasiveness.
- Cellular Biology
  Potential: R&D in public and private laboratories, *e. g.*, observation of morphological modifications of biological cells due to electrical or mechanical stimulation or through substance perfusions.
  Advantages: Noninvasive, real time, 3D observation of cellular morphological change.
  Current situation: Systems are incapable of invasive observations.
- Biochips
  Potential: Provision of in-production quality control for defects and dust on the substrate of biochips.
  Advantages: Noninvasiveness, high axial resolution, speed, 3D, and ease of use are key elements for users in R&D and the industrial environment.
  Current situation: The only available system is ellipsometry, which needs more than one minute per spot (biochips can have several thousands of spots).

Exhibit 3 (continued)

- Pharmaceutical Industry
  Potential: R&D in public and private laboratories and high-throughput drug screening and fundamental research in pharmaceutical companies. Lyncée Tec could sign a contract with a pharma company for a complete system able to recognize and analyze cancerous cells.
  Advantages: Noninvasive, and the high resolution in the axial direction allows the detection of the density of a cell's nucleus to characterize the pathology of cells.
  Current situation: Systems are incapable of invasive observations.

Additional life science segments with huge long-term potential are biochip readers and tomography applications. Additional technical developments are needed to address these applications.

Source: Company literature

Exhibit 4
Operation Plan

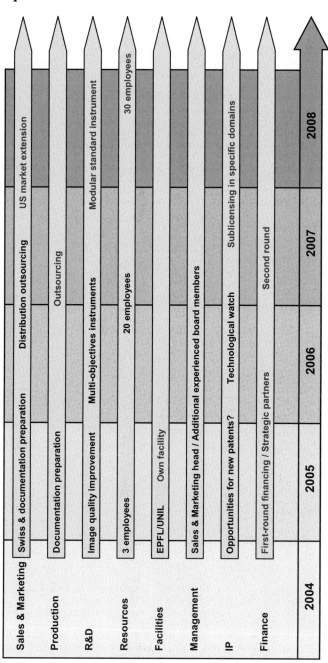

Source: Company literature

Chapter 3
Venture Financing

A fundamental characteristic of many entrepreneurial ventures is the imbalance between the resources currently controlled and those needed to capitalize on the opportunities. Few ventures truly face pure financing problems, but rather more complex "resourcing" issues, *i.e.*, how to gain access to the extensive collection of resources needed to succeed, such as management skills, distribution channels, networks, technology, and the like. In many instances, money will indeed provide access to those resources. But when there are funding constraints, where access to finance is not unlimited or is associated with huge costs, it becomes critical to use the fundraising exercise in a more creative manner, as a holistic approach to re-sourcing the firm.

Resourcing entrepreneurial ventures differs most from established businesses on at least two dimensions – risk and growth. Typically, entrepreneurial ventures share a comparatively higher exposure to uncertainty as they lack the ability of large firms to hedge risk in a broad product or project portfolio. Second, venture growth is one of the biggest consumers of resources and is typically closely re-lated to risk as it stretches the venture's and entrepreneur's capabilities to their limits and beyond. Attracting investors under these circumstances is reportedly the most challenging task for small organizations, in particular if the entrepreneur is inexperienced and without a track record demonstrating an ability to generate the returns commensurate with the risks involved in a new venture. This chapter will focus on how to develop a proper financing strategy for early-stage, higher risk ventures, which investors to target, and how to develop a compelling invest-ment case.

3.1 Financing Ecotypes and Resource Maps: Matching Needs with Funders

Venture financing is primarily about matching resources (and their providers) with the new venture's needs. As such, it is easy to see that an improper match will drastically hamper a venture's potential success or, more positively, a proper selection of financing will significantly improve a startup's success potential [1]. In this chapter, we will move way beyond the traditional debt/equity dilemma, drawing more comprehensive "maps" of relevant characteristics of various sources of financing. In a second stage, we will introduce the concept of financing ecotypes, *i.e.*, specific venture conditions that tend to be favored by each type of financing source. The ability to match the needs of ventures and the preferred habitats of investors is where art meets science in possibly the most creative fashion. Finally, we will look at the specifics of valuation in venture settings.

3.1.1 Debt Financing

Debt financing typically occurs through a loan from a financing institution such as a bank, credit union, or other commercial lender. These lenders operate on very thin margins and hence have very low tolerance for risk. While they primarily rely on the borrowers' cash flows for repayments, they also require special guarantees – known as collaterals – such as business or personal assets to ensure repayment if the firm proves incapable of making the required principal or interest payments. In practice, debt financing is rarely available to new ventures as their particular risk profile is simply beyond the scope of most capital providers. A new breed of lenders has recently appeared that is willing to supply loan capital to early-stage companies. These venture lenders play a hybrid role, taking the standard lender guarantees but also requiring upside exposure through warrants on the company's stock.

3.1.2 Equity Financing

Equity investors, by contrast, are much more adventurous and will provide the core of the new venture financing. These funds often come from the founders themselves, their family and friends, informal (angel investors) and formal (venture capitalists) equity investors, and established corporate entities. Equity capital does not come cheap because investors want to be compensated for the risk they assume: As residual owners in a startup, they are essentially unprotected and stand to lose all their money if the business sinks. Equity capital can also be more difficult to locate than debt financing, particularly when provided by informal investors, and can be demanding in terms of management's time. Equity investors are sometimes classified either as *strategic investors* whose investments are motivated by strong strategic concerns or self-interest or as purely *financial investors* focused strictly on the value creation potential of a deal as a self-standing entity.

In the next section, we offer a convenient map of the specific issues and requirements of each potential source of capital. These maps will provide the conceptual grounds for engineering optimal financing structures for early-stage companies.

3.1.3 Mapping Resources: The General Framework [2]

As we have seen, entrepreneurial ventures need a comprehensive set of resources, stretching well beyond financial means. Often, young and high-growth ventures will not be able to single out individual needs and source the appropriate resources but are faced with "bundled" offers. Hence, most startups follow strategies to source capital and noncapital resources together. For example, raising money from a reputable venture capitalist (VC) may not only provide some necessary financial resources but may also open the door to new channels of distribution. To analyze the quality of a bundled offer and determine the most appropriate mix of financial sources for any particular situation, resource maps have proved helpful. We will use a framework with 12 fundamental dimensions that are categorized in two generic categories: Financial and nonfinancial characteristics (*Table 3.1*).

Table 3.1 General Parameters of the Resource Maps Framework

Financial Characteristics	Nonfinancial Characteristics
• Cost of money	• Management skills
• Fundraising expenses	• Speed of decision
• Available funding	• Industry/product knowledge
• Collateral requirements	• Flexibility
• Risk-bearing abilities	• Other resources
• Reputation/signaling	• Control tendencies

3.1.3.1 Cost of Money

As classic financial theory implies, the cost of money is an opportunity cost driven by the risk profile of the project to be financed, not by the source of capital. This generic statement also holds true for early-stage companies. One caveat, though, is that many techno-ventures, because of their strong technology focus, often have access to heavily subsidized capital through government development programs and R&D grants. In effect, these programs lower the effective cost of capital for the firms. But it is important to understand that in this, as in all matters of life, there are very few free lunches; public money often comes with strings attached, and the actual cost (including reporting and the like) can be much higher than the nominal required rates of return might imply. In all instances, it is critical to understand the required rates of return of various sources of capital.

3.1.3.2 Fundraising Expenses

Raising capital does indeed incur costs, in the form of money, time, and effort. Therefore fundraising needs to be planned early, when enough cash and time are still available. The various sources of financing vary significantly in their requirements regarding things such as documentation, additional audits, and lawyer involvement, and thus incur very different fundraising expenses.

3.1.3.3 Available Funding

Different sources of funding typically differ in their ability to provide funds for large projects and repeated rounds of funding. The types of ventures we are focusing on in this book are typically high-growth ventures, which often need repeated cash infusions, otherwise called staggered financing. It is thus critical to integrate into the funding considerations the capital provider's ability either to fund the next rounds directly or to get access to the appropriate investors for the next (larger) rounds. One critical mistake many early-stage companies make is to overfocus on cheap capital for an early round (for example from some friendly angel investors) without considering that these investors may not be able to follow through on successive rounds and may not even be able to provide access to other funding sources.

3.1.3.4 Collateral Requirements

In order to raise money, one often needs to have money – or so the joke goes. Different investors will have different requirements in terms of the guarantees they will ask for from the founders. These often take the form of the dreaded "reps and warranties," in which founders are asked to certify a number of key company facts under threat of monetary penalties (facts they are sometimes unable to verify), as well as other forms of liens on the founders' assets.

3.1.3.5 Risk-bearing Abilities

Risk profiles vary between different types of investors, ranging from complete risk aversion to complete risk sharing. Besides accounting for the general exposure to risk, the portfolio component needs to be considered. Investors that own a portfolio of different projects with uncorrelated risks are more likely to benefit from hedging and thus are more likely to bear high-risk projects.

3.1.3.6 Reputation and Signaling

The entrepreneurial world is one that is best characterized by asymmetric information. Hence investors will look at any second-hand indication of quality and rely on it to infer as much as possible about the quality of a project. For example, the types of investors that have been attracted by a project will reveal much about its quality and can have a strong signaling character for followup rounds of financing.

For example, a high-tech startup that has neither a track record nor any proven prototype could well convince investors if it is able to show that venture capital fund X, a widely accepted expert in the field, has committed funds to the project.

3.1.3.7 Management Skills

Financial resources are core for any high-tech startup, but the accumulation of sufficient management skills to handle the growth requirements is equally important. Frequently new ventures lack the reputation and financial resources to attract the top-quality people they need. Some financing sources can, however, help alleviate management problems, either by direct intervention or by tapping a network or pool of superior managers.

3.1.3.8 Speed of Decision

The speed of decision making with which potential investors work or have to work (think of documentation requirements, *etc.*) can be strategically important for the entrepreneurial firm. A slow decision on financing can mean the difference between a profitable first-mover advantage and a permanent competitive disadvantage. Slow or convoluted decision making can apply to large corporate or institutional investors, in particular, where layer after layer of investment committees need to review each investment.

3.1.3.9 Industry/Product Knowledge

Another important dimension to consider when searching for investors is their in-depth understanding of the product and/or its markets. This expertise is absolutely critical for younger, technology-loaded firms, which often lack the means to acquire professional expertise through expert consultants.

3.1.3.10 Flexibility

With the uncertainty inherent in new ventures, the ultimate outcome of any project is likely to differ from expectations. Hence, it is worth evaluating up front the ability and willingness of potential capital suppliers to adapt terms and/or amounts to changing circumstances. Experienced VCs know that change is part of the process, if not *required*, to reach maturity. Some investors more accustomed to "stable" situations can find these changes disturbing and balk at accepting some of them.

3.1.3.11 Other Resources

Investors can provide not only capital, management skills, and product/marketing knowledge but also access to a comprehensive toolbox containing, for example, access to quality real estate locations, distribution systems, supplier networks, and the like. Each and every one of these resources could be critical to the venture and hence should be considered in the financing decision.

3.1.3.12 Control Tendencies

Whereas in large corporations control and ownership are typically clearly separated, this is less transparent in most entrepreneurial firms. Often the suppliers of funds and managers are the same, initially mitigating agency problems. In later rounds, when external investors are brought in, issues may arise when these investors start to require control rights, something many owner/managers can resent. When external capital is brought in, the rules of control will have to be adapted to reflect the fact that these players are investing in a business, not in people. They will not hesitate to remove owner/managers if they are perceived as standing in the way of superior value creation for the firm. This managerial expropriation, interestingly enough, is often to the benefit of the founders too (who remain as shareholders).

3.1.4 Resource Maps and Financing Ecotypes

Entrepreneurs have access to a large collection of financing sources, each exhibiting a very specific profile when it comes to the previously described 12 dimensions. We refer to these profiles as *resource maps*. Resource maps help decide, with respect to a particular venture and its development stage, which source of financing is likely to fit better than others. Over their life cycle, ventures typically face some standard phases of financing steps [3], as shown in *Figure 3.1*.

In the *seed* stage, financing typically corresponds to smaller amounts of capital to determine whether an idea deserves further consideration and investment. This stage usually does not involve production or sales but sets the stage for the proof of concept and early market validations. The *startup* stage entails commitment of more significant funds to an organization that uses the money for product development, prototype testing, test marketing, studies of market penetration potential, or management team recruitment. *First-round* financing would target a business not yet operating profitably but having demonstrated products and markets and the ability to manufacture and ship in commercial quantities. In this case, financing often supports the first major marketing efforts. *Second-round* financing usually supports expansion by investing in working capital or fixed-asset needs to grow the firm, which may possibly have demonstrated profitability. *Third-round*

Development
Information Availability

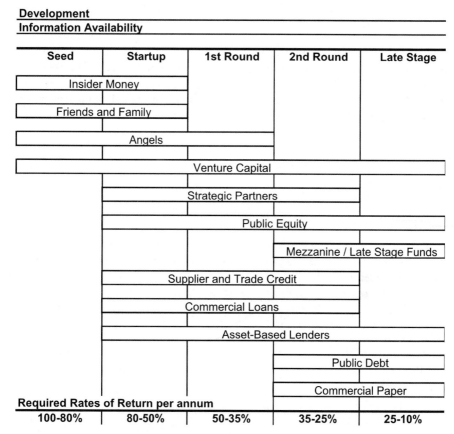

Figure 3.1   Financing Ecotypes by Stage of Development

financing provides capital to fast-growth companies with established profits that are as yet insufficient to finance further growth. Finally, *late-stage* financing, comprising everything from management buyouts or buy-ins to bridge, mezzanine, and replacement financing, provides resources for the restructuring activities of well-established firms, including preparation of cash-out or exit.

To determine the relevant mix of resources to be used, it is necessary to look at the resource maps, *i.e.*, the detailed "offering" of each major source of capital in terms of the 12-dimensional framework. We will use a 12-dimensional radial graph to facilitate the cognitive identification of the "best fit" source of financing. Each of the 12 fundamental characteristics described in the previous section will be ranked from 1 to 10, with 1 indicating that the characteristic is not present at all in the particular source of financing and 10 indicating an extremely strong presence. Furthermore, the graphs are organized so that their left half represents the *financial characteristics* and the right half the nonfinancial items. As sources of financing are not homogenous in their characteristics, a range of value for each dimension is represented with a low and high.

3.1.4.1 Friends and Family

Friends and family rely primarily on trust and friendship to deal with information asymmetries. However, they can also create tension, especially when things do not turn out for the best. Raising money from relatives certainly has advantages, but it can turn family gatherings into pretty acrimonious shareholder meetings. *Figure 3.2* presents the estimated ratings for friends and family investors according to the characteristics of the framework.

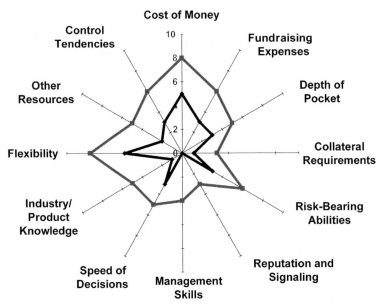

Figure 3.2 Friends and family

3.1.4.2 Angels or High-net-worth Individuals

Business angels or informal private equity investors are often former or current entrepreneurs, professionals (lawyers, consultants, *etc.*) or wealthy families. These individuals invest on a variety of grounds, for example identifying with the entrepreneur – a proxy for entrepreneurial thrills – or as an extra layer of diversification in an investment portfolio. Angels can bring very different levels of professionalism and resources to the table, from none to a high degree of industry knowledge and networks. *Figure 3.3* presents the estimated ratings for an angel investor according to the characteristics of the framework.

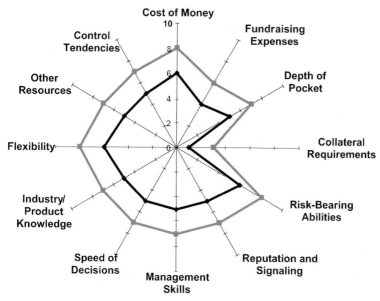

Figure 3.3 Informal Private Equity (Angels)

3.1.4.3 Venture Capital

Venture capitalists (VCs), or institutional investors, raise money from institutions, individuals, and even public entities to target equity investments in firms at all stages of development, from seed to buyouts and bridge financing. VCs share a professional approach to investments, running extensive due diligence and bringing internal and external expertise to the table. In addition, they are able to bear high risks and provide management expertise [4]. The presence of certain VCs in the capital structure of a startup can send a strong signal to other investors and stakeholders. For many technology and science ventures, venture capital provides a major source of capital. Because of this primacy we have included an extensive note on venture capital later in this chapter. *Figure 3.4* presents the estimated ratings for a typical VC according to the characteristics of the framework.

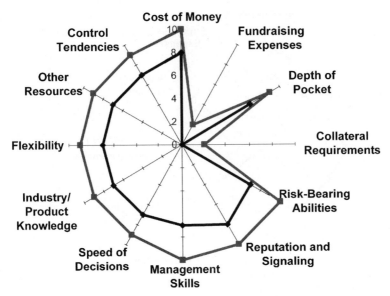

Figure 3.4   Professional Equity Investors (Venture Capitalists)

3.1.4.4 Public Equity

Public markets beat all other sources of capital when it comes to large-scale funding and efficient evaluation of management performance. Unfortunately, the process of going public is convoluted and expensive, requiring extensive disclosures and high costs associated with the use of reputed advisers. Going public also incurs ongoing disclosure costs for as long as the firm remains publicly owned. *Figure 3.5* presents the estimated ratings for the public equity market according to the characteristics of the framework.

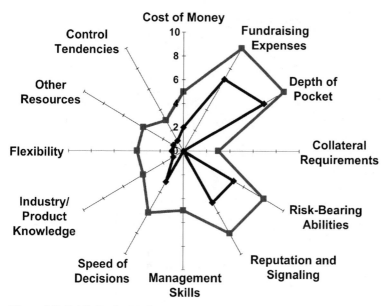

Figure 3.5 Public Equity Markets

3.1.4.5 Strategic Partners

Another source of financing can come from individuals or companies that are likely to benefit from your own success. This vested interest tends to make them more receptive to investing – and at better valuations, *i.e.*, lower required rates of return. Strategic partners can also provide invaluable information and access to markets and customers. Conversely, this type of investment carries the potential for complicated relationships as strategic investors get almost as much access to inside information as the owners and, thus, command over an inordinate leverage with the business. *Figure 3.6* presents the estimated ratings for strategic investors.

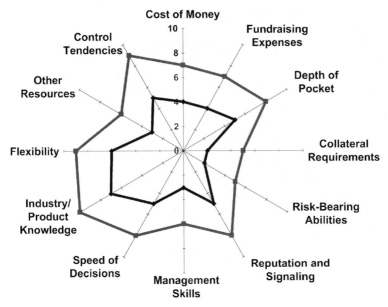

Figure 3.6   Strategic Partners

3.1.4.6 Trade Credit

While not a source of financing per se, trade credit can substantially lower a start-up's need for external financing. It is readily available and grows as the business itself grows. Typically, your suppliers have a direct interest in your business, since for them it represents a cash flow stream in the making. By providing normal trade credit terms or even expanding them, suppliers essentially give you the free use of their goods or services for the period of credit. On the negative side this source of financing often offers little other than cheap goods and services. Moreover, terms can be changed easily and excessive reliance on this source of financing can expose the business to sudden capital needs. *Figure 3.7* presents the estimated ratings for trade credit according to the characteristics of the framework.

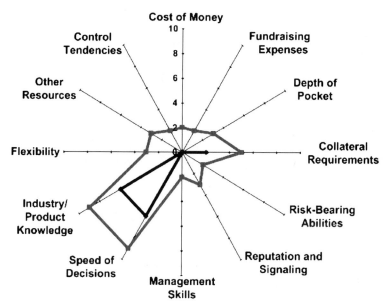

Figure 3.7 Trade Credit

3.1.4.7 Asset-based Lenders

In some cases, startup firms do have access to debt capital. Often, funding is used to invest in relatively standard assets, such as technical infrastructure or airplanes. As these assets have a well-defined resale value and markets, specialized lenders provide debt-based financing using the assets themselves as collateral for the loans. A whole new "venture leasing" industry has also emerged, to provide fixed-asset financing in areas previously regarded as unsafe. Some of these specialized lenders require warrants or stock options for the higher level of risk they assume. *Figure 3.8* presents the estimated ratings for asset-based lenders in our framework.

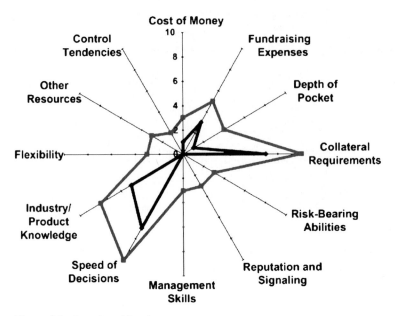

Figure 3.8   Asset-based Lenders

3.1.4.8 Commercial Loans

Besides asset-based lenders, banks and other financial institutions may also be willing to provide loans on the basis of other guarantees. Bankers analyze loans using the "3C" framework: cash flow, collateral, and character. The primary concern is usually whether or not the firm will be able to generate the cash flow needed to make the repayments. If the firm should fail to do so, a banker will check if collateral is available to secure the debts. If both criteria are satisfied, the banker will then typically rely on his or her perception of the entrepreneur's personality. In most instances, any one of these criteria will rule out many early-stage companies. *Figure 3.9* presents the estimated ratings for commercial loans according to the characteristics of the framework.

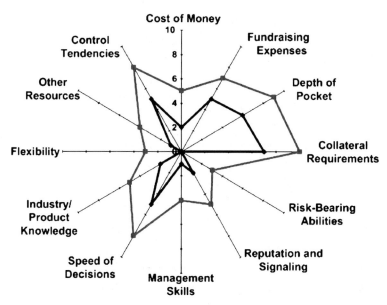

Figure 3.9 Commercial Loans

3.1.4.9 Government Sources of Financing

Since the development of a dynamic economy is often regarded as a political objective, many governments take an active role in financing entrepreneurial ventures. Interventions assume very different, often bundled, forms. They range from consulting schemes providing advice, to the adoption of legislation that is more supportive of entrepreneurs, to actual financing – at normal or subsidized rates. Typically, government funding is available under very specific circumstances and for very specific projects. Recent years have seen the increasing professionalization of government schemes and agencies. However, not all representatives have the incentive or skills to make certain business projects work: Their task, all too often, is to handle a government program administratively. Another potential issue is that time is typically not an issue in these programs, whereas it is for the entrepreneur. On the upside, however, government funding usually comes at very low costs, even close to zero, and some programs carry very high risks. *Figure 3.10* presents the estimated ratings for government financing according to the characteristics of the framework.

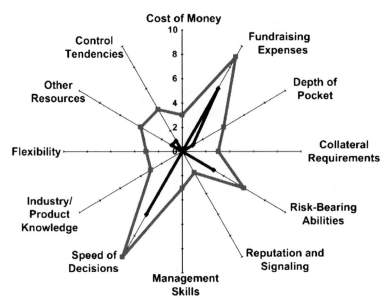

Figure 3.10 Government Financing

3.2 Business Valuation – Science or Art?

Knowing what type of financing to pursue, or what combination of sources, is but one component of venture financing as a discipline. Another critical skill for all entrepreneurs is the ability to understand the key drivers of business value. Business valuation refers to the assessment of the value of a company or part of it. Assessing and determining this value is critical to structuring the deals at the different stages of the venture's life cycle and eventually when attempting to harvest it [5]. Valuations are often required to support the key strategic activities of a company [6], such as strategic planning (where it is critical to evaluate the value contributions of alternative scenarios), negotiating strategic partnerships and alliances (to determine profit-sharing rules, for example), analyzing investments and acquisitions (in particular if they are to be financed with stock), establishing stock option incentive plans, and analyzing a potential flotation of the venture (IPO) or other forms of harvesting and exits.

Mostly, however, a would-be entrepreneur will face valuation for the first time when negotiating a first round of outside investments, and in particular when determining the percentage of equity that will be offered to these investors. Knowing different approaches and valuation techniques with their advantages and disadvantages will allow him or her to better prepare for the negotiation process.

3.2.1 Valuation Techniques [7]

Many valuation techniques have been developed; most of them are not necessarily geared toward small and privately held companies. The valuation process for these companies in the early stages of their life cycle is a difficult and highly subjective process. Difficult, because it typically has to account for a period of negative cash flows with highly uncertain future rewards. Subjective, because it is highly dependent on the context, perspective, and goals of the valuating party. Different stakeholders will use different assumptions and different methods. In general, no two valuation methods will produce the same results, and even the same method applied by two different experts is likely to lead to different outcomes. In this respect, valuation is very much an art form rather than a science.

The most common valuation methods can be categorized as (1) asset based, (2) comparables (or multiples) based, and (3) cash flow based. Researchers have also attempted to implement more sophisticated option-based valuation approaches to the world of startup firms but with very limited effects so far, mainly due to the inherent uncertainty in the data. Biotech and pharmaceutical firms, with their relatively structured development processes (through the FDA approval stages), have so far offered the most receptive grounds for these new approaches.

3.2.1.1 Asset-based Valuation Methods

One possible way of determining the value of a venture is to look at the underlying worth of the assets of the business. This method is often seen as providing a "liquidation" value for a firm, since its value would be equated with the net value of its assets if sold on the market today. Three common variations within this method use the equity book value, the adjusted book value, or the liquidation value:

- *Equity Book Value:* The equity book value method uses the balance sheet as the primary source of information. The main formula is:

  Net Worth = Total Assets − Total Liabilities

  However, it must be remembered that the book value is highly influenced by accounting practices, and it is most definitely not a valuation tool. For example, if reserves for losses on accounts receivable is too low for the business, it will inflate book value and vice versa. One of the risks of this method is that it misses the true market value of some assets that are not reflected in the balance sheet. The adjusted book value method proves more accurate for this.
- *Adjusted Book Value*: By using adjusted book value methods, the true market value for all balance sheet objects is taken. This is advisable, for example, when assets such as buildings or equipment have been depreciated far below their market value. This method also accounts for intangible assets such as patents or otherwise protected knowledge that has a market value. More tricky,

though, is accounting for elements that are not present in balance sheets at all, such as hidden liabilities.

- *Liquidation Value*: Instead of trying to work from the company's balance sheet, the liquidation value approach takes a more holistic view and tries to determine the value at which the whole company could be sold today, piecemeal or as a single entity. Whereas the adjusted book value implicitly works on the assumption that the business continues to operate, the liquidation value assumes the termination of activities and sale of the assets. The difference between the two methods would usually derive from the loss of value that is often associated with a "fire sale," *i.e.*, buyers driving prices down below estimated market values.

The various asset-based valuation methods outlined above share some strengths and weaknesses. Although they are simple and easy to comprehend and explain, they fail to account for the value of the ongoing business. In other words, they essentially provide the value of a "dead" business, which hopefully understates the value of the business "alive."

3.2.1.2 Comparables-based Valuation Methods

A second approach used by many investors is referred to as the comparables or multiples approach. It is based on "capitalizing" the estimated earnings of a firm, *i.e.*, its future business opportunities, by a multiple borrowed from comparable firms or transactions in the market. The underlying formula to start with in this case is:

Value = Earnings × Value Multiple

This involves multiplying an estimated, normalized earnings figure for the firm by a capitalization factor (such as price-earnings ratio) obtained by looking at similar companies or transactions in the market. The most classic version of the approach is the price-earnings multiple approach, where the normalized earnings before interest and taxes (EBIT) of the firm would be multiplied by the price-to-EBIT multiple of similar transactions of publicly listed firms. The approach rests on two fundamental components: What earnings base should be used and what is the most appropriate multiple to apply to it?

- *Normalized Earnings Base:* Two different types of earnings can be used to determine the value of a business: Historical or future earnings. Historical earnings are reliable but may understate the future earnings potential of a firm. Future earnings are just forecasts, *i.e.*, dreams, for all practical purposes. Both types of earnings offer valuable insights into the firm, so often a combination of historical and forecast numbers are used to determine a "normalized earning power" for the firm, *i.e.*, a reasonable estimate of the earning potential of the firm. That number is clearly full of uncertainties, so sensitivity analyses have to be run to understand its validity.

   Besides normalizing and standardizing the earnings, a decision also has to be made as to what type of earnings, such as profits before taxes, profits after taxes, operating income, or EBIT. Most early-stage companies are valued on the basis of EBIT or EBITDA (earnings before interest, taxes, and depreciation) to reflect their true operating performance. The argument is that interest is a result of financial strategy; taxes are heavily driven by the capital structure adopted (and hence not operational efficiency), and, finally, depreciation and amortization are pure accounting and tax decisions, again unrelated to the business fundamentals. Removing these "pollutions" creates more comparable numbers. However, consistency is key: Results can become totally meaningless if an after-tax earnings multiple is applied to a before-tax earnings base.

- *Earnings Multiple:* Earnings multiples are estimated by one of two methods. First, if comparable publicly listed firms exist, and they demonstrate sufficient liquidity so that the public share price can be seen as a proper market consensus about the firm's value, then the multiples that apply to these firms' earnings are good estimates to use for the private firm being valued. Unfortunately, finding comparable publicly listed companies for early-stage companies that are often pioneering technologies is often an impossible task. It is then possible to rely on private transaction data, *i.e.*, reported values on recently traded comparable companies (trade sales or new rounds of financing for these firms).

3.2.1.3 Discounted Cash Flow (DCF) Approaches to Valuation

DCF valuation methods use the present value of future cash flows, discounted at a rate appropriate to the risk levels of the cash flows, to determine a firm's value. They are based on the assumption that an asset is essentially worth what it is able to generate in the future in terms of cash flows, on a risk-adjusted basis. The advantages of DCF approaches over the asset-based and earnings-based methods are that they account for the time value of money and do not depend on comparison with similar companies. Conversely, these approaches are entirely based on the ability to make proper assumptions about the future cash flows and terminal value of a firm, a very difficult exercise with early-stage technology companies.

3.2.1.4 Real Options Approach [8]

One of the potential shortcomings of the DCF methods is that it may not properly incorporate certain real options built into projects, such as options to wait to make major investments, options to redirect applications, *etc*. These options have value for investors; the ability to stage the investments, *i.e.*, invest some money at the early stage with the option to follow on at a later stage in the venture's life cycle, makes it more appealing to get started with the investment process. Such an opportunity exhibits all the characteristics of a call option: It is a right, not an obligation, to make the investment. Entrepreneurial ventures – or some features of such ventures – can sometimes be valued as options, using sophisticated option valuation methods such as the Black-Scholes model.

However, there are some significant concerns associated with the use of option pricing methods. For one, they provide an undeserved sense of comfort: While the formulae may be better, the data they ingest remain highly uncertain, so the final result is still very unreliable. Technical difficulties also arise because of the "nested" nature of the options built into projects (*i.e.*, where one cannot be exercised before another). Often, it is thus easier to rely on Monte Carlo simulations than on analytical solutions to value the imbedded options.

Whatever the approach, it is fundamental to understand that there is no such thing as a "value" for an asset; hence we are shooting at a moving target. The objective of the various methods is to gain a better understanding of the value drivers and to provide rough estimates of value when required [9]. In other words, the critical elements are to:

- Identify critical assumptions and the interrelationships between the valuation elements that drive the valuation results and
- Develop realistic scenarios and identify sensitivities.

No single valuation technique is best at determining the value of a firm. Value is and will remain a function of individual perceptions and abilities to create it. Finally, one of the most critical determinants in the price paid in a transaction will remain the relative negotiation powers of the parties involved.

3.3 Cases in This Chapter

Instead of opening with a case, we begin this chapter with a comprehensive technical note focusing on the venture capital contracting process. *Venture Capital Investment Contracts: A Primer and Taxonomy* provides an overview of the main contractual features of venture capital contracts, explaining their *raison d'être* and their theoretical grounding and rationale. Contracts are by nature incomplete as it is impossible to foresee all contingencies. Nevertheless, contracting is an essential part of the venture capital investment process by which both the entrepreneur and the VC learn about each other's intents and formulate their agreements in writing. It is also a powerful complement to and support of the valuation process. In essence, contracts complement the inherent failings of the valuation process itself outlined above. Since value is so uncertain, it would be difficult for parties to come to common terms. A way to cope with the uncertainty is to accept it and make sure the investment contracts build in sufficient contingencies to cope what whatever future scenarios develop. A taxonomy is presented for linking contractual clauses to the respective types of actual or potential conflicts they address, and a number of observations relevant to venture capital contracting in general are also presented.

The first case in this chapter, *AVIQ SYSTEMS AG: Creative Technology, Innovative Financing*, effectively portrays the creativity necessary in entrepreneurial finance. It introduces Swiss-based AVIQ, a startup in the consumer electronics domain. Like many startups, AVIQ was the product of the visionary zeal of its founder – Rudy Kiseljak – an electronics engineer. In 2002 multimedia caught Rudy's imagination. As PCs and DVD players became commonplace, he started to believe that a combination DVD, CD player, video recorder, photo viewer, and TV tuner that also offered instant messaging would revolutionize in-house entertainment. He self-financed the development of the proof-of-concept machine but quickly reached the end of the line for bootstrapping: The company now really needs to raise €1.6 million to finance the production of the AVIQ A1 "media box" with the accompanying software. As the opportunity is too small to interest "classic" growth financiers, Rudy considers innovative and rather unusual financing techniques, most notably direct public offerings (DPO) and public product offerings (PPO).

The second case features *Tatis Limited*, a young Geneva-based startup company that is developing an integrated Web-based solution for managing preshipment inspections in emerging and developing countries – a key component of trade and customs management. The case investigates the challenges of business-to-government (B2G) activities and of developing a global partnership for solution implementations with a leading consulting firm, and it lays out Tatis' current fundraising considerations, which particularly focus on getting a leading VC on board. Tatis is at a turning point: It needs to define its future strategy, which will require additional investments and, hence, naturally calls for a valuation. A first approach is presented in the case entailing the question of the extent to which Tatis' strategic decision will influence the current valuation results. Another question to ask will be whether the startup has already identified all available financing strategies.

The ensuing three-case series will draw the student into the issues surrounding the valuation of a startup company in the biotechnology domain, one of the most speculative and volatile marketplaces. The center of attention is *Genedata*, a small Basel-based startup providing data processing software for the biotechnology sector. The startup is up for an independent valuation by Venture Valuation AG, a Zurich-based independent boutique firm that specializes in valuing high-growth-potential startups. Venture Valuation has been tasked by Novartis Venture Fund to help analyze and structure the next financing round at Genedata. The first case in this case series outlines Genedata's background, product portfolio, competitive positioning, and issues that it will need to deal with in the near future.

The second case in this series, *Venture Valuation AG: The Genedata Assignment*, reports the independent valuation of Genedata. Given a very tight deadline, Venture Valuation has to customize the evaluation process, using quantitative and qualitative data, to best meet its customer's needs. The case outlines the task ahead of Venture Valuation and requires the student to make assessments and calculations and finally come up with a valuation for the Genedata startup.

In the third part of this case series, *Novartis Venture Fund: Valuation Dilemmas*, we finally take the perspective of Novartis Venture Fund (NVF) in investing in Genedata. NVF must take into account the valuation analysis by both the startup and Venture Valuation AG. NVF uses the Venture Valuation inputs to determine how best to structure the next round of financing for Genedata. This case requires the student to discuss and analyze pricing, further financing requirements, and exit strategy for the fund.

References

1   Smith, Richard L. and Janet Kiholm Smith. *Entrepreneurial Finance*. New York: John Wiley & Sons, 2000, p. 23.
2   An extended version of this section can be found in: Leleux, Benoît. "Resource Maps and Financing Ecotypes: A Visual Approach to Resourcing the Entrepreneurial Business." *Babson Entrepreneurial Review*, Spring 1998.
3   Sahlman, William A. "The Structure and Governance of Venture-Capital Organizations." *Journal of Financial Economics*, 1990, *27* (2), pp. 473–521.
4   Muzyka, Dan, Sue Birley and Benoît Leleux. "Trade-offs in the Investment Decisons of European Venture Capitalists." *Journal of Business Venturing*, 1996, *11* (4), pp. 273–287; and Lerner, Josh. "Venture Capitalists and the Oversight of Private Firms." *Journal of Finance*, 1995, *50* (1), pp. 301–318.
5   Stevenson, Howard H., Michael J. Roberts, H. Irving Grousbeck and Amar V. Bhidé. *New Business Ventures and the Entrepreneur*. Boston: Irwin/McGraw-Hill, 1999.
6   Smith and Smith, *Entrepreneurial Finance*, p. 233.
7   For further reading we suggest: Stevenson et al. *New Business Ventures*, pp. 34–42; Smith and Smith, *Entrepreneurial Finance*, pp. 227–259.
8   Lerner, Josh and John Willinge. *A Note on Valuation in Private Equity Settings*. Harvard Business School Publishing, HBS 9-297-050, 2002.
9   Roberts, Michael J. *Valuation Techniques*. Harvard Business School Publishing, HBS 9-384-185, 1988.

NOTE 3-1
Venture Capital Investment Contracts:
A Primer and Taxonomy

Benoît Leleux and Frédéric Martel

Contracting in venture capital is a process by which both the entrepreneur and the VC learn about each other's intents and formulate their agreements in writing. Contracts are by nature incomplete as it is impossible to foresee all contingencies to be faced. Nevertheless, contracting is an essential part of the venture capital investment process and a powerful complement and support of the valuation process.

This technical note provides an overview of the main contractual features of venture capital contracts, focusing on their *raison d'être* and their theoretical grounding and rationale. A taxonomy is presented to link the contractual clauses to the respective types of actual or potential conflicts they address as well as present a number of observations relevant to venture capital contracting in general.

As much as possible, we elaborate on the actual effectiveness of the most common contractual clauses, *i.e.*, whether in practice they actually provide the level of comfort or benefits they are purported to and how they fare in typical funding negotiations. We also provide some insights on the relative strength of venture capital contracts in Europe *vs.* the USA.

Copyright © 2002 by IMD – International Institute for Management Development, Lausanne, Switzerland. Not to be used or reproduced without written permission directly from IMD.

1 Introduction

In the world of finance, early-stage financings (venture capital, angel financing, *etc.*) have always earned a very distinctive, and deserved, reputation as some of the more obscure, if not esoteric, dimensions of the field. The startup firms these financial investors cater to are known as much for the excitement and drive of their wizard-driven teams, the revolutionary technologies they hatch, and the dedication they generate as for their bad habit of failing in droves, burning cash as if there were no tomorrow, and ultimately not delivering the promised bounties, or only after excruciating delays and sufferings. So, how does one go about analyzing and providing financing to such "outliers" in terms of financial risk? The argument developed in this note is that to a large extent, the inherent valuation uncertainties are addressed through sophisticated contracting schemes that in effect (1) provide for "contingent repricing" through time as the venture develops, reallocating cash flow and control rights when need be; and (2) provide effective screening and incentivizing mechanisms, helping "smoke out" entrepreneurs with lesser-quality projects.

Financial resources are particularly difficult to come by for young growth companies because of the information asymmetries they face, often precluding them from raising capital from conventional sources (banks, *etc.*) and making raising capital very costly [1]. When family, friends, and angel investors become insufficient to fund the company development, startups turn to venture capitalists (VCs), professional risk investors who invest mainly third-party funds with an objective to maximizing financial returns.

According to Tyebjee and Bruno [2], the VC's investment process can be divided into five stages: (i) deal origination, (ii) screening, (iii) evaluation, (iv) deal structuring, and (v) postinvestment activities. This is described in more detail in *Figure 1* below, which also highlights the very critical juncture at which contracting stands. While much is known about how investors select investments and the nature of the

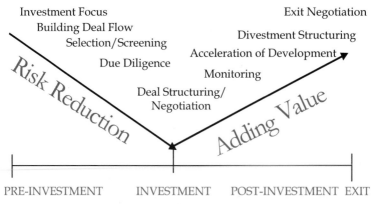

Figure 1 The VC Investment Process

postinvestment relationship [3], little is known about how the investment contracts are negotiated [4]. The purpose of this paper is to shed some light on aspects of the contracting process.

The contracting process is best described as a learning process whereby the various parties to the contract start divulging their true intent (generally known as "signaling"). Through contracts, VCs and entrepreneurs try to regulate a number of events and behaviors that have been extensively described in the literature. Before drafting a contract and including project specific contractual clauses, it is important to know the underlying potential conflicts that can arise during the life of the investment.

VCs translate their investment objectives in the contracts [5]: these include to (i) maximize their potential financial returns, (ii) provide downside protection against loss (risk mitigation), (iii) allow them to manage and control the investment if the company starts to diverge significantly from the business plan (contingency planning), and (iv) provide exit rights through an IPO or other value-maximizing transaction. To this end, VCs and entrepreneurs have found that a crucial element of contracts was the alignment of each party's objectives (goal congruence).

Contracts are the evidence of what was agreed upon at the beginning of a venture. Most VCs go back to contracts only to check on their rights. Contracts are mainly there to rule instances when events do not go well. This technical note focuses narrowly on venture capital contracting issues and the tools used to mitigate risks inherent in any venture; it does not attempt to cover other areas, such as the evaluation of potential investments and monitoring tools, that critically contribute to the success of new venture investments. A general taxonomy of venture capital contractual clauses is presented with the conflicts each is trying to address. Finally, as much as possible, a practical evaluation of the effectiveness of these clauses is presented.

The objective of this document is to facilitate a comprehensive understanding of venture capital contracting, in all its legal complexity and yet operational simplicity. Without an intuitive understanding of the goals to be achieved with such instruments, VCs are unlikely to reap the full benefits of their investing efforts.

2 Conflicts Addressed in Venture Capital Contracts

2.1 Introduction

Venture capital investments are more risky than investments in quoted companies in general [6]: informational difficulties, illiquidity, large investment sizes, long investment horizons, and high company-specific risks combine to explain the higher overall returns required *a priori* by VCs [7] and the existence of a less-than-perfect capital market with numerous potential conflicts [8].

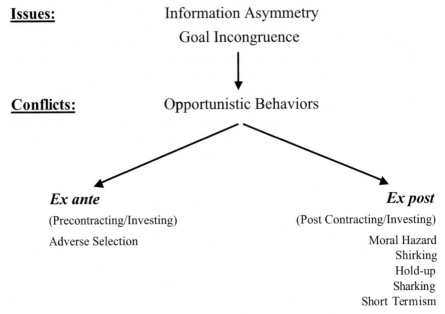

Principal-Agent Relationship

Issues: Information Asymmetry
                              Goal Incongruence

Conflicts: Opportunistic Behaviors

Ex ante   ***Ex post***
(Precontracting/Investing)   (Post Contracting/Investing)
Adverse Selection   Moral Hazard
                                                              Shirking
                                                              Hold-up
                                                              Sharking
                                                         Short Termism

Figure 2   Opportunistic Behaviors under Asymmetric Information and Goal Incongruence

In this section, we quickly review the types of conflicts present in VC/entrepreneur relationships, which are best described as typical principal-agent relationships under asymmetric information and goal incongruence. This situation can lead to opportunistic behaviors known in contract theory as adverse selection, moral hazard (including shirking), hold-up, sharking, and short-termism.

As prerequisites for any investment, VCs and entrepreneurs will try to reduce the information asymmetries and goal incongruence in order to mitigate the risks of opportunistic behavior both precontracting (*ex ante*) and postcontracting (*ex post*) (*Figure 2*).

2.2 Agency Relationship Theory

An agency relationship is born of a contract under which one or more persons (the principals) engage another person (the agent) to perform some service on their behalf, which involves delegating some decision-making authority to the agent [9]. For Orts [10], an agency relationship is a fiduciary relationship that results from the manifestation of consent that one person shall act on behalf of the other to act this way. In the VC-entrepreneur relationship, the VC is usually considered as the principal who delegates decision-making authority to the entrepreneur (agent), even if, as developed below, it is sometimes difficult to determine who acts as agent and who as principal [11].

Most academic writings on the relationship between VCs and entrepreneurs rely on a principal-agent model under which the VCs are principals and the entrepreneurs are agents able to take actions that harm the investors [12]. Under this view, the primary purpose of any contract between the two parties is "to align entrepreneurs' incentives with venture capitalists' goals." [13]. Obtaining goal congruence, then, is often the essence of contracting.

Alternative perspectives are obtained when assuming that the entrepreneurs are in fact the principals and VCs are agents of the entrepreneurs [14] or of their fund providers [15]. Although suggested in prior work [16], Gifford [17] actually developed the first formal model examining VCs as agents of entrepreneurs; the model notes the dilemma facing VCs in allocating their attention between improving current ventures and evaluating possibilities for new investments. The model acknowledges potential incentives for VCs to take actions that negatively affect the performance of the venture and entrepreneurs' and investors' wealth [18]. VCs also act as agents when they are active in structuring new financing rounds for their investee companies and motivating new potential investors.

Due to the fact that the agents act on behalf of the principal, two risk factors arise: (i) goal incongruence (*i.e.*, the agent pursues a different goal than the principal) and (ii) information asymmetry (*i.e.*, it is impossible for one party to know what the other party is doing or intends to do). Goal conflicts as well as information asymmetries are significantly present in the VC-entrepreneur relationship [19]. Manigart confirms Fiet's speculation that conditions leading to greater agency risk increase involvement of the VCs [20].

As the venture capital market is inherently imperfect, agency risks and business risks are likely to become important [21]. Agency risk refers to the risk that the ability of the entrepreneur is lower than expected (incompetence problem) or that the entrepreneur may take actions that are in his/her personal interest (opportunistic behavior) but that destroy value for the VC [22].

2.3 Asymmetric Information Theory

Information asymmetries arise whenever one party to a transaction possesses superior information compared to the other party. Inside information can be the source of opportunistic behavior, as contractors have the opportunity to influence their reward schemes to positively influence the rewards derived from privileged information.

Asymmetric information in venture capital projects is most evident during the selection of the projects (the entrepreneur usually knows more than the investor about the true prospects of the project) and the due diligence phase (the market knows more than the investor and the VC knows more about his/her ability to add value to the venture than the entrepreneur does). Practically, the VC is mainly concerned with the entrepreneur's quality and the potential of the venture, while the entrepreneur is concerned with the value added the VC promises to deliver along with his or her cash investment. Once the investment is made, information asymmetries may arise when the entrepreneur refrains from giving complete information on an investment to his investors. The VC is obliged to monitor his investments and obtain regular confirmations that the business is going according to plan.

2.4 Goal Incongruence

Goal incongruence describes the sometimes-conflicting objectives present within the VC-entrepreneur relationship [23]. *Ex ante*, the entrepreneur is keen on maximizing the premoney valuation of his project and receiving adequate funding and VC support while the VC is concerned with minimizing the premoney valuation, reducing the intrinsic risk of the venture investment. *Ex post*, the entrepreneur is interested in leveraging his/her network of investors to solve operational issues (*i.e.*, strengthening the management team, finding new clients, assisting in the globalization of the firm, *etc.*).

2.5 Opportunistic Behaviors

According to Williamson [24], opportunism is "self-interest seeking with guile." Opportunistic behaviors can occur *ex ante* (prior to making the investment) and *ex post* (after investing). The six main opportunistic behaviors described in the literature include (i) adverse selection, (ii) moral hazard, (iii) hold-up, (iv) shirking, (v) sharking, and (vi) window-dressing. Adverse selection occurs *ex ante* while the other five conflicts occur mainly *ex post*.

(i) Adverse Selection Problem

Milgrom and Roberts [25] define adverse selection as "the kind of precontractual opportunism that arises when one party to a bargain has private information about something that affects the other's net benefit from the contract and when only those whose private information implies that the contract will be especially disadvantageous for the other party agree to a contract." Sahlman and Amit, Glosten, and Muller both expressed concerns about adverse selection in the venture capital market [26].

From the entrepreneur's perspective, the adverse selection problem centers on the VC's competence at providing nonfinancial contributions to the firm. From the VC's perspective, the adverse selection problem lies in the difficulty of selecting projects with entrepreneurs able and willing to deliver their business plan's promised returns.

Two important elements used to mitigate the adverse selection process are (i) signaling and (ii) reputation. Both try to address the information asymmetries within a project at the precontracting stage and improve the quality of the information available to the contractors. A number of contractual clauses address the issue of adverse selection at the postcontracting stage. These clauses seek to protect the VC from the entrepreneur's possible opportunistic behaviors after the investment has occurred. Contingency clauses, vesting provisions, and the right to dismiss the management team members all address the adverse selection problem *ex post*.

(ii) Moral Hazard Problem

Holmström [27] described the moral hazard problem as "an asymmetry of information among individuals that results because individual actions cannot be observed and hence contracted upon." Moral hazard occurs when one party attempts to gain advantage within the present terms of the contractual arrangements, often due to lack of specification or enforceability of the contract [28]. According to Chan *et al.*, moral hazard in the VC-entrepreneur relationship can occur when the entrepreneur has control over his own ability but performs below this ability by, for example, not exerting enough effort (shirking) [29].

Barney, Busenitz, Fiet, and Moesel identify the moral hazard problem in VC-entrepreneur relationships as a problem of opportunism [30]. They assume that the contract between the two parties can best be seen as a governance device used to manage conflicts of interest between them and that the need to write and enforce contractual covenants is partly a function of the probability of opportunistic behavior occurring. Two kinds of opportunistic behaviors may arise: managerial opportunism (*e. g.*, managers engage in actions that reduce the wealth of VCs by spending too much money on research and development, taking out too high a salary, *etc.*) and competitive opportunism (*e. g.*, managers reduce the wealth of VCs by leaving the firm and starting a new competing firm or acting as advisors to competing firms).

(iii) Hold-up (Ex Post Expropriation)

Goldberg coined the "Hold-Up Theory" to describe the contractual worries of someone being forced to make follow-on investments, accept disadvantageous terms *ex post*, or seeing his investment being devalued by the actions of others [31]. The party that is forced to accept a worsening of the effective terms of the relationship has been held up.

Various strategies, such as staging, are available to mitigate the hold-up problem. For Neher, an important element of the hold-up problem that motivates the use of staged financing is the initial intangibility of the physical capital that is invested in the venture [32]. In contracting, several authors show that typical contracts and governance structures are seen as mitigating the hold-up problem when the investor is empowered [33].

(iv) Shirking (Not Exerting Enough Effort)

As stated by Holmström, shirking is "the problem of inducing agents to supply proper amounts of productive inputs when their actions cannot be observed and contracted for directly." [34]. For Orts, agents will shirk their responsibilities if given half a chance [35]. If not sufficiently monitored, they will not work as hard as they should, take too many breaks, or otherwise behave selfishly, lazily, or irresponsibly.

Alchian and Demsetz argued that shirking within a firm could be reduced to an efficient level by a principal (VC) who specializes in monitoring [36]. Assigning the VC's title to the residual earnings[1] of the venture would encourage the VC to engage in the "right" amount of monitoring, defined as the amount at which the marginal costs of monitoring equals the marginal benefits of shirking reduction. Any solution to the shirking problem in venture capital contracting, therefore, must allow for reallocation of rewards after the initial investment.

VCs also promise to perform value-added services but fail to perform up to the entrepreneur's expectations or later attempt to renegotiate this promise when the entrepreneur has reduced bargaining power. This is a problem of "incompetence" and may be significant in the venture capital community, but for Smith, this "free rider" problem does not appear to be material [37].

(v) Sharking (Conflicting Private and Corporate Interests)

Sharking refers to situations in which (1) agents divert assets that are "owned" by others in the firm and (2) principals take undue advantage of power held over agents. Orts describes four types of sharking [38]. First, sharking against a minority shareholder involves the unfair alliance of majority shareholders, corporate management, and the board to "freeze out" or "squeeze out" minority shareholders. Second, excessive executive compensation involves the problem of overreaching by entrepreneurs to inflate their compensation package by exerting undue influence on the board. Third, the enterprise may restructure the company to the detriment of creditors and employees. Fourth, financial reengineering of an enterprise by outside investors and new capital providers may result in a form of sharking against other participants in the firm, including shareholders, employees, and former creditors.

In venture capital the interests of the VC and the entrepreneur can diverge on many topics, especially, and most importantly, on the exit routes. Entrepreneurs

---

[1] Alchian and Demsetz (1972) define residual earnings as all earnings remaining after the entrepreneurs are paid for their inputs.

are seldom inclined to consider the financial exit alternative when it could result in having to relinquish control.

(vi) Short-Termism (Window Dressing)

At every stage of a company's financing, new information about the venture is released [39]. Sahlman describes how the entrepreneur may try to improve short-term performance reporting to make sure that the project gets refinanced when conditions improve [40].

To mitigate short-termism, contracts often rely on convertible debt instruments, which make manipulation less profitable because the VC will convert debt into equity if the firm looks "too good," which would reduce the entrepreneur's profit [41]. Furthermore, the entrepreneur can be made to commit to the projected financial figures and milestones presented in the business plan by staging the investment around performance and development milestones.

2.6 Conclusion

Principal-agency theory tackles three interconnected contracting issues: the existence of asymmetric information and goal incongruence between the contracting parties, which then leads to the potential for opportunistic behaviors. This section showed that the opportunistic behaviors could be mitigated by using appropriate contractual tools. Once these potential opportunistic behaviors are capped and goals are better aligned, VCs and entrepreneurs can enter meaningfully into long-term contractual relationships establishing the basis for their future collaboration during the venture. The next section addresses in greater detail five mitigation strategies commonly used against the conflicts described above.

3 Conflict-Mitigation Strategies

3.1 Introduction

VC investments are typically concluded with a set of contracts that include a stock purchase agreement, a certificate of designations (or restated certificate of incorporation), a shareholder agreement, and a registration rights agreement collectively referred to as "venture capital contracts." [42]. However, many believe that these agreements, while very standard [43], tend to be very one-sided in favor of VCs [44]. Smith even wrote that the explicit terms of most venture capital contracts do not adequately protect entrepreneurs [45].

Table 1 below summarizes five key mitigating strategies used in venture capital contracting. First, the improvement of the information quality at the precontracting stage through signaling and careful due diligence reduces adverse selection. Second, deal syndication reduces the cost of information gathering and postcontract monitoring. Third, stage financing reduces the risks inherent in ventures and makes

Table 1   Key Mitigating Strategies in Venture Capital Contracting

Mitigation Strategies	
Precontracting	Information quality improvement
Contracting	Deal syndication
	Stage financing
	Deal structuring
Postcontracting	Monitoring

financing cheaper. Fourth, the deal structure, especially the securities used and contractual clauses, can realign interests. Finally, monitoring can be used *ex post* to reduce agency risks.

3.2 Information Quality Improvement

At the precontracting stage, the costs of adverse selection can be mitigated through the improvement of the quality of the information available. Prior to investing, during the due diligence process, a wealth of information is usually gathered by both the investor and the entrepreneur. This information gathering serves to provide comfort to both parties that their objectives are aligned (congruent) and that they stand to profit from their cooperation. The two most important elements of the due diligence phase are (i) the reference checks on reputation and (ii) the signaling that occurs with respect to the discussions on potential terms to be included in the contracts and the type of security to be used.

(i) Reputation

Most VCs rely heavily on the entrepreneur's reputation (brand name) to gain reassurance about his/her ability to perform and deliver the expected rewards detailed in the venture's business plan. A reputational check is also done by the entrepreneur, who is concerned with being financed by investors who bring not only money but also expertise, networks, and some form of value added [46].

Black and Gilson show that talented managers are more likely to invest their human capital in a company financed by a respected VC because the VC's participation provides a credible signal about the company's likelihood of success [47]. Further, empirical studies suggest that firms backed by well-respected VCs are able to sell shares in an IPO at higher prices than other firms [48]. Gompers argues convincingly that young venture capital firms have incentives to grandstand, *i.e.*, they take actions that signal their ability to potential investors [49]. Specifically, they would tend to bring companies public earlier than older VCs in an effort to establish a reputation and successfully raise capital for new funds.

(ii) Signaling

Lazear shows that in a traditional principal-agent framework, contracts can also be used as screening devices if the ability of the entrepreneur is unknown by the VC [50]. By setting the entrepreneur's compensation as an increasing function of performance, the VC discourages less able entrepreneurs from accepting the contract. For signaling to happen, the VC creates "contractual scenarios" such as alternative forms of financing, which are costly for the entrepreneur to lie about. The type of contract the entrepreneur chooses will then signal to the VC the relevant information about quality (*i.e.*, contingency dependent clauses). Leland and Pyle, for example, show that an entrepreneur's willingness to invest in his/her own project can serve as a strong signal of project quality [51]. Later-stage ventures can also signal future prospects with their use of debt or commitments to dividends since those are costly commitments of cash flows that cannot be replicated by low-prospect firms [52].

3.3 Deal Syndication

VCs, upon finding a promising venture, will often send the proposal to other investors for their evaluation. Syndicating deals provide (1) a better comparison of opinions with other knowledgeable sources. As indicated by George Middlemas of Inco Securities, "If two or three other reputable VCs, whose thinking you respect, agree to go along, that is a double check to your own thinking" [53]; (2) a larger pool of capital for current and follow-on cash needs; (3) a mitigation of risk; and (4) more expertise and support for the ventures.

Experienced VCs primarily syndicate first-round investments to venture investors with similar levels of experience. In later rounds they can invite less experienced VCs. Usually, the ownership stake of the lead VCs frequently stays constant in later venture rounds to mitigate the risk of opportunistic behavior by the lead VCs. The discussion by Lakonishok *et al.* of money manager "window dressing" also suggests a possible rationale for venture syndication [54]. Syndication may be a mechanism through which VCs exploit informational asymmetries and collude to overstate their performance to potential investors (especially in very negative markets) by organizing a new round of financing with new investors at favorable terms.

3.4 Stage Financing

Stage financing refers to a method by which a company is funded in "stages", *i.e.*, is provided incrementally with money as it passes milestones. For Sahlman [55], staging is one of the most important mechanisms for controlling the venture, as part of the generic, two-pronged approach used: (1) designing investment contracts which materially skew the distribution of payoffs in favor of the venture investors and (2) an active involvement in the development process of the invested company [56].

While the commitment might be to fund the entire amount, the later stages of funding are contingent on the company's attaining set goals (milestones). The option to abandon is essential for VCs because an entrepreneur will almost never stop investing in a failing project as long as others are providing capital [57]. Stage financing can also refer to the fact that most growing companies need funding at various points in their development. At each point, the company solicits investors to fund the next phase of company growth. As described by Cossin *et al.*, staging of the investment is a mutually beneficial arrangement [58]. It gives the VC the option to reinvest or abandon the project. It also provides the investee with gradually cheaper funding, as the sources of uncertainty are progressively removed.

Typically, the early rounds of investment create collateral that supports the valuations in the later rounds. Neher characterized the optimal staged investment path and showed how various features of the venture affect it [59]. Each stage of financing can reduce management's percentage ownership in the company. As the company progresses, however, its stock should command a higher price in each successive stage. The reason for this is that the investor's risk decreases as the company succeeds and meets its goals.

Staging the commitment of capital also helps reduce the uncertainty typically surrounding small ventures. As time passes, the VC is able to gather more information about the team, the market, and the product, thus reducing major risks and uncertainties considerably [60]. At every stage, new information about the venture is released [61]. Moreover, by staging the commitment of capital the VC gains the option to abandon and to revalue the project as new information arrives. As Sahlman shows, this option in fact raises the value of the investment. This is the cornerstone of the theory of investment under uncertainty, in which the irreversibility of investment creates value for the option of waiting for better information [62]. Gompers [63] provides empirical evidence on the importance of staging venture capital in gathering information on financed projects, where he found that high-tech companies, with less tangible assets and higher risks, are subject to more frequent monitoring by VCs [64].

Gompers argues that staged capital investments minimize agency costs [65]. There is also an efficiency cost to financing the venture via multiple rounds of investment *vs.* a single round. Though efficient, financing the venture up front may be infeasible because the entrepreneur cannot commit to not renegotiating down the outside investor's claim once he has made his investment. Staging the investment over time also helps to mitigate this commitment problem.

3.5 Contracting

VCs structure their investments using a mix of various types of contractual securities and contractual clauses to mitigate their risks and maximize their potential return. Each type of security offers a different mix of property and control rights. Property rights define the possible claims on shares of the company and consequently on the ultimate cash benefits of the venture, while control rights influence

Table 2 Key Characteristics of Debt and Equity Financing Instruments

Characteristics of the Two Financing Instruments		
Associated Characteristics of Financing Instruments	**Debt**	**Equity**
Property rights	Strong	Weak
(*i.e.*, benefits)	(Fixed returns)	(Residual claimants)
Control rights	Weak	Strong
(*i.e.*, available recourse)	(Only on default)	(Continuous monitoring)
Use of price controls	High	Low
Use of behavior controls	Low	High

Source: Kocchar (1997)

more the behavior of the entrepreneurs or the venture and the available recourse against possible improper deeds.

Table 2 summarizes the different characteristics of debt and equity and their coverage of both control rights that regulate the monitoring and effective controlling of the venture and property rights that regulate the exiting or refinancing of the venture. Control rights are mostly effective in settling behavior-induced conflicts, while property rights appear more effective at settling outcome-induced conflicts. This theme will be further evidenced in the next section's taxonomy of contractual clauses.

Contractual clauses are included in contracts to assist parties to a contract in formalizing their initial agreement on issues that may occur during the life of the contract. New contractual clauses are continually designed to settle *ex ante* new issues that have arisen in venture relationships.

Venture capital deal structures vary from investment to investment, but most venture capital deals are structured using the same concepts. In particular, the securities used revolve around two main varieties: equity and debt. Each subtype of debt or equity carries certain characteristics more suitable for certain types of investments; the main issue in contracting is then to select the liability class structure with the most appropriate amount of control and property rights to stimulate the entrepreneur to achieve the objective set in the business plan.

3.5.1 Equity

Equity is the initial simplest security form of a company. Equity is created at the onset of a company's establishment and represents an ownership interest in a company.

Common Equity

In most cases, common stock carries the right to vote for directors and to vote on other matters affecting the company. It also entitles holders to receive notice of shareholder meetings and to attend them, as well as the right to review corporate records or receive financial reports. Common stock can pay cash dividends based on company earnings but only after preferred stockholders receive their dividends. Dividends are usually paid at the discretion of the company's board of directors but are the exception, not the rule, with most growing, privately held companies. Holders of common stock participate last in company liquidations. Company creditors, debenture holders, and preferred stockholders get paid in that order, before holders of common stock.

Common stock can be issued in one or more classes. One frequently used company structure authorizes the issuance of two classes of common stock that differ only in the number of directors each class is entitled to elect. This arrangement can be used to insure an investor a seat on a company's board of directors while at the same time insuring that control of the board will remain with management.

Preferred Stock

Preferred stock provides the investors some "preference" rights not accorded the common stock holders. From management's perspective, preferred stock can be sold at a substantial premium over the price of common stock. This can allow the company, for example, to continue selling its common stock at a lower price to management and other key employees without triggering potential tax liability for receipt of undervalued stock.

Convertible Preferred Stock

A typical convertible preferred stock entitles the investor to convert the shares into common stock at a predetermined formula and to vote the preferred stock on issues presented for shareholder vote. The voting rights conferred by the convertible preferred stock often include the right to elect one or more directors to the company's board of directors and to approve certain major decisions. Common "major decisions" include the issuance of additional shares of stock to others, the sale or merger of the business, the creation of new stock with preferential rights, and the change of the company's core business activity.

Venture capital investors use convertible preferred stock frequently. The stock's preferred status gives the investor a preference in the event of a company liquidation or sale. In growing companies, this preference usually entitles the investor to receive back his/her investment, and an agreed upon return, before other investors receive any proceeds from a liquidation or qualifying sale. This right is very important because many startups expect to be acquired. Upon a sale or liquidation, convertible preferred stock typically receives a return of its per-share purchase price before the common stock receives any payment. Beyond this point, however, liquidation preferences vary greatly. At one extreme, the convertible preferred stock may be fully participating (see below): after receiving the amount of its purchase price per share (sometimes even with interest), the convertible receives payment on an equal per-share basis with the common stock. This greatly dilutes the payout to the company's founders holding common stock, to the benefit of the VC. At the other extreme, the convertible preferred stock may only have the alternative of receiving payment as if it were converted into common stock instead of recovering its initial purchase price per share. This results in treating the VC similarly to the founders of the company, except with downside protection.

Convertible preferred stock is usually accorded protection against dilution from stock splits, stock dividends, and the future sale of the company's stock at a price less than that which the preferred stock was sold. Clauses that are often embodied in the convertible preferred stock issue are further detailed below. The inclusions of certain clauses, such as liquidation preferences, in the share purchase agreements and shareholder agreement are largely decided within the negotiation process between the VC and the entrepreneur.

Participating Convertible Preferred Stock

Participating preferred stock is a variant of convertible preferred stock. In a liquidation, participating preferred stock allows the holder to receive the amount he or she initially invested before the common shareholders and then receive a *pro rata* share of what's left after the liquidation preference is paid. Under most circumstances, the PCPS behaves more like a straight preferred stock or common stock rather than a position of convertible preferred [66]. However, upon liquidation or exit, investors receive both the principal amount of the preferred, as they would in an investment of straight preferred, and common stock. As a result, participating convertible preferred stock is better categorized as a position of straight preferred stock and common stock than as a position of convertible preferred.[2] In some

---

[2] Gompers (1998) describes participation provisions and refers to them as super priority provisions.

instances, the participating preferred does not receive a return of principal if the company's return is sufficiently high.

Redeemable Nonconvertible Preferred Stock

Redeemable nonconvertible preferred stock is similar to the convertible preferred stock except that the investor cannot convert. Redemption by the company of the preferred stock leaves only the investor's common stock outstanding. This permits investors, after having given the company a reasonable opportunity to otherwise create an exit, to liquidate their investment by redeeming 95% of their cost, with a return on common capital (*i.e.*, dividend), while retaining their equity position in the company through ownership of the common stock [67].

3.5.2 Debt

Debt, the second main type of financing instrument, is most often used in later-stage financing. For Trester (1993), debt should become more pronounced in situations where monitoring is easier and the probability of information asymmetries is lower. Indeed, debt contracts typically include weak control mechanisms but are strong property claims. As the control rights are low, debt contracts have few behavior controls over the actions of the entrepreneur.

Straight Debt/Loan/Bond/Note

The simplest form of debt is the loan. Loans can be made by an investor or, in some cases, a supplier. Debt enjoys seniority in the event of liquidation of the company: debt must be repaid prior to returning the nominal capital to the shareholders.

Convertible Debt

Venture capital financing is typically characterized by extensive use of convertible debt and stage financing. Cornelli and Felli show why convertible debt is better than a simple mixture of debt and equity in stage financing situations [68]. When the VC retains the option to abandon the project, the entrepreneur has an incentive to engage in window dressing or short-termism (to bias positively the short-term performance of the project in order to reduce the probability that the project will be liquidated). With a straight debt-equity contract, the entrepreneur will always engage in signal manipulation. With a convertible debt contract, such behavior reduces the likelihood of liquidation but increases the probability that the VC will convert debt into equity, reducing the entrepreneur's profits. Trester showed that for sufficiently high probabilities of asymmetric information, debt contracting might be neither feasible nor desirable [69]. Under high asymmetric information, equity contracting may be a better alternative precisely because of the lack of "foreclosure" liquidation rights. Preferred or convertible preferred stock may then be the better alternative because these types of security eliminate the foreclosure option while preserving some seniority in the event of bankruptcy. It is argued that this may be reason for the predominance of preferred stock in venture capital contracting.

Subordinated Convertible Debenture

Subordinated convertible debenture is a special form of convertible debt in which the principal and/or interest due under the debenture is convertible at the election of the holder into capital stock of the borrowing company at an agreed-upon rate of conversion. These debentures usually do not entitle their holders to any rights to vote on matters that come before company shareholders. Often, however, they contain contractual provisions that entitle the holder to receive financial reports or limit the amount or type of debt to which it will be subordinate. Because they are debt instruments, convertible subordinated debentures have preference over common stock and preferred stock in the event of company liquidation.

As financial capital is an uncertain but critical resource for all firms, suppliers of finance are able to exert control over firms [70]. Debt and equity are the two major classes of liabilities, with debt holders and equity holders representing the two types of investors in the firm. Each of these is associated with different levels of risk, benefits, and control. While debt holders exert lower control, they earn a fixed rate of return and are protected by contractual obligations with respect to their investment. Equity holders are the residual claimants, bearing most of the risk and, correspondingly, have greater control over decisions [71]. Each type of security is different in its degree of possible control over the company and possible cash-flow allocation from the proceeds of the venture. VCs and entrepreneur negotiate the use of securities that best fit their objectives and their views on the future outcome of the venture.

Each contract encloses specificities such as seniority in liquidation, *etc.* and allows the contracting parties to include relevant clauses to possibly enforce in the process of controlling and benefiting from the venture's unwinding.

As stated above, the use of specific contractual clauses forms a cornerstone of any mitigation strategy in venture capital. The complexity of the trade comes fully to light when one begins to examine in detail the agreements made between the VC and the entrepreneur. Sometimes, the piling on of multiple layers of different types of securities with different contractual clauses also results in a complexity only fully encompassed at the unwinding of the investment within the context of a trade sale or liquidation. These aspects will only be lightly addressed in this note. Section IV covers clauses in more details with a convenient taxonomy to sort them by the stages of the venture in which they can be enforced and linking them to the conflicts they potentially mitigate.

3.6 Monitoring

After the investment is made, VCs reduce agency risk by closely monitoring the entrepreneurs' companies. For this reason, VCs usually demand board seats and powerful control rights[3] VCs have to be efficient in monitoring the ventures to overcome moral hazard problems after the investment is made [72] and may achieve this by gaining specialist knowledge of specific sectors or industries [73]. Thanks to this monitoring, entrepreneurs are less able to get away with opportunistic actions that destroy company value. While Gorman and Sahlman and Sapienza both found monitoring to be heaviest in early-stage ventures [74], MacMillan *et al.* and Elango *et al.* found no relationship between the company's current stage and the involvement of its VCs [75]. In any case, one purpose of monitoring would appear to be to lower agency risks. Therefore, according to finance theory, the presence of mechanisms to monitor should lower the perceived variability of future performance and lower the required return [76].

4 A Simple Taxonomy of Major Contract Clauses

4.1 Introduction to the Taxonomy and Generic Propositions

Contract clauses are one of the key mitigation tools used in venture capital investments; in some instances, they can even turn not-so-successful investments into significant profits for the VCs. It is thus essential for investors to understand these clauses, the rationale for their inclusion, their potential for mitigation, and their ultimate interactions with other clauses in various project circumstances. To facilitate the organization of the presentation of these clauses, we rely on the contractual dimensions identified by Landström [77], namely the terms:

1. Affecting the management of the entrepreneur's company
2. Focusing on the monitoring of the company
3. Effecting changes in ownership and management
4. Directed toward "exit."

---

[3] Lerner (1994) cites Sahlman for proposing that "venture capitalists usually have several board seats and powerful control rights."

Items (1) and (2) are generally seen as control rights, while items in categories (3) and (4) relate more closely to property rights. To structure the review of these contract clauses and their impact on entrepreneur-VC relationships, a taxonomy is presented showing also the conflicts these clauses attempt to mitigate and the objectives they pursue. This taxonomy serves as the basis for a number of propositions:

> *Proposition 1:* It is convenient to split contract clauses between those settling the monitoring and control aspects of a venture (control rights) and those focusing on settlement of the exit of a venture (cash-flow or property rights).

> *Proposition 2:* Property-right clauses are virtually all concerned with outcome-based issues, while control-right clauses deal mostly with behavior-based issues.

Clauses are indeed induced either by certain behaviors (behavior based) or by certain outcomes in the investment process (outcome based). Further, Kaplan and Strömberg explain that clauses either affect the control capacity (control rights) of the investors and/or the income streams derived from the venture (cash-flow or property rights) [78].

> *Proposition 3:* Few property clauses actually attempt to mitigate conflicts linked to information asymmetries or agency conflicts as these conflicts tend to appear earlier in the life of the investment.

Behavioral conflicts tend to happen earlier in the life of the venture and necessitate monitoring and control mechanisms from the VC. Later in the life of the investment, either the management-entrepreneur has adapted its behavior to that of the VC or has been replaced by a more market aware management team. Investors then only worry about maximizing the liquidity event and have relinquished much of their control and monitoring activities to the operational management team. This is consistent with the theory developed by Jensen and Meckling [79].

> *Proposition 4:* Few clauses directly penalize incompetent VCs.

Most startup entrepreneurs are in dire need of funding as well as many equally important resources, such as the contacts and the networks available to the VC.

Entrepreneur-VC discussions often focus on the added value the VC can bring to the company in its efforts to secure market share, globalize, *etc*. However, once the VC has entered the venture, few contracting clauses penalize a VC who does not live up to the entrepreneur's expectations. Arguably the VC will be penalized if the venture fails, but no specific clause actually protects the entrepreneur if he/she is unexpectedly obliged to perform activities that the investors initially agreed to perform.

> *Proposition 5:* Most clauses tend to protect the investor (VC).

It can be argued that the largest investor in a project is the entrepreneur who puts in his time, energy, commitment, and idea to the benefit of the venture. However, in contracting, the bargaining power seems to lie mostly on the investor's side.

To organize the presentation below, we use an analogy to the actual process used by investors to develop the contractual frame of the investment. First, before making their investment, they will conduct thorough due diligence. Once an investment opportunity has returned a positive due diligence, negotiations start on the formal investment contract. The final contract will include postinvestment monitoring and control rights as well as cash-flow rights.

The sections are then organized into four phases:

- Clauses used during the preentry or due diligence phase
- Clauses used during the investment entry phase
- Postinvestment monitoring and control clauses – behavior based
- Postinvestment property / cash-flow clauses – outcome based.

The resulting taxonomy is presented in *Exhibit 1*.

4.2 Preentry/Due Diligence Provisions

Conducting a due diligence is an expensive process for a VC, so he/she will normally initiate a contractual relationship with the entrepreneur. The first documents signed usually include (i) a nondisclosure agreement; (ii) a guarantee to complete information; (iii) an exclusivity right for VC during the due diligence, also known as a "no-shopping" clause.

(i) Nondisclosure Agreement

The nondisclosure agreement (NDA) ensures that both parties can exchange information freely about their project. The intention for the entrepreneur is to convince the VC to invest while preserving his business idea, concept, intellectual capital, plans, *etc.* For the VC this serves to ensure that his discussions with the entrepreneur will remain confidential. Most VCs refrain from signing NDAs, in particular before receiving business plans, since it could constrain their ability to enter a particular space. Others limit the validity of the NDA signed to one year to avoid unlimited liability. In practice, most entrepreneurs must weigh the possible costs with not getting signed NDAs for their projects and possibly getting an investor interested in his idea. Clearly, the professional code of conduct implicitly abided by VCs precludes divulging confidential information: doing so would seriously damage their ability to conduct business in the future. Reputation is the most effective guarantee in this respect. Entrepreneurs should also understand that VCs do not necessarily make good entrepreneurs and are not in the business of starting businesses on their own. Also, if a business concept is likely to be damaged or copied by simply presenting it, then the concept probably does not constitute a valid opportunity in the first place!

(ii) Guarantee to Complete Information

The guarantee to complete information (GCI) clause ensures the VC access to all information available within the company during the due diligence clause. This clause is often included in the letter of intent (LOI) signed by the VC wishing to perform a due diligence. The LOI also often includes the exclusivity clause.

(iii) Exclusivity, or No-shopping, Clause

Lead VCs typically demand exclusivity on the investment opportunity while they do their due diligence. This is done (1) because due diligence is an expensive endeavor and VCs want to guarantee that they have a reasonable chance to complete the transaction; (2) to keep the entrepreneur from stirring the interest of other potential investors and potentially starting a bidding war for the deal; (3) to allow the VC to complete the due diligence process and in parallel start structuring his financing round syndicate with worthy coinvestors.

4.3 Investment Entry Provisions

4.3.1 Investment Representations and Warranties

Representations and warranties refer to the statements of facts, opinions, and estimates investors ask companies and management to put into writing as conditions for receiving the funding by the investors. These representations (reps) and warranties usually appear in the financing agreement or in separate "subscription agreements." Investment representations help complete a funding deal quickly and cheaply. Used in connection with a proper exemption from registration under applicable securities laws, these representations also help protect a company and management from serious securities liabilities. Reps and warranties should be drafted carefully; violations create a default that releases the investors from their obligations and subjects management and the company to liability.

4.3.2 Operating Covenants

Operating covenants refer to agreements between an investor and management that outline undertakings management agrees to perform after the investor provides funding. They are typically preconditions to the making of an investment. Operating covenants can also require management to achieve specified goals or to refrain from engaging in specified activities. Failure to fulfill the obligations of an operating covenant can result in severe penalties for management.

Affirmative Covenants

Affirmative covenants refer to the contractual provisions in a venture funding that the company and its management agree to fulfill after the funding has been completed. They can also relate to specific deal points that are to be performed after the closing of a funding deal such as management's undertaking to identify an additional board member or hire a suitable chief financial officer or other designated officer to round out management.

Negative Covenants

Negative covenants refer to operating covenants that prohibit a company from taking specified actions without the consent of the investor. They are a regular part of venture financings and are designed to protect the investor from future events that may dilute or undermine the value of his investment.

4.4 Postinvestment Monitoring and Control Provisions – Behavior Based

Immediately after the investment and until the next financing round, the VC will be most concerned with monitoring his/her investment and controlling the proper execution of the business plan. Investors have joined the ranks of liability holders in the company and have every interest to seek the success of the venture. VCs spend a majority of their time monitoring their investments [80]. Within the contract negotiated, VCs will have put in place a number of control mechanisms or rights to ensure their ability to monitor the venture. As outlined by Black and Gilson, VCs will only be willing to cede their control rights when the venture is successful and the entrepreneur has proven his management acumen and when it maximizes the firm's value [81].

4.4.1 Board Rights

Significant venture investors will require board representation, and smaller ones will seek at least observer status on the board. In addition, venture investors will typically insist that, as a matter of good corporate governance, the majority of the board be composed of experienced directors independent of management and that the audit and compensation committees of the board consist solely of independent directors [82].

Venture capital investors on the board want a role, or possibly veto rights, with respect to the approval of annual budgets, management hires and dismissals, and compensation matters, including employee share ownership and incentive stock ownership plans. As well, venture capital investors, in their capacity as shareholders, will want the right to approve changes in capital structure, sales of control, amalgamations, and substantial asset sales [83]. The lead VC in a round of financing will generally expect a seat on the board of directors. If there are multiple rounds of financing or VCs, there may not be enough board seats, or the board might get too large or be dominated by financial investors. In the first round, the founders will most likely be able to retain control of the board. If there is only one VC in a round, it is common for it to request two board seats.

4.4.2 Voting Rights

Voting rights force the company to get VC approval in order to take certain actions such as dividends, mergers, issuances of stock, and amendments to the company's articles of incorporation or bylaws. A VC's preferred stock usually votes as if it were common stock. The preferred stock usually has a number of votes equal to the number of shares of common stock into which it is convertible and may have the right to elect a disproportionate percentage of the board of directors, frequently a majority.

Preferred shareholders have some protective provisions that require a majority vote of their class. These provisions usually involve changing the certificate of incorporation in ways that affect the preferred shareholders' rights, privileges, and preferences such as liquidation preferences, issuing securities senior or equal to existing preferred stock and changing voting rights.

Founders and investors will usually have a written voting agreement or designate in the certificate of incorporation how many seats common shareholders can elect and how many preferred shareholders can elect. Consequently, control over the board can change as later rounds of financings occur.

4.4.3 Veto Rights

In addition to board representation, VCs also usually negotiate for veto powers over certain corporate transactions, including the issuance of senior or parity securities, charter or bylaw amendments that adversely affect their interests, mergers and other business combinations, disposition of corporate assets, changes in management salary or benefits, and changes to any employee stock-option program.

4.4.4 Voting Trust Rights

Voting trust rights transfer control from management to an investor if the company does not meet its goals. It is a technique that enables minority investors to step in and take control if things get out of hand. VCs can exercise their rights under the voting trust, vote the shares of the company, and elect a new board of directors favorably disposed to them. Voting trusts are increasingly rare. A more usual way for the VC to regain control in extreme situations when the entrepreneur has deviated sharply from the business plan and needs money is to refinance the company under conditions that are much more attractive for the VC.

4.4.5 Class-voting Requirements

Some VCs will ask for protective provisions for each series of preferred stock with each series voting separately. In theory, these protect each investor class from possible expropriation by another (later) class of investors, in essence giving them veto power over future decisions. The latter can also reduce management's flexibility by increasing the chances that any one investor will be able to block changes.

4.4.6 Financial Reporting Requirements – Information and Monitoring Rights

Shareholders with a significantly large bloc of preferred stock may also be entitled to certain information such as monthly financial statements, audited financial statements, and the annual budget. These information rights are usually part of the share purchase agreement.

4.4.7 Access and Visitation Rights

Investors will usually require, at reasonable times and upon reasonable notice, full access to all books and records of the company. They shall be entitled to review them and copy them at their discretion and shall be entitled to inspect the properties of the company and consult with the management of the company, subject to usual confidentiality obligations of the investors.

4.4.8 Right to Dismiss

A major criterion for VC funding is the quality of the CEO and the entrepreneurial team [84]. Thus some VCs will negotiate specifically the right to dismiss management if the company runs astray. Typically, that decision will require board approval. The right to dismiss will thus be effective only if the VC controls the board or can convince a majority of its members to take action.

The perceived importance of the CEO for the success of the venture leads many VCs to focus on the CEO if any dissatisfaction develops. Dismissal of an entrepreneurial team member is more likely to occur in poorly performing companies [85]; it has a negative impact on long-term venture outcomes [86]. However, following the VCs' replacement of a CEO, perceived venture performance improves [87]. Bruton, Fried, and Hisrich examined the right of the VC to dismiss key members of the management team [88]. They found that CEOs are usually dismissed over issues of strategy rather than on operational issues. In contrast, European VCs rarely have the contractual right to dismiss the CEO. The renowned Swiss VC, Peter Friedli, often mentions that he has never seen a CEO fired too early in any of the 140 ventures on whose board he has served.[4]

VCs usually have the right to repurchase shares at book value from managers who have been dismissed or who decide to quit. This right, often structured as a call on the managers' shares, acts as a very strong incentive for the management to behave in accordance with the VCs' interest [89].

4.4.9 Vesting Provisions

VCs often ask founders to subject their stock, and all common stock sold, to a vesting schedule, *i.e.*, the stock will be repurchased by the company if the employee quits within a certain timeframe. Vesting usually occurs over four or five years, with the first group of shares vesting after a full year (known as the "Cliff") and then a fraction vesting each month (or year) afterwards.

If the company does not adopt a vesting schedule until after the venture financing, the founders may want to start the vesting period at an earlier date, like the day the founders first acquired their stock or joined the company. Sometimes investors negotiate to make the shares of the founders and management subject to future vesting so that they will forfeit shares if they leave the employment of the company before the vesting periods expire.

---

[4] Peter Friedli's remarks during the Venture Capital Conference in Munich in February 2001.

4.4.10 Noncompete Clause

Founders and management are usually required to enter into agreements with the company that require their full-time attention to the company's business, protect its trade secrets, and prevent the founders and managers from going into competition with the company. The purpose of this clause is clearly to reduce any goal incongruence. The noncompete clause usually prohibits entrepreneurs from working for another firm in the same industry for some period of time after departure. These clauses are usually written into the work contracts of the employees and the company [90].

Noncompete and vesting provisions are aimed at mitigating the potential hold-up problem between the entrepreneur and the investor. The vesting provisions are more common in early-stage financings where the hold-up problem is likely to be more severe [91].

4.4.11 Key-man Insurance

Where the success of the company hinges on the knowledge and skill of specific key individuals, the VC will insist on life insurance for those individuals. VCs will be the beneficial owners of these life insurances. These will serve to protect their investment in the company they are funding.

4.5 Property/Cash-flow Rights – Outcome Based

The first theories of modern property rights come from the Coase theorem [92]. It explains that, provided property rights are allocated, the potential parties to a bargain will draw up an agreement, which exhausts their potential mutual gains, and hence an efficient outcome is guaranteed. This theorem has not received unambiguous support [93].

Consider the three standard outcomes in venture capital investments: (i) a follow-on round, (ii) a liquidation, or (iii) an investment exit. Each and every one of those outcomes is addressed in the property-rights sections of contracts. In addition, ongoing property provisions can also modify the shareholdings and project returns during the life of an investment without any of the three above-stated outcomes happening.

A property right represents the fraction of a portfolio company's equity value that different investors and management have a claim to. Measuring cash-flow rights is not trivial, however, because many of the cash-flow rights accorded to founders and management are contingent either on subsequent performance (through performance vesting) or on remaining with the company (through time vesting). For Kaplan and Strömberg, cash-flow rights, voting rights, control rights, and future financings are frequently contingent on observable measures of financial and nonfinancial performance [94].

4.5.1 Ongoing Property Provisions

Ongoing property provisions change the shareholding mix during the life of the investment as the project matures and milestones are reached or not. Often VCs build into their project incentive/penalty clauses to reward/punish managers/entrepreneurs who succeed/fail to achieve the targets set out in their business plan.

Revenue Participation

Revenue participation gives an investor the right to receive a percentage of the company sales or revenues. Revenue participation works like royalties. The company pays the investor an agreed-upon percentage of sales and deducts the payments as a business expense. The investor shows the payment as ordinary income. Unlike royalties, however, revenue participation usually grants a percentage of all sales instead of just those on a particular product.

Sharing a percentage of revenue with an investor can accomplish some interesting results. First, the investor may become less concerned with the company maximizing earnings (profits) because he is paid based on sales. Second, management's attention may be focused on maximizing profits by minimizing expenses, so that its projected profit margin can be realized after the participation payment is made. Third, the investor may be less concerned with controlling or participating in company affairs. He/she may not even require a seat on the board or audited financial statements, at least as long as payments are forthcoming and things are running smoothly. Fourth, the entrepreneur gives up no equity.

Revenue participation is not common, but it does have their advocates among investors. One of the reasons for using the revenue participation certificate is that matters of management perks are then of no concern. The entrepreneur is free to live well "on the company" and spend at will without fear of shareholder criticism. Investors will remain passive as long as every week, month, or quarter they receive the agreed-upon share of revenues.

Dividends Provisions/Preferences

Preferred stock may have a dividend payment set at some rate if the directors declare a dividend. Since startups rarely declare dividends, this provision is used mostly to boost the argument that preferred stock is more valuable than common stock for tax purposes and to allow the founders to purchase common stock at a lower price.

There are three major types of dividend preferences. First, "when and if declared" dividends are paid on a noncumulative basis but before any dividends are paid to the common-stock shareholders (companies always prefer the first type of dividend). Second, "mandatory cumulative preferred dividends" accrue whether or not the company has declared a dividend or has the money to pay such dividends. Lastly, "state-contingent dividends" must be paid on a certain date if no exit for the investors has been found by then. Dividend provisions are not common.

Options

Options are securities that entitle but do not obligate their holders to purchase (call)/sell (put) securities in the future for a predetermined price or a price determined by a set formula. Because they do not obligate the holder to purchase/sell shares, they cannot be used to ensure future financing for a company. Even so, growing companies frequently use them to attract and retain qualified people, to obtain concessions on loans, and to reward outstanding performance.

VCs usually support stock-option plans for employees as long as the plan is not too generous and is vested over a certain period. Options provide an attractive incentive for employees because they give them the opportunity to share in the value created by a company's growth. A company will want to reserve a certain percentage of a company's stock for new and existing employees. As a general rule, 10% to 30% of the stock after venture financing is set aside for options. Setting aside shares will dilute the control of both the investors and founders but will help the company attract the top talent needed to succeed. Options generally take four to five years to vest.

A company's board of directors is usually responsible for granting the options. Shares that can be purchased through an option are referred to as the option shares. The purchase price for option shares is referred to as the exercise price or strike price. Options with exercise prices higher than the fair market value of the company's shares are considered to be "out of the money." Options with exercise prices lower than the fair market value of the company's shares are considered to be "in the money." Option grants are usually for a fixed number of shares at a fixed price per share. Often, the option only becomes exercisable in the future or gives access to more shares over time. Exercise, or purchase of the shares, is effected by paying to the company the purchase price for the shares after the option becomes exercisable.

VCs sometimes bargain for options when they fund a company. They do so because options enable them to increase the upside potential of their investment without obligating them to purchase additional shares from the company. Options are sometimes used to provide investors with antidilution protection: they give investors the right to purchase securities in the future at today's price or another bargain, discount price. Lenders sometimes request options, or warrants, in consideration of making a loan or granting an interest rate or term concession.

ISOs (Incentive Stock Options) and NSOs (Nonqualified ISOs)

Incentive stock options are rights to purchase company securities, usually common stock, that are issued to company employees and others the company wants to hire. They are designed to attract new employees and motivate existing employees. They give employees the right to purchase company shares in the future at present-day prices. If the company is successful, and its stock increases in value, the employee can purchase shares of company stock in the future at a price far below the then fair market value of the shares. By creating this potential windfall, the ISO motivates the employee to work hard to make the company succeed. ISOs are issued under guidelines adopted by the company's board of directors.

Performance Options

Performance options are options to purchase company stock that become exercisable only in the event the option holder, usually an employee, meets a predefined performance objective. Employers, as well as investors, like performance-based incentives when they motivate employees to achieve measurable objectives and assist the company in creating value for its investors. However, performance options are used sparingly in companies that have expectations of going public. This is because of accounting issues that can force a company to record a charge against earnings when a performance option matures and becomes exercisable. If great enough, this reduction in earnings can make a company a less desirable candidate for a public offering. It can also cause a company to violate a financial ratio covenant in its loan agreements.

Current accounting standards can require a company to book a charge against its earnings when the option becomes exercisable (when the performance is achieved) if the fair market value of the option shares is greater than the option's exercise price. Growing companies usually plan for increasing valuations and set exercise prices in their qualified options at or below fair market value at the time of option grant. If the company's value increases as planned, a charge to earnings will occur when the performance is achieved at a later date.

Call Options

Call options entitle the company to require a shareholder to surrender shares of company stock in return for payment of an agreed-upon sum of money. Calls are usually contingent upon some future event, such as the passage of time or the accomplishment of some goal by the company.

Management might exercise a call and redeem an investor's stock if another investor is willing to pay a higher price for shares than the call price. Even without a ready investor, management might call an investor's stock if it believes the market value of the company's stock has risen to more than the call price. In each case, redeeming stock at a call price that is below market value can increase the value of the remaining shareholders' stock. The cost to the company is the cash required to pay the investor for the shares. In deciding to exercise a call, management must weigh the potential benefit of redeeming shares at below market prices against the company's need for the cash used to exercise the call and its ability to replace that cash when needed.

Calls are usually optional but can, by express agreement, be made mandatory. A mandatory call, however, is nothing more than a contract to redeem stock at a future date. Calls, which become effective after a set period of time, are used sometimes to force a convertible preferred stock investor to choose, after the time has elapsed, between converting the preferred stock into common stock or accepting the call payment in redemption of the stock interest. When used in this way, the call mechanism sets a time limit on the preferences granted in the preferred stock.

Earn-ups

Earn-ups are arrangements whereby an investor acquires most of a company's capital stock but gives management the opportunity to increase its stock ownership by managing the company successfully. Usually, management operates the company under a contract that allows it to control the operations of the company as long as the company meets specified goals. Earn-ups are less a financing device than they are a technique for attracting good people to manage a company. They are sometimes used in a leveraged buyout by the outside investor to entice existing management to remain with the company. If management can keep the company profitable and meet its goals, it often can earn a substantial equity position in the company. Earn-ups assign little value to management's role in putting together the funding used to purchase the company. Instead, they reward management for making the company succeed after the funding.

Warrants

Warrants are options with long maturity dates. They are sometimes attached to convertible notes and offered to improve the conditions of a deal for the investor. VCs are often keen on obtaining some form of warrants to improve the possible transaction being negotiated.

4.5.2 State-contingency Rights

Usually VCs place great importance on the company reaching the objectives it sets itself in its business plan. VCs commonly enforce state-contingency rights dependent on the achievement of or failure to reach certain predefined milestones set in the business plan by the management. Should the business plan's targets not be reached, VCs use state-contingency clauses to increase their share of the company, reduce their average cost per share, and increase their control.

According to Kaplan and Strömberg, state-contingencies are mostly used in first VC rounds as opposed to subsequent rounds [95]. VCs commonly write (and presumably enforce) contracts in which control rights are contingent on subsequent measures of financial and nonfinancial performance or output. The contingencies appear to be related to the performance measure that is most important to the investors and the company. Cossin *et al.* show that "contingent precontracting" for followup rounds is theoretically a better proposition than the simple "rights of first refusal" commonly found in many contracts [96]. Examples of contingency clauses are presented in the appendix.

4.5.3 Take-away Provisions

Closely related to state-contingency clauses, take-away provisions are agreements between an investor and management that entitle the investor to penalize management when the company does not achieve agreed upon results. The penalty is often the reduction in management's shareholdings or in its ability to operate the company independently. Take-away provisions are most prevalent in financings where the company sells a controlling interest to the investors and management contracts with the investor to maintain operating control so long as company results are acceptable. They are also common in earn-ups, leveraged buyouts, and other transactions where management's participation is predicated upon its achieving certain results.

4.5.4 Next Financing Round Provisions

Subscription Rights

A subscription right clause stipulates under which conditions a capital increase can be approved by the existing shareholders of a company. A typical subscription right clause will stipulate that a capital increase must be approved by at least a majority of the investors and necessitate a shareholder meeting to have taken place. Subscription rights also typically allow "*pro rata*" preferential subscription rights (or "preemptive right") to existing shareholders. Should existing shareholders renounce their subscription rights, the company must first offer these shares to other existing shareholders prior to offering them to a third party.

Antidilution Provisions

Antidilution provisions entitle an investor to obtain additional equity in a company, with or without additional cost – depending on the type of antidilution clause, when the share capital of a company is increased. Preferred stock usually receives antidilution protection against stock dividends, stock splits, reverse splits, and similar recapitalizations.

Because these provisions can result in protected investors receiving free shares of stock when a future funding is at a lower price, antidilution provisions disproportionately reduce the percentage ownership of shareholders who are not protected. These "unprotected" shareholders usually include the company's founders and management. Antidilution protections usually appear as provisions of a financing agreement, in the terms describing the conversion rights of a convertible preferred stock, in a warrant or option, in an investor rights agreement, or, finally, in a shareholder agreement.

Five main forms of antidilution provisions are used: (i) preemptive rights or rights of first refusal, (ii) "pay to play" provisions, (iii) "price protection" provisions, (iv) full ratchets, and (v) weighted average antidilution.

- *Preemptive or Investment Rights:* Preemptive rights, or investment rights, entitle existing shareholders to subscribe to a portion (equal to its percentage ownership of the company – *"pro rata"*) of newly issued shares or shares available for sale by the company or another shareholder to maintain his or her level of ownership. This right is a contractual one that expires upon the IPO. It can also be a right attached to preferred stock as set forth in a certificate of incorporation. Under extreme forms of the preemptive right, a company must first offer all the shares to the current shareholders rather than just *pro rata* shares.

  Venture capital investors will typically insist on rights of first refusal, including a preemptive right with respect to future issuances and a right of first purchase of the shares of selling shareholders. Venture investors will typically seek protection not only from dilution resulting from capital alterations but also from "economic" dilution such that both their shareholdings and conversion formula will be adjusted if the company issues additional equity in subsequent rounds of financing at less than the price paid by the venture investors. A typical economic antidilution clause gives the investors the right to obtain more common stock, without additional aggregate consideration, in the event that the company subsequently issues new common stock or common stock equivalents at a price below the effective "as converted" common-stock price paid by the investors based either on a full ratchet or weighted average formula.

- *"Pay to Play" Provisions:* The "pay to play" provisions require the preferred shareholders to buy their *pro rata* share in later down-priced rounds of financing in order to get any price and antidilution protection. If the preferred stockholders do not purchase their pro rata share, then their shares are converted to another series of preferred stock that has no price protection. This provision is designed to encourage all investors to help the company in difficult times. Such provisions, however, are atypical.

- *"Price Protection" Provisions:* "Price protection" gives the VC protection against later rounds of financing that issue stock at a lower price than he paid (called "down rounds"). The VC's valuation is an educated guess, and price protection allows the VC to make an adjustment should later rounds of financing prove the initial valuation incorrect. If the price in later rounds is lower, the VC receives additional stock to make up for the price difference.

- *Full Ratchet Antidilution:* Ratchets are powerful forms of antidilution whereby the investor is given additional shares of stock for free if the company later sells shares at a lower price. The number of free shares the investor receives is enough to make the investor's average cost per share (counting all of his purchased and free shares) equal to the lower price per share given to the later investor. What makes the ratchet so powerful is that the first investor is given these extra shares regardless of the number of shares purchased by the later investor.

Ratchets can also be tied to options or warrants. When they are, the investor receives extra shares when he/she exercises the options. When ratchets are tied to conversion prices, as in convertible preferred stock, the extra shares are received at conversion. Other types of antidilution provisions employ weighted average or other allocation mechanisms that give the investor fewer shares when the second investor purchases fewer shares. Full ratchet antidilution protection allows a VC who has invested in earlier rounds at a higher price than other VCs in later rounds to receive additional stock as if they had bought it at a lower price in the first round. The company issues those additional shares. Full ratchets are rarely used because they are widely viewed as unfair. Common shareholders suffer most of the dilution, and the later investors end up with a smaller percentage of the company than they bargained for. Additionally, a full ratchet occurs regardless of how small the later issuance is. However, there are some circumstances where a full ratchet is appropriate. If a VC discovers that the company is overvalued and needs more cash sooner than expected, the company might give a 6- to 12-month ratchet to assure the VC that it does not need to get later rounds of financing at a lower price. Additionally, as insurance against some future event occurring or not, such as getting a patent, ratchet protection might protect the VC if more money needs to be raised.

- *Weighted Average Antidilution:* Weighted average antidilution refers to a milder form of antidilution protection, which uses a weighted average formula to determine the dilutive effect of a later sale of cheaper securities and grants the investor enough extra shares for free to offset that dilutive effect. The weighted average antidilution often uses the following mathematical formula to set the conversion price:

> New Conversion Price = (Old Conversion Price) × (Number of Outstanding Shares Before Issuance + (Money Invested in Current Round/Old Conversion Price)/(Number of Outstanding Shares Before Issuance + Shares Issued in Current Round)

The result of this calculation gives a new price per share for the protected venture investor that is then divided into the monetary amount invested to determine the total number of shares to be received. The difference between this number and the number of shares the venture investor already owns is the number of new shares the venture becomes entitled to receive for free. This formula adjusts the conversion price based on the relative amount of the company being sold at the lower price. Some variations to the formula exist, and the most common variations involve counting options as either issued or unissued stock. Shares reserved for granted options are often counted as already issued, while those for options to be granted are not.

Employee Reserve

Quite often the company will want to exclude from the above antidilution protection a set number of shares to be issued to key employees as incentives. The VC is usually also highly in favor of this exclusion since the success of the company often hinges on the dedication of the company's employees. For a first-round venture financing, an amount equal to 10% to 20% of the outstanding shares will often be reserved for issuances under incentive stock option plans and other incentive stock issuance plans to the company's employees. These plans normally provide for a monthly vesting schedule over a two- to three-year period during which time the employee vests the right to purchase and retain shares in the company. The purchase price is usually established at anywhere from 10% to 20% of the preferred stock purchase price.

Antiredemption Rights

These prohibitions usually apply to sales of virtually all the company's shares or assets by the entrepreneurs and to increases in the number of authorized number of shares or directors. VCs are more favorably disposed toward mandatory redemption provisions, but even these are rare in venture capital agreements because the provisions are "not viewed as a realistic alternative" for a company with little or no cash inflow [97].

4.5.5 Exit Rights Agreement/Liquidity Agreements/Exit Provisions

Liquidity agreements or exit provisions allow a stockholder to convert his investment into cash or at least makes it easier for investors to get their money back. Liquidity agreements are common in venture capital financings, particularly when the VCs are uncertain whether the company will ever be able to offer its stock for sale in the public markets. Since many venture capital investments end up being less successful than the business plan predicted, these options can be very important to investors. The parties to investment agreements can consider a number of contractual devices designed to influence the VCs' exit from the arrangement. These include:

- Exit-control covenants, obliging management to use reasonable endeavors to facilitate exit before a certain date
- Demand registration rights that enable an investor to force a company to register his shares for sale to the public
- Liquidation provisions
- Put options, requiring management or the investee company to buy the VCs' shares if no exit is achieved within a defined period [98]
- Buy-sell agreements that enable an investor to force management to either purchase his shares or sell its shares to him
- Unlocking provisions
- Tag-along/cosale rights
- Take-along rights.

These rights as well as other variants are presented in greater detail below.

Convertible debentures also increase investor liquidity and provide an escape when a company is not living up to expectations. They do so by giving the investor the option of not converting his debenture into shares of the company's common stock. If the company does not do well enough to give the investor a "healthy" profit upon the sale of the common stock he could obtain by converting his debenture, the investor can choose to forgo converting and collect instead the interest and principal repayment on his debenture. The VCs must then ensure that, at the debt's maturity, the company will have enough cash to refund their investment.

The exit-rights agreement, if there is one, typically replaces the registration-rights agreement and may replace the stockholder agreement. It will contain the registration rights of the investors as well as any redemption or "put" redemption rights as to common stock or warrants and any "cosale" rights, all of which provide opportunities for the investors to obtain liquidity for, or "exit," the investment [99].

Registration Rights

Venture investors typically negotiate agreements that entitle them to require the company to register their shares for resale to the public. The purpose of the agreements is to provide the investor with a mechanism to sell his shares. In the USA, a registration right allows the holder to force the company to register the holder's security with the Securities and Exchange Commission (SEC). This allows the investor to sell his shares publicly. When a company goes public, the underwriters will often be unwilling to allow existing shareholders to sell their shares since it could undermine the new issuance of stock. Additionally, SEC rules may limit certain shareholders from selling shares, especially if the holder is part of the management or owns over 5% of a company, unless the shares are registered. Investors are likely to request one of three types of registration rights: (i) demand rights, (ii) piggyback rights, or (iii) S-3 rights.

- *Demand Rights:* Demand rights enable the investor to liquidate his investment by forcing the company to register his/her shares for sale to the public. Because this can increase a company's cost of raising equity in the future, the number of demand rights granted and the conditions under which they can be exercised are usually heavily negotiated issues. The outcome of those negotiations can seriously affect the value of a funding and the future success of a company. Among the points management frequently seeks to obtain in negotiations over demand rights are:

- *Limitations on the Number of Demand Rights*: The fewer demand rights, the better for management. Generally, a company will grant such demand rights only to specific investor groups.
- *Control over When the Demand Right May Be First Exercised*: Typically, managements seek to delay demand registration until at least twelve months after the company has conducted a company-initiated initial pubic offering. Alternatively, managements negotiate for a fixed period of time, such as three years, during which time the demand right may not be exercised.
- *Minimum Requirements to Exercise a Demand Right*: Usually, this translates into a requirement that a fixed minimum number of shares be offered for sale and that the proposed offering be of at least a certain minimum monetary size.
- *Rights to Delay or Convert a Demand Registration*: Managements routinely request the right to delay a demand offering for a fixed maximum period of time if the company's directors have commenced preparations for a public offering or believe a delay would benefit an offering because of market or other conditions. They also frequently request the right to convert a demand offering into a company initiated offering to give the company preference and control over the number of shares offered.

- *Piggyback Rights/Come-along Rights:* Piggyback rights allow an investor to participate in a registration already proposed by the company so as to include his/her shares in a public offering that the company voluntarily conducts for its own benefit. Unlike demand rights, piggyback rights do not entitle investors to require a company to conduct a public offering but, rather, allow them to include shares in a registration that is initiated by the company. Piggyback rights usually appear as provisions in a financing agreement, an investor-rights agreement, or a registration-rights agreement. Piggyback rights are common in venture financings. Except for certain fees required by various state securities law administrators to be shared by all selling shareholders, companies usually bear the cost of investors exercising piggyback rights.
- *S-3 Rights*: In the USA, S-3 rights are a variation of demand rights. They force a company to register the holder's stock on an S-3 form with the SEC. Companies with a "public float" over $75 million in securities held by parties other than management and shareholders with over 5% of the stock and which have been public for at least a year can use an S-3 form. An S-3 registration allows the company to incorporate by reference other documents it has filed with the SEC, so disclosure is much less burdensome. Usually, VCs get S-3 demand rights that are limited in time but not in the number of shares.

Redemption Rights

Three types of redemption provisions are commonly found: (i) call redemption rights, (ii) put redemption rights, and (iii) discretionary stock redemption provisions. However, redemption rights can also be structured to include such events as termination of employment, death and invalidity, bankruptcy, or breach of contract. Usually upon one of these events occurring the other parties to the contract enforce the redemption of the defaulting party's shares at a fair market price minus a preagreed discount.

- *Call Redemption Rights:* This right grants the company the right to redeem the preferred stock at a predetermined price at some point in the future if the preferred shareholders do not elect to convert their shares into common stock. Usually this right is only triggered if the company meets certain milestones and can serve as an incentive for the entrepreneurs.
- *Put Redemption Rights:* This right grants the preferred shareholders the right to require the company to purchase their shares at a predetermined price at some point in the future. Usually, this redemption price is the original purchase price plus some cumulative rate of return. From the VC's perspective, this "put" redemption is an exit provision for companies that are marginally profitable or, worse, among the "living dead."
- *Discretionary Stock Redemption Provisions:* The entrepreneur's only effective contractual protection against adverse selection is a discretionary stock redemption provision allowing him/her to repurchase the stock sold to VCs at a predetermined price. Suchman found that discretionary stock redemption provisions were common in weak, preprogrammed, and legalistic contracts, but they are not usually found in close or flexible contracts – the most common modern forms of venture capital agreements [100]. Indeed, discretionary stock redemption provisions are sufficiently rare in modern venture capital contracts that many VCs claim never to have seen them.[5]

Even if a discretionary stock redemption provision were proposed, VCs would be unenthusiastic. Such provisions limit a VC's ability to decide the form of his investment. VCs are more favorably disposed toward mandatory redemption provisions, but even these are rare in venture capital agreements because the provisions are "not viewed as a realistic alternative" for a company with little or no cash inflow. For example, if a company was contemplating the exercise of a discretionary right of redemption, it is safe to assume that the value of the company had risen and the VC would convert preferred shares into common shares.

---

[5] Gerard H. Langeler of Olympic Venture Partners observed: "You can divorce your spouse. You cannot divorce your investors." Remarks at the Venture Oregon '97 Conference (October 14, 1997).

Although the VC would remain interested in the company, many of the control mechanisms previously awarded to the VC would evaporate upon conversion.[6]

It is misleading to assume that the investor is going to gamble that the company's fortune will not improve, because if the company's fortune does improve, the investor's conversion right appreciates. Consequently, in venture capital deals, the corporation's option to redeem works primarily to force an investor conversion that will relieve the company of the burdens of interest (or dividend) payments and restrictive covenants [101].

The absence of discretionary stock redemption provisions from most modern venture capital contracts suggests that the costs to the entrepreneur of protecting against adverse selection through contract are greater than the risks. This in turn suggests that entrepreneurs have other methods of protecting against adverse selection. The most likely candidate for such protection is the market for reputation, discussed above.

Automatic Conversion Rights

Preferred stockholders in venture deals usually have the right to convert their preferred stock into common stock at any time. Preferred stock also often automatically converts to common stock upon the occurrence of certain events (usually at a ratio determined by dividing the initial purchase price by the conversion price). Such events include an initial public offering by the company or if a certain percentage of the preferred shares votes to convert. A company should try to get the preferred shares to convert as soon as possible to get rid of their special rights and streamline the balance sheet for the initial public offering. Usually a vote by the majority or a supermajority of preferred shares will force an automatic conversion. An entrepreneur should seek a majority requirement or the smallest supermajority requirement possible.

Automatic conversion at the initial public offering usually requires that (i) the IPO be firmly underwritten, *i.e.*, the underwriters must have committed to placing the entire offering; (ii) the offering raise a certain amount of money for the company, and (iii) the offering price must exceed a certain minimum, such as four times the conversion price of the preferred stock.

When the preferred stock converts, the inherent rights associated with it cease to exist. Some contractual rights, such as registration rights that force the company to register a shareholder's stock, usually survive. Other rights, like information rights (the right to receive certain ongoing company financial information) and preemptive rights (the right to buy stock the company issues), usually terminate upon the IPO.

---

[6] Halloran *et al.* (1998) observed that "corporations prefer redemption provisions exercisable at their option so that they can force the Preferred Stock to convert into Common Stock (and surrender its privileges) at some time in the future, even if the conditions for automatic conversions are not met."

Mandatory Conversion Clauses

The investor's right to convert his preferred stock into common stock is usually coupled with the company's right to require conversion in the event of an initial public offering or, less frequently, sale of the company. This is the mandatory conversion clause. The conversion price is usually equal to the purchase price of the preferred stock but is subject to downward revision (resulting in giving the investor more common shares upon conversion of the preferred shares) after a subsequent sale of stock by the company at lower prices.

Unlocking Provisions (incl. Buy-Sell Agreements)

Unlocking provisions enable investors to disengage their financial arrangements with management. A typical unlocking provision is a buy-sell agreement. Another provision creates a mechanism through which either management or the investor can force the other to sell its interest in the company to a third party or buy the stock of the other party. Typically, this arrangement allows either management or the investor to accept a third party's offer to buy the company's assets or the stock of both shareholders in the company. The other party must then either buy the first party's shares at the per-share price contained in the third party offer or sell his interest, along with the interests of the other shareholder, to the third party.

- *Buy-Sell Agreements:* Buy-sell agreements are contracts that entitle one shareholder to force others either to buy his stock or to sell him theirs. These contracts often appear as provisions in a financing or shareholders' agreement. Many investors insist on buy-sell agreements with the entrepreneurs of companies they fund. They view these agreements as a way to help them withdraw from a company that does not live up to their expectations. With a buy-sell, investors know they can either get their money out or get management's shares. With management's shares, they can usually replace management or have enough shares to sell control of the company to others. In these arrangements, two or more shareholders agree that if the company fails to meet certain goals, either may buy all of the other's stock in the company. By the terms of the agreement, whichever shareholder exercises this option must first offer to sell his stock to the other shareholder. The purchase price per share is usually the same for both shareholders (although sometimes the price is slightly lower for management). In this way, either shareholder can force the other out of the company but only at the risk of being bought out himself.

Take-along/Drag-along Rights

In cases where the investors control a majority of the company on a common stock equivalent basis, it is not uncommon for the purchase agreement to contain "drag-along" rights in favor of the investors. These give the investors the right to require management and other stockholders to sell (to be "dragged along") in the sale of all or substantially all of the company's stock to a third party, providing yet another full exit opportunity to the investors. Note that the drag-along might not be enforced by investors/VCs wishing to force the management to stay and work for the new shareholder. For management to exit, they need a tag-along right (discussed later). This is especially useful in cases of trade sales where the buyer might insist on detaining 100% of the company for the trade sale to happen.

Tag-along Rights/Cosale Rights

Tag-along rights are rights that require one or more parties, usually management, to include another party, usually an outside investor, in the opportunities they receive to sell their stock. They are also referred to as cosale rights. A cosale right is the right to sell its stock to anyone to whom a founder or another investor of the company proposes to sell its stock.

A cosale right is designed to keep the founders tied to the company and requires management members to share any future sale of their stock with the present investors. It usually allows an investor to sell his stock at the same time a founder decides to sell and replace some of the shares the founder has offered for sale. It is reasonable for a founder to resist a cosale right unless a substantial portion of all the founder's stock is sold. A founder can also insist on certain exemptions from a cosale right for personal liquidity reasons such as making a down payment on a house, paying a child's school tuition, sudden death, or termination of employment. Founders may ask for a reciprocal right from the VCs, but such a reciprocal right is rarely granted.

In practice, whenever a management shareholder is approached and has the opportunity to sell his/her shares of company stock, he/she must notify all investors and tell them the terms that have been proposed. Then the outside investors have an opportunity to sell some of their shares to the purchaser. If the outside investors elect to include some of their shares in the sale, the management shareholder will reduce the number of his shares being sold so that the purchaser acquires the number of shares he originally offered to buy. The number of shares the outside investors are entitled to include is usually a predetermined percentage of the total number of shares being sold. Often, that percentage reflects the relative number of shares held by the investors ("*pro rata*") and the selling management shareholder.

Liquidation Provisions

Liquidation provisions regulate the settlement of the proceeds accruing to the party upon a liquidation or bankruptcy.

- *Liquidation Preference:* A liquidation preference entitles the preferred shareholders to receive their original investment back and usually any declared or mandatory but unpaid dividends before the common shareholders receive any proceeds upon liquidation of the company. This provision is a downside protective measure for the VC. In reality, however, rarely is a company liquidated unless its debts to creditors exceed its assets. In this case, there is usually no money for any of the shareholders, preferred or common. Liquidation, however, is often defined to include mergers and reorganizations. In this situation, a merger "liquidation" would give the preferred shareholders the right to receive their original investment back if they elected not to participate in the merger. Again, this provision is rarely exercised since the preferred shareholders often have the voting power granted to them to preclude any merger or reorganization that they do not approve.

  If the company issues multiple rounds of preferred stock, then it needs to decide whether the earlier series of preferred stock get a liquidation preference over the others or whether they are treated equally or *pari passu*. A liquidation preference will usually add accrued and unpaid dividends, though very few startups pay dividends. However, some arrangements will have a "cumulative dividend" that is a mandatory annual dividend designed to build up value of the liquidation preference over time. Other arrangements will mandate that the liquidation preference increase by some rate every year.
- *Cumulative Preferred Dividend:* One common way of making the liquidation rights stronger is by giving the investor cumulative preferred dividends. Even though these are dividends, and strictly speaking do not have to be paid out, they will accumulate and be added to the liquidation claim.
- *Minimum Guaranteed Annual Return Clause*: On some occasions, usually when a trade sale is expected, the VCs will negotiate a minimum guaranteed return clause on their investment. This clause allows the investors to have utmost equity seniority at the time of exit. In the event of a trade sale, this clause is extremely powerful for the investor as it ensures him a minimum guaranteed annual return.

4.6 Conclusion

As VCs tend to try to use the same contracts over and over again out of legal-cost optimization and time-efficiency concerns, contracts tend to be relatively similar across the industry and are often referred to as "boilerplate" or "plain vanilla." VCs will add new clauses to their standard contracts as they encounter them in their coinvestments. This implies a certain peer-to-peer learning process. Finally, clauses are adopted more firmly as retrospective analysis shows their effectiveness.

As evidenced in this section, the higher the probability of opportunistic behavior, the more probability that additional clauses are included in the contract. Further, as the successes materialize and the valuations increase, companies get financed by increasingly sophisticated investors who invest larger amounts of money and need to spend the time and legal costs necessary to maximize their return while minimizing their risks.

A number of studies have been done on the potential effect of combining clauses in contracts and their potential effect in a venture. However, much remains to be analyzed on this topic. Furthermore, the use of multiple layers of contractual clauses in venture capital rounds of financing must also be examined as a crucial tool to maximize return for investors and minimize risks.

5 Conclusions

Most contractual clauses are included in order to avoid principal-agency problems by (1) enhancing goal congruence between entrepreneurs and investors, (2) reducing information asymmetry between entrepreneurs and investors, or (3) reducing the opportunity for opportunistic behaviors. The main purpose of the written contract is thus to reduce possible agency problems at the outset, rather than to build trust between both parties. It is striking that most clauses protect investors against adverse actions of entrepreneurs, rather than vice versa. Although the former are possibly more important than the latter, investors may also take detrimental actions toward the venture or the entrepreneur, e. g., by not giving further financing when needed or by not giving as much assistance as needed or expected by the venture. Clauses regulating these possible problems are seldom included in the contracts at this point.

Further studies could clearly concentrate more on the financial affect and mathematical modelization of venture capital clauses. A step in that direction is the real options modeling of venture capital contractual features in Cossin *et al.*, which includes a first analysis of the interactions between clauses and their valuation effects [102]. These results would have to be compared with empirical evidence and progressively built into the industry common practices.

Another interesting angle of analysis is that provided by Kaplan, Martel, and Strömberg [103]. The authors postulate that the long-standing performance differential between US and other international VCs (an astounding 10% per annum on average for early-stage investments over the last 20 years, with US VCs performing much better than their European counterparts) could be linked to weaker VC investment contracts. The authors find that:

- VCs investing in non-US deals have significantly weaker liquidation and exit rights, which lowers their ability to recover the investment in companies that perform below expectations. In addition, due to a much lower use of vesting provisions and antidilution rights, entrepreneurs on average receive less high-powered cash-flow incentives compared to their US counterparts.
- These differences seem to be somewhat related to the institutional environment. In particular, VC investments in common-law countries (Anglo-Saxon-inspired legal systems) exhibit contractual features less different from US VC contracts.

- Although contractual differences correlate with the institutional environment, they cannot fully explain these differences, since there are VC funds that are able to implement US contractual features in any institutional environment. More sophisticated VCs (larger, investing internationally, and/or having coinvested with US venture capital funds in the past) are more likely to write contracts similar to those of US VC funds. Moreover, VC funds are more likely to write US-style contracts when the amount of funds committed in the deal is large.

Taken together, these results suggest that the limited accumulated experience in venture capital markets outside of the USA, together with fixed costs of learning, are important for explaining the differences between international and US contracts.

Exhibit 1
Taxonomy of Contractual Clauses Part I

Control / Monitoring Rights	Behavior Based Clause	Outcome Based Clause	Protection for	Reduce Information Asymmetry	Reduce Goal Incongruence	Improve VCs Liquidity	Main Objective	Adverse Selection Problem (1)	Moral Hazard Problem (2)	Hold-up Problem (3)	Shirking Problem / Not Exerting Enough Effort (4)	Shirking Problem / Private vs Corporate Interest (5)
Due Diligence Provisions												
Nondisclosure Agreement (NDA)			Entrep.	X			Allow protection of info.	X	X			
Guarantee Complete Info			VC	X				X				
Letter of Intent (LOI)												
Ordinary Course												
DD Expense Paid by Investee Co.			VC				Reduce VC costs					
Exclusivity Right for VC During Due Diligence			VC				Avoid entrep. shopping around for best offer			X		
Preinvesting, Entry Provisions												
Company Representations (Reps) and Warranties	X		VC	X			Protect investor		X			
Investor Representations (Reps) and Warranties	X		Entrep.	X			Protect entrep.		X			
Affirmative Covenants	X		VC	X	X		Protect investor		X			
Negative Covenants			VC		X		Protect investor		X			
Right to Rescind Subscription Agre.			VC		X		Protect investor	X	X			X
Personal Guarantee of Entrep.		X	VC		X		Protect investor	X	X	X	X	
Confidentiality Clause	X		Entrep./VC				Protects confidentiality		X			
Patent Rights		X	Entrep.				Protect inventors		X			
Key-Man Insurance		X	VC				Protect VC against death of main employees	X				
Deal Structure												
Staging			both				Reduce risk, improve flexibility		X	X		
Syndication			both				Reduce risk, improve networking		X		X	
Choice of Security (Debt/Equity Mix)	X	X	both	X	X	X	Reduce agency cost	X	X	X	X	X
Postinvestment Control Provisions												
Statement of Goal and Vision	X		both		X		Improve VC control		X			
Use of Proceed Definition	X		VC	X	X		Improve VC control		X			
Board Rights / Control Rights	X		VC	X			Improve VC control		X		X	X
Board Voting Rights	X		both				Improves control		X			
Director Indemnity		X	Directors		X		Improves protection					
Voting Rights	X		VC				Improve VC control		X	X		
Veto Rights	X		VC				Improve VC control		X	X		
Voting Trust Right	X		VC				Improve VC control		X	X		
Class-Voting Requirements	X		VC	X			Improve VC control		X			
Information Rights	X		VC	X			Ensures proper reporting		X			
Inspection Right	X		VC	X			Reduce information assymetry		X			X

Exhibit 1 (continued)

Control / Monitoring Rights (continued)	Clause Type		Clause Objectives					Type of Conflict Mitigation				
	Behavior Based Clause	Outcome Based Clause	Protection for	Reduce Information Asymmetry	Reduce Goal Incongruence	Improve VCs Liquidity	Main Objective	1 Adverse Selection Problem	2 Moral Hazard Problem	3 Hold-up Problem	4 Shirking Problem / Not Exerting Enough Effort	5 Shirking Problem / Private vs Corporate Interest
Notice of Adverse Change Requirement	X		VC	X			Reduce information assymetry		X			
Financial Reporting Requirements	X		VC	X			Ensure proper reporting		X		X	X
Auditors	X		VC	X			Ensure proper reporting		X		X	X
Remuneration of Management	X		Entrep.		X		Motivate management					
Minimum Time Assistance by VC	X		Entrep.				Protect entrep from incompetent VC				X	
Maximum Debt/Equity Ratio	X		VC		X		Protect VC from increased risk / debt		X			
Shares in Safe Keeping		X	VC				Protect VC from entrep. tampering with certificates		X			
Confidentiality Clause	X		Both		X		Protect company against all its shareholders		X			
Press Announcement Restrictions	X		Both		X		Protect company from unapproved press coverage		X			
Future Shareholder Agreement	X		Both		X		Protect all shareholders from 1 not signing agreement		X	X		
Exclusivity, Noncompete Clause	X		VC		X				X	X		
Contract Closing Clauses												
Finder's Fee		X	Both		X		Protect shareholders		X			
No Revocation	X		Both		X		Protect shareholders		X			
Amendments	X		Both		X		Protect shareholders		X			
Remedies / Indemnification	X	X	Both		X		Protect shareholders		X			
Notices	X		Both		X		Protect shareholders		X			
Notification of Change	X		Both		X		Protect shareholders		X			
Binding Effect	X		Both		X		Protect shareholders		X			
Survival / Binding on Successor	X	X	Both		X		Protect shareholders		X			
Rights Assignement / Assignability	X	X	Both		X		Protect shareholders		X			
Separability of Agreements, Severability of	X	X	Both		X		Protect shareholders, if one		X			
Jurisdiction / Arbitration Clause	X	X	Both		X		Protect shareholders		X			
Governing Law	X	X	Both		X		Protect shareholders		X			
Duration or Termination Clause	X		Both		X		Protect shareholders		X			

Exhibit 1 (continued)
Taxonomy of Contractual Clauses Part II

Cash-Flow / Property Rights	Clause Type		Protection for	Clause Objectives				Type of Conflict Mitigation				
	Behavior Based Clause	Outcome Based Clause		Reduce Information Asymmetry	Reduce Goal Incongruence	Improve VCs Liquidity	Main Objective	1 Adverse Selection Problem	2 Moral Hazard Problem	3 Hold-up Problem	4 Not Exerting Enough Effort	5 Sharking Problem / Private vs Corporate Interest
On going Property Provisions												
Dividends												
Revenue Participations		X	both		X		VCs get agreed revenues, Entrep. Keep equity		X			X
Yearly Dividend Right		X	both				Assure minimum return					
Accrued or Cummulative Dividend Right		X	VC				Assure minimum return					
Sate Contingent Dividend		X	VC			X	Assure minimum return					
No Div. Without 75% Approval		X	VC		X		Ensures alignment of Objectives		X			
Management Incentives												
ESOP (Employee Stock Option Plan)	X		both	X	X		Motivates employees				X	
ISOPs (Incentive Stock Options) and NSOs (Non Qualified Incentive Stock Options)		X	both	X	X		Motivates employees				X	
Mgmt. Vesting Provisions	X		VC	X			Ensures Entry stays committed	X	X	X	X	X
Performance Options		X	both				Ensures mgmt commitment	X	X	X	X	X
State-Contingency Clauses		X	both		X		Motivates Entrep., Protects VCs	X	X		X	X
Take Away Provisions	X	X	VC		X		Ensures mgmt commitment	X	X	X	X	X
Earn-out Rights		X	both		X		Incentivise Mgmt / Protect VC	X				
Options												
Put Option/Warrant		X	VC		X	X	Ensures mgmt commitment		X	X	X	X
Call Option/Warrant		X	both				Allows the beneficiary to increase his share of the co.		X			
Next Financing Round Property Provisions												
Follow-on Investment by VC		X	Entrep.				Forces VC to participate in next round	X				
Discount Right on Follow-on Round		X	VC				Allows VC discount on next round (highly unusual)				X	
Milestones Plan for Follow-on Tranches		X	VC		X		Protects VCs against paying too high valuations		X		X	
Contingency Clauses		X	both		X		Protect against deviations from plans	X	X		X	X
Anti-Dilution												
Preferred Subscription Right / Preemptive Rights / Investment Rights		X	VC		X		Protect VC against dilution					
"Pay to play" Provision		X	Co.				Get current investors to reinvest					
"Price Protection" Provision		X	VC				Protect VC against down-rounds and dilution					
Anti-dilution Full Ratchets		X	VC				Protect VC against dilution					
Weighted Average Anti-dilution Ratchet		X	VC				Protect VC against dilution					
Employee Reserve		X	Entrep.		X		Incentivises mgmt.					
Anti-Redemption Right		X	VC		X		Protect VC against dilution, sale of assets, etc.		X			X

Exhibit 1 (continued)

Cash-Flow / Property (continued)	Clause Type			Clause				Type of Conflict Mitigation				
	Behavior Based Clause	Outcome Based Clause	Protection for	Reduce Information Asymmetry	Reduce Goal Incongruence	Improve VCs Liquidity	Main Objective	1 Adverse Selection Problem	2 Moral Hazard Problem	3 Hold-up Problem	4 Not Exerting Enough Effort	5 Sharking Problem / Private vs Corporate Interest
Exit-Liquidity Agreements and Exit Provisions												
Redemption Rights												
Redemption for Termination of Employment	X		VC		X		Allows shareholders to reclaim shares of mgmt leaving co.	X		X		
Redemption for Breach of Contract	X		both		X		Allows Inv. to reclaim shares		X	X		
Redemption for Death	X		both		X		Allows Inv. to reclaim shares			X		
Redemption for Bankruptcy	X		both		X		Allows Inv. to reclaim shares			X		
No Redemption												
"Put" Redemption Rights		X	VC			X	Provides an exit for the VCs from non performing investments.	X		X		
"Call" Redemption Rights		X	Co.		X		Incentivises the Entrep. To reach milestones			X		
Discretionary Stock Redemption Provisions	X		Entrep.		X	X	Protects Entrp from non supporting VCs	X		X		
Unlocking Provisions:												
Deadlock / Buy-Sell Agreements		X	both		X	X	Eliminates possible dead-lock situations			X		
Registration Rights												
Demand Registration Rights	X	X	VC			X	Assists VC in exiting					
Piggyback / Come-Along Rights		X	VC			X	Assists VC in exiting					
S-3 Right	X	X	VC			X	Assists VC in exiting					
Automatic Conversion Right		X	VC			X	Allows shareholders to convert shares into common stock at IPO or trade sale					
Mandatory Conversion Right		X	Co.				Force conversion into common stock at IPO time			X		
Share Transfer Provisions												
Intra-Company Transactions	X	X	both			X	Allows shareholders to transfer assets between comp./funds					
Right of First Refusal		X	both				Protect VC against new investors coming in					
No Sale Management		X	VC		X		Proctects VC against mgmt selling out					
Take-Along Clause/ Drag-Along Rights		X	both		X	X	Force the sale of all shares in trade sale			X		
Tag-Along Rights / Co-Sale Rights		X	VC		X	X	Assists VC in exiting	X	X			
Liquidation Provisions												
Right to Force Trade Sale or IPO		X	VC			X	Force a trade-sale or IPO			X		
Liquidation Preference		X	VC				Ensures first priority in liquidation, merger, sale		X	X		
Cumulative Preferred Dividend	X	X	VC				Ensures minimum return		X			
Minimum Annual Return Clause	X	X	VC				Ensures minimum return	X				

References

1   Brito, Paulo and Antonio S. Mello. "Financial Constraints and Firm Post-Entry Performance." *International Journal of Industrial Organization*, 1995, *13* (4), pp. 543–565; Petersen, Mitchell and Raghuram Rajan. "The Benefits of Lending Relationships: Evidence from Small Business Data." *Journal of Finance*, 1994, *49*, pp. 3–37; Manigart, Sophie. "Start-up Characteristics and Growth." Faculty of Economic & Applied Sciences, De Vlerick School for Management, University of Ghent, Working Paper, 2000.

2   Tyebjee, Tyzoon T. and Albert V. Bruno. "A Model of Venture Capitalist Investment Activity." *Management Science*, 1984, *30* (9), pp. 1051–1066.

3   Sapienza, Harry J. and J.A. Timmons. "The Roles of Venture Capitalists in New Ventures: What Determines Their Importance?" Academy of Management Best Papers Proceedings, 1989, pp. 74–78; Fiet, James O. "The Informational Basis of Entrepreneurial Discovery." *Small Business Economics*, 1996, *8*, pp. 419–430; Fredriksen, Øystein. "Venture Capital Firms Relationship and Cooperation with Entrepreneurial Companies." Thesis No 625, Department of Management and Economics, Linköping University, Sweden, 1997.

4   Steiner, L. and R. Greenwood. "Venture Capitalist Relationships in the Deal Structuring and Post-Investment Stages of New Firm Creation." *Journal of Management Studies*, 1995, *32* (3), pp. 337–357.

5   Hudec, Albert J. "Negotiating and Drafting Shareholders Agreements in Venture Capital Transactions." Davis and Company, Vancouver, BC, May 2000.

6   Schilit, Keith W. "A Comparative Analysis of the Performance of Venture Capital Funds, Stocks and Bonds, and Other Investment Opportunities." *International Review of Strategic Management*, 1993, *4*, pp. 301–320.

7   Manigart, Sophie. "The Survival of Venture Capital Backed Companies." De Vlerick School for Management, University of Ghent, Working Paper, September 2000.

8   Wright, Mike and Ken Robbie. "Venture Capital and Private Equity: A Review and Synthesis." *Journal of Business Finance and Accounting*, 1998, *25* (5/6), pp. 521–570.

9   Jensen, Michael C. and William H. Meckling. "Theory of the Firm: Managerial Behavior, Agency Costs and Ownership Structure." *Journal of Financial Economics*, 1976, *3*, pp. 305–360.

10   Orts, Eric W. "Shirking and Sharking: A Legal Theory of the Firm." *Yale Law and Policy Review*, 1998, *16* (2) 265–329.

11   Admati, Anat R. and Paul Pfleiderer. "Robust Financial Contracting and the Role of Venture Capitalists." *Journal of Finance*, June 1994, *49* (2), pp. 371–402; Barry, Christopher B. "New Directions in Research on Venture Capital Finance." *Financial Management*, 1994, *23* (3), pp. 3–15.

12   Smith, Gordon D. "Venture Capital Contracting in the Information Age." Reprinted from *The Journal of Small and Emerging Business Law*, 1998, *133*.

13   Gompers, Paul A. "Ownership and Control in Entrepreneurial Firm: An Examination of Convertible Securities in Venture Capital Investments." HBS Working Paper, September 1999.

14   Utset, Manuel A. "Innovation and Governance, High-Powered Incentives, Opportunism, and Venture Capital Contracts." Unpublished Manuscript, February 24, 1997; Cable, Daniel M. and Scott Shane. "A Prisoner's Dilemma Approach to Entrepreneur–Venture Capitalist Relationship." *Academy of Management Review*, January 1997, *22* (1), pp. 142–176; Gifford, Sharon. "Limited Attention and the Role of the Venture Capitalist." *Journal of Business Venturing*, 1997, *12*, pp. 459–482; Sapienza, Harry J. "When Do Venture Capitals Add Value?" *Journal of Business Venturing*, 1992, *7*, pp. 9–27.

15 Robbie, Ken, M. Wright, and B. Chiplin. "The Monitoring of Venture Capital Firms." *Entrepreneurship Theory and Practice*, 1997, *21* (4), pp. 9–28.

16 Sapienza, Harry J. "When Do Venture Capitals Add Value?" *Journal of Business Venturing*, 1992, *7*, pp. 9–27.

17 Gifford, Sharon. "Limited Attention and the Role of the Venture Capitalist." *Journal of Business Venturing*, 1997, *12*, pp. 459–482.

18 Manigart, S. and H. Sapienza. "Venture Capital and Growth." *The Blackwell Handbook of Entrepreneurship*. Eds. D. Sexton and H. Landstrom. Oxford: Blackwell, 2000, pp.240-58.

19 Amit, Raphael, Lawrence Glosten and Eitan Muller. "Entrepreneurial Ability, Venture Investments, and Risk Sharing." *Management Science*, October 1990, *36* (10), pp. 1232–1245; Chan, Yuk-Shee, Daniel Siegel and Anjan V. Thakor. "Learning, Corporate Control and Performance Requirements in Venture Capital Contracts." *International Economic Review*, 1990, *34* (2), pp. 365–381; Barry, Christopher B. "New Directions in Research on Venture Capital Finance." *Financial Management*, 1994, *23* (3), pp. 3–15.

20 Manigart, Sophie. "The Survival of Venture Capital Backed Companies." De Vlerick School for Management, University of Ghent, Working Paper, September 2000; Fiet, James O. "Risk Avoidance Strategies in Venture Capital Markets." *Journal of Management Studies*, 1995, *32* (4), pp. 351–374.

21 Sapienza, Harry J. and A. K. Gupta. "Impact of Agency Risks and Task Uncertainty on Venture Capitalist-CEO Interaction." *Academy of Management Journal*, 1994, *37* (6), pp. 1618–1632.

22 Admati, Anat R. and Paul Pfleiderer. "Robust Financial Contracting and the Role of Venture Capitalists." *Journal of Finance*, June 1994, *49* (2), pp. 371–402; Amit, Raphael, Lawrence Glosten and Eitan Muller. "Entrepreneurial Ability, Venture Investments, and Risk Sharing." *Management Science*, October 1990, *36* (10), pp. 1232–1245; Sapienza, Harry J. and A. K. Gupta. "Impact of Agency Risks and Task Uncertainty on Venture Capitalist-CEO Interaction." *Academy of Management Journal*, 1994, *37* (6), pp. 1618–1632.

23 Amit, Raphael, Lawrence Glosten and Eitan Muller. "Entrepreneurial Ability, Venture Investments, and Risk Sharing." *Management Science*, October 1990, *36* (10), pp. 1232–1245; Chan, Yuk-Shee, Daniel Siegel and Anjan V. Thakor. "Learning, Corporate Control and Performance Requirements in Venture Capital Contracts." *International Economic Review*, 1990, *34* (2), pp. 365–381; Barry, Christopher B. "New Directions in Research on Venture Capital Finance." *Financial Management*, 1994, *23* (3), pp. 3–15.

24 Williamson, Oliver E. *The Economic Institutions of Capitalism: Firms, Markets, Relational Contracting*. New York: Free Press, 1985.

25 Milgrom, Paul R. and John Roberts. *Economics, Organization, and Management*. Englewood Cliffs, NJ: Prentice-Hall, 1992.

26 Sahlman, William A. "The Structure and Governance of Venture-Capital Organizations." *Journal of Financial Economics*, October 1990, 27, pp. 473–521; Amit, Raphael, Lawrence Glosten and Eitan Muller. "Entrepreneurial Ability, Venture Investments, and Risk Sharing." *Management Science*, October 1990, *36* (10), pp. 1232–1245.

27 Holmström, Bengt. "Moral Hazard and Observability." *Bell Journal of Economics*, 1979, *10* (1), pp. 74–91.

28 Masten, Scott E. *Case Studies in Contracting and Organization*. New York: Oxford University Press, 1996.

29 Chan, Yuk-Shee, Daniel Siegel and Anjan V. Thakor. "Learning, Corporate Control and Performance Requirements in Venture Capital Contracts." *International Economic Review*, 1990, *34* (2), pp. 365–381.

30 Barney, Jay B., Lowell W. Busenitz, James O. Fiet, and Douglas D. Moesel. "Determinants of a New Venture Team's Receptivity to Advice from Venture Capitalists." Paper presented at the Babson Entrepreneurship Research Conference, Boston, MA, June 1994.

31 Goldberg, Victor P. "Regulation and Administered Contracts." *Bell Journal of Economics*, 1976, *7*, pp. 426–448.

32 Neher, Darwin V. "Staged Financing: An Agency Perspective." *Review of Economic Studies*, 1999, *66*, pp. 255–274.

33 Williamson, Oliver E. *The Economic Institutions of Capitalism: Firms, Markets, Relational Contracting.* New York: Free Press, 1985; Klein, Benjamin, Robert G. Crawford, and Armen A. Alchian. "Vertical Integration, Appropriable Rents, and the Competitive Contracting Process." *Journal of Law and Economics*, 1978, *21*, pp. 297–396; Hart, Oliver D. and J. Moore. "Property Rights and the Nature of the Firm." *Journal of Political Economy*, 1990, *98* (6), p. 1119; North, Douglass C. and Barry R. Weingast. "Constitutions and Commitment: The Evolution of Institutions Governing Public Choice in Seventeenth-Century England." *Journal of Economic History*, 1989, *49* (4) pp. 803–832; Shleifer, Andrei and Robert W. Vishny. "A Survey of Corporate Governance." *Journal of Finance*, 1997, *52* (2), pp. 737–783.

34 Holmström, Bengt. "Moral Hazard in Teams." *Bell Journal of Economics*, 1982, *13*, p. 324.

35 Orts, Eric W. "Shirking and Sharking: A Legal Theory of the Firm." *Yale Law and Policy Review*, 1998, *16* (2) 265–329.

36 Alchian, Armen A. and Harold Demsetz. "Production, Information Costs, and Economic Organization." *American Economic Review*, 1972, *62*, pp. 777–795.

37 Smith, Gordon D. "Venture Capital Contracting in the Information Age." Reprinted from *The Journal of Small and Emerging Business Law*, 1998, *133*.

38 Orts, Eric W. "Shirking and Sharking: A Legal Theory of the Firm." *Yale Law and Policy Review*, 1998, *16* (2) 265–329.

39 Sahlman, William A. "The Structure and Governance of Venture-Capital Organizations." *Journal of Financial Economics*, October 1990, *27*, pp. 473–521; Kaplan, Steven N. and Per Strömberg. "Financial Contracting Theory Meets the Real World: An Empirical Analysis of Venture Capital Contracts." NBER Working Paper No. W7660, March 2000.

40 Sahlman, William A. "Aspects of Financial Contracting in Venture Capital." *Journal of Applied Corporate Finance*, 1988, *1*, pp. 23–36.

41 Schmidt, Klaus M. "Convertible Securities and Venture Capital Finance." Working Paper No. 217, December 1999.

42 Halloran, Michael J. "Making Portfolio Company Investments." *Venture Capital and Public Offering Negotiation* 6-1 (Lee F. Benton *et al.* second eds.), 1998.

43 Dent, George W. Jr. "Venture Capital and the Future of Corporate Finance." *Washington University Law Quarterly*, 1992, *70*, pp. 1029–1036.

44 Sahlman, William A. "The Structure and Governance of Venture-Capital Organizations." *Journal of Financial Economics*, October 1990, *27*, pp. 473–521.

45 Smith, Gordon D. "Venture Capital Contracting in the Information Age." Reprinted from *The Journal of Small and Emerging Business Law*, 1998, *133*.

46 Smith, Gordon D. "Venture Capital Contracting in the Information Age." Reprinted from *The Journal of Small and Emerging Business Law*, 1998, *133*.

47 Black, Bernard S. and Ronald J. Gilson. "Venture Capital and the Structure of Capital Markets: Banks versus Stock Markets." *Journal of Financial Economics*, 1998, *47*, pp. 243–277.

48 Barry, Christopher B., Chris Muscarella, John W. Peavy, and Michael Vetsuypens. "The Role of Venture Capital in the Creation of Public Companies: Evidence from the Going-Public Process." *Journal of Financial Economics*, 1990, *27* (2), pp. 447–472; Megginson, William L. and Kathleen A. Weiss. "Venture Capitalist Certification in Initial Public Offerings." *Journal of Finance*, 1991, *46* (3), pp. 879–903.

49 Gompers, Paul A. "Grandstanding in the Venture Capital Industry." *Journal of Financial Economics*, 1996, *43*, pp. 133–156.

50 Lazear, Edward P. "Raids and Offer Matching." *Research in Labor Economics*, Vol. 8, part A. Ed. Ronald G. Ehrenberg. Greenwich, CT: JAI, 1986, pp. 141–165.

51 Leland, Hayne E. and David H. Pyle. "Informational Asymmetries, Financial Structure and Financial Intermediation." *Journal of Finance*, 1977, *32*, pp. 371–387.

52 Ravid, Abraham S. and Oded H. Sarig. "Financial Signaling by Committing to Cash Flows." *Journal of Financial and Quantitative Analysis*, 1991, *26*, pp. 165–180.

53 Perez, Robert C. *Inside Venture Capital: Past, Present, and Future*. New York: Praeger, 1986.

54 Lakonishok, Josef, Andrei Shleifer, Richard Thaler and Robert Vishny. "Window Dressing by Pension Fund Managers." *American Economic Review*, 1991, *81* (2), pp. 227–231.

55 Sahlman, William A. "The Structure and Governance of Venture-Capital Organizations." *Journal of Financial Economics*, October 1990, *27*, pp. 473–521.

56 Cossin, Didier, Benoît Leleux and Entela Saliasi. "Understanding the Economic Value of Legal Covenants in Investment Contracts: A Real-Options Approach to Venture Equity Contracts." IMD working paper, March 2002.

57 Admati, Anat R. and Paul Pfleiderer. "Robust Financial Contracting and the Role of Venture Capitalists." *Journal of Finance*, June 1994, *49* (2), pp. 371–402.

58 Cossin, Didier, Benoît Leleux and Entela Saliasi. "Understanding the Economic Value of Legal Covenants in Investment Contracts: A Real-Options Approach to Venture Equity Contracts." IMD working paper, March 2002.

59 Neher, Darwin V. "Staged Financing: An Agency Perspective." *Review of Economic Studies*, 1999, *66*, pp. 255–274.

60 Sahlman, William A. "Aspects of Financial Contracting in Venture Capital." *Journal of Applied Corporate Finance*, 1988, *1*, pp. 23–36.

61 Sahlman, William A. "The Structure and Governance of Venture-Capital Organizations." *Journal of Financial Economics*, October 1990, *27*, pp. 473–521; Kaplan, Steven N. and Per Strömberg. "Financial Contracting Theory Meets the Real World: An Empirical Analysis of Venture Capital Contracts." NBER Working Paper No. W7660, March 2000.

62 Dixit, Avinash K. and Robert S. Pindyck. *Investment under Uncertainty*. Princeton, NJ: Princeton University Press, 1994.

63 Gompers, Paul A. "Optimal Investment, Monitoring, and the Staging of Venture Capital." *Journal of Finance*, 1995, *50* (5), pp. 1461–1489.

64 Al-Suwailem, Sami. "Venture Capital: A Potential Model of Musharakah." Al Rajhi Banking and Investment Corp, Research Center, Riyadh, 1994.

65 Gompers, Paul A. "Optimal Investment, Monitoring, and the Staging of Venture Capital." *Journal of Finance*, 1995, *50* (5), pp. 1461–1489.

66 Kaplan, Steven N. and Per Strömberg. "Financial Contracting Theory Meets the Real World: An Empirical Analysis of Venture Capital Contracts." NBER Working Paper No. W7660, March 2000.

67 Rubinstein, Frederic A. and Marc H. Morgenstern. "Balancing the Deal, Private Placements – What they can mean to your Client." American Bar Association, 1995.

68 Cornelli, Franscesca and L. Felli. "How to Sell a (Bankrupt) Company." Previously circulated as "Revenue Efficiency and Change of Control: The Case of Bankruptcy," Rodney L. White Center Working Paper No. 18-98, Wharton School, University of Pennsylvania and FMG Discussion Paper No. 300, London School of Economics, 1998.

69 Trester, Jeffrey J. "Venture Capital Contracting Under Asymmetric Information." The Wharton Financial Institutions Center Working Paper, October 1993.

70 Stearns, Linda Brewster. "Capital Market Effects on External Control of Corporations." *Theory and Society*, 1986, B, pp. 47–75; Stearns, Linda Brewster and Mark S. Mizruchi. "Board Composition and Corporate Financing: The Impact of Financial Institution Representation on Borrowing." *Academy of Management Journal*, 1993, *36*, pp. 603–618.

71 Kochhar, Rahul. "Strategic Assets, Capital Structure, and Firm Performance." *Journal of Financial and Strategic Decision*, Fall 1997, *10* (3), pp. 23–36.

72 Admati, Anat R. and Paul Pfleiderer. "Robust Financial Contracting and the Role of Venture Capitalists." *Journal of Finance*, June 1994, *49* (2), pp. 371–402;
    Barry, Christopher B., Chris Muscarella, John W. Peavy, and Michael Vetsuypens. "The Role of Venture Capital in the Creation of Public Companies: Evidence from the Going-Public Process." *Journal of Financial Economics*, 1990, *27* (2), pp. 447–472; Lerner, Josh. "Venture Capitalists and the Oversight of Privately-Held Firms." *Journal of Finance*, 1995, *50*, pp. 301–318.

73 Hall, John and Charles W. Hofer. "Venture Capitalists' Decision Criteria in New Venture Evaluation." *Journal of Business Venturing*, 1993, *8*, pp. 25–42; Sahlman, William A. "The Structure and Governance of Venture-Capital Organizations." *Journal of Financial Economics*, October 1990, *27*, pp. 473–521; Manigart, S. and H. Sapienza. "Venture Capital and Growth." *The Blackwell Handbook of Entrepreneurship*. Eds. D. Sexton and H. Landstrom. Oxford: Blackwell, 2000, pp.240–58.

74 Gorman, Michael, and William A. Sahlman. "What Do Venture Capitalists Do?" *Journal of Business Venturing*, 1989, *4*, pp. 231–238; Sapienza, Harry J. "When Do Venture Capitals Add Value?" *Journal of Business Venturing*, 1992, *7*, pp. 9–27.

75 MacMillan, Ian C., D. M. Kulow and R. Khoylian. "Venture Capitalists' Involvement in Their Investments: Extent and Performance." *Journal of Business Venturing*, 1989, *4*, pp. 27–47; Elango, B., Vance H. Fried, R.D. Hisrich and A. Polonchek. "How Do Venture Capital Firms Differ?" *Journal of Business Venturing*, 1995, *10*, pp. 157–179.

76 Manigart, Sophie and Koen De Waele. "Determinants of Required Return in Venture Capital Investments: A Five Country Study." De Vlerick School for Management, University of Ghent. Working Paper, 2000.

77 Landström, Hans. "Riskkapitalföretagens resurstillförsel till små företag." Lund, Sweden: IMIT, 1990.

78 Kaplan, Steven N. and Per Strömberg. "Financial Contracting Theory Meets the Real World: An Empirical Analysis of Venture Capital Contracts." NBER Working Paper No. W7660, March 2000.

79 Jensen, Michael C. and William H. Meckling. "Theory of the Firm: Managerial Behavior, Agency Costs and Ownership Structure." *Journal of Financial Economics*, 1976, *3*, pp. 305–360.

80 Gorman, Michael and William A. Sahlman. "What Do Venture Capitalists Do?" *Journal of Business Venturing*, 1989, *4*, pp. 231–238.

81 Black, Bernard S. and Ronald J. Gilson. "Venture Capital and the Structure of Capital Markets: Banks versus Stock Markets." *Journal of Financial Economics*, 1998, *47*, pp. 243–277.

82 Hudec, Albert J. "Canadian Venture Capital Opportunities for Japanese Investors and Enterprises." Davis and Company, Vancouver, BC, May 2000.

83 Hudec, Albert J. "Canadian Venture Capital Opportunities for Japanese Investors and Enterprises." Davis and Company, Vancouver, BC, May 2000.

84 Fried, Vance H. and Robert D. Hisrich. "Toward a Model of Venture Capital Decision Making." *Financial Management*, 1994, *23* (3), pp. 28–37;
    Fried, Vance H., Robert D. Hisrich, and Amy Poloncheck. "Venture Capitalists Investment Criteria: A Replication." *Journal of Small Business Finance*, 1993, *3* (1), pp. 37–42.

85 Fiet, James O., Lowell. W. Busenitz, Doug D. Moesel, and Jay B. Barney. "Complementary Theoretical Perspectives on the Dismissal of New Venture Team Members." *Journal of Business Venturing*, 1997, *12*, pp. 347–366.

86 Busenitz, Lowell W., Douglas D. Moesel, and James O. Fiet. "The Impact of Post-funding Involvement by Venture Capitalists on Long-term Performance Outcomes." *Frontiers of Entrepreneurship Research*. Eds. P. Reynolds, W. Bygrave, N. Carter, P. Davidsson, W. Gartner, C. Mason, and P. McDougall. Wellesley, MA: Babson College, 1997, pp. 498–512.

87 Bruton, Garry D., Vance H. Fried, and Robert D. Hisrich. "Venture Capitalists and CEO Dismissal." *Entrepreneurship Theory & Practice*, 1998, *21*, pp. 41–54.

88   Bruton, Garry D., Vance H. Fried, and Robert D. Hisrich. "Venture Capitalists and CEO Dismissal." *Entrepreneurship Theory & Practice*, 1998, *21*, pp. 41–54.
89   Sahlman, William A. "The Structure and Governance of Venture-Capital Organizations." *Journal of Financial Economics*, October 1990, *27*, pp. 473–521.
90   Kaplan, Steven N. and Per Strömberg. "Financial Contracting Theory Meets the Real World: An Empirical Analysis of Venture Capital Contracts." NBER Working Paper No. W7660, March 2000.
91   Kaplan, Steven N. and Per Strömberg. "Financial Contracting Theory Meets the Real World: An Empirical Analysis of Venture Capital Contracts." NBER Working Paper No. W7660, March 2000.
92   Coase, Ronald H. "The Nature of the Firm." *Economica*, 1937, *4*, pp. 386–405.
93   Anderlini, Luca and Leonardo Felli. *Costly Coasian Contracts*. DAE working papers, 9704. Cambridge: Department of Applied Economics, University of Cambridge, 1997; Anderlini, Luca and Leonardo Felli. *Costly Contingent Contracts*. London: London School of Economics and Political Science, Suntory and Toyota International Centres for Economics and Related Disciplines, Theoretical Economics Workshop, 1996.
94   Kaplan, Steven N. and Per Strömberg. "Financial Contracting Theory Meets the Real World: An Empirical Analysis of Venture Capital Contracts." NBER Working Paper No. W7660, March 2000.
     Kaplan, Steven N. and Per Strömberg. "Financial Contracting Theory Meets the Real World: An Empirical Analysis of Venture Capital Contracts." NBER Working Paper No. W7660, March 2000.
96   Cossin, Didier, Benoît Leleux, and Entela Saliasi. "Understanding the Economic Value of Legal Covenants in Investment Contracts: A Real-Options Approach to Venture Equity Contracts." IMD working paper, March 2002.
97   Halloran, Michael J. "Making Portfolio Company Investments." *Venture Capital and Public Offering Negotiation* 6-1 (Lee F. Benton *et al.* second eds.), 1998.
98   Stroud, Helen and Koen Vanhaerents. "Venture Capital Exit Routes." Chicago: Baker & McKenzie, 2000.
99   Cole, Jonathan E. and Albert Sokol. "Structuring Venture Capital Investments." Pratt's Guide to Venture Capital Sources, Edwards & Angell, LLP, 1999.
100 Suchman, Mark C. "On Advice of Counsel: Law Firms and Venture Capital Funds as Information Intermediaries in the Structuration of Silicon Valley." Unpublished Ph.D. dissertation, Palo Alto: Stanford University, 1994.
101 Dent, George W. Jr. "Venture Capital and the Future of Corporate Finance." *Washington University Law Quarterly*, 1992, *70*, pp. 1029–1036.
102 Cossin, Didier, Benoît Leleux, and Entela Saliasi. "Understanding the Economic Value of Legal Covenants in Investment Contracts: A Real-Options Approach to Venture Equity Contracts." IMD working paper, March 2002.
103 Kaplan, Steven N., Frédéric Martel, and Per Strömberg. "Venture Capital Contracts around the World." University of Chicago Working Paper, April 2002.

CASE 3-1
AVIQ Systems AG: Creative Technology, Innovative Financing

Benoît Leleux and Inna Francis

Like many startup companies, AVIQ was the product of the visionary zeal of
its founder, Rudy Kiseljak, an electronics engineer based in German-speaking
Switzerland. In 2002 multimedia caught Rudy's imagination. As PCs and DVD
players became commonplace, Rudy believed that a combined DVD, CD player,
video recorder, photo viewer, and TV tuner that also offered instant messaging
would revolutionize in-house entertainment. It would allow users to interact easily with the new combined media – sound, video, Internet, and text.

He financed the development of the proof-of-concept himself but quickly
reached the end of the line for bootstrapping: The company really needed to raise
€ 1.6 million to finance the production of the AVIQ A1 entertainment media center
(EMC), with its user-friendly software. Rudy took to the road and made sales
pitches to banks, institutional investors, VCs, and angel investors to raise the
funds needed to bridge the gap to digital-convergence nirvana. Commercial
and investment bankers did not offer much hope, though; nor did private equity
houses. Even VC firms showed little interest. A direct public offering of shares
was difficult to engineer. But Rudy would not let go and came up with the idea of
a PPO, a public product offering: If he could convince just 1,000 customers to
commit about CHF 1,000 each and wait between two and four months for the
media box to be delivered, AVIQ's short-term capital needs would be covered and
the company would have the capital it so desperately needed to start the manufacturing process. But would this untested funding mechanism work?

Copyright © 2003 by IMD – International Institute for Management Development, Lausanne,
Switzerland. Not to be used or reproduced without written permission directly from IMD.

1 AVIQ Systems

As digital TV gained popularity, and people wanted to watch in more comfort in their living rooms, Rudy had the vision of offering the ultimate convergence between the Internet and the audio/video worlds by connecting hi-fi speakers, the Internet, and a PC to a TV set. Ordinary TV sets, with their wide screen, central location, and high-quality sound systems, would become the device of choice to display all sorts of digital content. AVIQ's mission was born:

> Forget the medium – enjoy the content!

Excited by the idea, Rudy and his staff worked feverishly to flesh out the prototype of AVIQ's A1 media center. It would integrate all home media entertainment into one simple-to-use media box supported by Media Manager software and Web-based services that would allow online access to an electronic program guide (EPG) and remote access to other Internet services.

> You can pause live TV and pick up where you left off, rewind live TV without missing anything, record one or two programs while watching a third, bypass commercials easily, and once the machine knows that you like to watch, say, Friends and your kids cannot miss the Tweenies, it will check the program schedules of all the channels you receive and automatically record every episode on any channel. Most importantly, the "record" function is simple to use. AVIQ A1 offers complete control over when you watch the programs you want.

The company motto was soon born:

> I watch when I want.

Apart from allowing you to watch your recorded shows, it would let you record, organize, and create play lists of all your music, as well as organize your digital photo slideshows and your videos for convenient access and play out. Rudy hoped that the Media Manager software would become a new standard for home media interaction (*see Exhibit 1 for AVIQ's key features*).

Rudy tried to convince the owners of local retail chains to provide space for him to demonstrate the wonders of AVIQ's multimedia system:

> It's too complex a concept for people to get their head around until they actually see and touch one. If you sit consumers down with one of these things, after 10 or 15 minutes, they're amazed at what it can do for them.

The company also actively approached the media for active product coverage in the most widely read magazines in the German part of Switzerland. Extensive negotiations were slow to yield results, though.

AVIQ's lead advertising and marketing channel was its website, where Rudy's 16-year-old daughter illustrated the essence of the product in a very creative way (*see Exhibit 2 for some descriptive cartoons from AVIQ's website*).

Rudy also planned to host an event once the product was delivered to the first 100 buyers to gain extensive media coverage that would help market the product.

1.1 Market Size and Developments

AVIQ's market research indicated that there were about 280 million TV households in the European target markets. The total accessible market size for an entertainment media center was estimated at over 7 million users. Rudy hoped to capture 500 customers by the end of 2003 and reach some 400,000 customers by the end of 2007.

These projections were based on the US experience in introducing a product with comparable characteristics: the personal video recorder – PVR. After a couple of years on the market, PVR sales reached 1 million units in 2002 and were predicted to grow to 40 million units by 2008.

The European home entertainment market was heavily segmented by price. The top segment included end users prepared to pay over € 3,000 per unit. The midclass market covered equipment priced at € 1,000 to € 3,000. The low-end segment, with prices below € 1,000 per machine, was by far the largest. AVIQ targeted the bottom segment and intended to price its product at between CHF 1,000 (€ 600) and € 1,000.

AVIQ planned to grow through rapid geographic penetration of new markets, starting in the German part of Switzerland and moving to nationwide and Europewide coverage, with possible subsequent expansion into the USA.

1.2 TV Viewers

AVIQ targeted two primary customer groups: technically savvy, early adopter consumers aged 20 to 35 who were setting up a household; and the 55-to-70 age group, who spent a lot of time watching TV. A secondary target group was consumers aged 30 to 55 – mostly replacement buyers. Ease of use was the overarching goal of AVIQ's service:

The 8-year-old kid or his grandmother must be able to pick up the remote control and set a recording or play back a program with the touch of a button without opening a user guide. Otherwise it's just going to remain a really nice niche piece of technology. We don't want that to happen. We want it to be the fundamental way that most people watch TV because it increases satisfaction to a level never seen before.

The AVIQ A1's value proposition was to be the simplest machine of its kind – a technology marvel that anyone could use in seconds and feel comfortable with. While it would not be considered a technological revolution, it would be a breakthrough in terms of ease of use. (*See Exhibits 3 and 4 for its key characteristics and competitive advantages.*)

1.3 Vendors of High-end Analog and Digital TV Sets

The Consumer Electronics Association expected 4 million digital TVs to be sold in the USA in 2003 and 10.5 million in 2006. Rudy expected that, increasingly, PVRs would be built into TV boxes. AVIQ hoped to capture part of the digital TV market by supplying its Media Manager software to digital TV vendors for a royalty fee.

1.3.1 TV Content Distributors

By gathering information on customer viewing habits, such as channel and content selection and channel-switching data, AVIQ also planned to offer day-to-day marketing input – raw or condensed – to the TV content distributors (cable or satellite operators and TV stations). This would provide an additional, recurrent knowledge-based revenue source for the company.

1.3.2 Competition

No single company had yet dominated the market with a similar product in Europe, but Rudy knew that it was just a matter of time, since the barriers to entry were limited. People were starting to wake up to the power of a hard disk sitting under the TV.

The TiVo PVR, a pioneer in the US market, was similar to the AVIQ A1. TiVo's results had been disappointing in Europe, and the company stopped selling its boxes in Britain in 2003, claiming margins on selling the "box" were below 3% – insufficient to stay in the market. Eventually TiVo managed to turn cash flow positive as the service and licensing revenues reached $13.7 million in the USA, with a 50% margin. TiVo clearly emphasized the services and advertising revenues and was expected to show a profit in the 2003 fiscal year, derived totally from these sources. TiVo's financial forecasts focused on the subscriber-installed base and service revenues, with forecasts of over 1 million users by year-end 2003.

Hughes, Samsung, and Philips offered set-top boxes (STBs) based on licensed TiVo technology, and Hughes was planning a high-definition TV (HDTV) with an embedded TiVo-based PVR. Toshiba was also planning a DVD/TiVo PVR combination product. These partnership arrangements drove licensing revenues for TiVo, as well as services fees, although some of the agreements required TiVo to subsidize manufacturing costs and share services fees, with an unknown impact on the bottom line.

Cable and satellite service companies, such as AOL Time Warner, Cox, and Canal Satellite in France, also saw clear opportunities in bundling PVR functionality into the premium subscription price.

Sony, another TiVo licensee, decided to create its own software and services offering for its popular CoCoon product line, which was expected to start shipping in the USA late in 2003 and in Europe by 2004. SONICblue had also been active in developing PVR technologies, but after selling its ReplayTV business unit, it filed for bankruptcy protection in March 2003. British Sky Broadcasting (BSkyB) was more successful and offered its Sky Plus service in Britain from August 2001. By January 2002 it had 65,000 subscribers. FAST Server's TVS200 became a main player in Germany, rapidly advancing its service offering to other markets. Telly offered similar services but, being a PC-based product, required a keyboard and mouse (*see Exhibit 5 for product comparisons*).

DVD-based products, set-top boxes, and PC-based products such as Microsoft Windows XP Media Center were potential substitutes for the personal recorder technology.

EchoStar, a direct broadcast satellite (DBS) service supplier, was estimated to have added 1.4 million subscribers since 2002. Of those, approximately 500,000 bought a Dishnet PVR receiver/PVR combo with 35 hours of video storage capacity for $349. The company advertised the ability to "zap commercials" by using a function that leaped forward 30 seconds at the touch of a button.

DirecTV offered a Sony satellite receiver/PVR unit with 35 hours of video storage, as well as the ability to watch one channel and record a second simultaneously for $250.

Meanwhile, Charter Communications was deploying a Motorola home theater system that included PVR capability. Panasonic and Pioneer were also developing PVRs. Panasonic had a cable-based PVR capability under development, and system PVR functionality was already available in iMagicTV cable system offerings.

The semiconductor industry was alerted to opportunities for surfing the new wave of digital convergence by adding value to silicon chips. Broadcom created a special purpose STB chip enabling PVR capabilities; TeraLogic offered a single-chip Cougar STB platform; and a variety of similar devices were announced. Portal Play, a company that offered turnkey silicon software, and a reference design for an MP3 jukebox, had recently announced that it was moving into video products.

Rudy knew that he needed to move fast and get the product to the market as soon as possible. He was ready to shift into high gear to accelerate his efforts. After some soul-searching, Rudy and his staff realized that although they had an exciting vision and some highly talented people, converting the dream to reality was going to take a lot of hard work and significant chunks of money. The central question on Rudy's mind was how best to fund AVIQ's way to the market.

1.4 Analyzing the Financing Alternatives

After conventional ways of funding had been explored without yielding the expected results, Rudy and his four partners met at his home near Zurich to consider the financing options left. The partners had already sunk CHF 300,000 in seed funding and now faced the classic financing crunch. They were highly motivated by the prospects the emerging market offered and the possibility of eventually striking it rich by selling to a strategic buyer or offering stock to the public. But in the meantime Rudy knew that he had very little leverage in bargaining with the banks, VC firms, angel investors, or even suppliers. Most of the commercial banks the company approached praised the product but had little interest in a CHF 2 million loan, claiming that the paperwork required would cost them as much as for a CHF 300 million loan and that the fees would not compensate for the bank's effort and risk, especially as there was no collateral to secure the loan.

The hard drive for the standalone media center was expensive and represented the bulk of the production cost. Resources were also required to pay for the continued development and upgrading of the software, to pay the suppliers of key parts in Taiwan, to deliver the media box components to Switzerland for assembly, and to pay the company's marketing and operating expenses. CHF 1 million was set as a short-term prerequisite for starting production.

As a last resort, Rudy studied a practice that some small businesses in the USA used to raise funds: the direct public offering (DPO; *see below*). He also heard about some Swiss entrepreneurs – Easy-Glider, Mystery Park, and Digital Logic – that had experimented with this idea of alternative financing by raising money in a public product offering (PPO) from enthusiastic customers, *i.e.*, early adopters prepared to back the company and wait for the benefits to be delivered later. Would the PPO route be more appropriate for Rudy, or should he resort to the more classic DPO? Whatever the option chosen, how would he structure the deal to benefit the company and its consumers?

1.5 The Easy-Glider's Fundraising Effort: A Swiss PPO Pioneer

The founder of Easy-Glider, Stephan Soder, a former professional Swiss figure skater, and his inventor team based in Oberglatt (northern Switzerland) were convinced that their slick-looking electric scooter could become the individual environmentally friendly transportation device of the future. Two years of hard work and five Easy-Glider generations later, the product was ready to take to market. Finding an investment banker in Switzerland willing to underwrite a stock offering under $100 million was impossible. To make things more difficult, the company had no collateral to speak of to secure a bank loan.

The entrepreneurs decided to go directly to their potential customers. After all, traditional bootstrap financing has always maintained that the best sources of finance are the people or companies with a vested interest in the success of the startup. Two months after making its offer to clients on its website, Easy-Glider had secured over CHF 600,000 in preorders, mostly from investors in California, where the product was most actively marketed. The investors were each asked to deposit CHF 1,000 into Easy-Glider's bank account, to be held in a trust until the minimum investment level of CHF 1 million was reached. Easy-Glider promised to deliver its product within two months of the time it could start production. It would return the money with interest if the minimum capital requirement was not reached. Investors would also be able to participate in a company-sponsored profit-sharing scheme: for every 20,000 Easy-Gliders sold, these early investors would receive CHF 1,000 back. The profit-sharing benefit would extend to the first 200,000 Easy-Gliders sold, after which the scheme would stop.

1.6 Direct Public Offering (DPO) Alternative

Small businesses with revenues under $25 million typically did not generate much interest from commercial or investment banks when it came to raising funds. As a result, DPOs rapidly gained popularity in the 1990s in the USA, where they were used successfully for pure startups as well as for firms that had existed and been privately financed for decades.

Small Corporate Offering Registration (SCOR) offered a simple procedure, approved by the SEC in 1989,[1] to raise relatively small amounts of capital from the public, without the administrative burden of a full-blown public offering and free of the damaging constraints on public advertising. If a firm did not raise the minimum it had set within the offering period, investors would get their money back, with interest.

A DPO could not only give a company the capital it needed but also be used as a clever marketing tool for its products and/or services. For example, a restaurant or brewery could sell its stock directly to its customers. Wit Brewery, then a three-year-old New York microbrewery, was too tiny to interest Wall Street underwriters, but the owners did not want to sell out to VCs either. They launched an initial public offering – all by themselves. The company put up a home page on the Internet to alert the public to the impending sale and printed notices on

---

[1] DPOs generally fall into three regulatory classifications. Regulation D Section 504 of the Securities Act of 1933 has become the most widely known, called the "Small Corporate Offering Registration" or "SCOR." SCOR allowed companies to raise up to $1 million every 12 months by registering with state securities administrations. The Regulation A offering increased the amount to $5 million but also required registration with the Small Business Office of the SEC. Finally, intrastate offerings typically had no ceilings. To qualify for SCOR, a company had to answer 50 questions about itself and its financial history and list principal risks to investors, the minimum and maximum amount of money desired, and the details. Once regulators approved the SCOR, the company had 12 months to raise the money.

six-packs of its Wit ale. The announcements invited customers to call or write for a prospectus and stock order form. The offering drew 3,500 investors, mostly beer connoisseurs. Some 900,000 shares were sold for $1.85 each, raising $1.6 million for the company, without a cent paid in fees to underwriters or brokers. It was clear that when people were involved with the business economically, they could become loyal and helpful supporters.

Since the issuing company handled all the paperwork associated with the issue as well as marketing the shares and cutting out most registration, brokerage, and legal fees, a DPO was a cheap alternative to a traditional IPO underwriting. The upfront costs of a typical IPO could represent up to 13% of the expected funds raised, a transaction cost that put it beyond the reach of many smaller firms. By contrast, DPO costs could be as low as 6% to 7%. Moreover, DPO investors tended to be loyal customers and hence did not demand immediate profits. "For shareholders, it's more like buying into a business than buying a stock," remarked Drew Field, a San Francisco lawyer and DPO specialist.

According to the Swiss Federal Banking Commission (SFBC), DPOs were not subject to prior approval by the Swiss regulatory authorities:

Issuing shares for public subscription, as part of an Initial Public Offering (IPO), is not subject to prior approval from the Swiss authorities, provided the issuer itself sold equity directly to investors (members of the public) in the framework of an "own issue". The same held true for public compliance with the provisions governing the issue prospectus. If newly issued shares were offered for public subscription, the provisions of the Swiss Code of Obligations (OR) concerning the issue prospectus had to be observed. As of 2003, it was not really clear if a prospectus written in accordance with the Code of Obligations could be validly published on the Internet: the SFBC was not responsible for monitoring obligations under Swiss civil law, with the responsibility for ruling on such disputes lying with the civil courts. (SFBC Regulations)

1.7 Public Product Offering Alternative

Rudy came up with the term public product offering (PPO) to cover the type of fundraising Easy-Glider used. The idea was to find a way to incentivize customers to make a prepayment for a product to be delivered later. The arrangement would entitle early customers to certain privileges (similar to those that retail chains or banks sometimes granted to their shareholders, such as discounts on merchandise or interest rate reductions). Rudy decided to offer his early customers free software updates (for a limited period) and a claim on the company's profits in the form of dividends or capital appreciation. In a sense, it combined features of the DPO with a product component. In this way, AVIQ's customers would share the risk and benefits of taking the product to market.

1.8 AVIQ's Investment Proposition

After mulling over the alternatives, Rudy and his team decided to give the PPO a shot. After a few long nights they came up with a powerful value proposition for their future customer-investors. In essence, they would offer the public the possibility to preacquire an AVIQ A1 machine with extra bonuses for the risk taken in backing a preproduction company. The firm's offering included the following elements:

Members of the public could transfer a fixed amount of CHF 1,000 to the company's bank account. By 31 January 2004, the "investors" would receive their AVIQ A1 machine. In addition, the first 100 investors (on a first-come, first-served basis, by date/time of investment) would receive:

- A free lifetime standard subscription to the Electronic Programming Guide
- Free software updates until the end of 2004
- The right to be heard: all feedback would be gathered directly and used to improve the A1's features.

Subsequent investors would receive, in addition to the A1 media center, free software updates until the end of 2004 and a delivery time between April and May 2004.

The company set the goal of showing first net profits within 15 to 18 months of the start of production and a minimum of 25% operating profit and 15% net profit after three years. Tough cost controls and subsistence salaries for the management team were also imposed. AVIQ's prospectus highlighted the analyses and forecasts assuming first product delivery in Q1 of 2004 and first profit in Q1 2005. By December 2006, the company planned to have 75 employees and €90 million in revenues.

The revenues from the PPO would be supplemented by regular product sales starting in March 2004 and by the sale of premium service subscriptions, accessories, and product support (*Exhibit 6*). To be able to finance the subsequent growth to break even, the company expected it would need to raise an additional €1.6 million of capital, in what it envisioned as a three-stage process.

Exhibit 1
Product and Services Concept

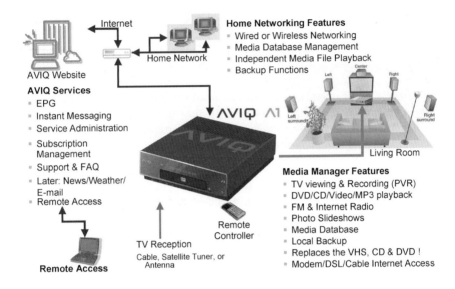

AVIQ Website

AVIQ Services
- EPG
- Instant Messaging
- Service Administration
- Subscription Management
- Support & FAQ
- Later: News/Weather/ E-mail
- Remote Access

Remote Access

Home Networking Features
- Wired or Wireless Networking
- Media Database Management
- Independent Media File Playback
- Backup Functions

Living Room

Media Manager Features
- TV viewing & Recording (PVR)
- DVD/CD/Video/MP3 playback
- FM & Internet Radio
- Photo Slideshows
- Media Database
- Local Backup
- Replaces the VHS, CD & DVD !
- Modem/DSL/Cable Internet Access

Remote Controller

TV Reception
Cable, Satellite Tuner, or Antenna

Watch TV

- A1 helps me decide what to watch when I want
- Manage interruptions (phone, kids,..)
- Pause Live TV
- Never miss my favorite show

Record TV

- A1 always knows what I like
- No need for TV magazines
- No need to worry what time or tape
- Simple one-touch programming, even remotely

Listen to Music

- Get rid of all CD clutter
- Enjoy hours of selected music
- Make my own party playlists

Watch Videos

- Watch DVDs, digital films, own videos
- Digitalize videos from VHS cassettes
- Watch while I record – time-shift

Listen to Radio

- Get international with Internet radio
- Categorize stations (jazz, rock,...)

View Photos

- Entertain family & friends
- All my photos at a touch
- Full digital quality

Send & Receive Messages

- Send messages to my home
- Receive messages from AVIQ

Internet & Home Network

- Use A1 as a media server
- Play music/videos on any PC at home
- Use PC for backup and editing

Source: Company website

Exhibit 2
AVIQ A1: Before and After

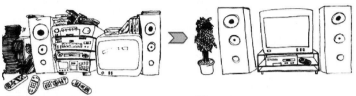

Before:

* Dozens of VHS tapes and CDs take up lots of valuable space

* Tapes need to be erased, labeled and ordered

* Three or four A/V devices (VHS recorder, CD, DVD player) take up even more space

After:

* Space savings by getting rid of many tapes and CDs

* Even more savings by removing VHS recorder, CD and DVD player

* Only one remote needed for all

* Cabling becomes simple and hassle-free

✓ A neat, compact, and great-looking audio/video system

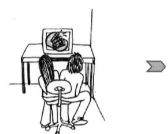

Before:

Uncomfortable to watch a movie or even listen to music in study

Small screen and plastic speakers...

After:

* Enjoy a movie while sitting on sofa

* Listen to favorite music on big speakers

* Use the PC to download and manage your material and stream it to the living room

✓ Enjoy more movies and music from your sofa!

Before:

* Difficult to see the small photos

* Only one or two people can follow your explanations

* Some of the group are not interested!

After:

* Show favorite vacation images or your new car

* Involve everyone in the group

* Run automatic or manual slideshows

* Play background music to the pictures

✓ Friends or family are much easier to interest and attract to view your pictures!

Source: Company website

Exhibit 3
AVIQ A1's Unique Advantages

The A1 always knows what I like.	I can call up "My Favorites" information and enjoy my favorite show at any time.
The A1 helps me decide what to watch.	I get comprehensive EPG information, reviews and pictures, 2 weeks' advance programming, search/favorite functions, rating of broadcasts, automatic recording of series, and "what's on now" information.
Anyone in my household can easily use it.	Ergonomically designed menus, a single remote control for the A1, the TV, and the sound.
My investment is future proof.	A1 can be upgraded online, it is networked and connected to the Internet.
I save space and money.	No more cassettes to buy, only one box to place, much less cabling, no need for a TV magazine.
And if I forget to program the recording?	The A1 is remotely programmable – you can program recording of an important tennis match from your office via the Internet.
Never again miss a favorite TV show!	Improve the quality of your free-TV-program enjoyment by deciding what to watch and when, regardless of the TV schedules. Start watching a show from the beginning, even if it's still recording! Effortlessly use the 14-day Electronic Program Guide with one-touch recording capability and easy-to-find favorite programs. Never miss the crucial movie moment if you are interrupted (phone, door, kids, *etc.*) – pause live TV! Rewind, instant-replay, and slow-motion live or recorded TV and videos; skip boring parts using the fast-forward function. Forget the hassle with videotapes and have your favorite shows recorded digitally.
All your video, music, and photo media in one place so they're ready to enjoy anytime!	Enjoy all your media using only ONE ergonomically shaped remote. Even works for your TV and hi-fi system! Make your own music playlists or listen to your favorite songs from any networked PC in the home, wired or wireless! Impress family and friends by showing the latest photo slideshow with background music. Save space and money by getting rid of VHS tapes and CDs – and the players as well. Replaces over 50 VHS cassettes.
Superb audio and video quality ensures total media enjoyment.	Selection of industry-standard video compression schemes for optimal viewing quality. Great sounding audio with proprietary surround features. Direct audio/video connections possible using active loudspeakers; alternatively uses an A/V amplifier.
Easy 3-step installation at home.	Just unpack, connect several cables, and hit Power on. The unit asks several simple questions about the location and user preferences. The unit downloads all necessary configuration information from the Internet, configures itself automatically, and is ready to go.
Advanced technology for long-lasting usability.	Server-based online software updates and upgrades. Selectable compression optimizes disk usage depending on program type. Industry-standard hardware ensures upgradeability – not a dead-end proprietary design.

Exhibit 4
Technical Characteristics

A1 preliminary technical data (Status September 03 – Subject to change without notice)	
Front connections	Video/S-Video IN
	Audio L/R IN
	USB Port
	FireWire (IEEE1394) Port
Rear connections	Video/S-Video OUT
	Audio L/R/C/Ls/Rs/Sub analog Line OUT
	Audio S/PDIF Digital OUT
	2 USB Port
	FireWire (IEEE1394) Port
	Ethernet Port
	TV IN/THRU (cable or terrestrial)
	FM IN
Recording	Video: MPEG1/MPEG2, 3 quality settings
	Audio: MP3, 4 quality settings
Playback	Video: MPEG1/MPEG2, DivX
	Audio: MP3
	Photo: JPG
Recording capacity	Depending on the selected quality settings, freely combined, up to 200 hours of video (equivalent to 100 standard VHS cassettes) or 1,600 hours of MP3 music (equivalent to 1,600 CDs) or 100,000 photos

Source: Company website

Exhibit 5
Competitors' Features

	AVIQ A1	KiSS DP-508	Telly	FAST TV	TiVo Plus
Price	CHF 990 plus €9.95 or €19.95/month	CHF 948	$899	€1,000–€1,500	From $299 plus $12.95/month
Hard disk recording	Up to 200 hours	80-GB drive	80-GB drive	Up to 200 hours	Up to 80 hours
Record from guide	x		x	x	x
Record by time and channel	x	x	x	x	x
Control live TV	x		x	x	x
Program guide data	x		x	x	x
Broadband connectivity	x	x	x		
Search by title	x	x	x	x	x
Season pass	x		x	x	x
WishList			x		x
Smart recording options	x		x	x	x
Suggestions			x	x	
Digital music and photos	x	x	x	x	x
Multiroom viewing	x	x	x	x	x
Remote scheduling	x		x	x	
Keyboard/mouse			x		

Source: Literature of individual companies

Exhibit 6
AVIQ Financial Projections

Revenue Streams [KEUR]	2003	2004	2005	2006	2007
A1 Media Box Average Street Price (EUR)		1'069	908	772	618
Based on # of units sold	500	10'300	52'400	141'900	272'100
AVIQ A1 FOB Revenue	330	7'329	32'439	76'425	124'566
Standard Services revenue	-	-	2'292	8'379	16'389
Premium Services revenue	-	-	611	3724	7'284
Repair, spares, accessories revenue	-	37	324	764	1'246
Training + Support revenue	-	73	649	1528	2491
Total services	-	-	2'903	12'103	23'672
Total spares, acc, training	-	110	973	2'293	3737
Sales Grand Total	330	7'439	36'315	90'821	151'975

Investment Overview [KEUR]	Q2 / 03	Q3 / 03	Q4 / 03	Q1 / 04	Q2 / 04
Founders	93	53	-	-	-
Investors	-	-	200	733	667
Total	93	53	200	733	667
Cumulative Total	93	147	347	1'080	1747

Source: Company prospectus

CASE 3-2
Tatis Limited

Benoît Leleux, Georges Haour, and Lisa (Mwezi) Schüpbach

B2G. Business-to-government solutions. While all eyes were on the B2C (business-to-consumers) and B2B (business-to-business) Internet applications, a quiet revolution was taking place in the less glamorous world of business-to-government relationships and the management of global trade relationships.

Governments were increasingly faced with the delicate task of balancing the need to regulate and control trade effectively while not stifling its development. Government controls were forced to reinvent themselves, shedding their old garb as rigid barriers to trade to reveal more up-to-date trade enhancement and facilitation characteristics.

Nowhere was the change more evident than in developing and transitioning nations. The urgency to bridge the gap with more developed countries, and the role that New Economy tools could play in the forced march to riches, had created a new market-driven fervor in the last months of the 20th century.

In the traditional time-consuming, paper-rich, and maddeningly complex world of customs declarations and import/export controls, the need for transaction automation was too obvious to ignore. But the obstacles to serving the multibillion-dollar trade were as blatant: There were as many conflicting regulations as countries to serve, entrenched competitors in the freight forwarding business, technophobic bureaucrats, *etc.*

> In one African country, for example, 24 sets of documents were required for each import. The customs clearing agent would simply sit in front of a typewriter and type them out!
>
> In Asia, it took six weeks to clear a container… it was really due to the fact that all those documents needed to be processed and collected.".

It seemed that the Internet would be able to offer the speedy, transparent, and efficient processing of trade flows that governments and the trading community sought. So who would unblock the bottleneck and free global trade of its 19th-century attire?

Copyright © 2001 by IMD – International Institute for Management Development, Lausanne, Switzerland. Not to be used or reproduced without written permission directly from IMD.

Tatis (Trade And Transport Information Systems) Limited, founded in Geneva by Marc-Henri Veyrassat in 1998, had been looking at those very solutions and hoped to be the savior of global trade. But in November 2000, its immediate future was in jeopardy if Veyrassat could not get organized quickly.

1 Global Trade Patterns: A Data Management Nightmare

Prior to founding Tatis, Veyrassat had spent over five years in southern Africa in charge of various government programs. In particular, he had designed and implemented an innovative Customs Revenue Enhancement Program for and on behalf of the Zambia Revenue Authority. This program offered, for the first time ever, online access to trade and fiscal data, thus facilitating and expediting the work of the customs department.

With his understanding of world trade and all its complexities (*Figure 1*), Veyrassat believed the current market opportunity and the ability to leverage on new services created a compelling investment opportunity with an appealing risk/reward profile.

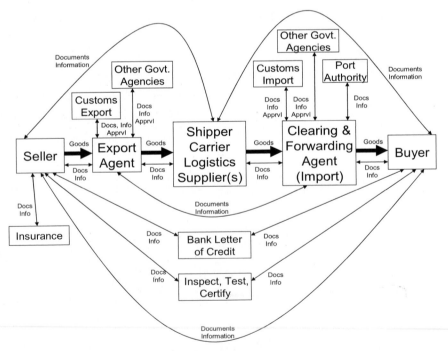

Figure 1   Administration Complexity in Global Trade Arrangements
Source: Tatis Limited company literature

Some of the problems Veyrassat had encountered included the fact that:

> ...many developing and transitioning countries, in Africa, Latin America, Asia, used a service called Preshipment Inspection to monitor and control goods being imported into the country. The service would provide an indication to customs on the valuation and the classification of the goods (pricing, quality, *etc.*). However, the information to customs was often not available in due time, the service was costly, it was perceived as an impediment to trade because another administrative layer was added to the import process: the service had to be called for by the countries of import, inspections had to be arranged, documentation had to be made available, and so forth. Because of the issue of enforceability, obviously there was money involved in the transactions, and therefore there was also a big issue of corruption at customs.

With the difficulty and complexity of dealing with bureaucratic and conservative governments as well as international institutions, which constituted the core customers for Tatis, Veyrassat needed to build the right partnerships to be able to penetrate the market and to have any influence there.

The key to the right partnerships was Tatis' SmartDocument™ line of products, which offered an innovative solution to managing the information necessary for the control of international trade transactions. (*See Exhibit 1 for a depiction of the underlying technology.*)

2 Tatis: Bringing out the Smarts

Building on the extensive experience of its founders and its management (*Exhibit 2*), Tatis set about developing tools that could be introduced rapidly and solve the most pressing needs of trade authorities. Since it would clearly not be possible to solve all issues related to international trade at once, the products were designed to be modular, scalable, and rapidly upgradeable to whatever new technology came to dominate information dissemination (*i.e.*, platform independent or at least compatible).

A first step in the process was the creation of the proprietary Trade Information System (TIS), which ultimately gave its name to the company. TIS was a system that allowed participants such as exporters, importers, shipping agents, banks, and government agencies to share information more easily, securely, and rapidly. It integrated electronic information from networks with the SmartSymbol™, a machine-readable stamp that carried pertinent information from documents.

SmartDocument™ technology took TIS a step further by incorporating e-business databases and publishing technologies. This created the core technology base on which to develop a whole range of trade, fiscal, and financial information-management products.

The SmartDocument™ technology was essentially an application server software designed to help manage and distribute digital information and secure paper documents. The key elements of the SmartDocument™ technology were:

- A proprietary software program to generate a SmartSymbol™ based on *information* provided by one party and *a digital signature* provided by another party – usually a government-appointed agency – and provide information about the document (the SmartDocument™).
- The SmartSymbol™ itself – a unique integration of a sophisticated 2-D bar code with a digital signature and a unique compression algorithm to provide strong protection, multilevel security and high data capacity.
- Information distribution through the Internet. SmartDocument™ information could be downloaded from Web sites or sent as an e-mail attachment or a fax message. The SmartDocument™ server could upload, process, and store data, as well as distribute and retrieve secure documents.
- The ability to fax paper copies of the printed SmartDocument™ documents, since they included a SmartSymbol™ that could be scanned on receipt to confirm authenticity.
- The ability to transfer the SmartDocument™ technology to other applications where secure information has to be shared between many different organizations without risk of compromising security.

The benefits of the SmartDocument™ technology were obvious.

- The distribution of information using the Internet enabled key information to be passed between different organizations with speed, ease, and security. The strong encryption used with the SmartDocument™ guaranteed data integrity. Furthermore, the Web platform allowed clients to use the system without installing dedicated software on their computers, hence allowing easy remote access.
- The system provided a seamless bridge between paper-based systems and electronic systems.

Tatis leveraged the key ingredients of SmartDocument™ to create the SmartFiscal™ Portal, an online, multitemplate, secure trade and fiscal document library.

Web "data corridors" formed the basis of the International Trade Portal. The Trade Services Procurement Portal linked with the SmartFiscal™ Portal and was designed to aggregate multiple trade services via data corridors (*e.g.*, logistics, multimodal transport, supply-chain management, trade finance, shipping) and facilitate trade transactions (*Figure 2*).

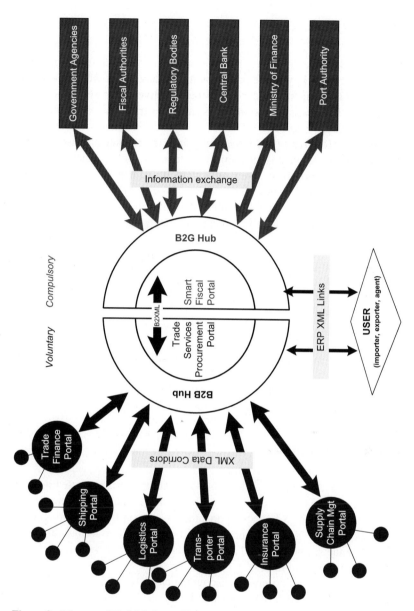

Figure 2 Diagram of Tatis' Sphere of Intervention
Source: Tatis Limited company literature

3 Market Entry Strategy

The priority for Tatis was identifying the most attractive markets for its products. Clearly, given the complexity of the problems it was tackling, it was not viable to launch globally: The launch would have to be selectively targeted. To provide a strong basis for that crucial decision, Tatis rated the trade blocks of the world on the basis of three key success factors: (1) the existence of known trade control problems (marks out of five), (2) ease of entry into the market (strength of contacts; marks out of five), and (3) availability of funding for a project and ability to pay. Tatis decided initially to enter only markets with a score of 13+. (*See Exhibit 3 for the trade block table and scores.*)

The steps taken for each geographical area were to:

- Identify the high revenue risk transactions where the introduction of Smart-Fiscal™ products would generate immediate interest.
- Demonstrate the product to local governments, including showing how to link control systems of neighboring governments (especially within trade blocks) and demonstrating the synergies that could be achieved.
- Get recipient governments to accept the product.
- Submit a project plan linking the product to revenue enhancement and good governance ideals.
- Obtain funding to implement a pilot project if full funding could not be obtained.
- Implement according to the plan agreed with government clients.

4 Financing Tatis Limited

Assuming it would receive early- and later-stage funding, Tatis projected its revenues would reach US$34 million in year 5 (2005). Based on these projections – developed with the help of a financial consultant – and on the conditions in the market, Tatis expected to generate a pretax profit of $13 million in that same year 5. Given this revenue scenario, a brief analysis yielded a current valuation of Tatis in the range of $15 to $16 million.

The valuation was based loosely on discounted cash flows – or in this case the profits before tax discounted at 10% over the five-year projections with no residual value. According to Tatis, a more precise and better valuation methodology could have resulted in a much higher figure. Tatis intended to offer 30% of the company's equity for sale, potentially yielding up to $4.5 million, which it would use as working capital to fund the following key activities:

- Market penetration through a joint marketing effort with its financial consulting company.
- Continued development of the SmartFiscal™ technology platform.

- Building organizational capacity and recruiting highly qualified personnel to fill key management positions. These personnel would then train and support the customers (investment boards, customs, central banks, bureau of standards, ministries, *etc.*) prior to handing over the operating systems.

Once operational, SmartFiscal™ projected it would be cash flow positive after two years. Exit strategies for investors would be offered within three to five years through a public offering, a recapitalization, dividend distributions, or the sale of the company.

An alternative for Tatis to secure a substantially larger capital injection (*i.e.*, more than $20 million) was to rework the strategic model to develop Smart-Fiscal™ portals for governments worldwide, hence taking the opportunity to move higher on the "development chain" (*i.e.*, to higher-level transition and developed economies). Fiscal authorities would have to authorize (or approve) green channel clearance for voluntary use of the SmartFiscal™ portal. In parallel, a generic trade services procurement portal could be developed for simultaneous implementation with each SmartFiscal™ portal. The revenue potential from this "high-end" business model would be significantly greater than from the "low-end" approach, as would the corresponding return on investment. But the risk was also higher, and adoptions were far from guaranteed by the new customers.

To further develop the range of SmartFiscal™ products, Tatis paired up with Scientific Generics, a leading European technology and investment firm. It was chosen for its extensive expertise in the core technologies (2-D and 3-D bar codes, data encryption and compression, *etc.*) underlying the SmartFiscal™ product line and a proven track record in the design, development, and roll-out of complex information services.

As early as January 1999, Veyssarat approached PricewaterhouseCoopers (PwC) to investigate the possibility of a strategic partnership with their Overseas Revenue Consulting Group (ORCG). This partnership could facilitate the promotion and deployment of the products. In particular, the London-based group specialized in advising governments and the private sector on customs and fiscal-related matters. It had assisted governments with restructuring revenue and tax systems and legal frameworks. Its extensive experience organizing and managing large-scale tax-reform projects provided the expertise to successfully implement Tatis' solutions. By November 2000, after some 600 days of discussion, no formal agreement had been concluded.

In early 2000 Tatis began discussions with Venture Partners, a leading early-stage technology venture capital group based in Geneva. Venture Partners would be providing a much needed capital injection and its extensive network connections and might also be able to help in recruiting top-quality managers to the venture. The reputational boost and signaling effect of such a high flyer investing in Tatis was also something that the company could use.

The objective for Tatis and its alliance partners was to identify opportunities to address international trade and fiscal issues faced by governments, businesses, and other parties such as banks and insurance companies and to find solutions within the context of the SmartFiscal™ suite of products. Tatis offered a fee-generating solution to known and potential future problems, with a direct benefit to clients and all the alliance partners.

By 2001 customs agencies would be able to license Tatis' solutions and operate them directly and independently with the assurance of ongoing technical assistance as needed.

When compared to the traditional alternatives, Veyssarat and Tatis had gone a long way toward unblocking the bottleneck and free global trade of its 19th-century attire. But could Veyssarat take the company to B2G stardom? In other words, would he be able to deliver the promised benefits to a large enough audience worldwide and in a profitable manner? The challenge was as high as the opportunity was compelling.

Exhibit 1
First Tatis Product (Underlying Technology to Be Incorporated into Trade-related Information Management Software Systems)

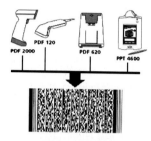

2-D data-rich barcode bridging the gap between
paper-based and electronic information

---

http://tatis.hesalog.com/servlet/DisplaySmartDoc?TransitID=1&XSLDocID=1 - Microsoft Internet Explo... ☐ ☐ ☒

File Edit View Favorites Tools Help

Register at mreport.port5.com

sm▲rtdocument

| Consignor (Exporter)
Exporter A
Alain Praz
1950 SION
SWITZERLAND | Date; Reference No.
Transport Reg. no. : 2001032108 | |
|---|---|---|
| Consignee
Importer A
Person A
Address A
SPAIN | Buyer (if other than consigee) or other address | |
| Notify/Delivery address | Country whence consignee | |
| | Country of origin
SWITZERLAND | Country of destination
SPAIN |
| Transport details
Method of transport : Truck
Route : MSA-NBO-NAM
From : GENEVA To : MADRID
Distance : 654km Duration : 7.0 | Terms of delivery and payment | |

| Shipping marks; Container No.
Number & kind of packages; Goods description
- -	Commodity No.	Gross weight 50	Cube
	1006.30.00	Net quantity 100.0	Value 10'000
	Place and date of issue; Authentication		

TATIS SmartDocument™ Technology and System
Source: Tatis Limited and Generics Group company literature

Exhibit 2
Tatis Management (as of Fourth Quarter 2000)

Marc-Henri Veyrassat, Chief Executive Officer

Marc-Henri Veyrassat is the founder and CEO of Tatis Limited. Marc-Henri spent over five years in southern Africa in charge of various government programs. In particular, he designed and implemented an innovative Customs Revenue Enhancement Program for and on behalf of the Zambia Revenue Authority. This program offered, for the first time ever, online access to trade and fiscal data, thus facilitating and expediting the work of the customs department. In a comprehensive audit report on PSI conducted by the World Bank (1993–1995), the Customs Revenue Enhancement Program in Zambia was analyzed in detail and praised for its impact on revenue collection in that country. Marc-Henri began his career in the Far East. He was educated as an economist at the universities of Lausanne (HEC) and Neuchâtel.

Mark Miller, Chief E-Business Officer

Mark has worked extensively overseas and has over 13 years of business experience in North America, western Europe, and Africa. He is a member of the Global 500 Future Business Leaders Panel. Mark's international experience includes diverse management and business roles, most notably as the contract executive for a high-profile government program executed by a Swiss multinational company, as an international business development executive, and as a management consultant. Mark holds a BSc in electrical engineering from the Massachusetts Institute of Technology (MIT) in Cambridge, MA, USA and an MBA from IMD in Lausanne, Switzerland.

Nicolas Haenni, Chief Technology Officer

Before joining Tatis Limited, Nicolas worked for the Geneva Stock Exchange and the Swiss Exchange (SWX) for several years. Among other projects in the systems area he set up a BBS (bulletin board system) for stock closing price downloads in the pre-WWW era that led to an Internet connection and to one of the first European stock exchanges to provide end-of-day market data. There he also coordinated software development for the Internet, assisted Internet site development teams, launched SWX's Financial Quote Server, and was responsible for the configuration and security of the SWX Web services. Nicolas received an MSc in management of technology from the MIT Sloan School of Management, Cambridge, MA, USA (2000) and a degree in computer science from the Swiss Federal Institute of Technology in Zurich, Switzerland (1985).

Exhibit 2 (continued)

Frédéric de Senarclens, Vice President, Business Development

Frédéric spent over three years managing various high-profile government programs in Ethiopia, Saudi Arabia, Brazil, and Angola. In Angola, in particular, he gained comprehensive experience in new business development with government clients and valuable expertise in international trade and fiscal-related issues. Moreover, Frédéric has gained experience in warehousing management through the monitoring of food aid programs in Ethiopia and Angola. He has a master's degree in environmental economics from the University of Sussex, England and a license in international relations from the Graduate School of International Studies, Geneva, Switzerland.

Source: Tatis Limited company literature

Exhibit 3
Trade Block Table and Scores (High Score = 5, Low Score = 0)

Area	Ranking				Visit priority
	Problems	Entry	Funding	Total	
APEC	3	2	5	10	2
Imports = $2,625 billion					
Exports = $2,497 billion					
EU	1	3	5	9	3
Imports = $2,232 billion					
Exports = $2,180 billion					
NAFTA	1	2	5	8	3
Imports = $1,420 billion					
Exports = $1,070 billion					
ASEAN	5	4	4	13	1
Imports = $299 billion					
Exports = $359 billion					
CEFTA	5	4	4	13	1
Imports = $134 billion					
Exports = $107 billion					
MERCOSUR	4	3	3	10	2
Imports = $83 billion					
Exports = $74 billion					
ANDEAN	4	3	3	10	2
Imports = $36 billion					
Exports = $43 billion					
COMESA	5	5	4	14	1
Imports = $45 billion					
Exports = $23 billion					

Source: Tatis Limited company literature

CASE 3-3
Genedata

Benoît Leleux and Atul Pahwa

Dr. Othmar Pfannes, chairman and founder of Genedata – a startup bioinformatics company located in Basel, Switzerland – was wondering what to do with the valuation data obtained for his company from Venture Valuation AG, a young financial services boutique specializing in the valuation of early-stage, high-tech startups. Up to this point he had always tried to stay clear of these issues and remain entirely focused on developing the next-generation analytical software products, but it was increasingly difficult not to jump into the debate. His main concern was to keep the company on track for the next round of funding and not to get distracted by valuation numbers.

Dr. Pfannes did not much care for numbers just as long as they signaled that the company was on the right track developing the next-generation analytical tools supporting the genomics revolution. In a sense, the valuation exercise would only confirm the strategic positioning of the company and the validity of its business model. On the other hand, curiosity pushed him to consider the different angles utilized by Venture Valuation AG to come up with the numbers. Maybe there would be some insights he could use to further improve the company's positioning for the next round of funding.

Copyright © 2005 by IMD - International Institute for Management Development, Lausanne, Switzerland. Not to be used or reproduced without written permission directly from IMD.

1 Company Background

In 1996 four employees of Ciba-Geigy Computational Biosciences group decided to start Genedata, a company focusing on developing and marketing computational solutions for pharmaceutical research and development. This happened at a time when Ciba-Geigy and Sandoz merged to form the pharmaceutical company Novartis.

Originally targeting European pharmaceutical companies, Genedata sold a mix of consulting services and software licenses to several leading European commercial and research institutions. Hoechst Marion Roussel, Bayer, Boehringer Ingelheim, and Novartis were some of Genedata's early customers. In 1997 revenues were CHF 500,000 and losses were CHF 37,000. By 1998, both revenues and losses increased to CHF 1,400,000 and CHF 50,000, respectively.

1.1 Product Strategy

Genedata focused on developing and marketing high-end software products that enabled pharmaceutical companies to process, store, and interpret huge quantities of experimental data in a highly automated workflow. The need for such software systems resulted from the application of so-called omics technologies (genomics, proteomics, metabolomics, *etc.*) in drug discovery that had been discovered in the 1990s and were then implemented in research processes.

One major differentiation for Genedata was that it provided not only software but also related professional services, including system support, scientific data analysis, and technical consulting. Customers had the flexibility of choosing the extent of the services they wanted to contract. The company had three major software products on the market:

- *Genedata Phylosopher*® for the analysis and management of gene and protein sequence and related data.
- *Genedata Expressionist*® for the analysis and management of gene expression, protein expression, and metabolic profiling data.
- *Genedata Screener*® for the management and analysis of high-throughput screening data.

(*See Exhibit 1 for Genedata's product range and Exhibit 2 for process flow analysis.*)

1.2 Key Competitive Strengths

Genedata's staff consisted of scientists with backgrounds in mathematics, statistics, physics, computer science, and molecular biology/biochemistry. Also, expertise in the discovery-related context enabled Genedata to specifically address customer needs. Core to Genedata's business concept has been a close collaboration with its customers. Although other providers of software existed, most of them were unable to meet specific needs or researchers or keep up the continuous development process.

Customers could also access Genedata's high-performance computing system with large genomic databases via the Internet. It was very important to perform correct analysis and to provide comprehensive graphical displays, which facilitates interpretation. Achieving high quality in analysis hinged on scientific quality of the algorithm.

Genedata's software facilitated easy integration in other analysis systems, and close relationships to scientists of leading research centers ensured the availability of the latest algorithms as well as the quality of the analysis and their efficiency.

1.3 Competitive Landscape

The competition in Europe consisted mainly of a limited number of smaller highly specialized bioinformatics companies (usually around 20 people), mostly of academic background. There had also been one major commercial enterprise (Lion Bioscience) that went public in 1999 and had been widely considered as market leader due to very expensive marketing, as well as a number of US-based companies with offerings similar to that of Genedata. They were strong competitors for Genedata in the USA, but not necessarily in Europe.

1.4 Industry Development and Trends

The pharmaceutical business had come under pressure as the time and costs for R&D of new drugs had exploded over the last few years. They hoped that through the application of new technologies (omics technologies) as well as automation and standardization of research processes they would be able to gain on efficiency of R&D: more drugs at smaller cost.

They therefore looked for bioinformatics companies as providers of computational solutions that supported them in becoming more efficient.

1.5 Staying One Step Ahead

To remain in a leading position, Genedata had to capitalize on its European foothold while preparing for growth in the USA and Asia. As it was expected that US companies would also prepare for entering the European market, Genedata needed to counter this with a strong marketing effort in the USA as well.

1.6 Crucial Factors

Genedata's core competencies were based on the specific knowledge and experience of its staff in pharmaceutical R&D. These assets were key for further success. Therefore the main risk factor was the loss of key employees that would be difficult to replace. The recruitment of additional highly skilled professionals would be required to support the services called for in the business plan.

Underlying technologies were also changing rapidly. It was crucial to stay at the edge of technology and science. Genedata's close contacts to universities had to be maintained and intensified.

Genedata had a clear competitive advantage with its intensive customer focus in the European market. This advantage would weaken if other already existing companies developed more customer-oriented software and also started to provide similar support.

1.7 Performance Scenarios for the Future

Not only was there the threat of American companies entering the European market with new innovative products, but also there was the fact that software could be copied and sold as long as there was demand, the production costs being very low compared with the cost of development. However, there was a high growth potential for Genedata in the future: if Genedata was able to develop real solutions according to customer demand, it would benefit from the huge copy-multiplier potential.

The development process of a software product in such close interaction with the customer was very promising and would enable Genedata to grow rapidly. Maintaining contact with universities and researchers represented an important basis to future development. Through the licensing of the software Genedata would be able to keep the customers' dependence high and assure future growth of its turnover.

Exhibit 1
Genedata's Software Products

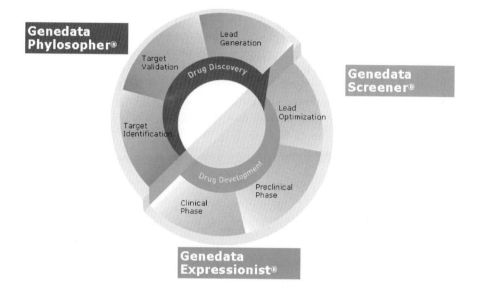

Genedata Phylosopher®

Genedata Phylosopher® is the first fully integrated computational solution that systematically analyzes gene and protein functions based on comparative genome sequence analyses. It is an invaluable tool in industrial research for the identification and characterization of potential target proteins, leading to a time- and cost-efficient prioritization of research projects. It has already been successfully applied in antibiotics research and, with the availability of more eukaryotic genome sequences, is positioned to become the standard for the prediction of gene function in plant, animal, and human genomes.

Genedata Screener®

Genedata Screener® is an integrated platform for high-throughput screening analysis. Thousands of plates can be visualized simultaneously, and robust dose-response curve fitting is performed in an automated and standardized fashion. Tailor-made for high-content and high-throughput screening data, the platform features advanced algorithms for data processing and quality evaluation. Only high-quality data are considered in downstream analysis, focusing valuable resources on the most promising leads.

Exhibit 1 (continued)

Genedata Expressionist®

Genedata Expressionist® is the most flexible industrially scaled system for the analysis and management of microarray data. It optimizes the value of data from leading technologies and spotted arrays through automated quality control and easy-to-use, interactive, graphical data mining tools. The system incorporates proprietary algorithms and supports all major data formats. Expression data can be linked to internal as well as external data sources for gene annotation, pathway information, and literature. The system is used for life science research with special applications in medical, toxicological, and microbial research.

Source: Company literature

Exhibit 2
Genedata Process Flow Analysis – July 1999

Source: Company literature

Exhibit 2 (continued)
Gene Expression Data Analysis

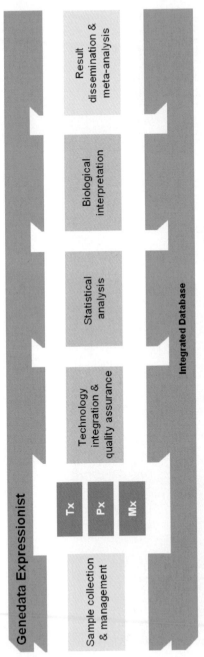

Tx = transcriptomics expression data, Px = protein expression data, Mx = metabolite expression data

Exhibit 2 (continued)
Genedata Graphical User Interfaces

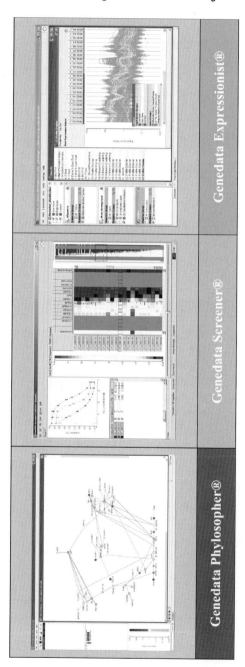

Source: Company literature

CASE 3-4
Venture Valuation AG: The Genedata Assignment[1]

Benoît Leleux, Victoria Kemanian, Atul Pahwa, and Katrin Siebenbürger

JULY 1999. Patrik Frei, CEO and founder of Venture Valuation, a newly created valuation boutique in Zurich, was thrilled with the news of his engagement in a project for Novartis Venture Fund (a corporate venture fund of pharmaceutical giant Novartis). At the same time, he was nervous at the realization of the new-found responsibility he had been assigned. This assignment was a breakthrough for his fledging startup firm and a unique chance to validate the methodologies he had developed to value high-potential, early-stage companies.

As he got off the phone with Dr. Jürg Meier from NVF, he could hardly believe his luck. This was their very first full-scale project for a major-league VC. The assignment was to value Genedata, a fast-growing bioinformatics company already in NVF's portfolio, currently seeking a new round of expansion capital.

The mandate was relatively straightforward: Venture Valuation would conduct an independent valuation of the Swiss startup company in order to help Novartis Venture Fund analyze and structure the next financing round for Genedata. For Frei, this seemed an innocuous-enough assignment, but he had to deliver a top-quality product. Time was not on his side though; he was given a very tight deadline, somewhat in conflict with the need to customize the evaluation process to this assignment.

With emotions ranging from elation to terror, Frei took his laptop and walked over to the nearby Starbucks coffeehouse, hoping to organize his thoughts on the valuation process that lay ahead of him with the help of a Venti Latte. "Make that shot of caffeine a triple," he said as he ordered his coffee.

Copyright © 2005 by IMD – International Institute for Management Development, Lausanne, Switzerland. Not to be used or reproduced without written permission directly from IMD.
[1] We would like to thank Patrik Frei at Venture Valuation AG for his contributions to this case study.

1 Company Background

In the spring of 1999 Patrik Frei founded Venture Valuation in Lucerne, Switzerland, while still finishing his studies at the University in St. Gallen (HSG), where his diploma research focused on valuing high-growth private companies. In the course of his research, he interviewed several venture capital funds in Switzerland and noted their strong interest in the valuation of early-stage, high-potential companies and their lack of structure about the process. VCs seemed to approach valuations with an *ad hoc* combination of experience, gut feel, fundamental analysis, and comparison with relevant matches. Frei was surprised by the relatively little academic research in the field of early-stage, high-tech valuations and the high level of reliance on industry and operational experience in the valuation process. He set out to design a conceptual "service model" that combined grounded valuation methods with the heuristics (*i.e.*, rules of thumb) commonly used by practitioners.

The newly founded Novartis Venture Fund (NVF) had shown great interest in his valuation work. After Frei made a presentation at one of their meetings in 1998, NVF fund managers asked him to conduct the valuation of a series of nine portfolio companies. On the back of this "proof of concept," Frei decided to formally register the company and pursue its development as a career. The company specialized in independent, third-party assessment, valuation, and monitoring of emerging high-growth companies in industries such as biotechnology, med-tech, high-tech, IT, and telecom.

To craft a product offering, Frei spent time identifying the reasons for companies or funds to conduct valuations. Valuations seemed to be triggered either by internal company issues or market-related factors. The need to raise fresh capital (in the case of a startup) or to invest (in the case of a venture fund) were obvious triggers. Less obvious ones included (1) a need to set objective bases for negotiations on equity allocation among project partners or other significant stakeholders; (2) periodic updatings of portfolio valuations, especially after market downturns or upturns; (3) increasing pressure to improve communication between fund managers and their limited partners; and (4) major corporate changes, such as initial public offerings (IPO), acquisitions, mergers, management buyouts (MBO), leveraged buyouts (LBO), or other business development transactions. Clients consisted mostly of venture investors and emerging companies raising money or embarking on major business development activities.

2 Venture Valuation Product Portfolio

2.1 ValuationReport™: The Star Product

Venture Valuation clearly positioned valuation as its core competency, with ValuationReport™ its flagship product. The report was a comprehensive analysis that combined soft valuation factors, such as an assessment of the quality of management and of the science and technology and corresponding intellectual property, with a comprehensive risk profile and standard valuation methods. The report also highlighted the critical success factors that drove company value in the long term.

2.2 ValuationRadar™: The Recurrent Revenue Engine

Unlike ValuationReport™, ValuationRadar™ was a monitoring instrument tailored to the venture capital fund's need to obtain a comparative assessment of their portfolio companies over time (*Figure 1*). Each company in the fund's portfolio was assessed using Venture Valuation's standard methodologies and common criteria, allowing funds to monitor portfolio value through a common set of valuation lenses. The radar was conceptually grounded in the classic balanced scorecard concept.

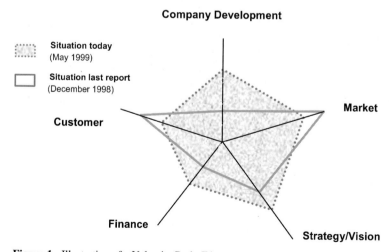

Figure 1 Illustration of a ValuationRadar™ report

2.3 Portfolio Performance Measurements: For Investors' Eyes Only

Venture Valuation also offered a portfolio performance measurement service compliant with the Performance Measurement Principles of the European Private Equity and Venture Capital Association (EVCA). This service was relatively popular with VCs since EVCA had recommended that its members use only independent third parties to conduct such valuations.

2.4 Workshops: The How-to

The company also organized valuation workshops upon client request. To offer a flexible client-specific service, Venture Valuation created a program based on progressive modules – eight in total – that could be individually adapted to the specific needs and backgrounds of participants.

2.5 Venformis™: Reporting Software

Valuation and reporting were closely related and linked issues. Thus, Venture Valuation decided early on to offer its investor clients a software tool allowing them to efficiently manage their portfolio reporting tasks (*Figure 2*). Portfolio companies

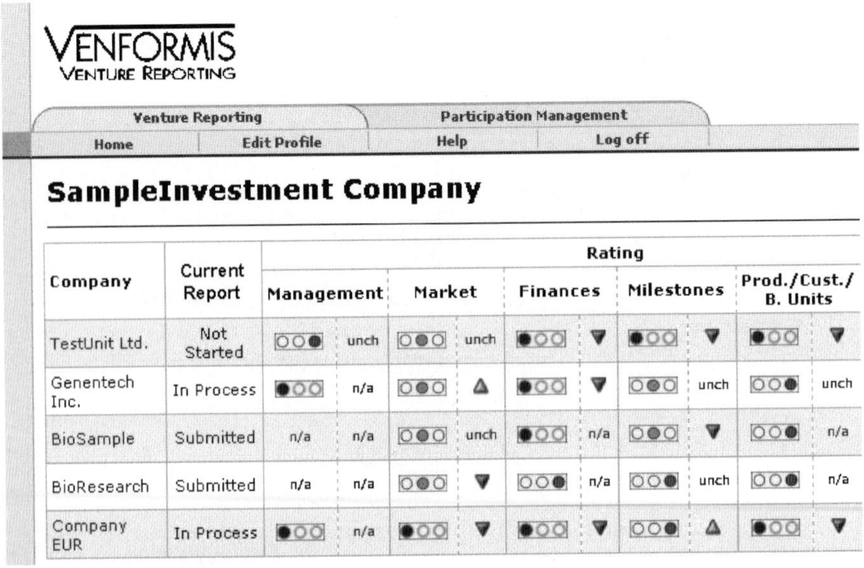

Figure 2 Example of Venformis™ report

shared their quarterly reports on a Web-based tool. All accredited investors had access to this reporting, which included automatic red flags (traffic lights) in case set targets were not met. This software allowed for a dynamic reporting on the evolution of the company as well as on the whole portfolio. Included in the package was also a convenient shareholder management tool.

2.6 The Venture Valuation Methodology

Venture Valuation's methodology was built on three fundamental building blocks:

1. A general assessment of the company, *i.e.*, an analysis of its vision and strategy, products and technology, the business model, management team capabilities and experience, financial situation, and stage of development. In this phase, Venture Valuation looked at the company's whole life cycle and the various investors' investment cycle. Factors specific to each company – like intellectual property and proprietary technology – were thoroughly analyzed.
2. A comprehensive risk analysis, which implied gaining a deep understanding of the company's specific risk factors and value drivers.
3. A detailed valuation exercise through several established methods, including at the very least an operations-based method (*e.g.*, discounted cash flow) and a market-based (*e.g.*, market comparables) approach.

The entire valuation process started with information provided by the company, such as business plans, strategy presentations, product descriptions, and market

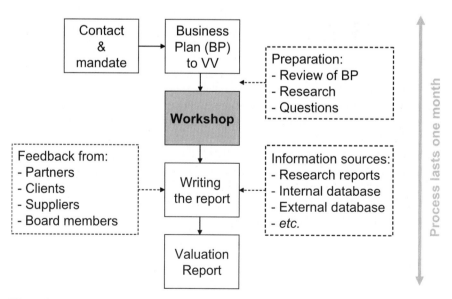

Figure 3   The Client Interface

and financial data, including forecasts for the next 5 to 10 years. After analysis of the information provided by the client, Venture Valuation usually organized a one-day workshop with management to gain a better understanding of the business, the market, and financial forecasts. The workshop was also an opportunity to get to know management and the team dynamics. The methodology is illustrated in *Figure 3*.

Based on the assessment, Venture Valuation provided the company with not only a valuation range based on different scenarios but also strategic recommendations, in particular around growth drivers. (*A more detailed overview of the valuation process is given in Exhibit 1.*)

3 The Genedata Project

When Frei received Dr. Meier's call, he immediately got in touch with Genedata's CEO to arrange for the company data and for setting up the workshop with management. He had only a week to compile the report and present the results to NVF.

Genedata was spun off in 1997 from Novartis, the global pharma giant created from the merger between Ciba-Geigy and Sandoz the previous year. The founders, most of them computational bioscience specialists formerly part of the Central Research Unit at Ciba-Geigy, decided to provide integrated solutions to the global bioinformatics market, focusing on data analysis applications for genomics research. Its offering differed from that of its competitors by the ability to provide an integrated service pack, including system support, software development, technology transfer, data analysis, and consulting, and specifically addressing the customers' needs. Genedata had also come away from Novartis with some important customers, including Novartis itself, Bayer, and Boehringer Ingelheim.

As a valuation exercise, Genedata provided a few special challenges. First the management team had to be convinced to share confidential information. This was not only about getting access to a "dataroom" but also obtaining an in-depth understanding of the ways and means of the company. An interactive, open interview, presentation, and QandA session were necessary. Second, it was a challenge for nonscientists to understand the highly technical business of Genedata. Patrik activated all his contacts to get hold of different research reports and industry analysis. He spent countless hours in the university library acquainting himself with the most obscure corners of bioinformatics, just to be able to ask the right questions at the meeting with the management team.

4 The Valuation Exercise

4.1 Soft Factors Analysis

Before analytics were brought to bear on the valuation exercise, it was key to obtain a solid footing in the soft, *i.e.*, qualitative, factors on Genedata's market and competition, drivers for growth, and strategic issues that would need to be addressed. Although Genedata had over a year of operational experience and growing revenues, it still needed to expand its customer base.

The soft factor analysis usually looked at the major factors affecting a startup company's growth and success rate, such as the quality of the management team, uniqueness of products and technology, and market and industry dynamics. The factors could be rated according to their relative importance to the company and thus served as an initial guidance to specific value drivers. For the Genedata analysis, Frei had come to the following conclusions:

4.1.1 Management

Genedata's staff consisted of scientists with backgrounds in mathematics, statistics, physics, computer science, and molecular biology or biochemistry. The team had superior expertise in these various technical areas which, combined, represented a very potent mix. On the other hand, business development and marketing skills were becoming critical for the company, and these areas were still in short supply.

4.1.2 Market

Competition in bioinformatics included four public companies and more than a dozen private ones. Geographically, Genedata focused on the European market (mainly Switzerland, Germany, and Austria), where it targeted commercial as well as nonprofit research companies. Genedata differentiated itself by providing an integrated service package, including system support, software development, technology transfer data analysis, and consulting. Most other players in the field were research institutions, with limited business expertise. US bioinformatics firms had so far shied away from the European market. A more serious threat came from manufacturers of biochips, small "laboratories on a chip" that would allow most bioinformatics processes to be carried out automatically, with very high speed and in a controlled environment.

4.1.3 Product and Technology

Bioinformatics companies, unlike biopharmaceuticals, had to cope with a volatile market, with shorter product lifecycles and high technology risk. Genedata's products and technologies were not formally protected by patents but instead relied on the unique intellectual capital and knowledge base of its scientists. The major concern for bioinformatics companies was to establish a strong installed customer base from which to evolve and tailor its offerings.

4.1.4 Advancement Stage

The company tripled its sales in 1998 and a subsidiary in Germany was established. A marketing and sales unit was under construction. On the negative side, Genedata's sales relied presently on a handful of customers. After thorough assessment of these value drivers and a number of meetings with Genedata's team, Frei put together the data for the various possible valuation approaches. He elected to combine a discounted cash-flow approach with market-based comparables, and he would cross-check the results with NVF's requirements for return on investment.

4.2 Discounted Cash Flow Approach

The discounted cash flow (DCF) method used the financial projections provided by the company as shown in *Exhibit 2* and valued these expected cash flows over the foreseeable future. To do this properly, Frei needed to determine an appropriate discount rate and the conditions for the terminal value calculation.

4.2.1 Estimating a Discount Rate

To estimate an appropriate discount rate for this early-stage, nonlisted company, Frei used a simple three-parameter risk-adjustment procedure, above and beyond the standard market risk premium. Starting with the prevailing market risk free rate (the rate obtained by the most creditworthy borrower, in general the government or State itself), Frei would first add a market risk premium (MRP), the long-term difference between the returns on the risk-free asset and a broadly defined stock market. Over the last 80 years, that spread averaged around 6% in the USA. He would then correct the result with three additional risk premiums – related to additional liquidity, value added, and cash-flow concerns (*Exhibit 3*).

1. The liquidity premium compensated investors for the illiquidity of investments in nonlisted securities.
2. The value-added premium took into account the value added that the new investor would bring on board. Indicators included the positive image of the investor, strategy support, participation in board of directors, IT support, R&D support, or short-term office space provided.
3. The cash-flow premium accounted for the uncertainty of cash-flow projections and reflected both experience and skepticism. It was linked directly to the three main "soft" factors presented above, namely management, market and product and technology.

These estimated risk premiums, and the corresponding calculated discount rate, are presented in *Table 1*.

Table 1 Estimation of Discount Rate for Genedata's DCF Calculation

Risk-free rate of return	3%
Systematic risk	6%
Illiquidity premium	11%
Value-added premium	1%
Cash-flow adjustment premium	17%
Discount rate (d)	38%

VCs often "ballparked" these return adjustments on the basis of the stage of development reached by the company to be valued. In a sense, stage of development proxied for the progressive reduction in the major risks faced by a developing business. These industry heuristics, or rules of thumb, are presented in *Table 2*. The various stages of development are explained in detail in *Exhibit 4*.

Table 2 Risk Premium Standards for Early-stage Companies

in %	Seed	Startup	First	Second	Late
Liquidity premium	20–30	15–25	11–21	9–19	7–17
Value-added premium	8–12	6–10	5–9	4–8	3–7
Cash-flow adjustment premium	25–35	19–29	15–25	12–22	10–20

4.2.2 Estimating the Terminal Value

The terminal value (TV) was the value of all future free cash flow (FCF) from the end of the explicit forecast period. For example, if explicit projections were made for the cash flows over the next five years, then the value of the cash flows from year six to infinity had to be estimated. To calculate the TV in year five, Frei used the following formula:

$$TV_5 = \text{normalized } FCF_6 / (d - g),$$

where

TV_5 = terminal value of the company in year 5,
d = discount rate,
g = estimated constant growth rate for year 6 onwards.

The DCF method was often used to value high-growth companies but suffered from some serious limitations. First and foremost, it relied on forecasted cash flows, with the understanding that startup companies faced a high risk of failure and/or of having to seriously deviate from the business plan. Startup cash flows were thus extremely hard to predict and highly dependent on a small number of events. Second, the choice of discount rate affected dramatically the final result, and it was fair to recognize that the discount rate estimated was at best an educated guess and at worst a random shot. Finally, because of the long lead time to steady state, the terminal value would inevitably represent a large proportion of the total value, hence be highly sensitive to the assumptions made there, such as the expected growth rate in perpetuity.

Taking these limitations into account, Frei would always run a detailed sensitivity analysis around the key drivers of the discount rate, often using a spread of ±4% around the base rate to get a better understanding of the sensitivity of value to these factors.

4.3 Market Comparables Approach

As a double check, Frei always ran market comparables valuations. The method required the identification of a number of public companies with business models as similar to Genedata's as possible. This valuation approach was based on analogies, assuming that the comparable companies had been properly valued and could thus serve as reference points and benchmarks for valuation. For example, if a comparable public company was valued at $1,000,000 with R&D expenditures of $500,000, for a price to R&D ratio of 2, then a "similar" private company with R&D expenditures of $200,000 would, by analogy, be worth around $400,000, all other things being equal.

Public company shares were traded relatively easily on a stock exchange, but private company shares were not, and it could be very difficult, if at all feasible legally, to find a new owner for them.[2] To recognize the illiquidity constraint inherent in nonlisted stock, a liquidity premium was traditionally applied that ranged from 10% to 30% depending again on the stage of development of the company.

Frei decided to obtain share price information at the time of the valuation from four key publicly listed competitors: Lion Bioscience (Neuer Markt/Nasdaq: LEON), Compugen (Nasdaq: CGEN), InforMax (Nasdaq: INMX), and Tripos (Nasdaq: TRPS). *Exhibit 5* provides some descriptive statistics on the sample of competitors that allowed the calculation of the following multiples:

Enterprise Value/Sales, or EV/S
Enterprise Value/Earnings Before Interest and Taxes, or EV/EBIT
Price/R&D expenses
Price/Employees
Price/Earnings, also known as the P/E ratio

where

P = Share price or market capitalization
EV = Enterprise value or equity market value + net financial debt
R&D = R&D expenses per year
NFD = Net financial debt or interest-bearing debt less cash, cash equivalents, and short-term investments.

As with the DCF method, the market comparables method resulted in a range of valuations, depending on the choices made for the multiples. In addition, the sample group of competitors needed careful analysis, in particular in terms of the similarity of business models, business environments, and growth potentials. An illiquidity discount would also have to be applied to reflect the private nature of the company being assessed.

4.4 The Venture Capital Method, or the Required IRR Approach

When VCs are involved in funding a venture, it is also essential to bring their perspective on the valuation exercise. Venture capital funds had shareholders, called limited partners, to whom they had to deliver returns (ROI or IRR) in line with the risk they assumed. Given the high risk of early-stage, technology-based investments, limited partners usually required rates of return in excess of 50% on each project (depending on the specific risk profile of the venture). This high required return – also called hurdle rate – was required to compensate for project risk, with more than half the projects often not bearing fruits, and the bulk of the fund returns

---

[2] These private placement securities are often referred to as "restricted" stock in the USA because of the stringent liquidity constraints imposed on them. They are often nontradable for years and only gain liquidity slowly over time.

often generated by the top 20% of all investments made. Since most of the value was realized at exit, either through a public offering or a trade sale, the venture capital method started with the estimated exit valuation and discounted it at the required return by the venture investor, to obtain a present value for the VC.

For Genedata, NVF had estimated that it should take around four years before a liquidity event could occur. To estimate the exit price, Frei fell back to the two valuation methods already used, DCF and market comparables. For the required rate of return, Frei used the same rate as for the DCF. Discounting the exit price provided a "valuation cap," the maximum price NVF would pay for the stock today. It served as the basis for defining pre- and postmoney valuations and the ultimate allocation of equity ownership in the round.

The VCs' required IRR varied according to the stage of development of the venture seeking funding, again a proxy for the risk exposure of the round. The round-to-round IRRs could thus vary from more than 100% for a seed round to as little as 15% to 20% for a late-stage growth round. In order to reach its overall IRR target, a VC would therefore not only have to estimate the current value based on its own IRR but would also have to take into consideration future financing rounds and potential dilution.

To calculate the increase in value required from round to round to realize the required rate of return of the investors, a simple formula could be used:

$$\text{Round Multiple} = (1 + \text{required IRR})^{(\text{months of cash}/12)}$$

As an example, with the 38% annual required rate of return deemed appropriate for Genedata's risk profile, and 24 months to the next round financing, the multiple comes out to 1.9×, implying that the valuation would have to close to double to earn the required rate of return. The valuation multiples were often used in a recursive mode (starting from the exit, moving back to the day of the proposed investment) to estimate the maximum value (postmoney) of the company at each stage that would ensure the required rate of return for investors.

In a next step, Frei realistically assumed that a liquidity event for Genedata could occur in July 2003 and that one more financing round was going to be needed before exit. He was not sure about the conditions for the next financing round, which would depend heavily not only on Genedata's development but also on the market environment for biotechnology companies. He would have to create scenarios for the next financing rounds with respect to premoney valuation and investment amount. With these assumptions, he could come to a more accurate value for Genedata. *Figure 4* presents his base case scenario, *i.e.*, a $20 million Series B round to take place in March 2001, 24 months after the Series A investment. This would be followed 28 months later by the IPO. With an estimate of the IPO valuation and with the B round multiple, it was easy to reverse engineer the maximum postmoney value at the end of the B round that would secure the required rate of return. From there, a "capped" premoney valuation for the B round could be estimated, a key negotiation element for the investor. A similar rational could then be applied backward to the Series A financing.

	Series A	Series B	IPO
Date of round	March 1999	March 2001	July 2003
Pre-money value (in $ thousand)	N/M		???
Investment	N/M	20,000	
Post-money value (in $ thousand)	???		
Multiples	???x	???x	

Figure 4 Exit-based Multiples Calculations

4.5 Bridging the Valuation Gap

Frei looked again at the data that lay before him. Genedata's management had helped him get a good hand on the key value drivers in the business. He had investigated competitors and their financials and come to grips with the market valuations for such companies. He had extensive discussions with NVF to understand their requirements. It was now time to assemble the data in a proper valuation report. He knew a number of issues would have a material effect on Genedata's value. The real question was how to prioritize those issues and evaluate what impact they would have on Genedata's value. What was expected of him, besides the valuation, was to provide the company's management with clear guidance as to how to fuel the long-term value of the company.

Frei did not need to be reminded that the assignment was a critical step in establishing the credibility of his boutique valuation services company. Would the business model have a large enough market to be profitable? Did he understand sufficiently his target clients' and customers' value propositions? As he settled in front of his computer to finish the report, Frei thought about the presentation he would give to the NVF in a few days, and what he would recommend that they do in the case of Genedata.

Exhibit 1
Venture Valuation's Process

Understanding the Business Model

Venture Valuation looked first at the elements of the business model and analyzed whether they were correctly geared toward achieving the company's financial and market objectives. The analysis of the business model generally incorporated four main aspects:

1. Revenue stream: Which revenue sources generated the company's sales?
2. Cost structure: Did the company have a specific cost structure different from other models?
3. Customer segments: What was the company's target audience?
4. Channels: How did the company plan to reach that target audience?

The audience and channel strategies strongly influenced the valuation and the risk assessment. They represent a crucial point for any company, but they become even more important for startups with limited marketing muscle and no track record or brand name, especially when engaged in a new technology or market.

Parallel to the analysis of the business model, the company was categorized under a *subsector* according to its activities. For Genedata, the subsector was bioinformatics. Other biotech subsectors included biopharmaceuticals, biochips, diagnostics, and monoclonal antibodies. The purpose of this classification was twofold:

1. Identify competitors and determine approximate market sizes. Competitors' data were used in the valuation process for the application of the market comparables method.
2. Gauge investor interest in the subsector and measure demand. This information could then be used to advise the startup on how to rethink the fundraising strategy.

With a clear understanding of the business model and a solid categorization, other qualitative aspects were assessed, such as management, market, and technology. A simple rating was applied to each of these factors based on the criteria outlined below. The rating would go from 4 (very good) to 1 (very poor).

The Qualitative Analysis

Management

The skills and experiences of the combined management team were crucial to achieve the goals set in the business plan. This part of the analysis gauged the management team's ability to lead the company in the direction set and assessed the financial incentives and the management support structures. Venture Valuation looked at management's track record, the *complementarity* of skills, and the experience levels. In addition, it analyzed the remuneration packages and ownership structure. Lastly, it surveyed the composition and level of involvement of the board of directors and/or advisory board.

Market Screening through Porter's Five Forces

To conduct the environmental analysis, Venture Valuation utilized Michael Porter's Five Forces Framework, through which it identified and analyzed barriers to entry, substitute products, and bargaining power of suppliers and buyers.

Frei believed that a systematic, proven framework allowed them to cover all the relevant market conditions that could affect the company, reducing the probability that important aspects failed to be captured. The result of this environmental scan provided a clearer perspective on the company's market potential and its chances to succeed in a given environment.

Science and Technology

In attempting to measure how long it would take the company to realize the value of its proprietary technology, Venture Valuation took into consideration intellectual property issues, the technological platform, the product pipeline status, the collaborative agreements, and the management of future discoveries.

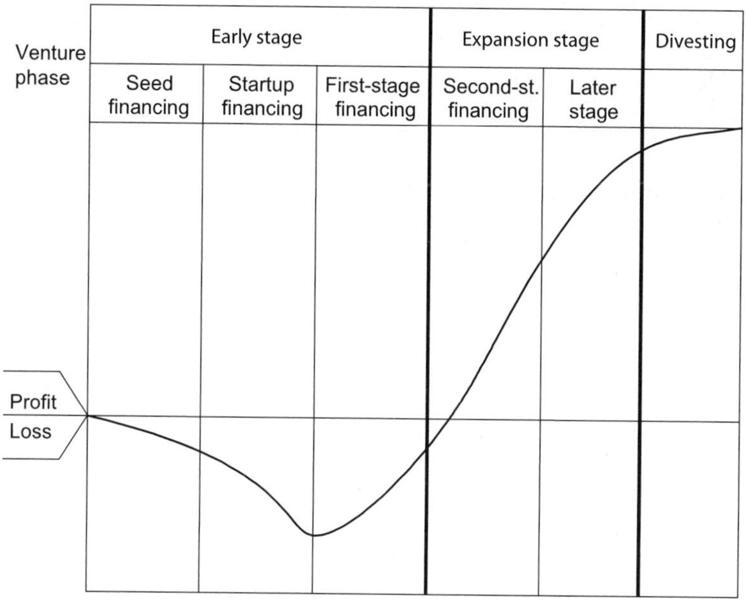

Figure 5 Typical development phases of high growth companies

Exhibit 1 (continued)

The Company Stage

The possible stages for a high-growth company were defined as follows: seed stage, startup stage, first stage, second stage, late stage, and exit. *Figure 5* shows the typical development of a high-growth company. These are described in detail in *Exhibit 4*. It is crucial to correctly position the high-growth company since each phase represents a specific risk situation and, thus, strongly influences the valuation.

The Quantitative Analysis

Defining Risks

The risks faced by a company were focused around operations, technology, liquidity, and the generic uncertainty of future estimates. These risks were also contingent on the development stage of the company. The proof of principle – the seed stage – often reduced the technology risk, while the proof of concept provided initial customer feedback and started to lower the market risk.

Valuation Methods

Venture Valuation applied, whenever possible, at least two different value perspectives when valuing a company: one *operations based*, the other *market based*. The operations-based perspective included methods such as *discounted cash flows*, *decision tree analysis*, or *real options*. These were purely based on the business plan and the company fundamentals.

The market-based approach included methods such as *market comparables* and *comparable transactions*. They depended on market trends and current prices.

Finally, the *venture capital method* presented a cross of both perspectives since the exit price could be established with either an operations-based or a market-based method. These various methods were then combined and weighted in an optimal fashion.

The final result of the process was a comprehensive report covering strategic and operational issues, including recommendations for management on "soft" factors and the calculations made using the different methods and assumptions.

Exhibit 1 (continued)
The Valuation Process Flow

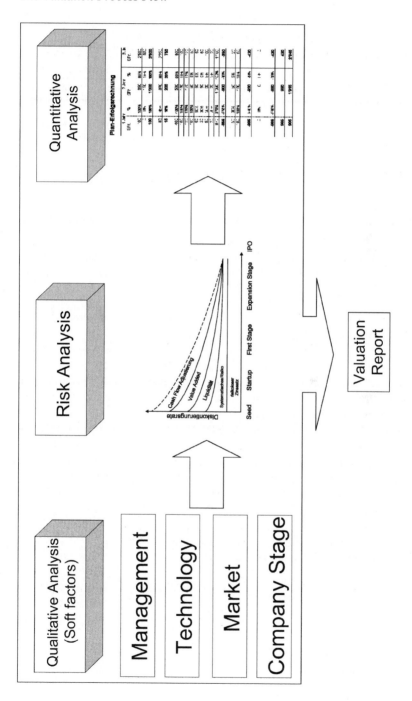

Exhibit 2
Genedata's Forecasted Key Figures

The table below summarizes the company's key forecasted financial and operational data, including its profit and loss account. Forecasts as of July 1999.

In thousand US$	2000	2001	2002	2003	2004
Sales	6,200	11,000	18,000	29,000	45,000
– Cost of sales					
– R&D expenses	1,400	2,000	2,500	3,500	4,500
– Other costs	2,250	4,000	8,000	11,000	16,000
– Depreciation	80	80	100	100	100
– Tax	20	50	1,850	3,600	6,100
Net profit	**2,450**	**4,870**	**5,550**	**10,800**	**18,300**
Employees	22	N/A	N/A	N/A	N/A
Change in working capital	517	400	583	917	1,333
Investments	80	100	100	100	100

Genedata had no net financial debt (NFD) at the time of the valuation. This means that equity value + NFD = enterprise value.

Exhibit 3
Evolution of the Risk Premium in Discount Rates [1]

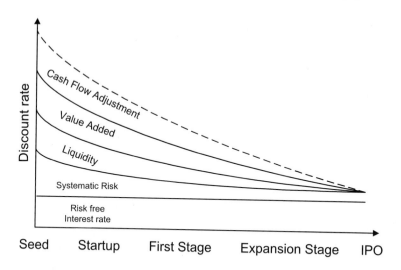

Systematic (Market) Risk Premium

Compensation for the risk in investing in large, publicly listed firms compared to a risk-free asset. This premium represents the historical spread in the returns between risk-free assets and stock markets in general. Depending on the time frame used and the reference stock market, this premium fluctuates between 4.5% and 8.5%, with 6% considered a safe bet.

Liquidity Premium

Compensation for the lack of liquidity (illiquidity) of investments in small, nonlisted companies. Indicators are the size of the firm, the transferability of the assets owned, the financial health, and cash flow, as well as the possibility of going public in the future. The premium ranges from 0% (pre-IPO private placement) to 30% for seed rounds in niche companies.

Value-added Premium

Compensation for the value added brought about by the new investor. Indicators include the positive image of an investor, strategy support, participation in board of directors, IT support, R&D support, and short-term office space. The premium ranges from 0% to12% depending on the stage of investment and the ability to add value. Top-tier VCs in the USA were shown by Tsu (2004) to invest at premoney valuations 14% lower on average than other lower-reputation investors.

Exhibit 3 (continued)

Cash-flow Premium

Compensation for the added uncertainty of cash-flow projections in smaller, early-stage firms, reflecting experience (in investors) and skepticism (in investee figures). It is linked to soft factors such as management, market, and product and technology. The premium ranges from 5% to 35% in "concept" seed rounds.

Exhibit 4
Stages of Development of a High-Growth Company

Seed Stage: *Conceptualizing*
No proof of concept, no product, incomplete team, no production, no sales, no earnings;"a man with an idea"
Discount rates: *in excess of 70–90%*

Startup Stage: *Launching*
Proven concept, beta version of product, marketing plan, production ready, company established, limited sales, net losses, capital bound for up to 8 years; "a team with a demonstrated opportunity"
Discount rates: *50–70%*

First Stage: *Developing*
Finished product, production established, sales building, still not breakeven, capital bound for up to 5 years
Discount rates: *40–60%*

Second Stage: *Accelerating Growth*
Established, profitable company that needs to extend product line and geographical reach, build global reach and supply chain; capital bound for up to 3 years
Discount rates: *25–40%*

Later Stage: *Restructuring for Value Added*
Established, profitable company that needs to restructure to survive, including relocating activities, integrating upstream, rolling up the industry, *etc.*, capital bound for 1 year
Discount rates: *25–40%*

Exit Stage: *Getting Ready for the Exit*
Established, profitable company, fully optimized and with maximum trade sale/IPO appeal
Discount rates: *20–30%*

Exhibit 5
Market Comparables (June 1999)

Market Comparables: Financial Aggregates

In thou-sands of USD	Market cap. (P)	NFD	Enter-prise Value	Sales	EBIT	Earn-ings	Em-ployee	R&D
Lion	65,988	−32,791	33,197	35,139	−50,766	−47,588	520	46,945
Compugen	38,604	−32,347	6,257	11,360	−19,019	−15,144	176	15,976
InforMax	16,928	−50,956	−34,028	18,528	−27,632	−27,459	223	10,192
Tripos	64,315	−16,010	48,305	52,331	1,062	3,185	267	9,406

Market Comparables: Multiples and Ratios

	Re-tained (Y/N)	EV/sales	EV/EBIT	P/employee	P/R&D	R&D/Sales	Earn./Sales
Lion	Y	0.94	N/M	127	1.41	134%	−1.4
Com-pugen	Y	0.55	N/M	219	2.42	141%	−1.3
Infor-Max	N	N/M	N/M	76	1.66	55%	−1.5
Tripos	Y	0.92	45.49	241	6.84	18%	0.1
Retained low		**0.55**	**N/A**	**127**	**1.41**	**18%**	**−1.4**
Retained mean		**0.81**	**N/A**	**196**	**3.55**	**97%**	**−0.9**
Retained high		**0.94**	**N/A**	**241**	**6.84**	**141%**	**0.1**
Genedata						23%	0.4

References

1 Concept based on *Harvard Business School Review*, June 21, 1989: "A Method for Valuing High-Risk, Long-Term Investments."

CASE 3-5
Novartis Venture Fund: Valuation Dilemmas

Benoît Leleux, Victoria Kemanian, Atul Pahwa, and Katrin Siebenbürger

In July 1999 Dr. Jürg Meier, Executive Director of the Novartis Venture Fund, the corporate venture fund created by the pharmaceutical giant Novartis, had carefully listened to Genedata's presentation of business results to date. Genedata, a provider of software solutions for drug discovery and development, was one of the fund's first investments and had shown strong results since its inception. Now, Genedata needed follow-on financing to maintain its current growth rate and expand its business development efforts.

The new business plan looked promising, and if it crystallized, the fund's returns would exceed initial expectations. Despite his early enthusiasm, Dr. Meier left the meeting with too many question marks and only a few days to make a decision.

Genedata was a compelling story, and Dr. Meier wanted to continue supporting the company, particularly at this stage when there was strong upside potential. Yet he was not convinced that the valuation presented by Genedata reflected the potential of contracts under negotiation, the market rollout of new products, and the corresponding risk factors they bore.

With that concern in mind, he decided to call Patrik Frei, CEO of Venture Valuation AG, a new financial service boutique in Zurich specializing in valuation of high-growth potential startups. An external valuation report would not only facilitate the decision-making process for the fund; it would also provide Genedata with a different set of lenses to look at their business. After listening to Dr. Meier's brief, Frei agreed to present a valuation report within one week.

Copyright © 2005 by IMD – International Institute for Management Development, Lausanne, Switzerland. Not to be used or reproduced without written permission directly from IMD.

1 Novartis Venture Fund

1.1 Origins

In 1996 the merger of Swiss pharmaceutical companies Sandoz and Ciba-Geigy gave birth to a new industry giant called Novartis. Later that year, the Novartis Venture Fund (NVF) was created to support new business initiatives in life sciences. In its early days, the fund very much acted as a vehicle to absorb some of the talent and scientific developments that otherwise would have been lost in the reorganization that followed the merger.

Mission Statement
The Novartis Venture Fund is based on the conviction that economic growth and the creation of new jobs can be achieved in the long run only if new entrepreneurial initiatives develop and promising ideas become a business reality.

Since its launch, the fund had made strong commitments to job creation, economic growth, and transparency across its investments and operating strategy. Over time, the fund evolved from being spin-off driven to a larger focus on external investments as the number of merger-related spin-offs diminished. Its mission and strategy made NVF a rare example of a sustained corporate venturing effort in Europe. Other pharmaceutical companies had established similar funds to access the ideas and technologies of entrepreneurial companies, but only a few had demonstrated staying power, with poor investments and the disappearance of IPO markets claiming the most victims. Contrary to traditional VCs, NVF explicitly favored scientific contributions and long-term commercial developments over high returns and quick exits. The strategy was unique, and early results seemed to support its validity.

1.2 Management

The management team of NVF came from both former companies, Sandoz and Ciba. (*See Exhibit 1 for short bios of the management team.*) The fund directors had solid groundings in science, most of them having held senior management positions in research and development globally in either one of the former entities. Their combined scientific and management experience clearly enhanced their ability to evaluate projects as well as to provide management support and advice to their investee companies.

However, the management team overall had limited experience in the venture capital environment. This was particularly critical when dealing with valuation issues, especially in the case of biotechnology companies with breakthrough technologies and long product-development cycles.

1.3 Investment Strategy

The fund invested exclusively in life sciences; it normally would not consider information technology, medical, or diagnosis devices. As a rule, NVF would seek to limit its equity stakes in any investee company to between 10% and 30% of the ownership and often took a coinvestor position. That left space for conventional VCs to become lead investors. In a sense, the fund sought to balance the potential pressure of being a majority investor with the ability to influence the long-term direction of the startup.

NVF insisted on representation on the board of its larger investments and actively supported companies in their startup and market development phases. Management maintained close contact with its companies through meetings, visits, and progress reports presented by CEOs. Given the strength and the scope of the combined funding activities, NVF played a major role in generating synergies among its companies. Its network in the pharmaceutical, technology, and venture capital sectors on both sides of the Atlantic allowed it to provide unparalleled access to resources and contacts. The combination of active involvement and leveraging of the portfolio supported their target of an expected rate of return from investments of 20%.

1.4 Track Record

NVF started operations in 1996 with an initial commitment of CHF 100 million (US$64.5 million),[1] entirely subscribed by Novartis. The fund received some 120 proposals in its first year of operation and invested CHF 28 million in 29 ventures. During its first four years of operation, NVF supported a total of 83 projects, for commitments in excess of CHF 96 million.

Several early investments of NVF had become success stories of their own, with products already reaching the market, large cooperation deals signed with leading pharmaceutical companies, or for obtaining regulatory approvals in record time. Genedata was one of these emerging stars, but other stars such as The Genetics Company, Genesoft, and Infinity all showed similarly impressive potential.

In terms of sector participation, pharmaceutical research led with 48% of investments, followed by life sciences instrumentation and services with 27%, and pharmaceutical development and marketing with 9%.

---

[1] Average exchange rate in July 1999. Source: Oanda.

1.5 Fund Structure

NVF had three funds under management in an effort to focus investments in priority sectors and geographies.

1.5.1 Spin-off Fund

The Spin-off Fund, which concentrated on investments in Europe, was set up to support, coach, and facilitate Novartis spin-offs worldwide. It mainly provided seed money to former or current employees intent on creating their own business. The underlying philosophy of the fund was to foster new ideas and technologies that in turn might complement and benefit the core businesses of Novartis.

1.5.2 Startup Fund

The Startup Fund, which concentrated on investments in Europe, focused on backing young, entrepreneurial startup companies mainly spun off from European universities. It participated with seed capital and typically continued supporting companies throughout further financing rounds. The fund concentrated on promoting scientific and technological innovation based on intellectual property.

1.5.3 BioVenture Fund

The BioVenture fund, with investments ranging from $0.5 million to $5 million, concentrated on investments in the USA. Created with the objective of generating long-term capital gains for Novartis, it invested in product- and platform-focused biotech, pharmaceutical, and healthcare companies at all development stages, but it favored later-stage participation in which it could become the lead investor.

The portfolios of each of the three funds were supposed to be managed independently. But since the funds were still reasonably small in terms of committed capital, its directors had a strong team approach and often exchanged views and ideas on their respective portfolio companies. Since its launch, the Spin-off Fund had been flooded with applications from former Sandoz or Ciba associates planning to create their own companies and eventually continue research started at Sandoz or Ciba. The Startup Fund was extremely successful, particularly in Switzerland, where the burgeoning biotech industry benefited from a culture and legal structure that facilitated the transfer of technologies from universities to startup companies. With a solid track record, NVF enjoyed a solid reputation among other investors, with the consequent frequent requests to coinvest in companies. Since its creation, NVF had received business plans at a rate of approximately one per day.

NVF relied heavily on a prestigious advisory board, which included prominent members from academia, including a Nobel Prize winner in chemistry, senior executives from the parent group, the fund managers themselves, and other investment banking luminaries. Despite the natural ties with the parent company, the fund acted reasonably independently, and Novartis management usually abstained from influencing investment decisions. Moreover, there was no obligation or commitment from the portfolio companies to cooperate with Novartis, although there were numerous opportunities to do so if companies shared a mutual interest. The network provided by Novartis acted as a unique, powerful resource for the fund and the portfolio companies.

1.6 Evaluation and Investment Process Closeup

The evaluation process employed by the fund consisted in a preliminary assessment to ensure projects were economically feasible and compatible with the fund's investment focus. Important aspects considered were the project's level of scientific innovation, quality and expertise of management, and ability to create economic impact, jobs, and added value to the community.

Upon a positive preliminary assessment, a meeting with the company was typically arranged and the funding application was given a more detailed appraisal. After a positive due diligence, depending on the amount of financial support required, the management team, investment committee, or advisory board made a final and irrevocable decision (*Table 1*).

Table 1 Evaluation Process

Type	Investment size (CHF)	Approval needed
Small projects	Up to 200,000	Chairman and management team
Medium projects	200,000–1,000,000	NVF investment committee
Large projects	1,000,000 and above	NVF advisory board

Even though new applications required most of management's time, the maturing of the funds' investee companies meant NVF increasingly participated in second and third rounds and follow-on financings. Genomics-based companies, such as Genedata, were rapidly advancing from the seed phase – in most cases defined by laboratory research activities – to a more advanced stage in which substantial additional investments were required to develop their technologies and products/services to market entry. NVF also assisted investee companies in setting realistic expectations on how long it would take to generate revenues.

NVF was pretty agnostic when it came to exits: most were expected to take the form of trade exits, even though IPOs were also considered. NVF, as a policy, always shunned exits that could hurt the investee company. Such concern also meant that NVF sometimes had to support a company through all rounds until an IPO, and hence that capital could be locked in for a longer period of time.

1.7 Genedata: A Longstanding Relationship

Genedata and NVF shared common roots. Both were offshoots of the Ciba-Sandoz merger. Genedata's management came from Ciba, while NVF's management came from both Ciba and Sandoz. Genedata began operations soon after the merger was announced and NVF was created. It took a few months for Genedata's management to actually contact NVF. Reluctant to give up equity, Genedata's management initially started looking for loans.

Dr. Meier recalled the hurdles in initial discussions with Genedata in October 1996. Neither of the sides had a predefined idea of the proper deal structure to apply or on alternative scenarios for the startup. NVF proposed a deal that involved loans as well as an equity stake. Genedata's initial reaction to that first offer was to require a call option on that equity. In response, NVF thought it would make sense for them to obtain a put option on their shares. In the end, NVF invested CHF 1 million in Genedata's first round, 50% in debt and 50% in equity, for a 25% stake in the firm's equity, for a premoney valuation of CHF 3 million. The founders and a few key employees had been the sole investors till this point.

By the end of 1998, Genedata's nominal (book) equity amounted to CHF 100,000. NVF invested in the capital increase of March 1999 a nominal capital of CHF 200,000. To address the delicate control issue, two classes of shares were created. Founders and staff were issued 1,000 type A shares, with a nominal face value of CHF 100. Class A shares had full voting rights. NVF invested in B shares (also referred as personal shares), with a face value of CHF 1,000. *Table 2* shows the resulting ownership structure.

Table 2 Ownership Structure

Type A nominal shares at CHF 100.00 (1,000 Shares)		
Founders	3 individuals	83%
Board		2%
Staff	First generation	3%
	Second generation	2%
Type B nominal shares at CHF 1,000.00 (200 Shares)		
Novartis Venture Fund		100%

There were a number of reasons explaining Genedata's reluctance to give away equity. The company was able to establish a client base generating revenues literally right after its foundation. Genedata achieved this feat with an above-average product technology and a clear customer focus. Its first and main product was a software to analyze genomic sequences, a key step in the drug-discovery process. The advantage of its product was that while the company could customize it to fit each client's specifications, it did not need to give away the know-how created in the process to the customers and could then accumulate it and capitalize on it later.

Within the following two years, Genedata continued building an impressive pharmaceutical client list, expanded by opening an office in Munich, and built an impressive product pipeline at various development stages. Keeping the rapid pace of development was critical to gaining and retaining a competitive advantage in a fast-moving, highly competitive market. And doing so required sufficient capital to recruit and retain world-class specialists and expand internationally.

Genedata planned to use the proceeds of a new round of financing to expand beyond its existing presence in Switzerland, Germany, and Austria into the USA and Japan – two major markets for the pharmaceutical industry. Many of its products in development had raised interest and secured commitments from current customers. This close collaboration with its customers, along with its proprietary software algorithms, remained among Genedata's most envied competitive advantages.

1.8 A Snapshot of Venture Valuation's Report

Dr. Meier and Dr. Gygax of NVF, curious to hear Venture Valuation's assessment of Genedata, accompanied Frei into the conference room. Beyond the specific valuation, they were keen to learn of the key value drivers (*i.e.*, the parameters most able to influence the value of their investment).

NVF had provided support for Frei's diploma thesis on "Valuation Methods for High Growth Companies," and he had subsequently developed a good rapport with NVF's directors. As Frei began his presentation, he walked Dr. Meier and Dr. Gygax through a comprehensive analysis, taking into consideration financial aspects as well as "soft" factors and important risk drivers. The report included a detailed analysis of the management team, markets for the company's products, and its technologies. It provided a detailed account of strengths and weaknesses as well as an industry analysis that drove the assumptions used as a basis for the valuation calculations.

In terms of valuation, the report presented a value range (under different sets of assumptions, with a value range of 5 corresponding to the most optimistic scenario) for Genedata according to three combined methods – discounted cash flows, venture capital, and market comparables (*Table 3*).[2] To aggregate the results from the various approaches, the first two methods were given a 45% weight each, while market comparable values were given a smaller 10% weight on the grounds that Genedata did not focus on an aggressive expansion strategy yet had the clear goal of an IPO. The calculations put a value range of about CHF 34.4 million to

---

[2] Specifics of Venture Valuation's methodology are explored in Case 3-4.

Table 3 Value Range for Genedata

Value spectrum		Value range				
In CHF million	Weighting	**5**	**4**	**3**	**2**	**1**
Discounted cash flow	45%	31.6	28.3	25.6	23.2	21.2
Venture capital	45%	51.7	48.7	45.9	43.4	41.0
Market comparable	10%	281.0	208.2	135.5	99.7	64.0
Average		**65.6**	**55.5**	**45.7**	**39.9**	**34.4**

CHF 65.6 million on the company. The mean value was CHF 45.7 million if calculated with a discount rate of 38% and the mean ratio from the market comparable method.

1.9 Decision Time

Dr. Meier knew that at this point in time the real question for Genedata was whether it would remain small and beautiful or build up momentum to realize its potential as a major player in its field. The dilemma Dr. Meier faced was a common one in the venture capital industry. Valuation of biotechnology companies had always been shrouded in much uncertainty, mainly due to market factors and the complexity of the technologies involved. This was particularly evident in early-stage biotechnology companies developing drugs, where much of the value contained in the initial phase of the project was based on the promise of developing a blockbuster drug. Valuation of those companies required very advanced valuation methods to reflect these inherent uncertainties, while in the case of Genedata – a service provider to these biotech and pharmaceutical companies – more traditional methods could be applied. However, a single method alone would be hard pressed to provide a fair value as most approaches had serious limitations depending on market conditions or the company stage.

The valuation report also identified key value drivers and risk factors for the company going forward. The report could also be used to allow comparisons among companies with similar technologies or shared with other investors. More importantly, it provided third-party, objective, and expert analysis of the company and its value, avoiding potential conflicts of interest.

1.10 The Task Ahead

With the data provided by Venture Valuation's report, Dr. Meier had clarified most of his concerns and made up his mind. Yet before offering a response to Genedata on the fund's decision to participate in the new round of financing for the company, he had to build a solid case for his own advisory board. On top of that, he needed to design a structure for a deal in case it would go through, work on the term sheet, and define the next steps. More importantly, he needed to put a cap on the transaction price. Other issues he needed to figure out included the timing of Genedata's exit, an estimation of the number of further financing rounds before exiting, and Genedata's position in the venture capital cycle.

It seemed like Dr. Meier was not going to get a lot of sleep that night; he had 24 hours to work on the deal specifics before he met with the advisory board and had to give a response to Genedata.

Exhibit 1
The Novartis Venture Fund Management Team

Dr. François L'Eplattenier, Chairman

Dr. L'Eplattenier received a Ph.D. in chemistry from the Swiss Federal Institute of Technology and held various research positions at Ciba-Geigy before becoming head of central research and later research and development of plastics, pigments, and additives. In 1988 he was nominated to the executive committee of Ciba-Geigy, with responsibility for research and development. François L'Eplattenier also serves as associate professor at the University of Neuchâtel and as board chairman or board member of several high-tech companies.

Dr. Jürg Meier, Executive Director

Before taking this position in 1998, Dr. Meier had a significant role in international research and development at the former Sandoz Pharma Ltd., including head of R&D of biochemistry in Austria, head of R&D in the United States, head of R&D worldwide, and president of Sandoz Pharma in Japan. Dr. Meier is also a member of the Swiss National Science Foundation, is involved in the Center for Technology Assessment at the Swiss Science and Technology Council, and heads the Spin-off Fund.

Dr. Rudolf Gygax, Managing Director

For nearly twenty years before joining the Novartis Venture Fund, Dr. Gygax held various positions in the management of research and development at Ciba, including process safety and high-technology materials research. Dr. Gygax studied physical chemistry at the University of Basel, Switzerland, and Stanford University, California, prior to joining Ciba-Geigy in 1978. Dr. Gygax heads the Startup Fund and also serves as a director on the boards of several startup companies in which the Novartis Venture Fund holds an equity stake.

Source: Company information and website

Chapter 4
Growing the Entrepreneurial Firm: Building Lasting Success

A promising opportunity has been identified, a skilled team assembled, a compelling business plan written, and the initial financing secured. Yet this is not the time to lean back – the journey has just started. Now, the venture is supposed to grow and leverage its first sales. Growth, however, entails new challenges and requirements. Only one in ten firms can measure up to these challenges and master employing more than ten people [1]. In fact, many firms die in their first two years, without ever having experienced "good times." Rather they have just endured as long as they could [2]. Most of them die due to a lack of resources, legitimacy, and coordination (liabilities of newness and smallness) [3]. So, how do the more successful companies actually manage to acquire legitimacy, status, and reputation? What distinguishes them from the less successful ones?

There are a myriad of factors that make a firm successful. This is no different for our set of science-based ventures. In this chapter we look at the leadership and management qualities of the new venture team and the organization-building capability required of the entrepreneurial manager. In addition, we will explore different growth strategies and highlight some challenges to be mastered along the way, which are also evident in the case examples included here.

4.1 Why Grow the Firm?

Having successfully incorporated the business, entrepreneurs find that scaling up becomes a major issue, although not all of them choose to scale up equally aggressively. Naturally, there are some decisive advantages linked to a larger firm size. From an operational viewpoint, growth is of immediate concern as it makes it possible to capture economies of scale. It also enhances the credibility of a venture and thus facilitates market access. It helps attract customers, more established business partners, and high-performing employees. Empirical evidence shows that company growth is positively correlated with market leadership, influence, negotiation power, and survivability [4]. And while bigger does not always imply better, there are clear benefits to reaching critical mass quickly in the commercialization of a new technology platform [5].

4.1.1 Measuring Growth

By and large, growth is measured in increase of revenue. However, it has been shown that firms that enjoy sustained revenue gains do not necessarily find a corresponding gain in shareholder value [6], which is regarded as a definition of quality growth. Evidently, there are factors beyond sales volume that spur growth. Not only the financial but also the strategic, organizational, and structural aspects of the company need to be considered [7]. In fact, companies also grow in the number of employees, clients, and projects, all of which contributes to the overall organizational complexity. Each aspect is accompanied by particular challenges for the entrepreneur.

4.1.2 Challenges to Growth, or Why Some Entrepreneurs Don't Scale

The ability to scale effectively is highly dependent on several managerial issues, ranging from partnering strategies and the establishment of sales channels to careful financial planning and deliberate management of cash conversion cycles. Ironically, providing managerial capacity to handle this range of issues is most likely the single biggest challenge facing the entrepreneur [8]. Even if those starting their own company have previously held senior management positions in larger corporations, the roles were probably not as all-encompassing as the role of an entrepreneur. When one is managing one's own company, suddenly the level of risk and responsibility (not only for the entrepreneur) reaches a different scale, imposing a significant emotional toll in addition to the extra hours.

Along with the growing operations, however, entrepreneurs soon face their next challenge. While initially all decision-making power is in their hands, an increasing number of tasks relate to follow-up and administration as opposed to initiation. Whereas in the startup phase they have been doing everything alone, the growth phase calls for sharing responsibility and relinquishing control. This can be a painful experience for entrepreneurs – no sooner have they developed a perspective on the different parts of the business and invested much personal time and energy in learning these roles than they need to delegate such tasks and often compromise on the way the company operates in the process.

Adding and integrating new employees can, in turn, destabilize a firm's operations. Yet human resource management is rarely the favorite activity of the average entrepreneur. Many of them have difficulty dealing adequately with supervising others' work and designing the right administrative structure and incentive system for the growing company. In addition, managerial capacity cannot simply be hired. New managers need time to socialize into a firm's culture, acquiring firm-specific skills and establishing trusting relationships with other members of the firm [9]. Meanwhile, some of the initial staff members who successfully advanced the original research and development may not be best equipped to drive application engineering, to provide customer service, and to coordinate market launch activities.

Another potential pitfall in growing a firm is the inability to deal with externally introduced change. Especially in high-tech and science businesses circumstances can change quickly, *e.g.*, new research results are published, new ways of solving a problem are developed, and the like. Primarily the entrepreneur needs to recognize and evaluate the kinds of "threats" that limit or put the venture's growth at risk. He or she needs to adapt the business plan and the human and financial resources accordingly. But inevitably, much uncertainty remains in casting a response to these developments, and the entrepreneur needs to persuade the employees to go along with the changes, all of which consumes significant time and personal energy.

In addition, there are a number of day-to-day challenges that constrain growth, including cash-flow management, price stability, quality control, and lack of capital. Securing sufficient financial resources is itself a recurring issue in any growing company, as we have shown in Chapter 3. In short, growth may seem as daunting a task as the prospect of not growing.

4.2 Strategies for Growth

Several strategies, broadly separated into internal and external growth, exist [10]. Whereas internal growth relies on capabilities and resources available inside the firm, external strategies use relationships with third parties to attain the goals.

4.2.1 Internal Strategies: Organic Growth

Internal growth strategies draw on efforts within the firm itself. They mainly include product-related strategies, but also geographic expansion.

4.2.1.1 Product-related Strategies

Many companies actively pursue the development of new and innovative products either to increase revenues from existing customers or to attract new customer segments. Product development is frequently the only way to create an initial foothold and remain competitive. Yet revenue growth can also be attained by improving existing products or services, which increases the value for the customer and the price potential [11]. Improving and modifying products is especially recommended if the company hopes to reach out to new customer groups. This can be accomplished either by customizing the product to very specific and usually smaller customer groups or by moving toward the mass market, which in turn often means simplifying features and making the product easier to use. However, entrepreneurs need to be aware of the technology adoption cycle technical products are subject to (*see box on Crossing the Chasm*). For example, selling a product into a laboratory environment for initial testing as opposed to selling it into a production environment for in-process quality control is rarely an easy transition. Volume growth cannot be expected to be smooth and steady.

Crossing the Chasm [12]
Just as a company follows a growth cycle, so do technologies and products. Knowledge of the technology adoption cycle is particularly important for technology- and science-based entrepreneurs so that they can realistically assess the business potential of their ideas, products, and services. Many of them start out with a unique and highly innovative product, which often only interests a very particular and small user group. However, we observe certain adoption patterns over time: Broader customer groups increasingly take an interest in the technology (*Figure 4.1*). The technology, in turn, needs to become more adapted to these broader user groups. Different requirements emerge with respect to product and marketing strategy; different functionality, distribution, and pricing strategies are needed. Mastering the transitions from a niche market player to mass-market company is frequently both a technical and a managerial challenge.

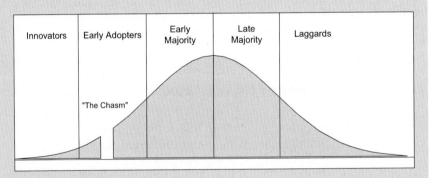

Figure 4.1 Technology Adoption Cycle

4.2.1.2 Geographic Expansion

Many entrepreneurial companies attempt to grow by expanding to additional geographic sites, hoping to reach new customers. They want to demonstrate the overall market potential and build a broader set of references. Making use of international distributors can seem like an easy route to build added volume. A key success factor for this strategy is a scalable business model, *i. e.*, a business model that is also suited to the new location. Yet the need for support for these channels and the added complexity is often underestimated, and any new business locations will typically need the same attention as the initial home market [13].

Table 4.1 highlights some of the advantages and disadvantages of internal growth strategies.

Table 4.1 Advantages and Disadvantages of Internal Growth [14]

Advantages	Disadvantages
• Enables the continued and more fine-tuned adjustment of operations to environmental changes • Provides control over product, quality, and processes • Preserves the organizational culture • Encourages internal entrepreneurship • Allows employees to be promoted from within, which provides a powerful motivational tool	• Slow form of growth • All new resources need to be developed alone, which is risky and expensive • Fewer opportunities for cobranding to build credibility • Failed investments are difficult to recoup

4.2.2 External Strategies: Partnering for Success

In light of the resource challenges, time lags, and uncertainty associated with internal growth strategies, smaller organizations often seek to boost company growth by linking themselves to third parties. This can occur either through mergers and acquisitions (M&As) or, more typically in the context of new ventures, through various forms of alliances or licensing agreements. Careful consideration of external growth strategies is vital for technology companies as it is almost impossible to have all the required resources in-house for ongoing product improvement and development [15]. A recent survey by PricewaterhouseCoopers revealed that CEOs of fast-growing companies increasingly regard external collaboration as critical to their company's success [16]. The two leading reasons for engaging in collaborations are to extend the customer base and to enter new markets or businesses. Many companies even entertain multiple partnerships at the same time, which has been shown to spur significantly higher growth rates [17]. On the flip side, many large corporations have indeed set up entire departments during the past decade to actively prospect for partnering and in-licensing opportunities. While the startup needs added resources, the corporation wants to leverage the technology base of science-based ventures to more fully utilize their own product development and marketing capacities.

4.2.2.1 Alliances

Alliances are long-term contracts that grant access to resources – either assets (technological alliances) or distribution capacity (marketing alliances) – that are otherwise available only through competition. They frequently focus on technical consultation and cooperative research. Other forms of alliance are educational workshops and seminars, contract research, the loan or exchange of employees, and lab visits to share information [18]. Alliances are attractive as they allow for the sharing of economies of scale, risk, and cost. They can provide access to foreign markets, facilitate learning, and speed up product development. Furthermore, they help to increase profit opportunities, defend the competitive position, and increase sales of existing products by providing security and legitimacy. On the downside, companies that are highly networked via alliances face financial and organizational risks, e. g., increased management complexity, dependence on the partner, and a loss of proprietary information, decision autonomy, and organizational flexibility. Moreover, if rapid changes occur, a contract can be stifling if it lacks the necessary flexibility.

4.2.2.2 Mergers and Acquisitions

Increasingly M&As are regarded as the ultimate means to grow a venture [19]. They are closely related, insofar as a single new business entity is created. However, in a merger two companies pool their resources on a supposedly equal footing, whereas in an acquisition, one firm has the lead and takes over another company. Mostly, entrepreneurial ventures are the target of acquisitions rather than the instigators. The next chapter describes some examples when discussing potential exit strategies. Occasionally, however, small ventures do grow by acquiring other firms. An example is the leading online networking platform openBC/XING, which acquired First Tuesday AG Zurich in February 2006 in order to strengthen its event-management capabilities.

The reasons for M&A activities are similar to those for entering into alliances: Access to resources and knowledge, or intellectual property (IP), access to new markets, the prospect of benefiting from economies of scale, and strengthening the competitive position. However, because of their contractual nature, M&As secure access rights much better than alliances do. Yet M&As carry the burden of the added effort of combining two firms, i. e., combining their operations, employees, cultures, and so forth. Frequently, this ties up resources for quite a long time.

4.2.2.3 Licensing [20]

We have already talked about licensing as part of the business model in Chapter 2, so here we will highlight only the aspects related to growth. From the licensor's viewpoint – especially if the company is rich in IP – licensing can be a highly effective source of income. It makes it possible, for example, to capitalize on IP that would otherwise be sitting on the shelf and of no economic value for the owning company. Licensing also opens up the potential to spread the risk and costs of developing technology. Another advantage is that it allows products to be brought to the market when the licensing company does not command the necessary resources or skills, as for example when entering foreign markets. However, licensing strategies are not without risk and many companies have regretted granting the wrong set of IP rights in hindsight.

Table 4.2 Advantages and Disadvantages of External Growth [21]

Advantages	Disadvantages
• Acquisitions can reduce competition • Access to new (proprietary) products, markets, technical expertise, or established brand names of the alliance partner or acquired company • Sharing in economies of scale • Diversification of business risk • Speed	• Potential incompatibility of top management teams when firms work together/ merge • Possible clash of corporate cultures • Incompatible operations • Increased business complexity • Loss of flexibility (*e. g.*, due to contracts) • Confidentiality issues and loss of control

4.3 Management Skills and Leadership

In the pursuit of growth, the entrepreneurial manager will eventually have to reassess his or her own managerial capacity and personal interests. New ventures typically employ the entrepreneur along with only a handful of people. As the venture flourishes, they will reach a natural limit in tasks they can or may want to handle. New hires, however, need to be funded, organized, and integrated into the firm's organization and culture, which requires resources and inevitably leads to a shift in the dynamics – and often the culture – of the firm. With around 20 or 30 people, *ad hoc* management of processes and responsibilities might work successfully. However, growing the number of employees beyond this threshold requires deliberate attention to the administrative structure of the firm.

4.3.1 Organizing Growth

The administrative structure of a firm defines its processes and the formal framework of decision making. It supports the effective and efficient handling of the tasks at hand [22]. Hence, developing the appropriate design is one of the critical tasks for the entrepreneur.

Generally, firms can be organized very differently, from almost no formalized structure to a highly mechanistic one. In less formalized, typically young, organizations there is often an *ad hoc* culture, which means that everyone does a bit of everything that comes up as urgent: There are no clear responsibilities and information usually flows freely – sometimes at the cost of being informal and erratic. Highly formalized firms, by contrast, have highly structured chains of command and rigid departmentalization with precisely defined roles and responsibilities, sometimes to the extent that innovativeness and flexibility are absent. Finding the right balance is vital.

Such balance, however, is usually the result of an evolutionary process, involving a fair amount of learning for the entrepreneur. He or she needs to define a clear structure and develop and implement a design that fits the corporate culture (*see box on Organizational Culture*).

Organizational culture [23]

Understanding the nature of an organizational culture is important since it is the culture that can indeed be a source of competitive advantage. Culture is the personality of an organization, *i. e.*, how it is perceived in the outside world and thus how customers and suppliers feel toward it and deal with it.

Organizational culture is formed by the beliefs, values, and behavioral norms shared among all members of the organization. It determines what employees perceive to be important and how they work, and it serves as a powerful guide for decisions and actions in the firm.

Dimensions of organizational culture

An organizational culture is difficult to describe explicitly. Mostly, we have only an intuitive understanding of it. It can be captured, though, via certain dimensions, ranging from low to high on an expressivity scale. These factors are the degree to which:

- Employees are encouraged to be innovative and risk-taking
- Employees are expected to pay attention to details, *i. e.*, exhibit precision and thorough analysis
- Outcome orientation prevails, *i. e.*, results are more important than the process that led to them
- Organizational decisions are people oriented, *i. e.*, they consider the effects on the team members
- Work is done in teams rather than individually
- Organizational members are cooperative (instead of competitive)
- Organizational actions emphasize stability and maintaining the status quo as opposed to growing.

Sources of organizational culture

A culture develops mainly as a result of what has been done before and the degree of success it has had. In addition, culture starts at the top. The main source of it is usually the founding entrepreneurs. Their values and initial attention to the organizational design form the blueprint for the organization, even beyond their involvement.

Culture is transmitted to organizational members through stories, rituals, material symbols, language, and physical surroundings. Employees are constantly exposed to these factors and thus assimilate a certain understanding and attitude. Maintaining a culture typically starts with hiring people with the right fit.

Founders may at times be tempted to project a desired company culture based merely on their own preferred ways of working – either explicitly or implicitly. For example, founders may prefer a flat organization without too many formal roles. Yet key employees who do not enjoy "founder's status" may have a legitimate desire for the establishment of formal roles, decision rights, and visible positions as the company evolves. They want to demonstrate career progression within the organization and toward the outside world.

4.3.2 Organizational Blueprints [24]

Research has shown that the evolution of different organizational forms is usually limited to a few very distinct blueprints that are based on the founder's mental model about the ideal organizational form for a technology venture. Typically, the form varies along three main dimensions: Basis of attachment and retention, criteria for employee selection, and means of control and coordination, each characterized by three or four distinct features (*Table 4.3*).

Table 4.3 Dimensions of Employment Blueprints [25]

Basis of attachment and retention	• Compensation ("money") • Quality of work ("work") • Work group as community ("love")
Selection criteria	• Skills • Exceptional talent/potential • Fit with the team or organization
Means of control and coordination	• Peer and/or cultural control • Reliance on professional standards • Formal processes and procedures • Direct monitoring

To create *attachment* to their firm, some founders like to establish a strong familylike community ("love"); many other technology-focused firms bind people via the excitement of working at the technological frontier; and yet others only see attachment occurring in exchange for salaries paid ("money"). When it comes to *selecting new employees*, some founders focus only on the efficient accomplishment of the tasks at hand and hence pick people according to their skills and ability to carry out the tasks immediately. Other founders stress the long-term potential of people to be able to move from project to project and solve tasks not yet envisioned. Finally, some founders emphasize values and cultural fit and hire people who are most likely to fit into the existing organization. For *control and coordination* in a firm, different models exist. Most common is an organization that involves peer and cultural control. Another model is reliance on professional control, which means that founders take it for granted that their workers will adhere to certain standards to which they have been "socialized" (*i.e.*, hiring only graduates from elite institutions). Other founders prefer to install a system of formal processes and procedures that control and coordinate work in their company. A fourth group, although small, prefers personal control by direct oversight.

Based on these three main dimensions and their characteristics, five distinct blueprint types have been identified, *i.e.*, five main models that characterize an organization (*Table 4.4*).

It is difficult to say which, if any, of the models promises more success. Each has advantages and disadvantages and is highly dependent on the founder's personality, preferences, and understanding of the world. However, what we can say is that organizations rarely change their blueprint on more than one dimension simultaneously and rarely move between blueprints. Generally, a change in the organizational blueprint – for example when there is a new top management team – is frequently experienced as quite destabilizing and adversely affects employee turnover. Therefore, selecting an initial blueprint that adequately suits the anticipated future strategy might be better than selecting one that is ideally suited to the current situation but likely to fail in the future. Being more aware about these choices can be a vital first step toward envisaging a platform for organizational growth.

Table 4.4 Typology of Employment Blueprints [26]

Employment blueprint	Dimensions			Typical statement
	Attachment	**Selection**	**Coordination**	
Star	Work	Potential	Professional	"We recruit only top talent, pay them top wages, and give them the resources and autonomy they need to do their job."
Commitment	Love	Fit	Peer/Culture	"I wanted to build the kind of company where people would only leave when they retire."
Bureaucracy	Work	Skills	Formal	"We make sure things are documented, have job descriptions, and pretty rigorous project management techniques."
Engineering	Work	Skills	Peer/Culture	"We were very committed. It was a sunk-works mentality and the binding energy was very high."
Autocracy	Money	Skills	Direct	"You work, you get paid."

4.3.3 Managing People

Managing a venture's growth is largely about managing people. In particular, entrepreneurs must be able to recruit, motivate, and retain high-quality employees [27]. Yet many firms with a more engineering-type approach – as can be observed in many science-based startups – report that organizational and HR aspects are of no concern when launching the firm [28]. Our cases illustrate that this can backfire, especially in a high-growth environment. Two capabilities can help an entrepreneur overcome some of the shortcomings. One key factor – especially in the early phases of a company – is to establish trust. Trust is set up by being credible, *i.e.*, doing what you say you will do, being open, honest, and direct. A second factor is the ability to resolve conflict. Especially, but not exclusively, in science-based startups, people are usually highly educated and they are used to working independently toward a set of objectives – often their own. Conflict will readily arise when resources are tight yet priorities are not shared. Hence, entrepreneurial managers need to be skilled in resolving differences, balancing viewpoints, building teamwork and consensus, and creating an atmosphere of trust and fairness [29].

4.3.3.1 Motivating Employees

At times, entrepreneurs may have difficulty accepting that not everybody is as self-motivated and goal focused as they are. Hence, entrepreneurial managers should start thinking about what makes their team members work and which types of jobs people actually find motivating. Understanding intrinsic motivation helps to incentivize people when financial resources are scarce. Corporate culture, the level of challenge of tasks, and the level of responsibility are possible enriching factors. Furthermore, people are motivated by the prospect of success. Goal setting is thus one of the most important tasks of a manager. However, goals need to be challenging and attainable, inspiring, and measurable. This is vital to be able to give employees feedback on their performance in relation to specific goals. Such feedback can, in turn, help overcome the threat of losing key people as it can work both ways. It helps to nurture a higher level of commitment and loyalty among employees and increases their personal involvement in the organization.

4.3.4 Other Necessary Management Competences [30]

Besides being good people managers, entrepreneurs need command over a whole range of other competences, mostly beyond their original education, technically specialized, and, often, cross-functional. *Table 4.5* indicates a list of the most common skills that startup managers need to bring with them or will have to learn over the course of their endeavor.

Table 4.5   Other Necessary Management Competences

Administration	Problem solving, communication, planning, decision making, project management, negotiating, managing outside professionals, personal administration
Law and taxes	Corporate and securities law, contract law, law relating to patent and proprietary rights, tax law, estate law, bankruptcy law
Marketing	Research and evaluation, marketing planning, sales management, direct selling, service management, distribution management, product management, new product planning
Operations/production	Manufacturing management, inventory control, cost analysis and control, quality control, production, scheduling and flow, purchasing, job evaluation
Financing	Raising capital, managing cash flow, credit and collection management, short-term financing alternatives, public and private offerings, bookkeeping, accounting and control, other specific skills
Technical skills	Unique to each venture

It is unlikely that many entrepreneurs will possess all the required skills in this list, let alone excel at them. This is particularly true for specialized knowledge such as patent law but also applies to a broad spectrum of general management tasks. Even if the venture can draw on a management team with complementary skills, the business will typically have to rely on experts. Outside experts can be a vital source of dialogue, yet they are rarely as attached to, or as familiar with, the company as the entrepreneur. Hence, being able to constructively work with their suggestions, balancing internal and external views, making use of advisers and board members' expertise without giving up control, is essential.

4.4 The Troubled Company

Most startups encounter their fair share of ups and downs in their evolution. The case examples in this chapter are certainly no exception to this reality of new venture formation, where growing a firm appears as a highly uncertain process. Changes in the political or economic climate, for example, can heavily affect target markets and the basic assumptions for running the business. A number of problems, however, can be traced back to inattention to strategic issues or poor financial planning [31].

4.4.1 Strategic Issues

Usually, a strategy has been defined early in the startup's life – on somewhat shaky grounds. It is unlikely to stay the same for years. Along the way, the entrepreneur learns from customers and competitors alike. Especially in high-tech and science businesses, circumstances can change quickly: New research results are published, new ways to solve a problem are developed, target markets shift due to enhanced product features and distribution capabilities, and the like. Likewise, entrepreneurs often realize only as they progress that their technology or product has the potential to fit a variety of other purposes beyond those originally identified. This calls for a careful reassessment of strategies over time and ongoing dialogue with key stakeholders.

Attending to the strategic issues frequently causes the entrepreneur multiple headaches. He or she is usually bogged down in day-to-day operations and thus not able to push the high-level planning. As long as the business is running smoothly, the need for strategic revision goes unnoticed and often problems have already grown quite a bit when the first signs can no longer be ignored. Unfortunately, this is particularly true when launching new products, as inevitably milestone setting and specifying sales targets is more of an art than a science – leaving plenty of room for interpretation in judging true progress. Eventually, board members and investors may start to deviate from company management in their assessment of the situation – or even differ among themselves. Maintaining an aligned view and ambition may become increasingly difficult, and ultimately it may seem impossible to change the way things are being done without making significant changes to the company's direction and management.

4.5 Cases in This Chapter

The first two cases in this chapter focus specifically on growth strategies. To begin with, we will revisit the sensor technology company IR Microsystems, which we already know from Chapter 2, where the setting up of the venture and the early definition of its business model were described. In *IR Microsystems (B): Taking Tunable Diode Laser Spectrometry to Market in 2006* we observe that the initial setup has not brought the envisioned success. In the four years since IR Microsystems began, the company has grown to eight employees and gathered some external funding. However, the initial product (or market) has just not taken off. In fact, company growth has been rather dissatisfying and investors are beginning to question their returns. The team has to make some serious decisions about how to move forward. So far, it has advanced several business models but has not managed to penetrate the market. In the (B) case, they are back to zero: A new product concept is to bring salvation. All efforts are refocused on tunable diode laser spectrometry, a laser gas detection technology with low-cost promise. Will the team be able to capitalize on their learning from the previous four years to establish this business? In trying to find an answer to this question, the case touches on several issues: IR Microsystems has to decide which strategy to pursue – building it by itself *vs.* looking for the "right" partners. It has to revalidate its market and rebuild its supply chain. A complex challenge awaits the team – balancing long-term growth with short-term cash generation. Will they be able to solve this dilemma?

IR Microsystems (C): Epilogue provides an overview of the key milestones in 2006 and the near-term strategy for the year to follow. Still, the critical challenges remain: To find sufficient development contracts, secure the financing, and resolve the perennial laser supply problem.

The next case, *AC Immune SA: Taking Research into Development*, examines the development of and a strategy for a biotech startup that performs research into cures for Alzheimer's disease. Following a successful B-financing round, the company's management identified three strategic tasks to take the venture further: (1) master the transition from a research organization to a development organization, given the existing resource constraints; (2) build a value-optimized product portfolio while keeping options open; and (3) adjust partner or licensing strategies to the latest financial situation to allow for maximum value creation. While discussing the case and how best to tackle the three challenges, students will be motivated to identify capabilities and qualities that help in leading a small venture through the stormy and uncertain terrain outlined in the case. It also provides opportunities to discuss external growth strategies.

To provide an update on AC Immune's development, we have included a recent press release, in the form of a handout. The news announces a megadeal the venture team was able to strike with renowned Genentech. The deal secures product development over the course of the following three years, a tremendous achievement underlining again the fine quality of AC Immune's experiences and dedicated management team.

Following AC Immune, we portray another venture in the biotechnology domain, *Covalys (A): Managing the Company's Growth and Development Strategy*. Basel-based Covalys develops and markets diagnostic kits that can be used in industrial and academic biomolecular research. Although the venture has met all its development milestones up to the point where the case begins, revenues are still below target. Key investors are increasingly concerned and have started putting more pressure on the management team to raise the company's value for a future trade sale or alternatively manage for a self-sustained and profitable company. CEO Christoph Bieri finds himself in the position of having to decide between two alternatives – and with the available resources, it is not possible to pursue both options half-heartedly. First, Bieri could invest in further developing, *i.e.*, enlarging, the product portfolio. This would allow for the diversification of risk and would generate future business potential, thus making Covalys more attractive to prospective buyers. Bieri's second option is to invest in the sales force by building one up directly under the company's control and by better educating the current distributors. It was hoped that this would spur the urgently needed revenues and build up legitimacy for the venture that is still suffering from not having a track record and proven technology. Covalys is fast approaching a strategic crossroads: Will it be able to grow and generate a return for its investors?

Again, we include a recent press release to give insight into the follow-up of the Covalys case, highlighting the strategic path actually chosen by the venture team

The company in the next case, *Technology Strategy at Dartfish*, is also facing the tradeoff between increased development and increased marketing/sales efforts. Introduced as InMotion Technology Inc. in Chapter 2, the company has since been renamed Dartfish and has grown to operate internationally. However, it has still not reached a level of profitability that satisfies its shareholders. The immediate issue for Dartfish's management is to decide how to move forward with its two distinct products. Dartfish's broadcast service business generates welcome visibility, yet the actual marketing effect is hard to measure. In addition the growth potential of the broadcast business is rather limited. Although the business is still sustainable, its future looks less promising, as market dynamics point to difficulties in recouping the technical development and maintenance efforts. By contrast, Dartfish's training solutions have high growth potential in terms of generating revenues and ultimately bringing the desired profitability. However, in order to succeed, the right product strategy, *i.e.*, a balance between feature development and marketing activities, needs to be found. Therefore, if the training application were to become the future of the company, it would also mean deciding about going into additional markets, first and foremost moving from the target clientele of high-end athletes to the mass market, *i.e.*, facing a classic crossing-the-chasm problem. In addition, Dartfish would need to decide how to go about its specialized training solutions, *i.e.*, applications designed exclusively for particular sports, such as golf or swimming. In essence, Dartfish's business model and the product strategy need to be revised.

In the ensuing three-case series, we follow the growth of Netcetera AG, a Zurich-based provider of software applications for networked computing environments. The series focuses on organizational evolution and people issues in growing a venture.

Netcetera (A): Hiring an External CEO? describes developments from the company's launch in 1996 up to spring 2000, by which time it employed 30 people. The firm's four cofounders envision further expansion, yet they realize that there is no time to handle strategic issues – management resources are completely taken up with winning new customers and dealing with the internal complexity caused by the increasing number of employees. At this point, they consider recruiting an external CEO to help address their challenges. They have even identified a candidate. While the case invites the student to evaluate the pros and cons of hiring an external manager to facilitate growth, it also illustrates the issues that occur in a fast-growing organization and the challenges facing the leader team.

Netcetera (B): Organizing for Sustainable Business Success picks up where the (A) case left off. It focuses on the development of Netcetera's organizational structure between 2000 and 2005. The cofounders had tried different formats to accommodate internal requirements (employee perspective) and the external economic situation. They have finally established a holding group and assumed more of a coaching role. Operations have to be extremely efficient in order for the company to maintain its position in the face of declining growth rates and slow incoming business. However, as the overall economy seems to be picking up, Netcetera's cofounders hope to resume growth. The case clearly demonstrates the need to question and, if necessary, adapt the organizational blueprint when growing a firm. It also provides opportunities to discuss the development of management roles, the targeted business model, and whether the organization is prepared to master the upcoming challenges.

Netcetera (C): Reflections and Outlook summarizes key learning points over the firm's ten-year history and provides an update on the aspirations of the founders. The case encourages critical discussion of the strategic direction and allows students to suggest their own strategy for further growth.

The final two cases showcase the somewhat dramatic rollercoaster experience of two ventures. First, we return to a company already introduced in Chapter 2. *Boblbee (E): Inventing the Urban Nomad* follows on from where the (D) case left off. Essentially, it depicts a company that has been overwhelmed by its own initial success. Boblbee's CEO Patrick Bernstein recognizes painfully that, having failed to be rigorous enough in choosing a branding and marketing strategy, he has maneuvered the company – and his dream – into serious trouble. Branding the product as a trendy fashion item could not support sales for long, but the company's distributor system has not enabled Boblbee to foresee these issues. Boblbee then took on the distributor role itself, including the entire excess inventory accrued, incurring high risks. The team had certainly paid its dues: Finding the right people to grow internationally, protecting itself from others infringing on its patents, and finding the right approach/strategy for collaborating entailed quite a lot of learning. Two years down the road, the company finds itself in a difficult selling position, as the market does not perceive the company as it wants to be perceived. It seems almost too late to rebrand. In addition, CEO Bernstein has to make some serious private career choices.

In *EndoArt: Creating and Funding a Medical Technology Startup (A)*, we learn about a venture that develops and commercializes medical devices for the telemetric control of the flow of bodily fluids. EndoArt had begun life in 1998 and since has encountered numerous challenges and setbacks in choosing, managing, and maintaining its technological approach. In particular, the venture team learned the hard way about choosing the business partners wisely. However, to date the venture has survived and brought out two very distinct products. The case leaves the cumbersome evolution at a point where EndoArt is running low in funding and has yet to launch its main product on the market, which requires the passing of time and resource-consuming approval procedures. How should the venture proceed toward success? There are several options to pursue – trade sale, development partnership, or VC-backed funding for further progressing on its own. The case provides a vehicle for reviewing pitfalls in the venture's development and discussing how, if at all, to prevent them. It points at the concept of uncertainty and could provoke discussion about chances and strategies of a small venture in a market dominated by big and affluent industry players.

As the cases in this chapter exemplify, growth eventually requires expanding internationally. This adds a particular challenge and its scope depends on a multitude of factors, among which the business model and the specific industry play an important role. To allow for a controversial discussion of the how-tos, chances, and limits of international growth in different industries and business models, we conclude this chapter with a contrasting case example – centering on *the* prototype of the Internet business. The *Google Search Engine and Advertisement* case highlights Google's staggering rise from a classic garage startup to the most popular Internet search engine today. On its way, the company has engaged in a high level of external partnering and acquisition activity. Will this strategy continue to support its strong competitive position – in particular against the omnipresent Microsoft, which is about to hit the market with its own service? In fact, both Google's overall strategy and its legacy business model are challenged. How should the company position itself to continue and even grow further?

References

1   Aldrich, Howard and Ellen R. Auster. "Even Dwarfs Started Small: Liabilities of Age and Size and Their Strategic Implications." *Research in Organizational Behavior*, 1986, *8*, pp. 165–198.
2   Aldrich and Auster, "Even Dwarfs Started Small."

3   Stinchcombe, Arthur L. "Social Structure and Organizations." *Handbook of Organizations*. Ed. J.G. March. Chicago: Rand McNally, 1965, pp. 142–193.
4   Barringer, Bruce R. and R. Duane Ireland. *Entrepreneurship: Successfully Launching New Ventures*. Upper Saddle River, NJ: Pearson Education, 2006, pp. 310–312.
5   Jolly, Vijay K. *Commercializing New Technologies: Getting from Mind to Market*. Cambridge, MA: Harvard Business School Press, 1997, p. 306.
6   Zook, Chris and James Allen. "The Facts about Growth." Bain Strategy Brief, 1999. http://www.loyaltyrules.com/bainweb/publications/written_by_bain_detail.asp?id=30&men u_url=written_by_bain.asp (viewed on August 7, 2006).
7   Mintzberg, Henry. *The Structuring of Organizations: A Synthesis of the Research*. Englewood Cliffs, NJ: Prentice-Hall, 1979.
8   Barringer and Ireland, *Entrepreneurship*, pp. 314–316.
9   Hay, Michael and Kimya Kamshad. "Small Firm Growth: Intentions, Implementation and Impediments." *Business Strategy Review*, 1994, 5 (3), pp. 49–68; Penrose, Edith. *The Theory of the Growth of the Firm*. New York: Oxford University Press, 1995.
10   Barringer and Ireland, *Entrepreneurship*, pp. 330–348; Coulter, Mary K. *Entrepreneurship in Action*. Upper Saddle River, NJ: Prentice Hall, 2000, pp. 284–286.
11   Barringer and Ireland, *Entrepreneurship*, p. 333.
12   Moore, Geoffrey A. *Chrossing the Chasm: Marketing and Selling Technology Products to Mainstream Customers*. Oxford: Capstone Publishing Limited, 1999.
13   Barringer and Ireland, *Entrepreneurship*, p. 335.
14   Barringer and Ireland, *Entrepreneurship*, p. 332.
15   Collins, Pete. "Partnerships Have Big Payoffs for Fast-Growth Companies." PricewaterhouseCoopers, *Trendsetter Barometer*. August 26, 2002. http://www.barometersurveys.com/production/BarSurv.nsf/vwResources/PR_PDF_Files_20 02/$file/tb020826.pdf (viewed on August 8, 2006).
16   Collins, Pete. "Alliances and Acquisitions Increasingly Important for Fast-Growth Companies." PricewaterhouseCoopers, Trendsetter Barometer. May 16, 2006. http://www.barometersurveys.com/production/BarSurv.nsf/vwResources/PR_PDF_Files_20 06/$file/tb060516.pdf (viewed on August 8, 2006).
17   Collins, "Partnerships Have Big Payoffs."
18   Collins, "Partnerships Have Big Payoffs."
19   Collins, "Alliances and Acquisitions."
20   Barringer and Ireland. *Entrepreneurship*, p. 343; Cardullo, Mario W. *Technological Entrepreneurism: Enterprise Formation, Financing and Growth*. Baldock, Hertfordshire: Research Studies Press, 1999, pp. 224–225.
21   Barringer and Ireland, *Entrepreneurship*, p. 341.
22   Coulter. *Entrepreneurship in Action*, p. 150.
23   Baron, James N. and Michael T. Hannan. "Organizational Blueprints for Success in High-tech Start-ups: Lessons from the Stanford Project on Emerging Companies." *California Management Review*, 2002, 44 (3), pp. 8–36; Coulter. *Entrepreneurship in Action*.
24   The entire section is based on the article Baron and Hannan, "Organizational blueprints for success in high-tech start-ups."
25   Baron and Hannan, "Organizational Blueprints for Success."
26   Based on Baron and Hannan, "Organizational Blueprints for Success."
27   Baron, Robert A. and Scott Andrew Shane. *Entrepreneurship: A Process Perspective*. Mason, OH: Thomson/South Western, 2005.
28   Baron and Hannan, "Organizational Blueprints for Success."
29   Timmons, Jeffry A. *New Venture Creation: Entrepreneurship for the 21st Century*. Boston, MA: Irwin, 1994, p. 216.
30   Timmons, *New Venture Creation*, pp. 219–222.
31   Timmons, *New Venture Creation*, pp. 598–600; Coulter. *Entrepreneurship in Action*, pp. 290–296.

CASE 4-1
IR Microsystems (B): Taking Tunable Diode Laser Spectrometry (TDLS) to Market in 2006

Benoît Leleux and the IMD MBA 2006 Venture IRM Project Team

By 2006, IR Microsystems (IRM) had eight employees on the payroll, all of them with technical/engineering backgrounds. The first major external funding had come in 2003 from a collection of investors, including a local VC fund, which was attracted by the growth potential of IRM's technological products.

Initially, the company had focused its resources on developing a low-cost solution (target price € 9,00) for infrared detection and gas/liquid analysis. However, sales did not take off and after a few years it was clear that IRM needed a new product. So the company refocused its efforts on the development of tunable diode laser spectrometry (TDLS), a laser gas detection technology with great promise. By 2006, pressure from investors to start generating serious revenues from TDLS, or at least to sign the first significant sales contract, was increasing.

How could the company move forward with this novel technology base? Could it capitalize on the learning of the last four years? The founders had been lucky to find investors who believed in them and the technology they had developed. It was now time to demonstrate that they could make good on their promises of market riches.

Copyright © 2007 by IMD – International Institute for Management Development, Lausanne, Switzerland. Not to be used or reproduced without written permission directly from IMD.

1 New Product Offering

From its original IR array detectors,[1] IRM had moved to offering what physicists referred to as laser spectrometry products. This advanced technology made it possible to detect quickly and without confusion (*i.e.*, cross-sensitivity) small amounts of a selected number of gases that had simple molecular structures including NH_3, O_2, CO_2, H_2O, HF, CH_4, C_2H_4, CO, N_2O, and H_2S. The range of detectable gases for this technology was driven by (1) physics, *i.e.*, only "simple" molecules and only gases in gases, but not in liquids, and (2) the cost of components, which were nonstandard laser devices (and thus required minimum orders of 1,000 units for economic viability). IRM's product – the microLGD gas detector – had almost completed its development phase. Five prototypes had been produced, and over 50 were going to be shipped to potential clients for testing in 2006.

The global market for gas detection was estimated at around US\$1 billion, with relatively low growth of 4% per annum and strong segmentation by gases and applications. The market was quite competitive; most potential customers relied on low-cost solid-state, electrochemical, and catalytic bead detectors.

IR Microsystems' offering was interesting because its detection solution could offer speed, functional safety (self testing and status control), excellent selectivity per gas, no recurrent calibration, and a lifetime of five years and more. This was a solution to some of the major drawbacks of other above-mentioned low-cost detection technologies that suffer from uncontrolled aging or poisoning, low lifetime, cross-sensitivities, and frequent need for calibration (high cost of ownership).

Over the years IRM had been able to obtain a number of patents on its measurement principles (*Exhibit 1*). This was perceived as the core know-how of the company, which could be leveraged in commercial products with, hopefully, serious profit potential.

Management forecast the target product cost to be \$50 to \$100 per unit if mass production could be achieved. There was, of course, a heavy scale effect on costs, mostly driven by the cost of the lasers. Several companies were already in the market with high-end laser gas analyzers for heavy industry applications like steel processing, waste incineration, *etc.*, where gas extraction and treatment before measurement in challenging environments represented a large part of the system cost. Laser gas spectrometry was thus generally perceived by the industry as high end.

IRM had the vision to make laser gas detection available to the safety industry at a cost that would allow it to compete with inferior low-cost detection technologies. This would be achieved by using only diffusive sampling and its intellectual property (IP), making it possible not to have a reference channel in the instrument.

This of course opened other opportunities in the medical and transportation industries. The target customers were the market leaders in the safety industry, such as Dräger, MSA, Honeywell, *etc.* Other target customers were companies

---

[1] The situation of the company in 2002 as well as its early technology development efforts are described in the case "IR Microsystems (A): June 2002" in Chapter 2.

working in fields of applications where the potential numbers of units/year would be high enough to be able to deliver cost competitive products, *e. g.*, for medical, transportation, or selected process control applications.

1.1 Challenge #1: The Business Model

Over the years, IRM had considered a number of business models and never settled on one for very long. At this point, at least five options were under consideration to allow it to move forward:

- Sell the company
- Sell part of the technology
- License the technology to a major OEM
- Manufacture the laser spectrometers in-house
- Outsource manufacturing, carry out quality control, calibration and testing in-house.

Qualitative analysis of the amount of funds required for each option and the effect of the decision on the valuation of the company led to the conclusion that only two of the alternatives had decent potential (*Exhibit 2*):

- Keeping the technology in-house but outsourcing the manufacturing of the laser spectrometers
- Licensing the technology to OEMs.

As a result of consultations and market research, management found that royalty fees for a license for similar products varied from 2% to 5% and the initial down payments would not exceed $200,000. They conducted a breakeven analysis to check the feasibility of both options based on the assumption that breakeven point would need to be reached in three years. Breakeven on the manufacturing option came at around 20,000 units, whereas it would take close to 500,000 units to break even on the licensing model. Considering the estimated market size for IRM's product was about $50 million a year, *i. e.*, approximately 300,000 devices a year, it would be virtually impossible to reach break even with licensing, even after capturing 100% of the market. In addition, the manufacturing option would allow the company to further develop the technology and keep the capabilities required to achieve continuous technological improvements in-house.

Outsourcing manufacturing was absolutely critical to avoid the necessary investments in equipment and personnel. On the other hand, small amounts could still be spent on licensing to enter markets not easily accessible for IRM (highly regulated, extremely price sensitive, distant geographies such as Japan).

1.2 Challenge #2: Keeping Focus

There was an endless list of markets and niches in which the product could be used and sold, and yet no material sales had been realized to date. Given the limited resources, it was key to prioritize the target markets and apply all resources to them.

Early market analysis revealed that the TDLS detection technology was applicable to a wide range of gases, but this did not drive economies of scope. In fact, for each gas sensed by IRM's microLGD the set of applications available was almost totally unrelated, thus giving hardly any overlap of client bases (*Exhibit 3*). The huge difference in client bases for each gas suggested a separate life cycle for each individual sensor, based on the specific gas and on the application of interest. Managing a product life cycle for more than ten gases with dozens of applications meant managing not just one product but a full portfolio of different products.

The level of complexity of such a portfolio, together with the limited resources available to IRM, suggested an early focus on markets where IRM was showing a strong technical advantage. And, once determined, IRM should concentrate all its efforts on responding to the specific customer needs for that market with a well-developed and focused product proposition. Two gases quickly stood out for which IRM's technological competitive advantage combined with strong growth potential: ammonia (NH_3) and hydrofluoric acid (HF) (*see Exhibit 4 for the technological competitive advantage and Exhibit 5 for the decision matrix*).

1.3 Challenge #3: Market Validation

A first-cut market validation for TDLS gas detection technology for selected gases (ammonia and hydrofluoric acid) was performed to gain insights into the way the TDLS technology was perceived, the purchasing criteria of buyers, their expectations of after-sales service, the overall size and segmentation of the market for gas detectors/sensors, and the regulatory environment (including certification). Two types of companies were contacted: competitors/potential customers in the industrial OEM space, and potential end users in industries such as pulp and paper, energy, and electronics.

As a result of this validation process, three themes were shown to dominate OEM behavior in this industry:

- The "not invented here" syndrome, which implied that licensing a technology from IRM would not be an easy deal to sell.
- The need to maintain existing quality and compliance certifications, which translated into the need for IRM itself to undergo an expensive certification process (ISO 900,* sensor- and country-specific, *etc.*) before actually selling its products to clients.

- The drive to lock in end users in order to uphold the OEMs' margins, which might lead to end user frustration, with OEMs being happy to create a "wall" between sensor makers and end users.

Based on these findings, a radically new segmentation of the end-user base emerged:

- About 75% of overall demand seemed to be driven by government-imposed regulations and was, therefore, very price sensitive. IRM, as a startup without production experience, was not well positioned to play a cost leader role.
- The remaining demand was driven by actual business needs such as safety. IRM seemed better positioned to address the needs of this segment, although certification was still an issue. This segment was more willing to pay a premium if it could perceive the benefits of IRM's product.

The conclusion at the end of the analysis was that IRM faced two major strategic options in the near term:

- Try to bypass the "wall" of OEMs by directly contracting with major end users with potential industry-scale demand and/or
- Identify, and then enter into alliance with, an OEM that would be ready to invest in TDLS and sell it as a part of its (turnkey) solution to its customers.

1.4 Challenge #4: Supply Chain Management

In April 2006 IRM's contact with a large manufacturer active in the transportation industry intensified. The company indicated an interest in IRM's technology to differentiate its own products from those of the competition. Its product was made on a large scale, and quality and functionality were "commoditized" in the current market. So a new gas sensing technology that did not need maintenance and reliably provided gas concentration readings fitted this customer's needs well. IRM hastened to provide this customer with a prototype device for a first gas that met its requirements. Extensive client meetings led to a better understanding of its technical requirements, the numbers of units needed and target delivery dates. The parties discussed plans to share IRM's development cost for a "3-gas-in-1" device and the potential sale of 5,000 to 10,000 units to be delivered by the third quarter of 2007. This triggered a reexamination of the supply chain arrangements needed to meet such a large order.

- The new triple-gas device and a total of about 50 components would have to be technically specified (R&D).
- Suppliers willing to accept IRM's hardly predictable R&D time requirements would have to be found.
- Assembly and testing capacity would have to be contracted.
- Various lead times within the supply chain would have to be coordinated, and manufacturing and assembly would have to be controlled and managed.

The task was highly complex and would have to be completed in parallel with the finalization of the R&D, setting up the "big-scale" supply chain and coordinating and controlling the supply chain once it was up and running. Problems in one area could easily derail efforts in another, dramatically delaying the whole process. As a consequence, the company decided to develop an integrated planning tool which would provide the opportunity to clarify and plan the challenge.

2 Next Steps

The year 2006 forced a number of key decisions on the company. First, finding an appropriate business model proved quite challenging. An evaluation of the different possibilities had shown that it was probably better for IRM to manage the manufacture of the gas detectors, as opposed to simply licensing the TDLS technology to an OEM. Second, it was hoped that focusing the sales effort on mainly two gases (ammonia and hydrofluoric acid) would avoid spreading the limited sales resources too thinly. The market validation of the TDLS technology had led to valuable new insights into the market and potential customers for IRM. Finally, the development of an integrated planning tool put the company on the right track for the eventual launch of a major production effort.

In the short term, IRM's priority remained to find the fastest way to cash. The plan to get there was to focus on a more limited set of gases and applications. The main challenge remained evolving from a "engineering" operation to becoming a fully fledged gas detector manufacturer.

Exhibit 1
Micro LGD Technology: Theory

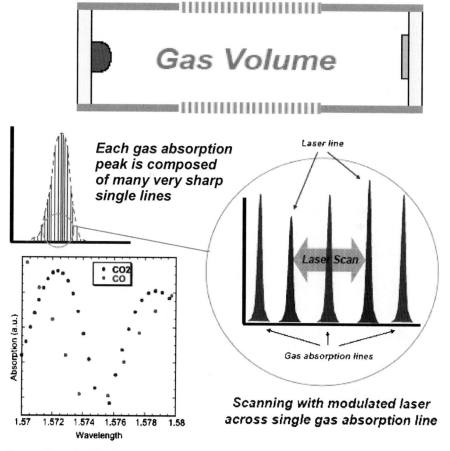

Source: Company literature

Exhibit 2
Challenges: Business Model Decision Tree

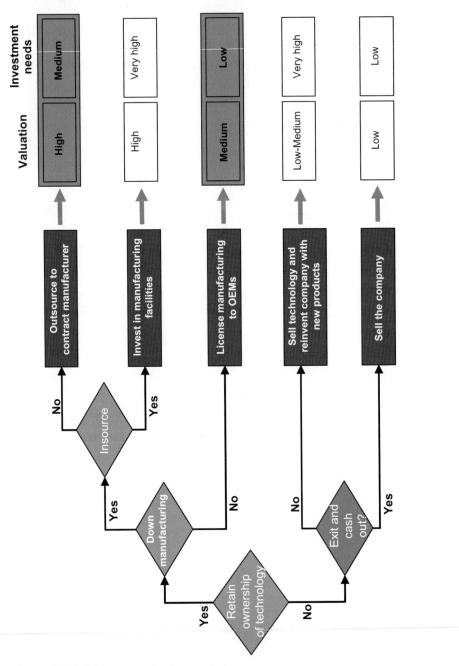

Source: IMD MBA venture project team analysis

Exhibit 3
Gases vs. Applications: No Client Base Overlap

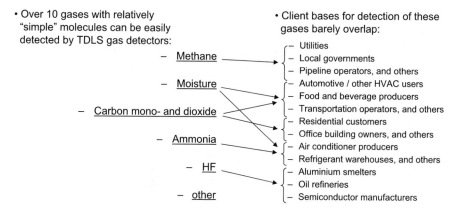

• Over 10 gases with relatively "simple" molecules can be easily detected by TDLS gas detectors:

- Methane
- Moisture
- Carbon mono- and dioxide
- Ammonia
- HF
- other

• Client bases for detection of these gases barely overlap:

- Utilities
- Local governments
- Pipeline operators, and others
- Automotive / other HVAC users
- Food and beverage producers
- Transportation operators, and others
- Residential customers
- Office building owners, and others
- Air conditioner producers
- Refrigerant warehouses, and others
- Aluminium smelters
- Oil refineries
- Semiconductor manufacturers

Source: IMD MBA venture project team analysis

Exhibit 4
TDLS vs. Other Sensing Technologies: Key Advantages

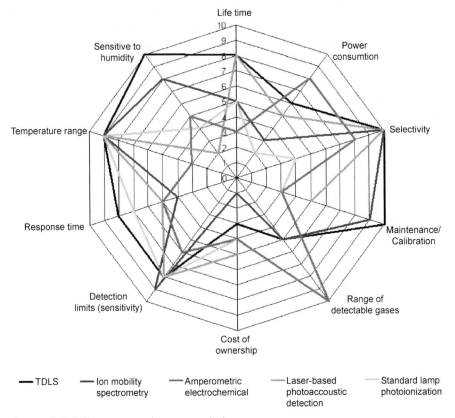

Source: IMD MBA venture project team analysis

Exhibit 5
Gas Decision Matrix

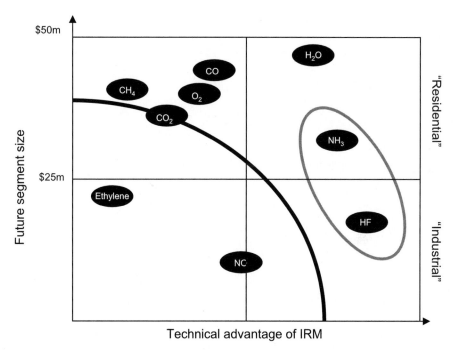

Source: IMD MBA venture project team analysis

CASE 4-2
IR Microsystems (C): Epilogue

Benoît Leleux and the IMD MBA 2006 Venture IRM Project Team

1 IR Microsystems' Outlook for 2006

In 2006 IRM's business model settled into the following mode:

- In-house: engineering, product design to OEM specs, quality control (calibration and testing)
- Outsourced: detector manufacturing (electronics, opto-electronics).

July 2006. IRM attracts key person with a 20-year extensive background in gas detection and safety instrumentation, as well as in-depth market knowledge as president of the company. IRM's new president built a gas detection company in Switzerland with several CHF million turnover. He then sold the company to one of the North American gas detection and safety market leaders.

IRM finds major interest from two of the three gas detection and safety market leaders. This is along the lines of IRM's near- and mid-term strategy decision to enter alliances with market-leader OEMs that will integrate TDLS technology into next-generation gas detection products.

IRM focuses on adapting solutions to concrete customer demands: NH3, O2, H2O, CO2, C2H4, combustibles.

September 2006. IRM needs to focus its resources and ceases small-scale production and commercialization of its microray and LVF spectrometer product lines (its original products). Production for orders of several hundred units can still be discussed on a case-by-case basis. Potentially the technology (with IP and know-how) is made available for a technology transfer.

October 2006. IRM continues to innovate and extends TDLS technology platform for detection of hydrocarbons using resonant photoacoustic laser spectrometry.

November 2006. Parallel negotiations with two OEM market leaders in gas detection and safety lead to a CHF 1 million development and product commercialization contract with one of them. Exclusivity for NH3, hydrocarbon gases.

Copyright © 2007 by IMD – International Institute for Management Development, Lausanne, Switzerland. Not to be used or reproduced without written permission directly from IMD.

Signature of a development and product commercialization contract with a French company for oxygen and moisture detection in the petrochemical industry.

IRM investors remain committed and secure financing until beginning 2007.

December 2006. Negotiations with OEM market leader in medical technology leads to term sheet for product development and commercialization.

Close cooperation with VCSEL as well as DFB laser manufacturers to surmount the critical issue of low-cost laser supply.

1.1 Key Elements of Near-term Strategy

Financing for 2007 will come mostly from development contracts, last quarter product sales, and some additional capital injections by investors.

Breakeven is expected in 2008, mostly out of product sales and some development contracts.

The critical challenges remain to find sufficient development contracts for 2007, secure financing, and resolve the perennial laser supply problem.

2 IR Microsystems Acquired by Leister Process Technologies

In a press release from its headquarters in Sarnen, Switzerland, Leister Process Technologies announced its acquisition of IRM on July 31, 2007, hailing it as an important milestone in Leister's long-term growth strategy:

> [...] "IR Microsystems diode laser based detection technology has great potential in the gas detection and gas monitoring market for a wide range of gases. We found their ammonia detector to show the best performance we have seen so far. It is an ideal extension to our activities in the NDIR (non dispersive infrared) gas detection market which we are currently serving with our black body IR source," says Thomas Hessler, Director of Leister's Axetris Microsystems Division.

> "The acquisition by Leister backs IR Microsystems by a highly performing industrial partner with a long standing. In the critical phase of product industrialization, this gives us an extremely important leverage and is equally beneficial to our customers," says IR Microsystems Co-Founder Bert Willing.

> Privately owned Leister Process Technologies (www.leister.com) is the market leader for hot-air and diode laser based equipment for plastic welding and industrial process heat applications. With its business unit Axetris Microsystems (www.axetris.com), Leister is giving OEM customers with MEMS based sensors and micro-optical components in industrial, telecommunication, medical and automotive applications. Axetris Microsystems operates its own 6″ and 8″ MEMS foundry running processes for its products and external customers. [...]

CASE 4-3
AC Immune SA: Taking Research into Development

Ralf W. Seifert and Jana Thiel

LAUSANNE, Switzerland, May 11, 2005 – Today, AC Immune, a Swiss biotech company developing innovative therapies against Alzheimer's disease, announced the successful completion of its round B financing. CHF 21 million was raised from private and strategic investors. Two-thirds of the invested capital was contributed by new investors.

It was one of only five financing rounds for European biotech to achieve more than CHF 20 million in the first quarter of 2005. The news ended months of hard work and difficult negotiations for cofounder and CEO Andrea Pfeifer and CFO/COO Armin Mäder.

> Actually, our negotiations became easier when we started asking for €15 million instead of €10 million. The higher amount was more attractive to investors, as it would allow us to take at least two projects to clinical trials, and thus raise the company's valuation. The other advantage was that for the next three years management would be able to fully concentrate on operational tasks other than raising money.

The financing was also proof of the investors' trust in the company's technological edge, its intellectual property (IP), and its management capabilities.

Shifting gear, the management team needed to ponder its next strategic moves for the future. CEO Pfeifer envisaged making AC Immune one of Europe's leading biotech companies for neurological ailments. The challenges and tasks ahead were soon identified:

- Master the transition from a research to a development organization.
- Build a value-optimized product portfolio while keeping options open.
- Adjust the partner or licensing strategy to the latest financial situation for maximum value creation.

Copyright © 2005 by the alliance for technology-based enterprise (IMD, EPFL and ETH Zurich). Copyright permissions are handled by IMD, Lausanne, Switzerland.

1 The Opportunity

The pharmaceutical industry, and specifically its biotech sector, offered attractive market opportunities. These were, however, accompanied by high risks. Developing a pharmaceutical product was a lengthy, highly uncertain, and expensive process. Usually it took 10 to 16 years, and the total costs could amount to $800 million [1], if one took into account the number of compounds that had to be developed to get one product to market. (*See Exhibit 1 for more information on the drug development process.*) To compensate for these risks, companies aimed to develop "blockbusters" – drugs with sales in excess of $1 billion.

One of these blockbuster opportunities was offered by the Alzheimer's disease market. Alzheimer's – an age-related disease – was one of the most expensive ailments in society. Its diagnosis, subsequent treatment, and care costs were exceeded only by those for cancer and heart disease. By 2005, an estimated 18 million people worldwide suffered from dementia, of which Alzheimer's was the most common form. Given the aging population, this number was expected to nearly double by 2025 [2] In 2005 none of the marketed Alzheimer's drugs was able to provide a cure; the symptoms could only be mitigated. (*See Exhibit 2 for additional information on Alzheimer's disease.*)

> Alzheimer's is a terrible disease – think of parents no longer recognizing their children. The necessity and usefulness of our research cannot be doubted. Likewise, the market is large. Whoever first presents a breakthrough technology does not need to question the economic success – there will be a significant return.

AC Immune SA set out to seize this opportunity.

2 AC Immune SA

AC Immune SA was incorporated as a private company in 2003. Its mission was finding and commercializing therapies for previously untreatable diseases such as Alzheimer's and Cancer. The startup was located in the Scientific Park (PSE) of the Ecole Polytechnique Fédérale de Lausanne (EPFL), leveraging the excellent conditions of the Lake Geneva region – a leading biotech area in Europe. Research laboratories were maintained at the Institut de Science et d'Ingénierie Supramoléculaires (ISIS) in Strasbourg, France.

The venture's distinguished founders and its scientific advisory board headed by the French Nobel Prize laureate Dr. Jean-Marie Lehn provided a true hallmark. In addition, AC Immune was managed by an experienced leadership team. CEO and cofounder Dr Andrea Pfeifer had previously been the head of Nestlé's Global Research. She had 15 years of senior management expertise in international R&D.

My experiences from working with Nestlé have proved invaluable. I don't believe that we would otherwise be as successful as we are today. It is fundamental to know how to set up effective business structures. Being able to build on a broad personal network is also crucial. If we already knew a potential partner, we got an appointment within a week. Where we had no prior personal contact, they never got back to us.

Pfeifer was joined by CFO/COO Dr. Armin Mäder, who had six years of experience in project management and business development from his work for UBS investment bank. (*See Exhibit 3 for more detailed information on the AC Immune team.*)

AC Immune had started with € 2.1 million in seed money from private investors and international business angels. In its first two years of operation, the company focused primarily on delivering proof of concept for the immunotherapy, which was accomplished by early 2005. The CHF 21 million raised would now allow these results to be taken to clinical trials. It was hoped to complete phase I/IIa of clinical trials of the two vaccine products by 2007. Market entry was expected for 2010.

AC Immune had successfully taken its first steps in a market full of risk and uncertainty. Obviously, the right strategies had been chosen to make this venture exceptional. AC Immune was repeatedly reputed to be one of Switzerland's top (biotech) startups.

3 The Venture's Assets

AC Immune's two prime assets were its technology, including IP, and the experience of its management team.

It is hard to tell which exactly sets you apart, whether technology or management. But I have seen so many companies run into difficulties along the way – and it was always good management that got the company out of danger.

With respect to business expertise, both the company's leadership team and its board of directors were well balanced. Besides Pfeifer, whose leadership background was undisputed, special entrepreneurial experience was added by the venture's CSO, Claude Nicolau, who had previously been involved in two startups.

Our CSO has this rare ability to see a test tube and, at the same time, a product.

The team was completed by Mäder who contributed financial and additional business strategy expertise. However, once the right management team had been assembled, it was back to technology issues and the strength of AC Immune's IP position, since without it there would be no startup.

It is this cutting edge which enables you to compete and attract interest in the market.

The originality of AC Immune's technological approach stemmed from the combination of both immunology and chemistry platform technologies.

The decision regarding our technological approach was no easy run. It was heavily discussed, and eventually we decided on the one which would presumably generate the most value.

AC Immune's first platform, supramolecular antigens for immunotherapy, aimed at destroying the body's immune tolerance to its own protein. This was achieved by generating antigens that led to active (vaccine) and passive (antibodies) immune therapies. Proof-of-concept testing had successfully demonstrated behavioral improvements in transgenic mice. The company planned to humanize its leading Alzheimer's antibody and to enter toxicology tests with at least one of its programs, both in 2005.

The immunotherapy approach was complemented by a second platform focused on chemistry. AC Immune investigated small molecules, called Morphomers™, which were able to create drugs capable of modifying the state of a targeted protein and render it harmless. The conceptional work on the Morphomers™ platform had started in January 2004 and was expected to reach a first candidate preclinical evaluation in 2006.

Besides focusing on Alzheimer's, AC Immune developed vaccines against multi-drug-resistant cancers (MDR1) (*Figure 1*).

AC Immune's technology could be leveraged further, later on. The venture considered future collaborations to develop active immunization for other pathologies such as Parkinson's disease and to investigate diagnostic methods for Alzheimer's. Diagnostics, in particular, provided its own business opportunity. Currently doctors had to rely on a list of signs rather than a tool to determine whether or not a patient suffered from Alzheimer's. Developing reliable and cheap diagnostics would grow the overall Alzheimer's drug market by identifying a larger number of patients at a much earlier stage.

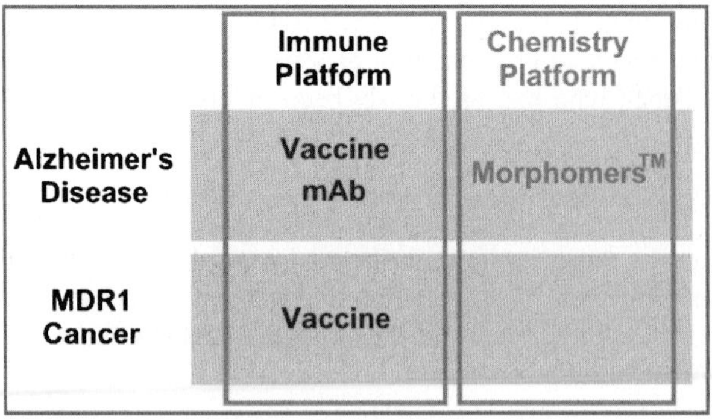

	Immune Platform	Chemistry Platform
Alzheimer's Disease	**Vaccine mAb**	Morphomers™
MDR1 Cancer	**Vaccine**	

Figure 1   AC Immune's Product Portfolio
Source: AC Immune (2005) Company Presentation, IMD, May

Creating options is key to our business. This is biotech – X percent of projects fail. As a "one-compound" company you will not survive if a project doesn't turn out as planned, and you certainly won't find a serious investor.

From an investor's perspective, balancing risks was a major issue. If a chosen path did not prove effective during clinical trials, the whole venture would be at stake. One of AC Immune's advantages was the flexibility of its platform approach: Different products building on the same technological foundations could be investigated and developed with low additional investments.

4 Protecting Assets

One of the first questions any potential investors raised was that of protecting the company's IP.

In biotech it is imperative to protect your approach. The consequences of neglect would be devastating. By the end of the day you need to market your product and if you cannot do so because of a fuzzy IP position, your entire work has – economically speaking – been in vain.

Several patent applications protected AC Immune's platform technologies. All patents and patent applications were the exclusive property of the company, except for its MDR (multidrug resistance), which was coowned by the University of Reims and AC Immune.

Filing patent applications was frequently accompanied by a thorough external patent review – an activity usually carried out by professionals. AC Immune relied on a US-based patent expert, with whom Pfeifer had previously worked.

When it comes to your patent application, you don't want to save money. You want the best possible attorney to work for you – which is what we did.

5 The Competitive Landscape

In light of the research results, we are positive that we have a sound basis for the protection of our intended business goals. Yet our life depends not only on a thorough patent review but also on timing. Hence, we have to watch our competitors closely. Today, we know our competition almost better than some of the large players active in this field.

Competitors used different research platforms, aiming either to mitigate symptoms or to develop cures (*Table 1*). AC Immune focused on cures. Its proclaimed differentiator from the competition was that by addressing very specific and well-defined targets, the potential product would cause fewer side effects.

Table 1 Competitive Landscape Based on Technological Approach (May 2005)
Source: AC Immune 2005, Company Presentation, IMD, May

	Target		**Main Players**	**Status**
Symptoms	Neurotransmitter		Large pharma	Main group of drugs on the market
	Anti-inflammatory molecules		Diverse group	Limited success
Potential cure	α-secretase activators			Via PKC activation
	β and γ–secretase inhibitors		Large pharma	No clinical candidates (β), potential side effects (γ)
	Tau		Biotech	Potential target for treatment
Cure/ treatment	Antiamyloid immunotherapy	ACI	Elan/Wyeth, Eli-Lilly, Novartis/Cytos, Roche/Morphosys	Potential for first real treatment
	β-sheet breakers	ACI	Biotech and large pharma	Existing clinical molecules

As of 2005, 15 companies were identified as being in direct competition with AC Immune. In a first assessment of their competitive strength, these players could be positioned along two dimensions: their technological platform and available resources [3] (*Figure 2*).

The viability and strength of the *technological platform* was determined by the time that would elapse before the potential drug reached the market. Competitors were ranked on a scale from 1 (research in preclinical phase) to 5 (moving the compound into Phase III clinical trials soon). The second dimension described to which extent *available resources* would allow the potential drug to be launched in the market. Again, competitors were ranked between 1 (seed financing or difficult financial situation) and 5 (major pharmaceutical or top biotech company).

The competitive space defined in this way could be split into three tiers: A, B, and C. Tier A companies (Sanofi, Neurochem, and Neurosearch) commanded both the required financial strength and a significant technological platform. Tier B companies lacked strength in one of the two dimensions but were serious players nonetheless. Tier C included companies at an early stage of research, carrying a high technological and financial risk.

One of the market players would eventually come out with the first product – it was just a question of time.

> We do not fear being a year late. First of all, our technological approach is outstanding, and secondly, the Alzheimer's market is a $4 billion market which offers attractive profit opportunities even for the second or third entrant. Moreover, as soon as the first is out there, large pharmaceutical companies will need to catch up quickly, which makes companies like ours even more attractive.

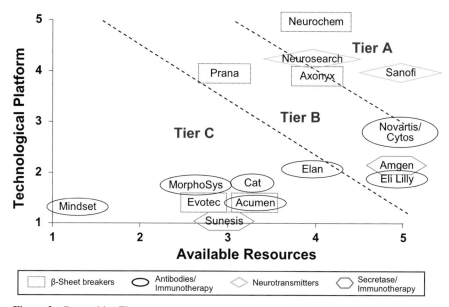

Figure 2 Competitive Tiers
Source: Saral *et al.* (2004) "AC Immune" MBA 2004 Start-Up Project, Final Presentation, IMD.

6 Industry Characteristics [4]

Traditionally, pharmaceutical companies had followed an integrated strategy that included development, patenting, manufacturing, promotion, sales, and distribution – doing all of these themselves. But the settings had changed. Due to the increasingly heterogeneous research approaches and capabilities that were subsequently required, companies were no longer able to cover all of these value steps on their own. Outsourcing of services as well as in-licensing had become a legitimate option and an increasingly strategic choice to speed up the development process and mitigate risks. This had opened opportunities for a growing number of smaller and specialized players.

The biotech scene in particular had enjoyed considerable growth over the years. In 2003 US biotech turnover was $17 billion. The importance of biotechnology was also highlighted in the new drug approval statistics – already half of all new drugs approved were biopharmaceuticals, *i.e.*, drugs created through genetic manipulation of living organisms (*Figure 3*).

The biotech industry was diverse, hosting three groups of players: (1) dedicated biotech firms, (2) traditional pharmaceutical companies marketing drugs which were developed either internally or by biotech firms, and (3) a specialized tier of firms serving both the pharma and the biotech industries with platform technologies that could speed up the drug discovery process. Pure biotech companies had

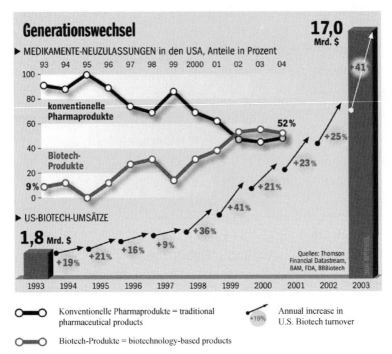

Figure 3 Share of New Drug Approvals in the U.S. and U.S. Biotech Turnover Industry Statistics: 1994–2003
Source: Der SPIEGEL, No. 15, May 2005, p. 86

only recently grown larger and profitable. Prominent examples were Amgen ($3.5 billion), Genentech ($1.7 billion), Serono ($1.3 billion), and Biogen ($1.0 billion). Yet compared to the pharma giants, these firms were still small and constrained by limited resources.

> These current market specifics result in two issues: First, it will be impossible for a startup to launch a product alone. Even a company as large as Serono needs a partner in the US to market its product because it is still too small. On the other hand, big pharmaceutical firms depend on the acquisition of promising approaches from external sources to complement their own R&D efforts.

In essence, both sides of the market were growing: More biotech startups were created (supply side) and, at the same time, there was a higher demand from big pharmaceutical companies (demand side). Nonetheless, gaining access to these big players had become increasingly difficult. Longer due diligence and negotiation processes meant deals took longer to sign – usually up to three years from the first contact with the first company to the signing of the actual contract with the nth company.

7 Developing Options in a Dynamic Environment

> Knowing that a company such as ours could not go all the way alone, we had to discuss licensing and partnering options.

In early 2004 AC Immune was on the verge of signing a high-value deal with one of the major pharmaceutical companies – a huge opportunity for the small venture. But reality took a different route. The potential partner, becoming entrenched in merger and acquisition activity, unexpectedly withdrew from negotiations for reasons beyond AC Immune's control.

> There are things which you cannot control. Our seed investors lost considerable faith in us in this process – even though it was not our fault. But you have to deal with it and carry on.

The experienced management team had succeeded in keeping parallel options open, one of which eventually played out. In November 2004 AC Immune was able to sign a collaboration agreement with a major European pharma company for compound testing in its core area of research. The agreement helped to build a track record and reinforce the venture's credibility. It led to first revenues, although they were too limited to sustain the venture's operations in the long run.

> In the end it comes back to being able to finance your business. We learned that one of the first questions asked by any potential partner is whether or not you are still financed.

As is typical for startups, manpower was scarce. AC Immune did not have resources to look simultaneously for both investors and business partners. Thus, by fall 2004, management decided to concentrate on independent financing for the time being – eventually raising a staggering CHF 21 million in May 2005, proving that their strategy had been right.

> Thanks to the funds raised, there is no immediate pressure on us. We are in the very comfortable position of having enough time to carefully plan our next steps.

With the company freed from financial worries, the collaboration and licensing discussion regained momentum. In addition, issues such as optimizing the product portfolio and refocusing the business orientation emerged.

8 The Next Strategic Moves

8.1 Master the Transition from a Research to a Development Organization

AC Immune's original setup required a strong focus on research to justify, through a proof of concept, further investments in the company's infrastructure. Having successfully accomplished that, AC Immune needed to shift the focus from its exclusive research orientation toward development. Infrastructures and systems to meet the demands of product submission, approval, and, eventually, launch had to be built. This was clearly a mind-shift issue. Yet, it was also a question of availability of adequate resources and capabilities. Sooner rather than later AC Immune would need more experience in pharmaceutical industry dynamics. Negotiations had started to add a new board member with a strong pharmaceutical background.

However, it was unlikely that all necessary knowledge and capabilities would ever be found in-house. A small company was simply not able to employ and occupy all the required specialists full-time. At this time, the market increasingly offered outsourcing opportunities which the venture could use to acquire additionally needed competencies [5].

8.2 Build a Value-optimized Product Portfolio While Keeping Options Open

Maintaining two technological platforms allowed risks to be balanced and offered innovative product opportunities. Yet it was financially impossible to drive both platforms forward at the same speed. AC Immune needed to concentrate all its forces on bringing one product to market, and it looked like the antigen platform was the closest to reaching this goal. Nonetheless, the chemical approach held huge potential.

> Scientists agree that our chemical approach is very powerful. There are only a few companies that are able to compete. We do own a core competency there, but right now we are not able to extract the entire value.

Just recently, AC Immune had been approached by a prospective buyer interested in acquiring the entire chemistry platform, including all IP and subsequent rights, for an undisclosed amount. Once again, the venture faced a decision on balancing risks against resources. Without the molecules the venture's product platform would still offer several opportunities. Cancer and Alzheimer's treatments could still be pursued. On the other hand, AC Immune would give up a competency that could become a major asset in future.

8.3 Adjust the Partner or Licensing Strategy to the Latest Financial Situation for Maximum Value Creation

Balancing risks against resources also required due consideration of *timing* and *scope* with regard to licensing and partner strategies. To actually capture value – *i.e.*, to generate cash flow and eventually profits – different options were discussed:

1. *Out-licensing*: Out-licensing would give external parties access to the venture's IP for specific purposes, which could be either an application of a research platform or a specific therapeutic function. Out-licensing would have the advantage of minimizing future risks while keeping ownership of the core technology and being able to extract higher value from licensing to different firms. On the other hand, AC Immune would risk both a reduction in future business options and a lower company valuation due to risk transfer to the licensee.

2. *Full Partnering*: The successful negotiation of a high-value partnership agreement would allow for expertise sharing and integration. AC Immune could benefit from this option by extending its product development expertise, generating a higher firm valuation through milestone payments, and finally earning credibility in terms of launching a product in the market. On the other hand, the venture faced the risk of being exclusively "chained" to one partner. In addition, the intrinsic research risk would remain with AC Immune.

Naturally neither alternative was free of risks, and the expected payoff of each would have to be evaluated in detail. Generally speaking, the further a company was able to carry its product through clinical trials, the less risky it became to potential partners, and the higher the potential return based on company valuation could be.

Round B financing had reached the bank accounts, and there were many paths that could be followed to develop the venture. Decisions had to be taken to make the most of the opportunity. The race was on.

Exhibit 1
Pharmaceutical Research and Development Stages [6]

The road to a new drug launch was long and winding. The overall discovery and development process could take 10 to 16 years (*Figure 4*). In addition, there was a large degree of uncertainty in the pharmaceutical R&D process. Fewer than 1 out of 25 compounds (drug substances) that entered clinical trials actually made it to market and brought a return on investment. This meant it was crucial to maximize the commercial value of every registered drug – before competitors caught up.

New drugs were given 20 years of patent protection in Europe and the USA, running from the first patent application, which was usually filed in the early discovery phase. The protection period included time-consuming and heavily regulated clinical testing that took place before a product could be considered for approval. National authorities, such as the Food and Drug Administration (FDA) in the USA, handled the registration of new drugs, which required substantial preparation and documentation.

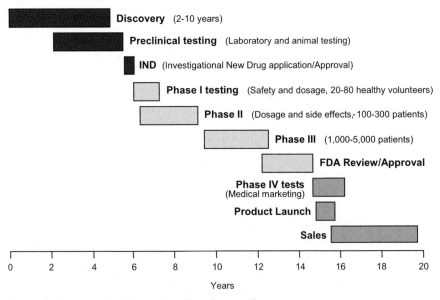

Figure 4 Pharmaceutical Research and Development Stages
Source: Seifert *et al.* (2004) "Outsourcing patterns in the pharmaceutical industry," IMD.

Exhibit 2
Alzheimer's Disease: Background Information [7]

Alzheimer's disease (AD) is an irreversible brain disorder that results in memory loss and in behavior and personality changes. The disease is mainly age-related; symptoms appear mostly after age 60.

In its early stages, AD frequently goes unrecognized and is brushed off as forgetfulness. As the disease progresses patients begin having trouble with daily activities such as driving, shopping, and following instructions. The symptoms become worse until patients are eventually incapable of looking after themselves and need constant supervision. In the later stages, all cognitive abilities are lost, and some patients may experience hallucinations, delusions, and uncontrolled outbursts. On average, patients live for 8 to 10 years after they have been diagnosed with AD.

AD damages nerve cells in the brain by inhibiting the chemical messaging system and thus halting communication between the cells, which eventually die. The underlying cause of this behavior is not yet completely known. Essential research continues to advance understanding.

Diagnosis of AD is usually made using tests for cognitive impairment, sometimes but not often cross-checked by brain scans. In general, these studies rule out other causes of dementia, such as seizures and cerebral bleeding, but they are not able to identify AD. The only was to test conclusively for AD is through an autopsy.

Figure 5 highlights some of the key platforms that were pursued in the search for a cure for Alzheimer's. By 2005, the vast majority of research into a cure and diagnostics was at the premarket stage.

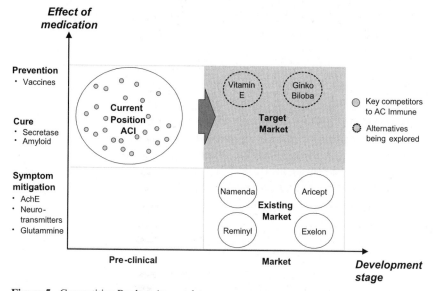

Figure 5 Competitive Product Approaches
Source: Saral *et al.* (2004) "AC Immune" MBA 2004 Start-Up Project, Final Presentation, IMD.

Exhibit 3
AC Immune Management as of May 2005

Andrea Pfeifer, PhD – CEO (founder)

As former head of Nestlé's Global Research, Andrea Pfeifer managed a group of more than 800 people. While at Nestlé, she led the development of the first Functional Food, LC1, and the first cosmeceutical product in a joint venture with L'Oréal, Innéov Fermeté. Dr. Pfeifer also cofounded the Nestlé Venture Capital Fund, a € 100 million life science corporate venture fund.

Andrea Pfeifer is an international expert in biotech and a professor of biology at the University of Lausanne. She completed her studies and doctoral work in Pharmacy and Pharmacology at the University of Würzburg, Germany, and continued with postdoctoral studies in molecular carcinogenesis at the National Institutes of Health, Human Carcinogenesis Branch, in Bethesda, MD, USA. She has published more than 200 papers and abstracts in leading scientific journals. Andrea Pfeifer is an alumna of the Mastering Technology Enterprise (MTE) program, jointly run by IMD, EPFL, and ETH Zurich.

Claude Nicolau, PhD – CSO (founder)

Dr. Nicolau is a visiting professor of medicine at Tufts University in Boston, USA. His main area of work is the study of membrane involvement in pharmacological processes at a molecular level. He reported the first gene transfer and expression in a live animal and has developed the technique of protein electroinsertion into cell plasma membranes. Dr. Nicolau has been working on immunotherapy for Alzheimer's disease and other "conformational" diseases for several years. He has published more than 210 scientific papers and has been awarded 30 patents to date. Claude Nicolau is also one of the founders of Alantos AG, Heidelberg, and has been Chief Scientific Officer of that company for two years.

Armin Mäder, PhD – CFO/COO

Dr. Mäder has six years of banking experience with UBS Investment Bank, the last few years being spent in the bank's equity business. He has worked in business development, business, and project management. Besides doing financial modeling work, he delivered several time-critical multimillion-dollar projects for the bank and was one of the key drivers behind UBS IB's securities business-to-business strategy in Switzerland. Dr. Mäder holds an MBA with a focus on finance from IMD in Lausanne. He also holds a PhD in molecular biology and biophysics from ETH Zurich, where he worked on the first high-resolution X-ray structure of the nucleosome core particle (NCP).

Ruth Greferath, PhD – Director of Biology (founder)

Dr. Ruth Greferath is a coinventor of conformation-sensitive therapeutic antibodies. She obtained her PhD at the Molecular Biology and General Pathology Laboratory of the Albert Einstein Universität in Ulm, Germany, in 1994. Postdoctoral research fellowships at the Max Delbrück-Centrum Berlin and at the Center for Blood Research Laboratories (CBR-Labs), Harvard Medical School in Cambridge, MA,

Exhibit 3 (continued)

followed. In 1998 she became a research fellow in the Laboratoire de Biotechnologie des Membranes, Ecole Supérieure de Biotechnologies de Strasbourg (ESBS), Université Louis Pasteur in Strasbourg, France. Since 2000, she has been a research scientist at the Institut de Recherche contre les Cancers de l'Appareil Digestif (IRCAD) in Strasbourg, France, working on the uptake of allosteric effectors for haemoglobin in RBCs.

Board of Directors

Martin Valesco, Chair: Internationally known and successful entrepreneur and business angel.

Hans-Beat Gürtler: Investor and industry expert, former head of Novartis Animal Health.

Detlev Riesner: PhD, International leader in Prion and Alzheimer's Research, founder of Qiagen, which makes tools to extract and purify nucleic acids for use in DNA sequencing, genomics, and drug research.

Andrea Pfeifer: CEO (see previous page for biographical details).

Scientific Advisory Board

J.M. Lehn, Chairman: 1987 Nobel Prize laureate in chemistry, professor of molecular chemistry at the Collège de France in Paris and the author of more than 700 scientific publications.

Adriano Aguzzy: International leader in Prion clinical research, head of the Institute of Neuropathology at the University Hospital Zurich Medical School, Switzerland.

Roscoe Brady, M.D.: Lasker Prize Winner NIH, Bethesda. International leader in the therapy of metabolic diseases.

Curtis C. Harris, M.D.: Chief of the Laboratory of Human Carcinogenesis, National Cancer Institute, Bethesda, and clinical professor, Division of Clinical Oncology, Georgetown University School of Medicine, Washington, DC.

Fred van Leuven: PhD, Dr. Sc., Katholieke Universiteit, Leuven. International leader in Alzheimer's disease research.

M. Monsigny: Professor of Biochemistry at the University of Orléans, France.

I. Rosenberg: PhD, dean of Tufts University's Gerald J. and Dorothy R. Friedmann School of Nutrition Science and Policy. Previously he led the Jean Mayer US Department of Agriculture Human Nutrition Research Center on Aging for 15 years.

Exhibit 3 (continued)

E. M. Voisin Lestringant: Expert in biotechnology and regulatory affairs/FDA.

Claude Nicolau: CSO (see previous page for biographical details).

Detlev Riesner: Also on the board of directors (see above).

References

1   "Tufts Center for the Study of Drug Development Pegs Cost of a New Prescription Medicine at \$802 Million," News Release 2001-11-30, Tufts Center for the Study of Drug Development, http://csdd.tufts.edu.
2   http://www.adi2005.org/ (2005-06-13).
3   Saral, Defne, Olivier Fortin, Harjeet Grewal, Anders Jepsen, Francesco Lissandrin and Roman Tarnovsky. 2004. "AC Immune" IMD MBA 2004 Start-Up Project, Final Presentation, IMD, June 6.
4   For further information refer to: Seifert, Ralf W., Carlos Cordon, and Maria Wilhelmsson. 2004. *Outsourcing Patterns in the Pharmaceutical Industry.* IMD case number IMD-6-0263.
5   Specific information on outsourcing options within the pharmaceutical industry can be found in: Seifert et al. 2004. *Outsourcing Patterns in the Pharmaceutical Industry.* IMD case number IMD-6-0263.
6   Seifert, Ralf W., Carlos Cordon, and Maria Wilhelmsson. 2004. *Outsourcing Patterns in the Pharmaceutical Industry.* IMD case number IMD-6-0263.
7   "2003 Progress Report on Alzheimer's," ADEAR – Alzheimer's Disease Education & Referral Center, http://www.alzheimers.org/.

CASE 4-3 Handout
AC Immune and Genentech: Exclusive Global License Agreement

PRESS RELEASE THURSDAY, DECEMBER 7, 2006.

AC Immune, Ltd. enters into an Exclusive Global License Agreement and Research Collaboration with Genentech, Inc. for Development of Anti-beta-amyloid Antibodies for the Potential Treatment of Alzheimer's Disease and Other Human Diseases.

Ecublens/Lausanne, Switzerland, December 7, 2006 – AC Immune, Ltd., a Swiss Biotech Company developing innovative therapies against Alzheimer's Disease, today announced that it has entered into an exclusive global license agreement and research collaboration with Genentech, Inc. for the development of anti-beta-amyloid antibodies for the potential treatment of Alzheimer's Disease and other human diseases. Beta-amyloid constitutes an important target for disease modification of Alzheimer's disease and AC Immune has developed conformation-specific antibodies against this protein generated by its SupraAntigen™ Technology.

Under the terms of this agreement, Genentech will make an upfront payment to AC Immune, with the potential for a total of over $300 million in payments upon successful completion of clinical and regulatory milestones for Alzheimer's and additional applications. Upon commercialization of a product, Genentech will pay AC Immune royalties on net sales of AC Immune's antibodies in the field of Alzheimer's or other human applications. Genentech will also provide funding for a multiyear collaborative research program and will cover all development and clinical costs of the lead antibody and subsequent antibody candidates.

"Out-licensing one of our lead programs to Genentech achieves a major business objective for the company and represents a significant milestone in the 3-year history of AC Immune," said Prof. Andrea Pfeifer, CEO of AC Immune Ltd. "Genentech is an excellent, and our preferred, collaborator for the antibody program due to its expertise in development and commercialization of antibodies."

"The terms of the deal will provide AC Immune with a solid financial basis beyond the next three years to develop other programs at full speed," said Dr. Armin Mäder, CFO of AC Immune.

AC Immune's proprietary SupraAntigen™ Technology resulted in the development of an anti-beta-amyloid antibody program with a selected lead antibody that has been shown to be highly active in animal disease models for Alzheimer's disease. The lead antibody is conformation specific and induced the proposed transition from an insoluble to the soluble form of the plaque forming beta-amyloid protein. This event directly correlated with memory improvement.

"AC Immune's technology has proven very effective for the generation of conformation-specific antibodies that have potential for the treatment of conformational diseases," said Dr. Andreas Muhs, CSO of AC Immune.

"We are thrilled that Genentech has taken such an interest in our Alzheimer's antibody program," said Martin Velasco, chairman of the board of AC Immune. "We are hopeful that in collaboration with Genentech we may develop a potential treatment for Alzheimer's disease for the benefit of millions of patients worldwide."

CASE 4-4
Covalys (A): Managing the Company's Growth and Development Strategy

Ralf W. Seifert, Jim Pulcrano, and Rebecca Meadows

> Looking back is always thrilling for me: Four years ago, there only was a piece of paper, a business plan; two years ago a lab somewhere in the Basel countryside; and now we are selling our products all around the globe.
>
> *Covalys CEO, Christoph Bieri*

SEPTEMBER 2006, WITTERSWIL, SWITZERLAND. Covalys's CEO, Christoph Bieri, sat at his desk contemplating the meeting he had just had with Tom Gibbs, his head of marketing and support. Gibbs had emphasized the need to increase sales significantly by changing the distribution from exclusive distributors to building up its own sales team. An hour before that, however, Bieri had met with Andreas Brecht, Covalys's CTO, who had asked for more resources focused to be on broadening the company's technology portfolio.

Bieri knew that the company was facing a key developmental balance point. The company had certain strengths and needs. At the same time, some key investors had already indicated that they would like to see an exit shortly, which would require management to put more focus on structuring the company as an even more attractive trade sale opportunity. Pursuing a trade sale in the short term would also mean actively identifying and soliciting potential acquirers. But preparing for such negotiations would consume significant time and energy as well. Bieri pondered whether it would be possible to do all of these tasks simultaneously or whether they would have to concentrate on one above the others. If so, which one?

If the company spent time and money building up its sales team, would this be of any value to the potential trade sale buyer, which might have its own channels already? Equally, if there was not a sufficiently attractive offer for a trade sale in the mid-term future, the company needed to be able to sustain itself, and increased sales would naturally be needed to attract employees, investors, and prospective buyers. Making further investments to enlarge the technology portfolio would make the company an even more attractive acquisition target, but could Covalys's management justify ongoing technology investments without driving sales first? It was a Catch-22 – or, put differently, the typical challenge for most entrepreneurs.

Copyright © 2006 by the alliance for technology-based enterprise (IMD, EPFL and ETH Zurich). Copyright permissions are handled by IMD, Lausanne, Switzerland.

1 Company Background

In July 2002 a diverse group of men with varying scientific, industry, and business backgrounds founded Covalys. They were Christoph Bieri (CEO), Kai Johnsson (director), Andreas Brecht (CTO), Maik Kindermann, Nils Johnsson, Patrik Sidler, and Henri Zinsli (chairman):

> The start-up aimed to develop, manufacture and market research kits for use by industrial and academic biomolecular labs in drug development and life sciences research.

Covalys's SNAP-tag technology was based on research conducted by Kai Johnsson in his protein engineering laboratory at the Ecole Polytechnique Fédérale de Lausanne (EPFL). Bieri, when he learned about Johnsson's research findings, was inspired by the potential to sell tools to the growing life sciences research community worldwide and engaged him to further develop and commercialize the technology.

> Covalys strived to deliver a crucial "shovel" for the 200,000 or so scientists digging in the labs for the "gold nugget" – a new discovery, a new drug, or even a Nobel Prize.

In 2002, during the process of forming the company, Covalys was awarded an exclusive worldwide license to develop and market the technology by EPFL. With a loan from the Swiss Start-up Foundation, it was able to prepare its operations, to support patent protection activities, and to pay the expenses (but no salaries) during the fundraising period. Subsequently it was awarded the CTI Start-up Label, which made it eligible for research grants provided by the Commission for Technology and Innovation.

In October 2003 three Swiss venture capitalists (VCs) and an undisclosed private investor provided substantial financing to Covalys. The VCs included Venture Incubator Partners, BV Group Private Equity, and Erfindungs-Verwertungs AG (EVA); they were later joined by Novartis Venture Fund. The company launched its operations shortly thereafter in Witterswil close to Basel, Switzerland. This allowed it to be close to potential customers in the Swiss German part of the country as well as those in French-speaking Switzerland and up the Rhine valley.

In March 2006 the company closed a C round of financing, bringing its total funding to CHF 10.7 million. The C round funds – from three of the original VCs and several private investors – were used to continue operations and further drive the market penetration of the company's tools for protein research. In June 2006 the company's key patent, filed back in 2001, was officially granted in Europe.

By September 2006 Covalys had grown to a headcount of 17 people including, in addition to the CEO, 12 researchers involved in technology and product platform development and 4 people predominantly focused on marketing. Successful senior executive recruits included Tom Gibbs (from Cytion SA and before that Molecular Devices Corp.) and Michel Crevoisier (from Rohner AG, and previously Bachem). The team was supported by a board of directors with extensive entrepreneurial and technical experience. The chairman, Zinsli, had wide pharmaceutical industry experience, as well as having founded a number of other science-based startups. Kai Johnsson provided technical direction, while both Diego Braguglia and Daniel Kusio had extensive experience in international commercial operations (*see Exhibit 1 for Covalys's organizational structure by 2006*).

2 The Life Sciences Industry

The life sciences R&D industry was rather diverse, but one commonality was the need for tools to conduct the research. Key tools included:

- Instruments
- Reagents
- Informatics/computational

Customers included research scientists in industry (*e. g.*, pharmaceutical, agricultural, and biotechnology companies) and academia/universities. As Bieri explained, the industry was fragmented and applications were diverse, making it difficult to define the exact market size. The total market was estimated to be worth about US$ 25 billion, while the core area Covalys targeted was conservatively estimated at about $ 140 million.

The industry consisted of several large companies and many smaller ones. The technology tool provider industry as such evolved continuously, with at least 24 companies currently operating in the area. Smaller companies were frequently acquired by the larger companies, seeking access to technologies that had been proven in the marketplace. Between 2002 and 2004 alone, the deal value of mergers and acquisitions amounted to over $3 billion. Likewise, average organic growth was healthy at 7.2% (*Exhibit 2*).

2.1 Covalys's Product

The startup's product helped scientists improve the efficiency of research into proteins and their function. Covalys's SNAP-tag technology was a self-labeling multipurpose protein tag that had a broad range of applications, from cell imaging to protein-interaction studies. The technology was based on the ability to create protein and tag combinations that bind strongly (covalently). This is useful in scientific research, where it is often necessary to isolate particular cellular components (such as proteins and antibodies) so that they can be studied.

Covalys's technology was simple and easy to use while being highly specific. It simply required the combination of a specific protein and a tag – both of which could vary in functionality. The tags could be fluorescent dyes or compounds such as biotin.

The strength of the SNAP-tag technology was that "one tag does it all." This increased flexibility, provided a good platform for later product and application development, and reduced labor. The company aimed to develop collaborative arrangements with other institutions and companies to further develop the SNAP-tag technology and its applications. (*See Exhibit 3 for more details.*) As Brecht noted:

> This technology for sure will be standard in most laboratories in the mid-term future due to its versatility, breadth of applications and ease of use.

2.2 Competing Firms and Technologies

Over the past few years a number of larger players had introduced competing products. These products either mimicked or were technically similar to Covalys's SNAP-tag technology, or the technologies enabled the same experiments (*e.g.*, fluorescent proteins *vs.* SNAP-tag for fluorescence microscopy). In 2006 conventional protein labeling products and fluorescent proteins dominated the market in both the academic and industry customer segments. Despite similarities in application or technical design, each of the competing technologies was protected by one or more patents.

In principle, Covalys had a strong patent. It was prepared and filed by EPFL and granted in Europe in June 2006. However, while EPFL was developing the technology and filing the patent, one of the large scientific research product providers, Promega, was developing a complementary technology known as HALO-tag. After Covalys had launched its product in 2004, Promega introduced the HALO-tag in 2005 and supported it with a significant marketing push in the subsequent years. This complementary technology did not infringe Covalys's patent – it used a slightly different approach but was still effective as a substitute supporting more or less identical applications.

A number of other established research product providers operated in the field. Those with very similar technologies included Invitrogen, with its Lumio-tag (a small binding peptide), and Active Motif, with a self-labeling DHFR tag. Some of the other large companies, such as Fisher Scientific, also had products that could be seen as substitutes for Covalys's products, although technically they were not similar to the SNAP-tag technology. (*See Exhibit 4 for more details on these companies.*)

Covalys sold a "tool" rather than a service or a product license. Its technology was open access, with no license required to use it, facilitating fast and easy access for customers in any application area they desired. Covalys then made money through consumables, just as computer printer manufacturers make their money by selling ink rather than the hardware.

Bieri commented:

> It is a platform – with the one technology you can do many things. We give away the plasmid (the gene or building plan for the SNAP-tag) for free, no strings attached. Once the customer has made the investment of fusing SNAP-tag to his gene of interest, he needs to come to us for the patent-protected consumables. It is intriguing from a commercial angle – you have recurring sales of consumables which are easy to protect in a platform that costs nothing to provide.

Competing companies typically provided their products to customers through a license agreement, which involved conditions outlining under what applications their technology could or could not be used. While this was also true for the vast majority of Promega's product suite, the one exception was the HALO-tag technology, which was offered as open access – following the same sales model as Covalys. In general, larger companies supported a broad product portfolio and offered bundled solutions as well as complementary services. In addition, they could leverage off an established direct marketing and distribution channel. That said, some of their marketing efforts also benefited Covalys indirectly as they helped establish the market for labeling technologies.

3 Shaping Covalys's Growth and Development Strategy

Covalys had a strong management team and excellent technical expertise. The challenge was how to leverage its current position to create and nurture sustained growth and development. The next meeting between the board of directors and the executive management team would focus on discussing the right balance between further technology investments and increased sales and distribution efforts:

3.1 Building a Portfolio of Technology

In 2006 Covalys had more than 20 products in its portfolio – all of them leveraging the same platform technology. The challenge for Covalys, though, was whether it had a large enough product suite to be of interest to potential acquirers and whether it was credible in having successfully established a new technology platform.

To complement its product portfolio, Covalys pursued a number of strategies, ranging from internal development to research licensing arrangements with institutions such as EPFL and research collaborations with companies in complementary areas. It also actively advanced further developments of research projects in co-operation with Kai Johnsson's laboratory at EPFL. The types of collaborations included:

- *Cross-licensing agreements/research collaborations* – Covalys and other companies shared components of their technologies and developed further applications, which were then cocommercialized. Covalys had a collaboration of this nature with Cisbio International and Gene Bridges.
  The collaboration with Cisbio and its homogeneous time-resolved fluorescence (HTRF) technology resulted in the development of the SNAP-vitro HTRF kit. The collaboration with Gene Bridges and its Red/ET recombination technology (allowing the precise engineering of large DNA sequences) resulted in the development of DNA engineering kits that could be used to generate SNAP-tag fusion proteins.
- *Comarketing/product bundling* – Covalys's product was sold together with another company's to yield greater customer value. Covalys partnered with Pierce to combine the SNAP-tag with its microtiter plates, and as a result Covalys gained access to the market leader's client base. Covalys also comarketed with Gene Bridges.
- *Exploratory technology evaluation* – This is typically done with large pharmaceutical companies.

Most recently, Covalys had collaborated with Kai Johnsson and his team at EPFL's protein engineering laboratory to develop a new ACP-tag. The ACP-tag was particularly suited to specifically labeling the external (extracellular) portion of membrane proteins, which would complement the existing product range well. As Bieri noted:

> Some easy-to-develop products could significantly increase the attractiveness of the technology as a platform. However, pursuing many of these "low-hanging fruits" may result in significant resources and distraction of management.

Clearly, these collaborations enhanced the company's competitive market positioning, and the pipeline of further products and labeling technologies also provided a safety net in case one of the technologies failed in the hands of the customers. Yet they also took up a significant amount of time and resources: On average, the product development cycle took about 8 months, with complex applications taking 12 to 15 months.

3.2 Marketing, Sales, and Distribution Strategy

Covalys was very active in marketing, producing numerous publications and informational documents for distribution at trade shows or in-laboratory demonstrations.

Many academic articles in leading international journals showcasing the use of the SNAP-tag technology helped build credibility further. Yet R&D scientists were typically cautious about switching technology and applications. Gibbs explained:

> Scientists are conservative in utilizing new technologies, as scientific research requires replicability of findings, and the introduction of a new technology can initially cause a significant delay – despite ultimately being faster, easier and cheaper.

Some scientific customers were reluctant to deal with Covalys because they already had well-established contacts and a track record of good performance with competitors such as Promega and Invitrogen. Others were wary about working with a startup company and new tools that did not have the backing of an established player because continued availability of the technology could potentially be at risk.

On the direct-sales side, Covalys leveraged off existing distributors in key markets such as North America, Europe, Switzerland, China, and South Korea. These distributors sold Covalys products alongside the other products already listed in their catalogue and paid Covalys a set price for the kits – selling them on to their customers at a marked-up "list" price at their own discretion.

However, the distributor model meant that sales of Covalys products depended heavily on the willingness of the distributors to market them. Covalys found that its SNAP-tag kits were selling slowly, partly because distributors did not have the necessary time and skill to sell the product, but also particularly because the product did not "sell itself" as it was so novel. In addition, Covalys's distributors would also be competing with many of the startup's competitors, such as Promega, Fisher Scientific, and Invitrogen, which had their own extensive and committed distribution networks. On the other hand, having established a worldwide network of distributors had given Covalys global visibility within just 12 months. Gibbs said:

> Recruiting distributors was a faster, much cheaper and easier way to have a global reach than building up our own sales force. At this early stage of the technology cycle this was for sure the right thing to do – the question is how to go on.

Covalys's target markets included research institutes, universities, and private companies. In the public sector, customers largely depended on government or other external funding. This segment was fragmented and volatile, depending on whether research grants were awarded or not. In the private sector, a major target client group was R&D scientists from both pharmaceutical companies and biotechnology and agricultural enterprises. This segment represented large-volume customers who would typically bargain more aggressively on price and impose more stringent quality standards. Given that Covalys's business model was to charge for consumables – consumables that were somewhat price transparent – this could have significant implications for the startup's bottom line, no matter how innovative, disruptive, or valuable its SNAP-tag technology.

The conservatism of scientists and the strong market share of companies such as Promega and Invitrogen meant that, for Covalys, marketing and "on-the-ground" sales people would be a vital resource for informing scientists about their products and the potential use and value of the technology. The challenge for Covalys was whether it should – and could – increase sales by creating a dedicated sales team to promote and distribute its products worldwide. If Covalys established its own sales team, this would involve considerable on-costs and commitment, whereas with the distributor model the setting-up costs were insignificant.

3.3 Increasing Shareholder Value

Covalys had a number of investors – both established VC firms and private investors (*see Exhibit 5 for details*). They invested with a view to exiting from the company through a trade sale three or four years after the initial investment. Although Covalys had done well to date and achieved all the milestones set, revenues were lower than expected. Bieri commented:

> With our means and timelines, it is difficult in this marketplace to generate substantial revenues just one or two years after market launch.

He added:

> Our investors supported us because they believed in the potential to sell both the technology and eventually the company ... [and] while they are open to selling the company, they are not pressed and are potentially willing to invest further.

Indeed, the most recent financing round underlined the investors' commitment. The challenge for Covalys was how to best utilize these funds to create the greatest value for the company and its investors. Resources would have to be carefully managed to balance the need to both create revenues and build an attractive product portfolio. Bieri did not want to invest significant funds in creating a sales force, particularly as costs would be high and there was an estimated lead time of four to five months before the first revenues would be received from each sales representative. But the distributor approach was not successful enough. Despite a decision in 2004 not to have a sales force, the company was now reconsidering that choice.

For Covalys to be attractive for a trade sale, Bieri knew that the acquiring company would want to

> ...see that the technical risk is gone and the product can achieve revenues. However, the companies don't want to buy positive cash flow, they want to buy potential. They would buy a company that has a number of products that had proved that they could be sold even if the company is still making losses.

Bieri summed up the dilemma:

> If you thought of this as a platform technology with outstanding commercial potential, then a potential acquirer would only want to see the technical risk reduced, so we would

then invest in more products, but if they were looking at it as an "average" technology, then we would have to work on the marketing and sales and prove that the products can generate significant revenues.

4 Outlook

Despite the last financing round, Covalys had only limited means and each month there were expenses to pay. If the company was not able to find a trade sale partner, it would need to increase revenues enough to cover costs and continue running. As Bieri noted:

> Burn rate is always an issue – there is always the concern as to whether we will still be able to pay salaries in three or six months from now.

Bieri knew that the company had done well to date – within a short period it had developed a great technology with much potential, achieved some market penetration, and secured financing from leading VCs. But Bieri also knew that the company was fast approaching a strategic crossroads, and the competition was also on the move. Bearing in mind the limited resources, Bieri grappled with the key question of whether he should focus efforts on developing a greater suite of complementary technologies or on expanding the sales and distribution of Covalys's product.

Exhibit 1
Covalys's Organizational Structure

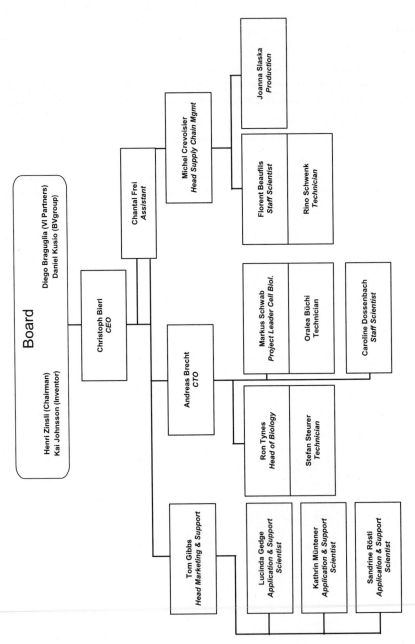

Source: Company literature

Exhibit 2
Market Size (Revenues 2004 in $ million)

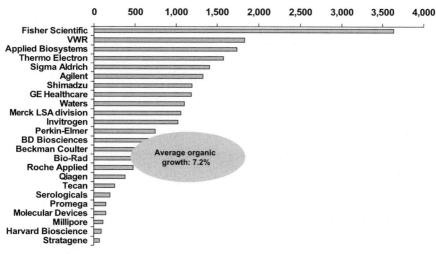

Source: Company presentation based on SEC data

Exhibit 3
Covalys's Technology

The applications of the SNAP-tag technology were diverse and included:

- Labeling of proteins inside living cells (known as SNAP-Biotin)
- Labeling of proteins inside extracts
- Immobilization of proteins on surfaces without the need for purification
- Detection of proteins on a gel (known as SNAP-vista)

The SNAP-tag was sold in kit form. There were a number of different kits for different applications, including:

- SNAP-vitro HTRF: labeling of donors and acceptors for HTRF binding assays
- SNAP-vitro fluorescent: labeling proteins with fluorescent probes
- SNAP-capture: immobilizing proteins on beads or a surface
- SNAP-cell: labeling proteins in live cells (using biotin or fluorescent elements/ fluorophores)
- SNAP-vitro: labeling proteins in solution.

Source: Company presentation

Exhibit 4
Covalys's Key Competitors

Promega Corporation

Promega is a large privately owned worldwide company that provides products and support for scientific research. It offers over 1,450 fully integrated products that assist researchers in exploring gene, protein, and cellular interactions. The key technology areas covered include:

- Nucleic acid purification
- Human identification
- Bioluminescence
- Coupled *in vitro* transcription and translation
- Cell biology.

Promega's annual revenues in 2005 were over $176 million and the company serves over 125 countries.

Invitrogen Corporation

Invitrogen, formed in 1987, is a NASDAQ-listed company that provides life science technologies for disease research, drug discovery, and commercial bioproduction. Invitrogen has a diverse product family, many from acquired companies including Molecular Probes, BioReliance, Gibco, and Dynal Biotech. Invitrogen is headquartered in the USA, has eight manufacturing locations, operations in more than 20 countries, and distributor relationships in 50 countries worldwide. It provides more than 25,000 unique products and services. The key technology areas covered include:

- Functional genomics
- Drug discovery solutions
- Detection technologies
- Bioproduction
- Cell culture
- Biological defense systems.

Invitrogen's annual revenues in 2005 exceeded $1.2 billion.

Active Motif

Active Motif is a US-based company that develops cell and molecular-biology-based research tools to assist in determining the functions, interactions, and regulation of proteins. Active Motif was formed in 2000 and in 2003, 2004, and 2005 was named one of San Diego's fastest growing companies in Deloitte and Touche's Technology

Fast 50 program. In October 2005 the company was ranked 88th fastest growing technology company in North America on the 2005 Deloitte Technology Fast 500 and named as one of America's 500 fastest growing private companies by *Inc.* Magazine. Active Motif has offices in Europe, Japan, and the USA and is represented throughout the rest of the world by distributors. Active Motif's technology categories include:

- Cell biology
- Assay kits
- Fluorescent detection
- Antibodies
- Cell extracts
- Recombinant proteins
- Molecular biology
- Biocomputing
- AlmaKnowledgeServer (a combination of biocomputing and wet lab biology provided as a result of an acquisition of TimeLogic).

Fisher Scientific International Incorporated

Founded in 1902, Fisher Scientific is a Fortune 500 company that has approximately 19,500 employees worldwide. Overall, it provides more than 600,000 products and services and has over 350,000 customers. Fisher Scientific has a number of different divisions – many of which were formed from acquired companies – all focusing on providing products and services to the scientific community. They include:

- Fisher Global Distribution Services: A supply-chain group that provides custom services and sells a portfolio of scientific products from a range of suppliers (private labeled, internally developed or exclusively sourced products) as well as Fisher's own proprietary brands.
- Fisher Scientific Products and Services: Manufactures and distributes products and services to entities conducting scientific research including drug discovery and development and process control and basic R&D.
- Fisher Scientific Research: Distributes a wide range of laboratory equipment, instruments, chemicals, biological products, and consumable supplies.
- Fisher Scientific Products: Manufactures a broad range of laboratory equipment and consumable supplies for scientific research applications.
- Fisher Biosciences: Develops and manufactures a wide range of products and services for the general chemistry and life science areas. Pierce Biotechnology is also part of the division. It offers reagents and kits for protein purification, protein detection and quantification, protein sample preparation, protein labeling, and protein interactions.
- Life Science Research: A global provider of innovative technological services for life science research and drug discovery.

Some of the other divisions include Fisher Safety, Fisher Science Education, Cole Parmer (instruments and supplies focusing on fluid handling), Fisher BioProcess Production, Fisher Microbiology, Fisher Global Chemicals, Fisher BioPharma Services, Fisher Healthcare Products and Services, Fisher Healthcare, Fisher Molecular Diagnostics, Fisher Clinical Products, Fisher Immunodiagnostics, Fisher Clinical Diagnostics, Fisher Diagnostic Technologies, and Fisher Hamilton.

Annual revenues for 2005 amounted to $5.6 billion.

Exhibit 5
Covalys's Venture Capitalists and Investors

Venture Incubator is a Swiss-based VC firm with CHF 101 million under management. It specializes in supporting university-based spinoffs as well as other promising startup companies with capital, coaching, consulting, and networks. It was established by McKinsey and Company and the Swiss Federal Institute of Technology in Zurich in 2001. VI investors represent ten blue chip enterprises from industry and finance.

BV Group is a Swiss-based VC firm that invests between CHF 1 million and CHF 15 million in companies located in Switzerland and bordering countries for:

- Innovation
- Growth, expansion, and acquisition
- Succession financing
- Management buyout and buy-in
- Restructuring.

Its particular focus is medical technology, IT and telecommunication technology, and industrial/material and process technology.

EVA (Erfindungs-Verwertungs AG) is a private enterprise that provides assistance to investors, innovators, and entrepreneurs. It focuses on the life sciences area (chemistry, biochemistry, molecular biology, and pharmaceuticals) and provides support with:

- Translating the initial discovery/idea from the R&D phase to a product that is ready for market
- Intellectual property commercialization and protection
- Setting up a new company (legal and organizational)

EVA is owned partly by private individuals and partly by the cantonal banks of the city of Basel and Basel County. EVA can inject some seed funding and, rather than taking a fee for its service, it also takes equity.

Novartis Venture Fund, with CHF 175 million under management, invests predominantly in the field of health sciences, in projects that demonstrate entrepreneurial and innovative spirit in future-oriented areas.

CASE 4-4 Handout
Covalys: Expansion Plans

After the decision point outlined in the (A) case, we continue to follow the progression of Covalys SA. To complement our perspective, we interviewed one of Covalys's investors, Diego Braguglia. He commented:

> We decided to promote the CTO – who is fairly sales and marketing driven – to CEO and also to retain in the company only the people who had the mindset of generating revenue.

He explained further:

> This was necessary, but not sufficient. What we are planning is to hire senior experienced sales and marketing people and focus on selected regions like Europe and the US.

Covalys formally announced the changes in the following press release.

PRESS RELEASE OCTOBER 3, 2006.
Covalys announces intentions to expand direct customer interactions in North America and major European countries

WITTERSWIL, SWITZERLAND 3RD OCT 2006 – Covalys Biosciences AG today announced that SNAP-tag and ACP-tag products will be available for sale directly from Covalys in North America and Europe (excluding Austria, Slovakia, and the Czech Republic).

"SNAP-tag and ACP-tag have undergone extensive development and have established themselves as a great way to label and immobilize proteins for drug development and life science research" said Andreas Brecht, Chief Executive Officer. "Market awareness of our products has increased significantly over the last eighteen months leading to adoption of the technology. Now is the time to increase our interactions with customers by supplying directly. In synergy with our local distributors we will further improve the level of support and accessibility."

CASE 4-5
Technology Strategy at Dartfish

Christopher L. Tucci and Jana Thiel

By February 2006, the shareholders of Dartfish, a provider of enhanced broadcast footage and sports training applications, were puzzled. A review of the 2005 figures had revealed that sales of the training applications fell short of expectations, and the business value of the broadcast service was not yet fully understood. Dartfish was operating on a breakeven basis but had not generated sufficient returns so far, which was even more surprising as the firm enjoyed a very positive brand image.

Dartfish's sports training applications were widely used by professional athletes. Over 60% of all contenders at the current 2006 Winter Games in Torino, Italy, had used Dartfish. In addition, Dartfish was also present on TV screens in many countries during the Olympics. It had developed a proprietary technology featured by many European and US-based TV stations in their enhanced broadcast footage.

However, the Swiss firm was still struggling to define its path to economic value. Anton Affentranger, executive chairman, summarized the situation:

> Already in early 2005, after reviewing the results of 2004, the board introduced a very aggressive growth budget and planned to revise the firm's strategy. A year later, however, we have grown substantially but neither met the aggressive budget nor truly launched a new strategy. The pressure from our shareholders is thus increasing. We are required to produce a credible strategic scale-up plan by the end of the first quarter of 2006.

Affentranger and his team had to answer the following questions:

- How to capitalize best on Dartfish's assets? Should both business lines – broadcasting and training software – be kept?
- Which strategy should the company potentially pursue to grow the training applications market?
- How could product development and support be managed more efficiently and effectively?

Copyright © 2006 by the Ecole Polytechnique Fédérale de Lausanne (EPFL). All rights reserved.

1 The Technology

Dartfish was primarily recognized for televised footage, which was based on two proprietary technologies: SimulCam and StroMotion [1].

SimulCam allowed two video sequences to be superimposed in one picture frame (*Figure 1*) [2]. By simultaneously replaying the action of two athletes who had competed on the same terrain but at different times, it could make otherwise imperceptible differences obvious. All that was needed was a digital camera equipped with sensors that transmitted information about position and capturing angle to a computer, together with the images. Sophisticated algorithms would then process the information by filtering the moving subjects from the background, aligning the imaging parameters, and finally blending the subjects into clear pictures – all in the space of one minute.

The second technology, *StroMotion*, created a sequential imprint of a subject as its action developed over time (*Figure 2*). StroMotion divided a movement into a series of pictures from which the user could quickly pick individual snapshots and repackage them into a new image series.

Both features were complemented by a set of other nonproprietary tools, developed over time, and included in Dartfish's sports training applications – the second business pillar of the Swiss firm.

Figure 1   SimulCam Example

Figure 2   StroMotion Example

2 Company History

Dartfish began life in the Audio-Visual Communications Lab (LCAV) at the École Polytechnique Fédérale de Lausanne (EPFL) in 1998. It was founded soon after the first TV stations had started to use the lab's innovative technology in their coverage of major skiing events [3]. The startup – then called InMotion Technologies Inc. – gathered CHF 1 million in seed capital, which it used to further develop and mature its offering. In October 1999 it entered a second business by launching a software package – today called DartTrainer – that allowed professional athletes to capture and evaluate their movements.

The broadcast and training applications were driven forward in parallel, which consumed another CHF 2 million of venture capital over the course of the following year. In 2000 Dartfish formed a partnership with NBC Quokka Ventures to provide content for the online coverage of the Summer Olympics in Sydney, Australia. At the same time, the small venture also developed and implemented video training solutions for associated members of the US Olympic Committee (USOC). Both partnerships boosted Dartfish's credibility, and by the end of 2000 it had raised CHF 11.5 million from several investors.

The establishment of an office in Atlanta, GA, USA, in February 2001 facilitated additional partnerships with several sports organizations across North America. A few months later, a subsidiary in Korea followed. As a result, the penetration rate of the training software increased significantly, making it a viable business within the young firm. To underline the shift from a technology-centered to a product-centered business, and to enhance brand recognition, the venture assumed its new name, Dartfish.

2.1 Changing Strategy

In 2002, however, the firm found itself at a crossroads. Would it be able to continue or would it have to close? It had certainly gained huge momentum when NBC used the broadcast features in the coverage of the Winter Olympics in Salt Lake City. But the corresponding development efforts and more employees – up to 60 – had cost a total of CHF 8.5 million, against too little income. The venture was in urgent need of funding as well as a strategy for generating a satisfying return from its technology.

Dartfish was able to arrange an injection of fresh capital in the summer of 2002. Over the course of 2003 the company then substantially cut back its burn rate. All efforts were focused on the growth segment of sports training applications. As a result of both – cost reductions and the focused approach to the sports market segment – the firm broke even in the first quarter of 2004.

By the end of 2005 Dartfish boasted an installed base of over 15,000 applications in over 150 countries. About 20 employees were working at Dartfish's headquarters in Fribourg, Switzerland, and a headcount of roughly 30 was spread over its subsidiaries in the US, Japan, Korea and France, and the various distribution channels. As Martin Vetterli, cofounder and EPFL professor, noted:

> Dartfish has come a long way. Eight years ago, it was just a PhD project – that means only a PhD student could run the technology. Today, however, our TV presence makes us globally known among sports enthusiasts and our training solutions enjoy high popularity worldwide [4].

Despite all the progress, however, the venture was still hovering around break-even, and its shareholders were wondering how to better capitalize on the existing markets and the tremendous potential in the technology.

3 Application Domains and Business Models

3.1 Selling a Service: Broadcasting and Media

The broadcast business was Dartfish's foundation. At its core were SimulCam and StroMotion, two powerful tools adopted by major TV stations around the globe to enhance broadcasts of sports such as skiing, soccer, swimming, and ice skating. By providing content value, Dartfish had claimed a unique position in the sports broadcast domain. Affentranger commented:

> Actually, there is no direct competition. No one does exactly the same as we do. Nonetheless, we offer only two out of many possible tools – or toys if you like – that broadcasters look for to spice up their footage. In the end, we compete with all other "toy producers" [5].

The broadcast business was founded completely on service delivery. Typically, TV producers contracted a fully equipped truck and experienced technicians from Dartfish who operated the various features and platforms (*Exhibit 1*). For postproduction and similar purposes, Dartfish sold its product DartStudio, which included the prominent features StroMotion and SimulCam. This business model had worked well over the last few years but naturally restricted the number of live events Dartfish could cover. Another challenge was posed by the specific characteristics of the TV entertainment market, as Vetterli pointed out:

> TV is not driven by technology. It is driven by advertisement. Hence, the broadcasters like to think you should pay for being played just as if you launched a commercial. This is out of the question for us – our financial position does not even allow us to think about it [6].

The fading novelty of SimulCam and StroMotion made broadcasters (especially smaller ones) hesitant about funding these high-tech gizmos. On the other hand, Dartfish was constantly required to keep up with technical developments and improvements, which was costly. Dartfish tried to keep investments low and targeted efforts toward satisfying a limited set of requests that would pay off quickly. Furthermore, it served only key broadcasters who were perceived as entertainment leaders.

3.1.1 Strategic Considerations

Broadcasts, specifically the Olympics, brought wide exposure, but Dartfish's management team was still struggling to identify the true value for the firm. Television did not pay the bulk of the bills, generating just 15% of the company's turnover. But the public exposure most likely helped fuel sales of the sports training software DartTrainer, although the effect could not be measured directly.

However, there was immense knowledge included in SimulCam and StroMotion. The question arose as to whether all the potential had been exploited so far and if commercialization via broadcasting was the only possible opportunity – or even the best one. An ironic twist was that the two flagship technologies on which the broadcasting business was based were increasingly less important for Dartfish's training solutions.

3.2 Selling a Product: Training Software and More

When it first entered the training business, Dartfish had just packaged its SimulCam application into a standalone software suite. The product ran on a standard PC, requiring only a connected video camera to capture the movements and positions. The footage could then be replayed, analyzed, and compared with external reference models or earlier records. This content could be enriched by graphic, oral, and written commentary.

DartTrainer had first been geared toward elite athletes, whose successes spoke volumes. Athletes who trained with it had won about 20% of all medals at the Athens Olympics in 2004 and an astounding 54% of all medals at the Winter Games in Torino. By 2006, over 5,000 top level athletes used DartTrainer, which dominated the top professional market.

Over time, the suite grew along with customer requirements and Dartfish started customizing it for particular sports, i. e., bringing out specific editions for golfers or swimmers. Along the way, Dartfish also recognized that its customers had very different requirements when it came to product features, as Vetterli noted:

> After a while, we realized that our typical user did not need all the functionality we provided – in particular he did not need our flagships SimulCam and StroMotion. Rather, we had to go back and improve the user interface. Through easy-to-use software, we have meanwhile managed to make an impact.

In 2006 DartTrainer was available in three versions, ranging from low-end to high-end suites, selling for CHF 3,000 on average. However, the cheaper "low-end" versions *without* the renowned SimulCam and StroMotion features were enjoying increasing popularity. Their competitive differentiator was more the user interface than sophisticated algorithms. According to Affentranger:

> …[W]e struggle with positioning our low-end solution on the market. We do not want to cannibalize our high-end products by offering a $49 suite that could possibly squeeze out parts of our high-end segment [7].

Consequently, Dartfish still served only a fraction of the potential market for training software (*Figure 3*). Pricing and distribution were still primarily geared toward professional athletes, trainers, and other sports experts such as professional organizations and committees.

However, observations revealed a trend toward professionalizing sports – increasingly amateurs wanted to train "like the pros." More opportunities thus awaited Dartfish if it went "down-market" by developing the broader but more price-sensitive amateur business.

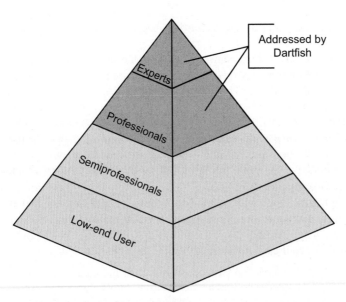

Figure 3 Dartfish's Market Coverage
Source: Company literature

3.2.1 Venturing into Other Domains

Building on its knowledge in sports training, Dartfish had ventured into *sports education*, which enjoyed broad success in the UK. By 2006, over 350 secondary schools had adopted the video-based solution to improve and enhance their physical education classes. The success had primarily been induced by a government-funded program to increase the level of IT equipment in schools. Nonetheless, Dartfish hoped to extend the business to other countries, although a replication of the success was unlikely to happen quickly.

Dartfish's technologies were also able to serve needs in *physical medicine*, where DartTrainer was used to support rehabilitation therapies.

Another lucrative market had been identified in fixed infrastructure solutions. One example was the video-feedback system at the Veltheim Driving Center in Switzerland. Here, using Dartfish's application, the driving of individuals following car-safety courses was recorded and superimposed on that of the instructor for immediate comparison. Potential customers included about 350 safety centers all over Europe and the USA, approximately 90 test tracks at manufacturer sites, and 100 motor-racing circuits worldwide. The application could also be used in ski domes, indoor cycling tracks, and tennis courts.

Dartfish had also made advances in the *industrial test environment* (crash tests, control of automated processes). It had initiated a couple of promising pilot projects. Affentranger commented:

> However, the issue with all this is: Dartfish is perceived as a sports company, not as a medical, nor an industrial or any other company for that matter. It might even be that we are only recognized for SimulCam and StroMotion, even though we have, in fact, developed a bunch of other applications with great business potential. Our primary concern thus is if and how we can possibly reposition ourselves [8].

3.2.2 Strategic Considerations

There were many options for exploiting Dartfish's knowledge in training software, but which were the most viable ones, and what should the company do first? Dartfish could either diversify into all possible businesses, hoping to reap profits from each segment's "early adopters," [9] or it could grow by broadening its existing markets, in particular the various sports training segments, and try to exploit the "low-end" business (*Figure 3*). However, entering mass markets was no walk in the park. It revealed surprising challenges, as Jean-Marie Ayer, cofounder and managing director in Europe, recalled:

> Our product is high-tech loaded, which requires a certain level of user affinity. I remember once we went to a professional golf player to teach him how to use our program. Soon, however, we realized that we first had to teach him how to start his notebook [10]!

Mass markets would also require distribution to be set up in a different way. Although Dartfish had moved away from selling exclusively via sports associations toward selling directly to athletes, it still needed a different strategy to penetrate the mass market. Furthermore, the market for "low-end" training software was not free of competition, although Dartfish did not need to fear the contest. It was in constant close contact with coaches and athletes, and the training suites were frequently updated based on their feedback. Nonetheless, cheaper solutions existed – a problem pointing back to the right positioning and pricing of Dartfish's low-end product suite.

On the other hand, diversifying into different application domains, or even only into different sports segments, posed a challenge in terms of organization. Currently Dartfish followed a divisional approach. It was split into the companywide sections of R&D, marketing, sales, sales support, finance, and operations. The only exception was the US business, which had its own marketing and sales organization. Addressing a variety of businesses, however, would most likely require organizing and coordinating the firm's product management differently.

4 Where to Go from Here?

Although Dartfish faced an abundance of opportunities to choose from, it lacked the financial and human resources to fully tap their potential. With a carefully elaborated strategy, however, it might be able to justify an aggressive budget and raise the necessary funds to eventually grow the company to the satisfaction of its shareholders.

Affentranger and his team (*Exhibit 2*) discussed all the options openly, aware of the crossroads at which they stood. Within the next month, they needed to finalize the strategic approach that would shape the firm's future.

The discussions boiled down to the fact that Dartfish could, on the one hand, continue covering the "expert markets" with its cutting-edge technology and a rather small business built around that. On the other hand, the firm could tackle very large business opportunities in the "consumer" or "student" markets, but the products and marketing strategies would still need to be developed. Questions to answer were:

- How could the company capitalize best on its assets? In particular, should both business lines – broadcasting and training solutions – be kept, or should the venture instead concentrate on training and try licensing or selling the broadcast business?
- Which market strategy should potentially be pursued for the training applications market? Should Dartfish diversify into different application fields or go "downmarket" – or even try both?
- Should the Dartfish vision evolve from a software provider to an ASP (application service provider)?
- In any case, how could product development and support be managed more efficiently and effectively?

Exhibit 1
Applications and Technical Platforms for Broadcast Business

DartCaster	• Live platform used during a sporting event.
	• Allows speedy and timely production of SimulCam composite footage, available for broadcast immediately after the second athlete's performance.
Fishbowl	• Hardware and software interface to transfer data from external camera to the DartCaster.
	• The fundamental component that delivers all kinds of special effects in real time.
DartCalib	• Software to calibrate lenses in studio and on site to deliver further special effects in real time.
DartStudio	• Portable postproduction platform that allows the creation of composite footage from tapes or cable network in a postproduction environment aiming at *ex post* analysis.
	• Includes SimulCam and StroMotion.
DartWeb	• Identical to DartStudio but with an interface to edit Internet pages.

Source: Dartfish "Overview Business Description," company brochure, May 2005

Exhibit 2
Profiles of Dartfish's Management Team and Cofounders

Anton Affentranger (business degree in economics)
Executive Chairman and CEO
Affentranger has extensive experience from various business leadership positions in Europe, Asia, and the USA. He was a member of the executive board of UBS, partner and CEO of Lombard Odier and CIE, and CFO of F. Hoffmann-La Roche.

Jean-Marie Ayer (PhD in economics)
Cofounder and Managing Director Europe
Prior to his Dartfish engagement, Ayer worked for large corporations in the high-tech sector (ABB, Swisscom).

Serge Ayer (PhD in computer science, EPFL)
Cofounder and Chief Technology Officer
Ayer is the lead inventor of the SimulCam technology and the lead engineer for the various products at Dartfish. Previously, he worked for IBM Research and the Laboratory for Audiovisual Communications at EPFL.

Victor Bergonzoli (business degree in economics)
Cofounder and US General Manager
Previously Bergonzoli was CFO of Sterling Cellular Company (Swisscom investment) in India and also worked for Crédit Suisse.

Ron Imbriale (master's degree in mathematics)
Executive VP of Sales, Dartfish USA
Imbriale gained experience as a senior partner of Venture Partners AG, in charge of US investments. He spent 25 years as senior sales and marketing executive at Tektronix, Cadre and Cayenne – all global software companies.

Daniel Morand (Business degree in economics)
CFO
Previously, Morand held various finance positions in banking and at Swisscom.

Emmanuel Reusens (PhD in computer engineering, EPFL)
Cofounder and Head of Broadcasting
Before starting on the Dartfish journey, Reusens worked as senior software engineer for Logitech's scanner and video business.

Philippe Schroeter (PhD in computer sciences, EPFL)
Head of Sales Support and Director of Training
Schroeter was multimedia program manager at Swisscom before joining Dartfish.

Ivan Pilet (engineer in microtechnology)
Head of Operations and Director of Quality
Pilet was senior project manager for SIG, Switzerland.

Martin Vetterli (Professor of audiovisual communications, EPFL)
Cofounder and Serge Ayer's PhD supervisor
Vetterli continues to support the venture regarding intellectual property decisions and as a technology advisor to management and the board of directors.

References

1   SimulCam, StroMotion, and DartTrainer are trademarks of Dartfish.
2   For image and video examples, visit the company website: http://www.dartfish.com.
3   Further information on Dartfish's early history is given in the case InMotion Technologies Ltd. by Olivier Courvoisier. ECCH no. 804-008-1, 2000.
4   Vetterli, Martin. Digital Image/Video Processing at the Audio Visual Communications Laboratory, Presentation, EPFL, May 23, 2005.
5   Affentranger 2006.
6   Vetterli 2005.
7   Affentranger 2006.
8   Affentranger 2006.
9   Refer to Geoffrey A. Moore: Crossing the Chasm. New York: HarperCollins, 2002.
10   Ayer, Jean-Marie. Company Presentation, MoT Event "Innovation and Entrepreneurial Clusters, EPFL, August 25, 2005.

CASE 4-6
Netcetera (A): Hiring an External CEO?

Ralf W. Seifert and Jana Thiel

Netcetera, a Swiss company that provided software applications for networked computing environments, had flourished since it was founded in 1996. Customers returned regularly, projects grew in size, and headcount increased. Yet the cofounders felt pressured to decide whether their current management approach was the right one to take the company to the next level.

By April 2000, Netcetera employed 30 people and was planning to recruit an additional 20 new employees by year end to keep up with the ever-increasing volume of projects and customer demands. Although the impressive growth was welcome, it was starting to place a strain on Netcetera's integration capabilities and knowledge transfer since the business was founded on the expertise of its people. Mike Franz, cofounder and head of business services, summarized the situation:

> Our growth over the last months, especially the integration of new employees and the development of larger projects, has absorbed all our resources and left strategic issues undone. In addition, an internal assessment has revealed that the quality and efficiency of our projects has dropped, even though our customers are still satisfied.

This situation jeopardized Netcetera's bright business prospects. Action was required to sustain its growth and to lay the foundation for a larger organization, while preserving the company's cultural integrity and efficiency in project completion. The board and cofounders had started to discuss whether to recruit an external CEO to strengthen Netcetera's management. A first candidate was soon identified; he had strong operational and process expertise based on 20 years in a variety of positions in large firms. Netcetera had to decide:

- Was an external CEO the solution to its challenges? Was the timing right? What were the alternatives, if any?
- Should Netcetera go ahead and appoint this candidate or should it broaden the search? How much time would it take for a new CEO to be fully operational?

Copyright © 2006 by the alliance for technology-based enterprise (IMD, EPFL and ETH Zurich). Copyright permissions are handled by IMD, Lausanne, Switzerland.

1 Netcetera

Netcetera was established as a limited company in January 1996 – after 18 months of deliberation – by four cofounders, who met while studying for engineering or science degrees at the University of Zurich and the Swiss Federal Institute of Technology in Zurich (ETH Zurich). In May 1996 Mike Franz joined as a fifth partner. The cofounders had all acquired knowledge in software design and Internet technologies, which built the foundation for their new software development company. Two silent shareholders were involved at the outset to provide business expertise and advice; they became board members later on. After carefully assessing the market, Netcetera concentrated its efforts on developing secure transaction systems, mostly for financial institutions. Over the years, the company acquired particular expertise in processing large quantities of data. Soon the venture was among the leading providers of market and stock data systems, and few companies equaled its competency in credit card processing.

Benefiting from the Internet boom of the late 1990s, Netcetera had quickly prospered. In 1999 it reported a turnover of CHF 5 million, and 2000 promised a closing result of almost double that amount. In terms of headcount, Netcetera had almost doubled the figure each year, and it expected to have more than 50 employees by the end of the year. (*See Exhibit 1 for data on turnover and employees.*)

However, the cofounders realized that this level of growth meant that Netcetera risked losing sight of its newfound strategic direction and not having the time to implement it properly.

2 Netcetera's Development 1996–2000

2.1 Choosing the Location

The startup initially took offices in Wildegg (canton Aargau, Switzerland), some 30 minutes outside of Zurich. It shared space and resources with an existing company, which helped Netcetera get off the ground. Soon, however, it became apparent that a downtown location would offer distinct advantages. Given the nature of Netcetera's business, customer proximity turned out to be key to winning accounts and providing responsive service. There were also recruitment benefits: Zurich was home to a large pool of highly qualified software engineers, who preferred the city's attractive work environment and were reluctant to move. The decision to relocate Netcetera to downtown Zurich came naturally.

2.2 Developing the Business Model

Netcetera specialized in the development and design of customized application software rather than licensing and configuring off-the-shelf products. The applications built mainly on open-source technologies such as Java/Unix. Typically, a Netcetera team was called in when a standard application like an ERP system[1] needed special additional functionality. The business concept was to provide turnkey solutions with a predefined scope, time frame, and budget and to excel at project management and customer service. Netcetera usually did not engage in "body leasing," *i.e.*, renting out specialists.

Early on customers recognized the cofounders' ability and commitment to *building* applications, and carefully chosen new employees complemented these skills. Soon customers were inviting Netcetera into adjacent business areas, *e.g.*, upfront requirements engineering and, later, into running the products it had built. Increasingly, Netcetera accepted the challenge and started to run applications for its customers, as Franz explained:

> One of our advantages, which finally led UBS and Credit Suisse to run systems with us, is that we spare them the hassle of bringing built software into the run environment.

2.3 Building the Customer Base

Back in 1996, only a few companies fully anticipated the vast opportunities afforded by the Internet, and even fewer made the necessary investments in the required technology. Financial institutions, however, were among the first to pursue these opportunities, and Netcetera thus focused heavily on financial solutions, soon attracting companies such as UBS and Credit Suisse as steady customers. (*See Exhibit 2 for project examples.*)

Although the economy was thriving in 1998 and 1999, new business did not come on its own. Netcetera quickly learned that winning new customers was hard work. Once there, however, customer retention was a matter of fulfilling expectations – and Netcetera proved to be good at this. Its employees were trained to identify requirements early on and carefully track customer satisfaction throughout a project. Doing a good job there frequently led to subsequent business and saved having to spend a lot of effort and lengthy sales cycles on acquiring new customers and accounts. Franz noted:

> This is Switzerland, still a small country where promotion is by word of mouth. Moreover, people tend to change employers, and if they had success with you in their previous job, they might remember and come back to you. This is exactly what happened to us.

---

[1] ERP stands for Enterprise Resource Planning – a complex business operation which is frequently supported by enterprisewide integrated IT systems (ERP systems) such as SAP, PeopleSoft, or Baan.

In late 1999 Netcetera relied almost entirely on the banking industry for its customer base. Moreover, nearly 50% of turnover was generated exclusively from one large customer – an issue that would need more attention once the CEO question had been resolved.

2.4 Creating a Netcetera Culture

Success in a service-oriented business was all about people. The venture had to rely on highly skilled and motivated employees. Hence, company values and culture played a crucial role in complementing monetary incentives, especially since financial gain was not a prime motivator for those seeking to work for the company, as Franz explained:

> Ambitious IT projects and challenging programming tasks drove people more than the extra dollar at the end of the month.

In the early days, Netcetera salaries were comparatively low and the company did not pay a 13th month installment (a Swiss-wide practice). It granted only 20 days of vacation, instead of the usual 25. Only after the business skyrocketed in 1998 were concessions made by introducing a generous bonus system. However, Netcetera had to offer additional incentives to sustain employee motivation and to compete for much-sought-after IT specialists: Company workshops, free drinks and snacks, regular sports events, and other benefits of this type contributed to a "complete value package" designed to set Netcetera apart from its competitors. It succeeded. Employee motivation and retention were not issues.

Yet in an increasingly "empty" labor market, recruitment had become more difficult and started to restrict the venture's growth plans.

2.5 Organizing Internal Process Quality and Efficiency

Organizational growth soon required extensive changes in procedures and standards. Keeping track of projects and their documentation was among the first challenges Netcetera faced. A database system was quickly installed to collect and manage all project records. The tool was then extended to be a performance-tracking instrument and had subsequently been Web-enabled. Yet the management team felt that they had not fully leveraged all the potential to govern and control the firm's operations and reassure its customers.

Internal communication was a second issue that had emerged over the years. As the team reached about 15 people, messages suddenly failed to come through the way they used to – coffee breaks were no longer sufficient to connect and the problem naturally grew worse over time. To bridge this gap, regular staff meetings were initiated, which helped to enforce the Netcetera community, but they were not truly effective in supporting new employees to adopt tacit standards. A first internal project – THEMAS (The Netcetera Manual of Style) – had been launched a few months earlier, aimed at standardizing and communicating shared processes, but it had yet to gain full support and yield measurable results.

2.6 Advancing the Strategic Approach

Business schools wouldn't like us: During our first years, there was no strategy at all. We just tried to have fun while doing our job and seizing the opportunity of the moment.

Mike Franz

The Netcetera team had understood how to capitalize on the thriving market of the late 1990s. However, the vast opportunities attracted others as well, and if Netcetera wanted to maintain its share it needed to step up to the next level. In April 1999 – admittedly more at the behest of the external shareholders than on their own initiative – the management team had engaged in their first ever strategic development plan, STEP '99. Three main issues were identified:

First, based on Netcetera's position and the potential business opportunities, there was no reason why the company should not grow to 80 or even 100 people within the next few years. However, until then, business had developed in reaction to external influences rather than being proactively directed. The ambitious growth plans required more focused management dedicated to stabilizing processes, quality, and efficiency.

Second, the cofounders intended to leverage the expertise and knowledge inherent in Netcetera's applications more effectively. Traditionally, each application was tailored to specific customer requirements, limiting reuse possibilities. To raise the level of reusability, they envisaged that software licensing could play a more prominent role.

Third, they needed to explore opportunities for business diversification. In response to increased customer demand, plans were outlined for extending the business model toward strategic IT consulting. Consulting, however, required competencies that Netcetera did not possess at that time.

A year after STEP '99 was drawn up, a review revealed that Netcetera was not even close to any of these goals. The company had become bogged down in daily operations and was simply too busy to advance these more strategic issues. All management resources had been completely taken up with acquiring and managing projects and integrating new employees. Moreover, all the cofounders, *i.e.*, management members, continued to be actively involved in project work. The management team confessed that they lacked the execution capabilities to implement STEP '99. This called for a change of approach.

3 Changing the Management: Hiring a New CEO?

Early on the cofounders had followed a pragmatic approach to managing the company, sharing responsibilities and tasks among them. When Netcetera started business, Joachim Hagger had been the most experienced cofounder in terms of leadership and management background. Hence, he became CEO. However, the position was understood as *primus inter pares* and essentially each cofounder would have been just as capable. Over time, individual preferences crystallized and management positions were defined more carefully. Four years after it was founded, Netcetera was split into two focus areas – development and services (*Figure 1*) – headed by two and one cofounders, respectively. Meanwhile, one of the cofounders had left the team due to differing opinions about Netcetera's future and the road to take.

> We basically had no prior management experience. These last years brought a fair share of learning for all of us.
>
> *Mike Franz*

Their strategic struggles led to an open discussion between the cofounders and the board on how to boost their management proficiency and the growth of the company. As a result, hopes rested on acquiring external (professional) expertise and, since none of the cofounders was overly concerned with positions or titles, it did not take them long to agree to look for a new CEO.

A first candidate was soon identified. Being about 10 years older than the cofounders, he had gathered 20 years of experience in various management positions in large companies (*e.g.*, serving IBM for more than 10 years). He had specifically developed strong operational and process expertise as COO for his last employer, a Swiss-based medium-sized industrial company. However, he had not worked in a particularly technical position before.

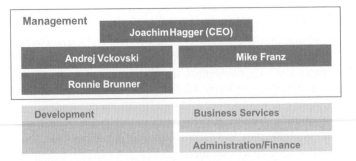

Figure 1   Netcetera's Organizational Structure in Early 2000

The negotiations had been successful; the candidate was keen to take up the challenge. Now it was up to Netcetera to make the final call.

- Was an external CEO the solution to their challenges? Was the timing right? What were the alternatives, if any?
- Should Netcetera go ahead and appoint this candidate or should it broaden the search? How much time would it take for a new CEO to be fully operational?

Exhibit 1
Turnover and Employee Figures 1996–1999

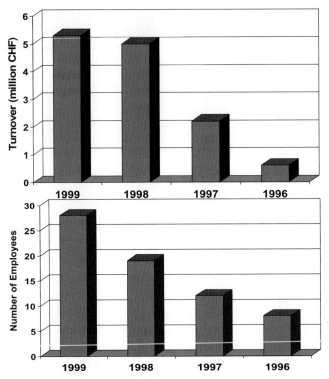

Source: Company literature

Exhibit 2
Reference Projects

UBS

- Since 1997 – ongoing
- Support in several projects related to financial market data
- Creation of a central data integration platform for a centralized and UBS-wide market data service known as "UBS Quotes"

Association of Swiss Cantonal Banks (Verband Schweizerischer Kantonalbanken VSKB)

- Since 1998 – ongoing (maintenance)
- Design, programming, and ongoing maintenance of a tool set that allows users to perform calculations and comparisons in financing, financial security, and tax issues

Public Administration – Swiss Federal Archive (http://www.dodis.ch)

- Since 1997 – ongoing (maintenance)
- Creation, regular maintenance and updates of a Web interface for the Swiss Federal Archive, which
- Contains pivotal documents in regards to Swiss Foreign Policy
- Makes documents electronically available for the public to enable reconstruction and foster understanding of Swiss history

Source: Company literature

CASE 4-7
Netcetera (B): Organizing for Sustainable Business Success

Ralf W. Seifert and Jana Thiel

The Netcetera cofounders appointed the external CEO as of May 2000 to drive an aggressive growth strategy. The aim was to potentially take the company public. Under his lead, Netcetera improved both internal communication and quality standards. The resulting boost to the firm's operational capabilities and project efficiency prepared the ground for further revenue and headcount growth. However, a lack of cultural match and internal support for the external CEO became obvious as he took charge. As he and the cofounders could not find a common denominator, he left the company after only eight months.

Netcetera's management structure returned to its previous two-division setup, with Andrej Vckovski, one of the cofounders, then assuming the CEO position previously served by Joachim Hagger. Further restructuring followed within less than a year: By February 2002, Netcetera was split into five divisions to bring more management responsibility to senior employees. The new structure, however, did not prove to be as effective as envisaged and eventually threatened profitability.

In early 2004, Netcetera's organizational form was replaced by a holding structure, legally named Netcetera Group AG, to govern the original company and new subsidiaries. This move was intended to help them resume growth as the economy picked up. During 2004 a new IT consulting subsidiary was established, and the original Zurich-based software business became a subsidiary alongside the others. Meanwhile, the market in 2004 called for continued improvements in operational efficiency, and one year into the new group structure significant questions about the level of its success remained:

- Was the new group structure the right setup to enable further growth and to weather dynamic market changes?
- How could Netcetera management improve efficiency and thus profitability within this new organization?
- How could Netcetera continue to maintain and further improve employee retention?

Copyright © 2006 by the alliance for technology-based enterprise (IMD, EPFL and ETH Zurich). Copyright permissions are handled by IMD, Lausanne, Switzerland.

1 Netcetera's Learning Points from the External CEO Decision

- *Successful leadership builds on trust.* In an environment of knowledge workers, the majority with university degrees or similar higher education, trust and personal chemistry are essential.
- *Don't underestimate the cultural match.* A company builds on its history and common understanding of values and procedures. External managers need to understand these factors in depth to gain respect and solicit support.
- *Cohesion in the core management team is critical.* The core management team needs to follow joint objectives to create the necessary cohesion. A lack of cohesion requires increased effort to align business interests and decision making – a resource- and time-consuming process.

Mike Franz summarized:

> Back then, we might simply not have had the capability to successfully integrate someone external in such a prominent position. If we had to do it again, we might try introducing the change in a step-wise way. If an external person was to be named CEO, we might have him or her work in another management position within the organization first.

2 Trying a Participative Approach to Organization

With the departure of the external CEO, Netcetera was back to square one – at least in the eyes of its employees. People's perception was that the company did not offer true development options internally, though it enjoyed good market growth.

This perception had been nurtured over time, as manifested by the creation of the first two subsidiaries back in 2001 and 2002: Netcetera Skopje (Macedonia) was set up in 2001 as a "nearshoring" outpost to address the current labor shortage in the Swiss IT market and to preserve competitiveness in a margin-squeezed environment. Despite having a local management team (holding shares in the business), customer contact was managed exclusively through Zurich. The second venture, Eveni – again a limited company – followed in 2002 and focused on providing Web-based conference and event administration services. The new ventures were led by professionally experienced CEOs, with the Netcetera cofounders taking seats on the board and holding the majority of the shares. Yet for the rest of the employees – particularly more senior people – these developments did not immediately translate into a rich variety of internal career options. Mike Franz noted:

> Increasingly, we felt the pressure: If someone wanted to excel in software development, the range was open, but for those leaning toward management positions, chances were limited.

Remedy was sought in a new organizational approach, effective as of February 2002. The new arrangement consisted of five divisions headed by the cofounders (*Exhibit 1*). For each domain, senior employees were promoted to support operations and assume management responsibilities. Franz explained:

> At that time, we seemed to have found the perfect solution by building a huge management team to which everyone would contribute. Pretty soon, however, we realized that this concept was not going to work. Everybody wanted to act as manager and too many people were telling how things should be done rather than taking the initiative.

About nine months later, management efficiency had deteriorated drastically and threatened Netcetera's profitability. Meanwhile, 2002 had been a difficult year for winning new customers, and 2003 did not look as if it would be any easier. The cofounders had learned the hard way that assigning "too many chiefs" to one task frequently led nowhere – a luxury they could no longer afford.

As the market slowed down, those who were promoted still found themselves spending 80% of their time on their daily projects, and the whole attempt failed to gain acceptance in the end. Meanwhile, the concept had put a strain on the company culture by constructing a barrier between "those who had made it and those who had not." The motivation and excitement levels of Netcetera's software engineers diminished noticeably. Enthusiasm was further dampened by the declining innovativeness in the overall market, reflected in less challenging projects and decreasing staff bonuses. Netcetera's "complete value package" for staff had started to crumble. The problem needed to be addressed.

Until now, Netcetera had experienced almost zero fluctuation in staff levels, despite a competitive market environment for skilled IT personnel. In 2003, however, the first people started leaving the firm. Franz noted:

> When you start losing people who are eager to progress in their careers, you are probably losing those you would have actually rather liked to retain.

Of course, the departures made way for new ideas and talents, but, as the cofounders were fully aware, business was bound to senior and knowledgeable staff. Hence, they set about remotivating employees while ensuring operational efficiency. Of course, this had to be accomplished in the context of ongoing market changes and Netcetera's strategy for growth.

3 Adapting to Market Developments

To date, Netcetera's customer acquisition strategy had not significantly changed from previous years. Existing accounts and project followups still constituted the majority of revenues with a maximum of 30% of new client business each year.

And although Netcetera served 40 active accounts, 8 of them contributed 80% of its turnover. Going forward, this picture would have to change.

In terms of industry coverage, Netcetera had already developed vigorously in recent years – leveraging competencies acquired in the financial sector in other industries (*Exhibit 2*). It now also served customers such as the Swiss Federal Railways and the European Space Agency (*Exhibit 3*) in addition to its stronghold in the financial sector. This constituted an attractive opportunity but also translated into an added strain on employee competence development.

Meanwhile, Netcetera's headcount had grown to 70 employees, and it would remain at that level through 2004. Turnover was likewise expected to level off, after falling from CHF 12 million in 2002 to CHF 11 million in 2003. (*See Exhibit 4 for data on headcount and turnover.*) Times were a bit less rosy than in the early days, yet nobody had had to leave Netcetera, as Franz commented:

> It is a question of culture, ethics and morale. We deliberately chose to keep everyone on board, despite the economic slowdown and the decline in our chargeable rates. It is important to uphold the commitment towards employees also in difficult times – hence we tightened our belt.

Even though profits had been squeezed in favor of retaining employees, Netcetera was still yielding positive returns. However, time was running out with respect to creating an efficient organizational setup and managing employee expectations.

By 2004, the lion's share of Netcetera's business – about 80% – continued to be custom-tailored software solutions for networked environments. An additional 15% to 20% was contributed by software maintenance and service fees, while less than 5% was generated through consulting services. Netcetera made efforts to enforce the consulting segment in the future. The objective was to gradually expand business along a defined Plan-Build-Run value chain reflecting the business process cycle of its customers (*Figure 1*).

The ambition to boost consulting revenues had been around since 1999, but Netcetera had not yet made much progress. It now pinned its hopes on Metaversum, another limited company under Netcetera's holding roof, which started

Figure 1   Netcetera's Business Process Coverage

operations in May. It focused on business process management and strategic planning of IT solutions. Furthermore, it was regarded as a testbed for a business model that would presumably replace Netcetera's current approach one day, as Franz explained:

> The market development will force us to change. Within five years, the Netcetera developer of today will have ceased to exist. Increasingly, we will employ specialists and project managers, meanwhile the pure programming is done elsewhere.

3.1 Organizing for Sustainable Business Success

Besides preserving motivation and efficiency, any new organizational approach also had to accommodate a targeted expansion of the company. Netcetera now ran three businesses outside its software engineering core, and the cofounders were not hostile to the idea of adding more ventures somewhere down the road.

Eventually, the complex organizational setup that had been in place since 2002 was discarded. Instead, a holding structure was adopted, which would host all present and future businesses and provide groupwide services such as finance and administration. Mirroring the existing subsidiaries, the original Netcetera, *i.e.*, the core software business, was turned into another group branch: Netcetera Zurich (*Figure 2*). Over the course of 2004, a new management crew was formed out of seasoned employees who would take over the new Netcetera Zurich and hence direct this business completely separately from the cofounders. The latter planned to retreat to the holding level from where they would provide strategic coaching and continue to engage in account management activities. Franz noted:

> It is a great moment when the company you founded enjoys sustainable development so that you can step back and let others grow.

Netcetera Group AG [Holding]
Dr. Andrej Vckovski **Joachim Hagger** **Mike Franz** **Ronnie Brunner**

Netcetera Engineering Zürich	Netcetera Engineering Skopje	Metaversum AG	Eveni AG
70 MA / 1996 / Zürich	**15 MA / 2001 / Skopje**	**2 MA / 2004 / Zürich**	**2 MA / 2002 / Zürich**
Software Engineering *Finance* *Aerospace* *Transportation*	Software Engineering *Off-/Near-Shore* *Services* *Medical Resource* *Planning*	IT Consulting *Business Process Mgmt* *Buiness Engineering*	Event Management ASP *Online Services for* *Conferences,* *Conventions, Events,* *Exhibitions*

Figure 2   New Netcetera Structure as of 2004

However, the market was still tight and did not leave much room for experimentation.

- Was the new group structure the right setup to enable further growth and to weather dynamic market changes?
- How could Netcetera management improve efficiency, and thus profitability, within this new organization?
- How could Netcetera continue to maintain and further improve employee retention?

Exhibit 1
Organizational Structure Between 2002 and 2004

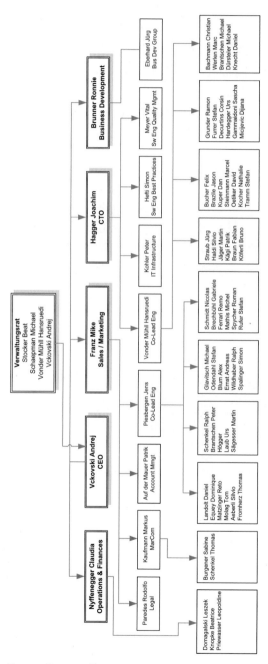

Source: Company literature

Exhibit 2
Competencies and Industry Coverage

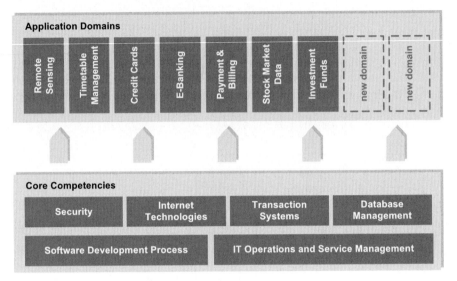

Map of Competencies as of 2004

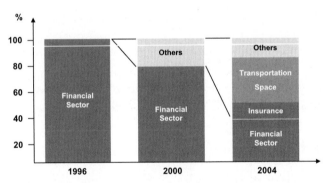

Development of Industry Coverage 1996-2004
Source: Company literature

Exhibit 3
Industry Coverage and Reference Projects

Insurance – eSure CSS (http://www.cssag.ch)

The project comprised the creation of a modular Web solution for the online purchase of insurances including payment. The platform specifically featured:

- The integration of online and offline contracts in the back-office environment and
- A shop-within-a-shop for brokers and partners.

Transportation – Swiss Federal Railway: FTE Pathfinder (no public access)

"Pathfinder" is a browser-based platform used by the majority of the big European railway companies and contracted through the Swiss Federal Railway. It supports the coordination of all European cross-border connections (passenger and freight traffic). The project involved (1) process-oriented planning over several steps and numerous organizations in various countries and (2) developing a wide variety of support functions like searching for and copying of planning dossiers, and extensive messaging within the closed user group.

Finance/Banking – myAccount Viseca Card Services SA (http://www.viseca.ch)

The project comprised the development and subsequent hosting and maintenance of a self-service platform (B2C) for Viseca, a major credit card issuer in Switzerland (with approximately 30% market share in Switzerland, 1 million cards issued). The application featured:

- Reporting services (list of transactions, invoices),
- Major use cases (change of address, card limit, *etc.*),
- Subscription to secure payment technology (3-D Secure).

Source: Company literature

Exhibit 4
Turnover and Headcount of Netcetera Group AG 1996–2004
(incl. Forecast 2005)

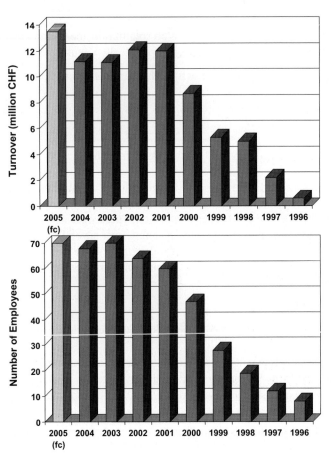

Source: Company literature

CASE 4-8
Netcetera (C): Reflections and Outlook

Ralf W. Seifert and Jana Thiel

Another year had passed. It was February 2006. The year marked Netcetera's tenth year of business, prompting cofounder Mike Franz to reflect on its journey to date.

> With hindsight, Netcetera surely ran the gamut from uncontrolled chaos to an over-controlled and overprocessed enterprise. Today, however, we are more balanced: The structure and tools in place, such as our monthly cockpit, allow us to track and maintain performance while keeping administrative effort in check.

Netcetera's new organizational setup, effective since early 2005, had gained broad acceptance among the company's employees. It had also had effects on the operational side, raising the internal profit rate considerably. The turnover in 2005 had been over CHF 13 million, exceeding expectations by 15%.

Netcetera was finally running to the satisfaction of its cofounders, who summarized their learning in terms of the organizational setup.

- *Structure needs to be transparent, with clearly defined responsibilities and information policies.* The distribution of power and responsibilities needs to be supported by a clear, written, understood, and "lived" charter. Governance is to be clearly formulated at the outset of a reorganization.
- *Flat hierarchies are not always better.* Decision making in less hierarchical structures needs more discipline and mutual understanding than in classical organizational structures. Not every subordinate appreciates working in flat hierarchies.
- *When growing to a larger organization, transition management requires careful timing and execution and, most importantly, communication.* The first attempt to transfer management responsibility from the cofounders to employees had almost put the business at risk. Only with the right timing and improved strategic approach was success achieved.

Copyright © 2006 by the alliance for technology-based enterprise (IMD, EPFL and ETH Zurich). Copyright permissions are handled by IMD, Lausanne, Switzerland.

1 New Projects

At the end of 2005, two new large-scale projects had been signed for 25 and 50 person years, respectively, bolstering revenues for the next two years. For example, Netcetera would develop the entire new routing system for the Swiss Federal Railways (SBB), which would further boost the company's visibility and credibility to perform large-scale IT programs and projects in business-critical and increasingly international contexts.

2 New Objectives

In general, the IT market had changed for the better, and Netcetera's cofounders had every intention to leverage the positive climate. They had introduced an updated strategic plan, envisioning the company's development until 2008:

- *Grow headcount by 10% to 15% annually,* in particular through the group ventures Netcetera Skopje, Metaversum, and Eveni.
- *Diversify projects geographically and into new application domains* to lessen the dependency on particular regions or industries. Geographically, the Middle East was targeted next, where business connections had been established by early 2006. As for the industries, possible sectors were blue chips, e-health, and security.
- *Grow financially, i.e.,* pass the mark of CHF 20 million turnover per year, and increase operating profit annually by 1% to 2% of turnover to sustain the self-financing strategy. However, as external money would grant more operational freedom, Netcetera also aimed to intensify investor relations (*i.e.*, increasing transparency and publishing performance figures on a regular basis).

3 New Challenges

Netcetera's cofounders did not rule out the creation of additional portfolio companies. The envisioned growth, however, required continued increases in efficiency and the development of the business model. The challenge was to clearly position the company in the market and to strengthen its success factors.

In particular, the cofounders had to work out:

- How to *better exploit synergies between the portfolio companies, i.e.*, realize up- and cross-selling opportunities within the group and thus also increase the acquisition efficiency.

  How to *develop the business model.* In particular, the cofounders aimed to increase the revenue share of "people-independent" business such as licensing, service level agreements (SLAs), and transaction-fee-based business (ASP business as in Eveni).

CASE 4-9
Boblbee (E): Inventing the Urban Nomad

Benoît Leleux, Joachim Schwass, and Anna Lindblom

Bernstein commented on the night in the hotel room in Munich when he, Blanking, and Bonnier sat together after the ISPO fair:

> We had planned our pricing on an agent basis but everybody wanted to be a distributor. We were unsure about the market standards so we used "a finger in the air" to determine the new prices. When we came home we discovered that there were a few problems with the pricing – we had calculated the distribution margin too low, but we were able to solve that. [*See Exhibit 1 for pricing strategy.*]

Using distributors meant that margins would be lower, but it also meant that there would be less risk involved for GAAB, as the distributors would carry the inventory. (*See Exhibit 2 for the distribution model.*)

GAAB also decided to do a horizontal market introduction in a number of countries. Bernstein commented:

> Our goal had always been to be global. We all had international experience and it felt natural to us. Our ambition was to become big, and by staying in Sweden that wouldn't happen.

In addition, they hoped that being "first to market" would give them access to key players in key markets and thus increase their chances of creating barriers to entry and building a number of strong key accounts in various European countries and the USA.

Copyright © 2006 by IMD – International Institute for Management Development, Lausanne, Switzerland. Not to be used or reproduced without written permission directly from IMD.

On their return from ISPO in Munich, the team immediately went to work. Blanking set about improving the harness with all the tips he had gotten during the fair. Bernstein sent letters to the 84 distributors that had shown an interest in Boblbee, asking them to submit a five-year business plan to GAAB. Roughly half responded. The distributors were a mix of large, small, established, and new. GAAB picked 20, mostly ones that were like GAAB itself – small, new, and entrepreneurial. Bernstein said:

> We did not want to become a small company in some big company's portfolio. Every time we went to see a big company we had to wait to see the person in charge – we did not mean anything to them. The smaller companies on the other hand were hungry and energetic and they were always keen to see us. We saw ourselves in them.

GAAB stipulated that each distributor would have to buy at least 500 bags in order to be listed as a distributor – this was the "cost of entry." Through these initial contracts, GAAB managed to sell 10,000 bags. In November 1998, two months after ISPO, the company began shipping backpacks to distributors and retailers.

Boblbee would initially be offered only through a select group of high-quality sports and leisure, computer hardware, and design stores and outlets. The distributors would have exclusive rights in their own territory. For customized or specialized versions of the backpack, the distributors acted as nonexclusive agents. This gave GAAB maximum flexibility and the added potential to sell directly to corporations and government institutions.

Bernstein established a 12-month rolling forecast model. The distributors had to make a 12-month sales estimate and commit three months in advance with a fixed order. GAAB used the estimates to make continuous sales projections for the upcoming periods and to plan the fiscal year.

By 1999, GAAB had sales and distribution partners for Boblbee in 23 countries on four continents. The company also opened its US Boblbee Web site for sales. The following year European customers were also able to use the Internet to purchase their Boblbee products.

1 Marketing Strategy

GAAB had to keep marketing expenses to a minimum and therefore used guerilla tactics and "smart deals" with various partners. The marketing mix for Boblbee consisted of trade shows (or fairs), event sponsoring, poster material, public relations (PR), and product placement. The goal was to create demand in the market and use market pull, as opposed to pushing the products to the customers. A slow increase would also be good for the learning curve in the manufacturing facility.

1.1 Trade Shows

GAAB went back to the ISPO winter fair in Munich in February 1999. The response was less enthusiastic than it had been the previous year. But the CEO of Pitti Immagine invited them to join Uomo, the fashion fair it was sponsoring in Florence, Italy, in May. He told Bernstein that it was the most important fashion fair in the world and that Boblbee would be a perfect match with its urban and trendy image. Bernstein commented:

> I remember thinking to myself "fashion? We produce sports goods!" But since sales to customers at The Stadium had been lukewarm and it seemed that targeting sports people was not the right thing, we decided to test the fashion route – so we went to Florence.

GAAB again had tremendous success at the fair, and several big department stores – among them Neiman Marcus, Bloomingdales, and Saks Fifth Avenue in New York – expressed an interest in selling Boblbee products. In 1999 Boblbee was featured at 14 different international trade shows.

1.2 Sponsoring

GAAB sponsored a number of sporting events, including the Red Bull Free Skiing event in Alaska and the World Cup of Windsurfing in Zandvoort, the Netherlands. Event participants wore the Boblbee backpack, and GAAB later used the photo material in Boblbee's marketing campaigns and brochures.

1.3 Product Placement

In 1998 GAAB started cooperating with the Propaganda agency in Geneva for various product placement opportunities in international movies and TV series. In 1999 Robin Williams wore a Boblbee in the movie *Bicentennial Man*, and in 2000 the backpack appeared in the movie *Charlie's Angels*.

1.4 Public Relations

Boblbee had been able to generate significant interest in the fashion, sports, and IT media. It was mentioned in magazines and newspapers such as *The Wall Street Journal*, *Vogue* (US edition), *Wallpaper* (an upmarket designer magazine), *GQ Magazine*, *Time Out*, *Orange Magazine*, *Axiom*, *The Daily Telegraph*, *Maxim*, *Arena Homme plus*, *The Guardian*, *Boi Chicago*, *Nylon Magazine*, and *Details Magazine*. Boblbee also featured prominently as one of four "trend" products in the Sport-Scheck catalogue, which was distributed in Germany, Austria, and Switzerland.

1.5 Display System

The display system that had been developed for the first ISPO fair in 1998 turned out to be a huge crowd pleaser and a great fit with Boblbee's image. It was unique, high-tech, and readily accepted by key retailers around Europe. In addition, GAAB was also able to save costs by using parts from Spotlight. In 1999 the display system was installed in various department stores, including Harrods in London.

2 Trouble Begins

By the end of 1999 Boblbee was distributed in 25 countries around the world. The main markets were Japan, Italy, France, and Sweden. The turnover was SEK 24.8 million (€ 2.5 million). (*See Exhibit 3 for financial statements.*)

By 2000 Boblbee was sold in more than 800 fashion outlets around the world. GAAB also changed its name to Boblbee AB, reflecting the company's focus on the Boblbee backpack.

But troubles were slowly creeping up on the company.

2.1 Problems in Distribution Channel

The Boblbee distributors had projected they would sell 140,000 backpacks in 2000. However, Bernstein knew they would never be able to sell this many and that such improbable projections were a clear flaw in the 12-month rolling budget model. Boblbee reduced the projections by 40% and ordered 56,000 bags. But the problems ran deeper. According to Bernstein:

> Our distributors never told us there was a serious problem with sales. The distributors were also smart – they believed that sales would pick up and then they wanted to be there to ride the wave. They did not want to lose the contract with us so instead they kept quiet. The bottom line is we were riding without a clear reading of the market because our distributors were keeping mum.

The distributors – encouraged by sales in 1998 and 1999 when trendsetters and designers had bought the backpack as soon as it reached the shops – had ordered large quantities of stock, but the repurchase rate after initial pipeline filling was weak due to the slow sell-out of the product. Many of the distributors had taken loans in order to buy the stock and hence were encountering financial difficulties. Bernstein commented afterwards:

> Italian shopowners were crazy about our backpack when it first came out as it was new, cool and different. But two years down the road they were looking for the newer "new thing" – the next Boblbee. In retrospect it was probably wrong to go with the fashion outlets, and the first marketing brochure was just "too much," too trendy. My Italian

distributor should have warned me about this – that this is how it would end. He must have known. We did not see the problems, nor did we listen carefully enough because we were too far away from the customers. We should already have seen it when we launched the bag in The Stadium in 1998 – but we were blinded by all the praise. In addition, I was so overworked with Spotlight and Boblbee that I had little idea about what was going on at the end of the chain.

As a result, Boblbee got stuck with unsold inventory worth more than € 1.2 million (SEK 12 million). The company was forced to take over the distributor role, with distributors becoming agents and Boblbee carrying the whole inventory risk.

2.2 International Adventures

2.2.1 USA

In 2000 Boblbee set up its US subsidiary, Boblbee Inc. Since Boblbee was not completely sure about how to market the backpack – was it a fashion accessory or a sports item? – it decided to handle distribution in the USA itself. The company hired a former manager of IKEA in Paris (who also invested € 22,000 in GAAB) and sent him to the USA to set up the sales network. He was treated as an expat and received all the usual perks – house, car, *etc.* Boblbee invested around € 700,000 (SEK 6 million) in the expansion into the USA, but the results were unsatisfactory. Bernstein noted:

> The decision to send our own man to the US was thought through and discussed with the board. It wasn't a rushed decision. He was hired because he was good. The US was too big to give to a single distributor. His job description was to build a network of independent distributors. If we decide to open a business in the US I won't sit here in Sweden telling people what to do. One has to be close to the market – I also had too much on my plate – I couldn't have done it even if I'd wanted to. But he could not get the business going – it turned out to be a total cash drain and we had to go our separate ways.

2.2.2 Pirate Copies in Korea

What started as an incredible story in late 1999 ended in disaster a few months later. A Korean, Mr. Lee, traveled all the way to Torekov with his translator to buy 1,200 Boblbee backpacks and paid the € 80,000 (SEK 720,000) in cash. Bernstein recalled:

> The story was crazy. This Korean calling me and then coming here and buying a whole container and paying in cash. It was unbelievable. I took all the money with me to the local bank and sat there and counted it. I went to Korea to do an inspection a few weeks later. A couple of weeks later I got an e-mail from one of Lee's employees who had been fired and he told me that I was being cheated. It turned out that they had made cheap copies of the mold behind our backs and were selling a bad copy of Boblbee under the brand name Ergo.

Boblbee sued the Koreans and a long court battle began in Australia. In the end, Boblbee's patents turned out to be solid, and the company eventually won, but it was a drain on cash and energy.

3 The Crisis

In 2000 HB Capital, a venture capital firm, came on the scene. Via a directed capital increase, it invested € 3.85 million (SEK 35 million), giving it a 34% stake in Boblbee AB. For Blanking, Bonnier, and Bernstein HB Capital's investment meant that they got some money back in their pockets since the venture capitalist (VC) decided to buy stocks from the three of them. The three partners then each had a market share of 10%, and the other 40% or so was held by various private investors (in several countries) who were all known to at least one of the partners. In the end the founders received € 200,000 each. Bernstein noted:

> It was a smart move on HB's part, as we felt so much better afterwards. We had gotten some of our investment back and the salary we had started to draw in 1998 was secured. Some of the financial worries we all had lifted a bit.

The money was spent partly on Boblbee and partly on Spotlight and helped to stabilize the company. But it turned out to be temporary relief – the company was burning serious cash via the buildup of inventory and an expensive cost structure in relation to the actual revenue stream.

In 2000 Boblbee's revenues reached € 3.3 million (SEK 30 million), with Japan being the strongest contributor. But the company lost € 825,000 (SEK 7.5 million) (*see Exhibit 3 for financial statements*). And the fundamental problems started to become obvious. Bernstein said:

> We should have seen it – I should have seen it. But there was so much going on with Boblbee and Spotlight, and on top of it there were serious tensions in my marriage. I needed a stronger board – someone should have supervised me and told me to slow down, step back and look at the situation. I have a background in branding – but I was not able to define a clear target group for Boblbee. What is Boblbee? The distribution strategy also had many flaws. The point in 2000 when we took over the distribution risk is the point when things really started to go bad – it was the end of the good years.

4 Product Development and Repositioning

The first marketing campaign had positioned Boblbee as trendy and urban – but Boblbee realized that being seen as a trendy product was not sustainable. In order to get away from fashion, the company moved to develop the Boblbee Sport Concept in 2000. Various add-ons and sports accessories were added to the product range in order to increase usability for athletes and sports enthusiasts. (*See Exhibit 4 for more of the sports range.*)

By the end of 2001, the product assortment had grown from two in 1998 to over 200 (*Exhibit 5*).

Marketing was directed to reposition the backpack as sporting gear, targeting mountain biking and mountaineering in particular. Several brochures were released with a clear sports focus, and the original marketing brochure was toned down, moving away from the purely trendy angle. Each outlet (fashion and trend) also got its own marketing brochure. But the repositioning was almost too late, as Bernstein noted:

> In the big stores we had earned a reputation for being hard to sell, and we had not done enough branding up until then.

4.1 The Urban Nomad

At the beginning of 2001 Boblbee finally got its act together and defined its target market as the "urban nomad," a phrase coined by the company itself (*Figure 1*). Blanking thought of himself as an urban nomad, a person who saw mobility as

Figure 1 Urban Nomads

a part of life and where age did not matter. Urban nomads felt a strong sense of belonging to the urban lifestyle but tried to balance it with an active outdoor life (skiing, biking, and hiking) in their leisure time. They were mostly sophisticated professionals, hard working, quality conscious, and interested in functional products and designs. They were open-minded extroverts.

Urban nomads used their bikes or rollerblades to get to work as they wanted mobility and freedom; they needed a backpack that would protect their laptops, mobile phones, and other electronic devices. On weekends the backpack could be used in the mountains.

From a segmentation standpoint, the company saw most of its customers as being between 25 and 36 years old (41%), but with significant tails on both ends: 36% of customers in the 16- to 25-year-old bracket, and the remaining 23% being 36 years old or more.

4.2 Brand Segmentation

A new, clear, and concise brand segmentation was also developed during 2001:

- *Boblbee Limited*: The range would feature new, innovative surface treatments and would be updated regularly to follow seasonal trends. It would be sold through the more fashion- and design-oriented partners.
- *Boblbee Executive:* A selected range of the best-selling colors would target retail partners in fashion/design and travel/luggage.
- *Boblbee Sport*: The sports range would be based on the Megalopolis monocoque hard shell, with the interior and harness adjusted for optimal fit and comfort for outdoor activities.

4.3 Distribution: Looking for Deals

In 2001 Boblbee realized that it had to be more open in its strategy in order to reach a wider market. The company started to use more direct sales to corporations (B2B) and made sales to companies such as Ericsson, Giant, Yamaha, Ducati, Nokia, Framfab, McKinsey, Compaq, and Siemens.

There were also a few more potential deals on the horizon. The first was with Carrefour, the French global supermarket chain. Boblbee would get one shelf meter for its products in 780 shops, a deal worth € 264,000 (SEK 24 million). Bernstein had shaken hands and verbally agreed on the deal with the people at Carrefour. Blanking started to develop a new harness, new packaging, and new colors for the hard shell that would be sold exclusively at Carrefour since the retail chain did not fit their differentiation strategy. The product development cost € 16,500 (SEK 150,000). But the joy was short-lived for Boblbee – the deal failed due to human error. The person in charge at Carrefour forgot to enter Boblbee's exact information into the company's computer, which meant that Boblbee missed out on the opportunity to participate in Carrefour's internal sales fair, a requirement to get into the shops.

The company pursued a few other deals. The most promising one was the shop-in-shop concept, BOB Shop, inspired by a Japanese distributor (*Exhibit 6*). Boblbee used 3-D software to allow customers to select the desired color and harness combination and see the result directly on a computer screen in the shop (*Exhibit 7*). The backpack would then be assembled while the customer waited. In summer 2001 Bernstein visited Galeries Lafayette to sell the concept, equipped with Blanking's drawings showing what a shop-in-shop might look like (*Exhibit 8*). Galeries Lafayette really liked what they saw and offered Boblbee 12 square meters (approx. 36 square feet) on the fourth floor of their flagship store in Paris, right by the escalator next to Nike.

The BOB Shop's opening day was… September 11, 2001. An ominous sign.

The only people at Galeries Lafayette in the afternoon were the employees – there were no customers! Our introduction totally flopped due to the terrible events in New York! The world changed overnight and the shops stopped buying anything that wasn't seen as safe – they did not want to buy new brands any more.

Psychologically it was very hard for me. I started to wonder if I could have done anything different. I started to go over things that I should have done better. It was a downward spiral and I became less pushy, less driven. I ended up handing in my resignation. But the board turned it down – they wanted me to stay. That did me a lot of good: It gave me new energy to keep going.

5 The Way Forward

In 2002 Boblbee initiated a strategic change toward a two-tier product portfolio, with both the classic hard-shell product line and a new "soft-shell" range (*Exhibit 5*). The decision proved to be strategically sound as in 2003 the soft-shell line accounted for most of the sales and helped the company to get quick cash.

Boblbee adopted the slogan "*Boblbee provides smartly engineered solutions for individual mobility*" and made changes in the sales and distribution system in order to get closer to its markets.

5.1 Distribution Strategy

Even if the BOB Shop concept had gotten off to a less than thrilling start, it soon turned around and picked up rapidly. By 2003, BOB Shops had opened in Las Vegas (*Figure 2*), Paris, Hamburg, Amsterdam, Tokyo, Osaka, Seoul, Bangkok, Prague, and Malmö.

Figure 2 BOB Shop in Las Vegas

E-bags also started to sell Boblbee bags through its e-portal. The largest challenge Boblbee faced in 2003 was replacing the small distributors with a new distribution network (*Exhibit 9*).

5.2 Alliance with Nike

In February 2002, Boblbee signed two contracts with Nike – one patent and license agreement and a service agreement. Bernstein recalled:

This was potentially the light at the end of a dark, long tunnel. I mean Nike – it's huge!

Figure 3 EPIC – The First Nike Product

The first Nike product, EPIC – for which the companies received two design awards – was released in May 2003. The process of developing the Nike linkage was long and painful. The initial contacts had been established as early as February 2001 during ISPO, when Bernstein met with the General Manager AMG of Nike Europe. The meeting was followed with numerous e-mails, phone calls, and meetings over the next 18 months. Blanking also started to design a full portfolio of hard-shell bags and backpacks specially for Nike.

5.3 Bernstein Resigns as CEO

During the summer of 2003 Boblbee hired an external consultant. He saw Bernstein as the biggest problem and wanted to rebuild the company into a small Swedish player. For Bernstein, this was a hard blow. Not only was he one of the cofounders, but he had also invested his soul in the business and had always seen Boblbee as an international brand and company.

For Bernstein, the Spotlight and Boblbee launches had been an emotional roller-coaster that had almost cost him his marriage, his savings, and his future pension. In fall 2003, he went on a retreat to Dalarna in northern Sweden to recharge his batteries. When he returned, he resigned from his position as CEO. He commented on his resignation:

> We needed to find new solutions and I had no energy to redo the whole thing with distributors, *etc.* Instead I resigned as CEO and took over the role of strategic development manager. I also stayed as a director on the board. In the CEO position, there were too many administrative things taking up my time and I wanted to be able to focus fully on the market and finding a way forward for Boblbee. So I chose to step down, which is something I should have done already in 2001. Jonas [Blanking] took over as interim CEO.

5.4 Boblbee up for Sale

The Nike alliance helped Boblbee get some much-needed cash, but it was not a long-term solution. Furthermore, after Nike started new discussions about patents, Boblbee bought back its patent and license agreements in 2003 – Bernstein had had enough. One of his main tasks as strategic development manager was to find a buyer for Boblbee. He sent a letter to all shareholders telling them that the company either had to redo the distribution network entirely or consider selling.

In late fall 2003 Bernstein met Anders Petersson – the CEO of Thule, the Swedish producer of roof boxes for cars and trucks. Thule was very interested in possibly taking over Boblbee. For it, Boblbee was a "roof box on the back" – the fit was good. Blanking had made a few sketches of what the concept might look like and they were impressive (*Exhibit 10*). After many meetings, handshakes, and promises the deal fell through after Thule's board decided to sell the company to Candover, a London venture capital firm in July 2004.

Bernstein continued the search for a buyer in the USA, with Tumi and Puma the hottest prospects. Tumi turned the offer down because it thought Boblbee was too small, Puma because Boblbee had been too closely associated with Nike.

5.5 More Strategic Alliances

In 2004 Boblbee established several new strategic alliances:

- A product development and sales agreement was signed with Yamaha under which Boblbee's current products would be cosmetically adjusted to fit Yamaha's motorbikes and scooters (*Exhibit 11*).
- A cooperation agreement was signed with Volvo Car Corporation for its new concept car. Volvo agreed to work with Boblbee products within the Volvo merchandising portfolio (*Exhibit 12*).
- A partnership with Poseidon, a Swedish diving equipment manufacturer, was initiated when Boblbee signed a letter of intent to assist in the development and supply of specialized diving products.
- XM Satellite Radio ordered specially designed bags for its product range as a part of a signed license agreement.

5.6 Leaving the Dream Behind

In September 2004 Bernstein made another radical decision – he decided to leave his position as strategic development manager. He would, however, stay on the board. Blanking took over as interim CEO and Bonnier became chairman of the board. Commenting on his decision to walk out of the company he had put his whole soul in, Bernstein said:

> I felt fantastic. I was relieved and I felt like I could finally breathe again.

Bernstein set up his own coaching and consulting company and started working for Boblbee on a commission basis, trying to sell the bags to shops in Germany, Switzerland, and Austria. The board gave him a prepayment of € 30,000 so that he had something to live off. Over the next few months he drove 70,000 km and managed to convince 75 shops to take on Boblbee. Bernstein was 47 years old – and he had gone from being a general manager at Unilever to being a bag sales person. But he wanted to do it to prove the case, show commitment to Boblbee, improve the company's financials, and find a solution for them all.

In March 2005 Bernstein stopped his operational work for Boblbee, but continued to work with corporate clients. He said:

> The company's short-term future was secured. I needed to earn some money again. It was time for me to move on.

***

Bernstein and I (one of the case writers) sat in a rustic restaurant in a small village in southern Germany in September 2005, where we had decided to meet due to its strategic location roughly halfway between Edenkoben and Lausanne. He fingered the glass of red wine in front of him and looked at me:

How could it have gone so wrong?

He answered his own question:

> The product was great – is great – I still believe so. We got lots of PR. We clipped the articles and put them on the wall. But we really overestimated the effect of the PR – we were blind and deaf! Now I find myself thinking that yes we got a lot of PR, but if you consider how much it was per country, then it was rather insignificant. When you read an article about a cool product in a magazine or on TV, you quickly forget about it. PR doesn't build a brand! For me, it was especially frustrating as I was supposed to be more or less an expert in branding. That was where I got started and at Unilever that is what I was responsible for! I cannot believe that we could not build the brand correctly.
>
> Another problem was that we had no clear target group. The "urban nomad" concept was good – but it took us three years to come up with it. Our sales strategy was very unclear.
>
> We also didn't pay enough attention to user needs. I must say that I also like to use bags that have wheels – rolling a bag is nice. A backpack can be heavy and if it is hot out there you get a sweaty back. Perhaps we should have been more critical of our own product before we set off on this journey.
>
> Another major problem was that we did not have a strong board. I had no one with international brand building experience to coach me or to challenge me, to tell me to step back and take a deep breath and look at the situation again. When people just nod their heads, then I go for it. Having been on the fast track in companies such as Unilever, I was very confident in myself and my abilities. Maybe too much so. I always believed – especially when I was younger – that I was good enough to do everything on my own. I needed a stronger board. This was something we discussed many times, but we never managed to find experienced people ready to join the board.
>
> Then there were problems with Spotlight. Taking on Boblbee when we had Spotlight was a mistake – there were absolutely no synergies between the two. Spotlight had been on its way – it was actually working – but then Boblbee came and sidetracked us. And both needed a lot of capital injection, time and effort. For me it meant I had to focus on two things at the same time.

Listening to Bernstein, it was clear he had dwelled on these questions many times and learned a lot in the process. He was now ready for a new challenge as the CEO designate at Biffar International.

Exhibit 1
Pricing Strategy (in € Rounded)

	People's Burden	Box Jellyfish	Megalopolis
Consumer price (incl. VAT)	131	142	164
VAT	26	29	32
Net sales per unit	105	114	132
Price to retail	59	66	73
Net sales Boblbee AB	59	66	73
COGS	33	35	37

Source: Boblbee internal information

Exhibit 2
Boblbee's Distribution Model, 1999–2000

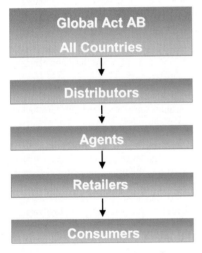

Source: Boblbee internal information

Exhibit 3
Boblbee Financial Statements, 1999–2004

Profit and Loss

	1999 [*]	2000 [*]	2001	2002	2003
	(SEK '000) *SEK 1 =* *€0.11*	*(SEK '000)* *SEK 1 =* *€0.11*	*(SEK '000)* *SEK 1 =* *€0.11*	*(SEK '000)* *SEK 1 =* *€0.11*	*(SEK '000)* *SEK 1 =* *€0.11*
Sales Total	**25,122**	**29,463**	**18,575**	**24,485**	**19,554**
Cost of goods sold	(16,300)	(18,555)	(11,118)	(12,476)	(12,268)
Gross income	11,339	29,894	7,457	12,009	7,286
Gross margin	35%	38%	40%	49%	37%
Operating costs			**(23,245)**	**(17,732)**	**(10,902)**
Rent and maintenance	(573)	(1,255)	(2,181)	(2,264)	(2,298)
Administration, patent, office supply	(1,461)	(3,797)	(4,157)	(4,021)	(1,097)
Marketing/Sales	(3,231)	(6,592)	(8,444)	(4,395)	(1,283)
Personnel	(3,593)	(7,699)	(8,463)	(7,052)	(6,224)
EBITDA	**(36)**	**(8,004)**	**(15,788)**	**(5,723)**	**(3,616)**
% of net sales			−85%	−23%	−18%
Depreciation	(1,235)	(2,444)	(3,460)	(2,962)	(1,838)
EBIT	**(1,271)**	**(10,448)**	**(19,248)**	**(8,685)**	**(5,454)**
% of net sales			−104%	−35%	−28%
Financial income/cost	(319)	861	299	(4,237)	(3,653)
Income after financial items	**(1,590)**	**(9,587)**	**(18,949)**	**(12,922)**	**(9,107)**
Result participating in associated company	(989)	(3,239)	(5,348)	(231)	–
Income before tax	**(2,579)**	**(12,826)**	**(24,297)**	**(13,153)**	**(9,107)**
Tax	(11)	–	–	3,449	–
Income after tax	**(2,590)**	**(12,826)**	**(24,297)**	**(9,704)**	**(9,107)**

Source: GAAB internal company information

---

* In 1999 and 2000 the income statement and balance sheet were consolidated for both Boblbee and Spotlight. Sales and specifics to Spotlight have been left out on purpose – contributions were insignificant.

Exhibit 3 (continued)

Balance Sheet

	1999	2000	2001	2002	2003
Assets	(SEK '000) SEK 1 = €0.11	(SEK '000) SEK 1 = €0.11	(SEK '000) SEK 1 = €0.11	(SEK '000) SEK 1 = €0.11	(SEK '000) SEK 1 = €0.11
Copyrights, patents, and equipment	5,478	5,308	3,726	2,027	976
Machinery and equipment	707	1,869	1,237	856	442
Shares, receivables in associated companies	2,127	5,579	231	0	0
Inventories	9,462	13,670	12,977	10,111	7.086
Accounts receivable	5,770	6,723	3,945	3,673	1,607
Other current receivables	1,446	1,994	688	3,885	3,764
Prepaid expenses and accrued income	1,022	1,459	563	304	648
Cash and bank	205	8,820	723	857	1,284
Total assets	**26,877**	**45,422**	**24,090**	**21,713**	**15,807**
Liabilities and equity	(SEK '000) SEK 1 = €0.11	(SEK '000) SEK 1 = €0.11	(SEK '000) SEK 1 = €0.11	(SEK '000) SEK 1 = €0.11	(SEK '000) SEK 1 = €0.11
Share capital	703	976	976	976	1,078
Reserves	13,300	47,843	54,883	21,248	27,805
Retained losses	(4,337)	(7,130)	(27,048)	(9,934)	(20,649)
Result of the year	(2,590)	(12,826)	(24,297)	(9,704)	(9,107)
Shareholders' equity	7,076	28,863	4,514	2,586	(873)
Long-term liabilities	3,828	7,250	8,696	7,227	6,082
Subordinated loan	5,000	5,000	5,000	5,000	1,595
Accounts payable	5,863	0	3,657	4,599	4,891
Other short-term liabilities	268	2,679	1,015	1,083	2,485
Accrued expenses and deferred income	983	433	1,208	1,218	1,627
Total liabilities and shareholders' equity	**26,877**	**45,422**	**24,090**	**21,713**	**15,807**

Source: GAAB internal company information

Exhibit 4
Product Development for Sports Range

Source: Boblbee internal information

Exhibit 5
Product Development

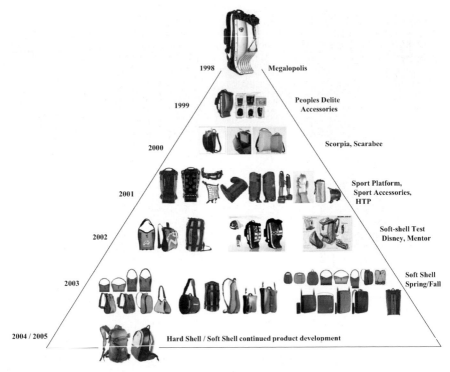

Source: Boblbee internal information

Exhibit 6
BOB Shop

Source: Boblbee internal information

Exhibit 7
BOB Shop Software

Source: Boblbee internal information

Exhibit 8
BOB Shop Sketches

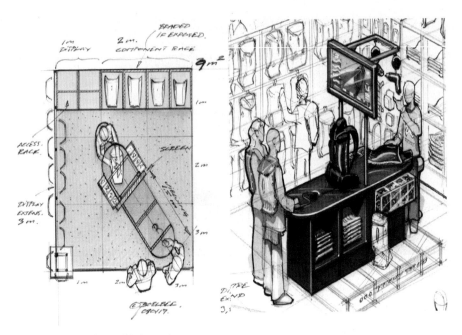

Source: Boblbee internal information

Exhibit 9
Boblbee's Distribution Model, 2001–2003

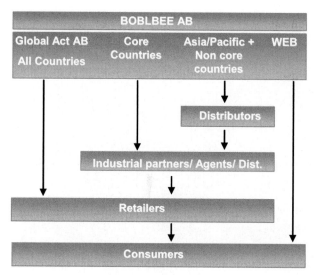

Source: Boblbee internal information

Exhibit 10
Strategic Alliance with Thule

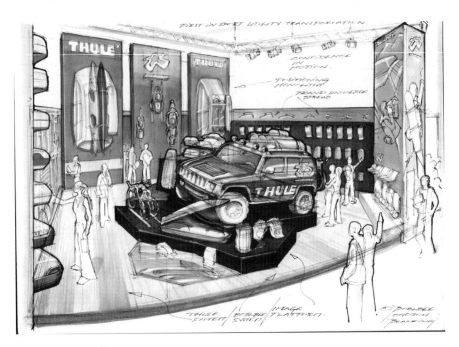

Source: Boblbee internal information

Exhibit 11
Strategic Alliance with Yamaha

Source: Boblbee internal information

Exhibit 12
Strategic Alliance with Volvo

Source: Boblbee internal information

CASE 4-10
EndoArt SA: Creating and Funding
a Medical Technology Startup (A)

Ralf W. Seifert and Jana Thiel

LAUSANNE, SWITZERLAND, NOVEMBER 2004. Nikos Stergiopulos, professor of biomedical engineering and chief scientific officer (CSO) of EndoArt SA, felt drained. He had just learned that the prospective strategic partner for his business was pulling out of the deal. The last few months had been extremely trying – indeed, looking back, the last six years since he had started the company had been a bit of a roller-coaster ride. The dull November weather did nothing to improve his mood.

Four weeks earlier, the prospective partner – a large US player – had sent eight people to conduct an in-depth due diligence of EndoArt, a medical technology venture specialized in the telemetric control of bodily fluids. As a result, they had given their green light to go ahead to further explore a collaboration, which could eventually lead to an acquisition. Stergiopulos was excited at having successfully passed the partner's scrutiny. Now the partner had unexpectedly dropped a bombshell: Having lost an intellectual property (IP) case a few days earlier in a completely unrelated business area and sentenced to pay heavy damages, it was concerned about any further technology acquisitions. Stergiopulos was frustrated:

> The option of EndoArt entering a collaborative agreement with this major player in the medical technology field had just vanished. And, although the circumstances were beyond our control, the damage to the team's motivation, investor relationships and EndoArt's reputation had been done.

EndoArt had raised and spent some CHF 15 million to date. Being out of money now, it needed to make a decision:

- Should it look for another trade sale partner, which might end in a long search considering the latest event?
- Or should it try to enter a development partnership with another company, making itself even more dependent on the actions of others?
- Or was it best to invest additional venture capital money to press on with product development and force the market entry of the product currently in its pipeline?

Copyright © 2006 by the alliance for technology-based enterprise (IMD, EPFL and ETH Zurich). Copyright permissions are handled by IMD, Lausanne, Switzerland.

1 Company Overview

EndoArt SA was a medical technology company whose business was based on innovative stent[1] implants and telemetric[2] devices to control the flow of bodily fluids. Since its inception in 1998, EndoArt had been established in the Scientific Park Ecublens, next to the campus of the Ecole Polytechnique Fédérale de Lausanne (EPFL). By the end of 2004 it occupied a whole floor with labs and technical equipment. Its headcount was 15, including the management duo of CEO Philippe Dro, who had some 15 years of experience in several pharmaceutical businesses, and CSO Nikos Stergiopulos. Whereas Dro had joined the venture only in February 2004, Stergiopulos was a cofounder and the main inventor of the fundamental technology. He currently headed EPFL's Laboratory of Hemodynamics and Cardiovascular Technology.

EndoArt was financed by venture capitalists (VCs), who had endowed funds of almost CHF 15 million in total. It held 16 patents defending its exploitation rights in Europe and the USA. The company had worked closely and repeatedly with patent professionals on both sides of the Atlantic to ensure the protection of its IP and in turn to prevent any allegations of infringement that competitors might potentially put forward.

By 2003, EndoArt had started commercializing one device for the treatment of cardiac malformations in infants. In early 2005, it would start testing a telemetrically controlled gastric band for the interventional treatment of morbid obesity – a serious disease that was spreading in the Western world. However, the initial starting point had been a different one.

2 Evolution of the Firm

2.1 Getting Started

EndoArt had its origins in an invention made by Stergiopulos back in 1994 in the EPFL laboratories where he had conducted research into stent technology. The first stent had been introduced into the human body in 1986 in France, and in 1994 the device was approved for use in the USA. Stents had quickly become the leading way to treat patients suffering from blocked coronary or peripheral arteries. A persistent problem, however, was the reclosure of arteries (known as restenosis) after stenting, which affected about 20% of patients and required subsequent surgical intervention. With the idea of inserting a diffuser into a conventional stent

---

[1] A stent is a tubular device, made of biocompatible metal or plastic, that is inserted into a blood vessel, duct, or canal in the body to hold it open.

[2] Telemetry is the science and technology of the automatic measurement and transmission of data by wire, radio, or other means from remote sources.

to increase the friction of blood in the area and thus reduce restenosis, Stergio-pulos felt he had discovered a significant improvement on all existing approaches (*Figures 1 and 2*).

To many people, the concept was counterintuitive and Stergiopulos was called crazy more than once. Nonetheless, he hoped to sell his idea to a bigger company to help develop the concept. His efforts in this regard were, however, not success-ful. Instead he was told to develop the technology on his own and come back when he had a proof of concept. In 1998 – almost four years after the initial idea – he eventually opted for his own company.

Stergiopulos was able to convince Christian Imbert, a well-known pioneer in the stent domain, to join him. Imbert had been the principal engineer in the devel-opment of the first stents and was experienced in managing medical technology projects. His expertise predestined him for the role of CEO. Yet before the duo could think of assigning any duties, they had to go in search of funding. Luckily, EndoArt was awarded the Prix de Vigier[3] in 1998, endowed with CHF 100,000 – exactly the capital stock required to incorporate a limited share company in Swit-zerland. But equally important was the sudden media coverage EndoArt enjoyed after receiving the award and the network connections that opened to the nascent venture.

Founding the company was only one part; another was to fund the proof of con-cept of their innovative device. Imbert and Stergiopulos soon met with several interested VCs such as the Banque Cantonale Vaudoise, Centre Patronal, and So-finnova Partners. Sofinnova was to become EndoArt's lead investor, not only

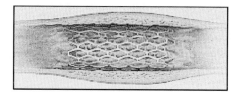

Figure 1 Conventional Stent, Implanted in a Vessel

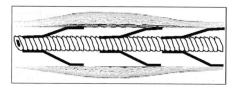

Figure 2 ARES® Stent, Implanted in a Vessel

---

[3] An annual award, granted by the Swiss Fondation WA de Vigier, to support entrepreneurs (http://www.devigier.ch).

granting funding but also providing valuable expertise in the life sciences domain. By March 1999 the young enterprise had raised CHF 1.8 million in capital and was valued at CHF 4.7 million. Its board consisted of experienced members, bringing considerable management proficiency in the medical and pharmaceutical business to support the company on the promising road ahead.

2.2 Broadening the Product Portfolio

The capital raised was sufficient to begin animal tests for the stent devices. The results looked positive. EndoArt could rightfully hope to reap a share of the stent business, which offered huge opportunities with an estimated total market volume of about $2 billion by 2001. Substantial revenue was looming. Yet the entrepreneurial enthusiasm and creativity of Stergiopulos's team did not stop here. They hoped to take their business to the next level by redefining the application:

> We were probably the first to think of actually *managing* blood flow, so we thought we could become *the* blood flow management company. At that time we thought we had hit gold and started ambitiously extending our research portfolio.

EndoArt broadened its research into a hemodialysis application with which it hoped to control blood flow in dialysis patients. The new activities, together with the fact that the stents had gone through preclinical testing, rapidly consumed all resources. Thus, just one year after the first financing round, the management team had to look for more funding. By July 2000, with a mere CHF 30,000 left in the bank, Stergiopulos and Imbert were able to close the big deal: Remarkably, all first-round investors followed up, and this time they financed a total of CHF 13.8 million. EndoArt was then valued at CHF 32 million. Asked what had made these vast amounts possible with just one potential product in their hands, Stergiopulos commented:

> Surely instrumental in this process was that our stents had proved effective in a short time and our product portfolio addressed blockbuster markets. But to some extent, we were certainly just lucky. Back then people thought that even a turkey could fly.

More comfortably endowed then, EndoArt's laboratories concentrated fully on research and product development. They had to proceed with clinical tests of the stents to get them approved for the European market and potentially the USA, where regulations were far more time and resource consuming than in Europe.

Meanwhile, work also continued on EndoArt's blood-flow application. Soon, however, the researchers had to write off about CHF 700,000 in development costs for their hemodialysis product. The project had to be stopped because of unsolvable issues having to do with material compatibility. Nonetheless, the original approach could still be used in a different way – in pulmonary artery banding (PAB), for example. PAB was used to treat congenital heart defects in small

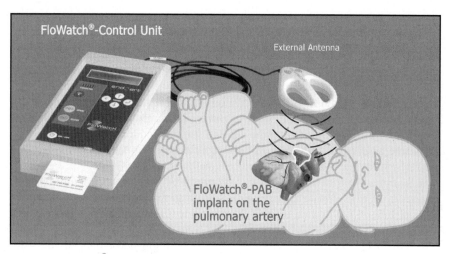

Figure 3 FloWatch® in pulmonary artery banding of a neonate

children by placing a synthetic band around the child's artery to restrict excessive blood flow (*Figure 3*).

EndoArt's approach of controlling the band telemetrically was a true novelty and granted authorization for trials in humans as early as June 2002. The first surgery was carried out at the university hospital of Lausanne (CHUV) in a one-month-old infant. It was a total success. As a result, in addition to its ARES® stents for both coronary and peripheral[4] application, EndoArt possessed another very promising technological platform: the telemetric control of blood flow, which would later be named FloWatch®.

2.3 FloWatch® – Telemetric Flow Control

In FloWatch®, the venture possessed a unique and proprietary approach to controlling an implanted synthetic band – as, for example, in PAB – telemetrically through the skin. Typically, adjusting the band to an optimal diameter caused problems and often required another risky operation. EndoArt's device, by contrast, did away with the need for further surgery as it was equipped with an electrical micromotor that operated wireless and battery free. It received all operational commands remotely from an external control unit.

---

[4] Peripheral refers to arteries in the legs.

2.3.1 Market Potential

Besides treating cardiac defects, the FloWatch® principle could be used in many other applications such as gastric bands for morbidly obese patients or for incontinence treatments. Whereas the cardiac application addressed a niche market with an estimated potential of $20 to $50 million per annum, obesity was clearly a more attractive target. As the health of increasing numbers of people in the West was at serious risk from obesity, the demand for interventional surgery to implant adjustable gastric bands (AGB) had risen significantly over the last few years. The market size for these applications was estimated at $2 to $4 billion at the time – with a strong trend toward growth. (*See Exhibit 1 for information on the basic principles of AGBs.*) The two remaining applications in incontinence treatment added another $200 million each to the prospects. (*See Exhibit 2 for an overview of the different applications and their market potential.*)

Although its unique and proprietary approach would make EndoArt the only player in cardiac banding, the business would be rivaled by larger companies in each of the other domains. The AGB market, in particular, was heavily competitive and included large and renowned players such as Johnson and Johnson and Inamed.

2.4 First Dip on the Rollercoaster

EndoArt's resources did not extend to driving both the ARES® and FloWatch® applications at the same speed. The decision on how to allocate resources to get the most out of its portfolio was eventually made for the company: Over the course of 2002, a sensation was looming in the stent market. Johnson and Johnson, one of the world's largest manufacturers of healthcare products, was finishing up clinical trials of a new type of stent, which, once implanted in the body, released an antirestenosis drug. All tests showed tremendous improvements in preventing restenosis. In April 2003 Johnson and Johnson's stent finally received FDA approval.

> Of course, we knew about their developments for quite some time. Most experts, however, had believed that tissue in-grow would restart as soon as the release of the drug declined. However, clinical trials had actually demonstrated a lasting effect.

The success in the field was immediate and rather devastating for EndoArt. From then on, the entire industry focused on drug-eluting stents, which had gained a market penetration of about 90% by the end of 2004 and commanded three times the price. EndoArt had to reevaluate its business opportunities and decided to shift its attention to the banding application, putting its stent development on hold.

> Of course, our investors were not overly happy at that time. But what could we do? We had to be pragmatic, so we quickly shifted all resources to our blood flow technology and tried to go full speed for a marketable product there.

2.5 Choosing a New Target Market

At the time, the FloWatch® technology had already passed its first trials. Fast progress was also facilitated by the excellent conditions in the venture's environment. In particular, Switzerland proved to be the right place for finding partners capable of further developing the micromotor technology needed for the remote-controlled bands.

With the shift in focus, however, a new decision was called for: Which of the several potential banding applications – cardiac, obesity, or incontinence – was to become the focal point? Based on its previous experience in interventional cardiology and on the competitive situation, the EndoArt team wholeheartedly chose to enter the cardiac banding market for treating neonates with congenital heart defects.

> Why did we choose the cardiovascular application? Well, as a small company we felt right with a niche market at that time. And we were a cardiovascular company: This was where we had gathered all our experience. Hence, we finally decided to license out the gastric banding to a French company to further develop and market the concept – against royalties.

EndoArt drove the pulmonary banding applications at full speed: soon it was allowed to carry out a first test on a human baby. The surgery was performed in Lausanne. Its success gave way to further clinical trials, conducted all over Europe. Impressive results were shown in each of the 49 implantations. Compared to conventional banding, the total costs of the new technology were comparable. The price premium on EndoArt's banding was easily offset by the lack of complications and by the convenient remote adjustment after the intervention, rendering reoperation unnecessary. By June 2003, EndoArt had obtained CE marking[5] for its FloWatch®-PAB. The application was built on a unique technology, with no immediate competitors. All seemed to be back on track.

2.6 Stranded Again

With the cardiac application ready for the European market, a fresh look at the opportunity revealed that the business was not as big as was hoped and needed. The total number of potential patients in Europe, i.e., newborns with this type of heart failure, would not sustain the company and product development. The market could be extended by going to the USA or Asia, but it would take a much bigger effort to release the device in these regions.

---

[5] CE marking demonstrates that a product complies with the essential requirements of the relevant European health, safety, and environmental protection legislations. It ensures the free movement of the product within the EFTA and European Union (EU) single market (total 28 countries).

EndoArt's management team started reconsidering the obesity application. The French licensee had meanwhile put the obesity application on hold due to a shift in strategy, so EndoArt decided that it wanted to regain the exploitation rights and took the French company to court for not having progressed the concept to the agreed level. Yet, having closed the contract without the help of a professional lawyer, the Swiss startup team had missed a detail that would turn out to be very important:

> At the time, we hadn't realized that the contract specifying the place of jurisdiction to be in France would turn out to be a major headache for a small, foreign start-up company. Our case had to be settled by the "tribunal de commerce" at the defendant's place – a small village in France. This court was non-professional and manned by local citizens. All the burden of traveling and interfacing with foreign law suddenly fell on us, adding to our cost. With hindsight, we should have paid much more attention – also to the worst case scenario – before signing this contract, but of course we were originally all too excited about the deal.

In September 2004 EndoArt lost the case. Yet it got back the rights to the obesity application, reaching a settlement with the French company. At the time, EndoArt's accounts showed a balance of CHF 500,000.

2.7 Trying to Convert Technology into Cash

One of the few options available at this point was a trade sale or a large collaborative agreement. In the summer of 2004 negotiations had already started, and by October EndoArt had the opportunity to enter discussions for a potential collaboration with a large US company. In the course of due diligence, two different lawyers – one based in the USA – had been checking on EndoArt's IP to build credibility. Both lawyers had assured them that there was no reason to fear any litigation. However, in a twist of fate, the buyer lost an IP battle against a physician in the USA and, despite the fact that there was no connection with EndoArt's business, got nervous and finally backed out – leaving the team high and dry.

> EndoArt has never been sued for patent infringements. Indeed, we had solicited several studies from independent patent lawyers in both the US and Europe, which unanimously asserted that our IP is airtight. But then again, there is not much to gain from a company like ours. When negotiating with medical giants, however, IP becomes one of the biggest issues. These guys need to pay a lot of attention to it because they are seen as attractive targets to go after.

3 Turning the Tide

This was where Stergiopulos was sitting now: The opportunity to enter a large collaboration or even to sell had just vanished, and EndoArt was burning cash each day.

Stergiopulos did not envy CEO Dro, who had taken over only a few months earlier. EndoArt's investors were banking on Dro's industry experience and network to save the company. Only a few options were left, and a decision had to be made quickly:

- The first possibility was for EndoArt to look for a trade sale, which would mean another round of due diligence with an uncertain outcome – and all the while the company would be spending money that might be better spent on advancing its applications. Furthermore, the fact that EndoArt was running out of time was unlikely to do an evaluation any good.
- A second opportunity was to approach larger companies for a development partnership instead of a trade sale. This would, however, require sharing exploitation rights and – as EndoArt had learned – would also depend on a partner with the right level of commitment to its project.
- The third possibility was to try to raise fresh venture capital with which EndoArt could then further increase company value by pressing ahead with product development and pushing market entry. This approach, however, would cause credibility issues for the team, and it would lead once again to a time-consuming and uncertain negotiation process. Then again, who would gladly invest if the better part of the money was to end up in reparations? If it were to happen at all, it would mean further dilution of the owners' equity.

The EndoArt team had accomplished much over the past seven years: They had developed an innovative and comparatively inexpensive stent technology, the exploitation rights of which the company still owned. With the cardiac banding, EndoArt even had a marketable product, and the gastric banding would likely receive approval for the European market within a couple of months of the tests starting. However, despite all his usual optimism, Stergiopulos could not help wondering whether EndoArt would be able to last.

Exhibit 1
Adjustable Gastric Banding: Basic Principles and Competitive Position

Conventional Hydraulic Banding

The most common adjustable gastric banding currently available is based on hydraulics: An inflatable silicone band divides the patient's stomach into a small pouch above the band and a larger one below. This small gastric pouch limits the amount of food the patient can eat at a time and also promotes a feeling of fullness after eating only a small amount. The band's diameter, and thus the opening between the two parts of the stomach, can be controlled by injecting or removing a saline solution from the band through a tube connected to an access port (reservoir). This access port is placed under the skin of the abdomen during surgery.

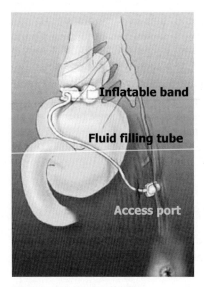

Figure 4   Hydraulic Banding

These conventional devices, however, can have complications:

- Access-port complications (infection, leakage)
- Band slippage/pouch dilatation
- Discomfort for the patient when the fluid is injected
- Pain due to the access port

Exhibit 1 (continued)

EndoArt's New Principle

EndoArt's telemetrically controlled device followed the conventional bands in terms of treatment, implantation, and removal. Yet it controlled the diameter of the band through a remote-controlled micromotor (*Figures 5 and 6*). Because of its features, EndoArt's device had definite advantages:

- No risk of leakage or infection
- No discomfort due to puncture and access-port size
- Outstanding ease and precision of adjustment

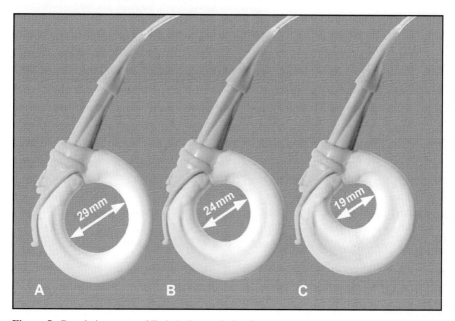

Figure 5   Regulation range of EndoArt's gastric band

Exhibit 1 (continued)

Figure 6 FloWatch®
components – external
control unit and band with
internal antenna

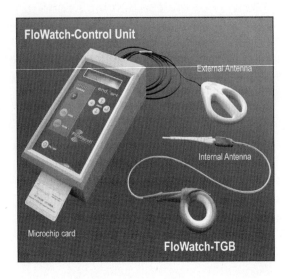

EndoArt's external control devices were activated through patient ID cards, which
ensured that only doctors could perform the necessary adjustments.

Competitive Applications on the Market

All currently approved applications were built on the hydraulic principle. Johnson
and Johnson's bought-in Swedish Adjustable Gastric Band and Inamed's LAP-
BAND® were the most widely used, and only Inamed's was approved for the US
market.

Source: Company literature

Exhibit 2
Market Potential and Entry Timing of EndoArt's FloWatch® Applications (Plan as of November 2004)

	Market potential (EU + US) mil. $	Development stage and competition	Competitive position	Launch date
Obesity	2,000 – 4,000	Hydraulic gastric banding (Inamed, JandJ, *etc.*)	+++	EU: 2007 US: 2009
Fecal incontinence	100 – 200	Hydraulic sphincter (AMS) Clinical Europe Q2 2005	++++	EU: 2007 US: 2009
Urinary incontinence	150 – 250	Hydraulic sphincter (AMS)	++	EU: 2009 US: 2010
Cardiac	20 – 50	No other player, proof of concept – CE mark obtained	++++	EU: 2003

Source: Company literature

CASE 4-11
Google Search Engine and Advertisement

Christopher L. Tucci, Guillaume Basset, Guillaume Gay, Sébastien Grisot, Bertrand Rey, and Aurélien Schmitt

It was October 18, 2006, and Eric Schmidt, the CEO of Google, was wondering how to react to the latest Microsoft assault aimed at attracting customers to its own search engine, Live Search, and probably away from Google.

According to surveys, 80% of websurfers used Microsoft's webbrowser, a large number of them performing their websearches directly from their home page or toolbar. If Google did nothing to counter this, its future would be clearly compromised. The google.com search engine accounted for 80% of Google services used, and that represented an astounding 97% of its revenue. A major cut in this revenue without some kind of diversification would certainly be the beginning of the end.

Copyright © 2007 by the Ecole Polytechnique Fédérale de Lausanne (EPFL). All rights reserved.

1 Creation

In January 1996, Larry Page was a PhD student in the Computer Science Department at Stanford University. In collaboration with Sergey Brin (another PhD student at Stanford), he started a research project in the area of "People, Computers, and Design." [1]. The two students thought that existing search engines ranked websites incorrectly by only considering the number of times a key word requested appeared on a webpage. They thought that the analyses of the relationships between websites would give better results.

One day in the Stanford University library, the two founders were looking for a name for their future company, convinced that their project was revolutionary. There, they found a book called "Mathematics and the Imagination," in which they saw the term "Googol." Then they remembered the old story:

> One day in 1938, Edward Kasner, a mathematician, was wondering what to call the number made of the digit one followed by a hundred zeros. His nine-year-old nephew was visiting him that day, and he decided to ask the boy. The answer was as surprising as it was childish, as the boy immediately said "gogol," which was later spelled "googol." [2].

The objective of the two students was to create something monumental, and they thought that a number made up of the digit one followed by a hundred zeros represented fairly well the greatness they aspired to for their company. That is how the project "Google," which was first called "BackRub" (because the ranking system was based on back-links), was launched. Their first server was an internal server of Stanford and the first domain of Google was google.stanford.edu.

September 15, 1997 was the first important date in the history of Google: that was the day that Brin and Page registered the well-known domain name google.com.

In the early days of the company, the founders searched for financial and moral support, often coming up empty-handed. Then one day, they met one of the founders of Sun Microsystems and former Stanford PhD student, Andy Bechtolsheim. He was impressed by the possibilities offered by the students' product, and he decided to participate by investing 100,000 dollars. Page and Brin counted on their friends and family as well, and they managed to get up to a million dollars in seed funding. That was how Google Inc. was created in 1998, and at the end of that year, their new basic Website treated around 10,000 requests a day and referenced around 25 million pages.

In September 1998, they formally incorporated the company, Google Inc., and became a full-fledged, prototypical startup in a small garage in Menlo Park, CA. The two students wanted their new firm to be competitive, but that was not possible without more resources. Therefore in March 1999, they decided to move slightly south to Silicon Valley in order to be a little closer to the action.

It seemed that everything happened quickly after that as the Google search engine got off to a good start. Users liked its design, which was simple and clean, and which was not filled with the useless visual distractions other search engines sported. Google became more and more widely adopted, with requests greatly increasing in a few months. At the beginning of 1999, the number of requests reached 500,000 a day and it was up to around three million in August 1999.

In order to provide more efficient services (with high storage capacity and high-speed servers), the company decided to sell advertisements. The key point was that when a request was sent, the user received the results, but in addition the search engine simultaneously sent advertisements closely linked to the keywords of the request. To add this content to the results while keeping the same loading speed, the Google engineers decided that the ads should be text-based.

In June 2000, Google was the first websearch engine to reference half a million webpages. A partnership agreement was then signed with Yahoo! [3], and the company started focused advertisements related to user requests. At the end of 2000, the Google Toolbar could be downloaded.

Realizing they had something huge on their hands, the founders and board of Google decided in March 2001 to appoint Eric Schmidt, chairman of Novell, as Chief Executive Officer of Google to focus on the further development of the company. Later that year, on September 4, another important date in the history of Google occurred: the US patent describing the ranking mechanism was granted. It was assigned to Stanford University and quoted Larry Page as the inventor. After that Google experienced even more explosive growth, making it the number one search engine on the web (*Exhibit 1*). They also launched the Google Catalog service.

The company continued to grow rapidly and steadily. The attractiveness of its success and organization culture allowed it to hire the best engineers (such as a developer of Mozilla) in order to diversify their activities. In May 2002, the Google Labs service, a laboratory for science and health, was launched. That same year, Brin announced the creation of a French subsidiary and of another service, Froogle, which was to be used for shopping.

2 The Technology and Original Idea

Google was the only company focused on creating the best search engine ever developed. Larry Page, cofounder, said their product understood "exactly what you mean and gives you exactly what you want" [4]. *Exhibit 2* shows the working principle of that concept. A user sends a specific request to the Google website and obtains answers corresponding to his or her exact request in a few seconds.

Moreover, the strength of the development teams was their continuous desire for innovation. Current limitations did not prevent them from developing new products. A consequence of this spirit was the well-known PageRank algorithm:

While other search engines only classified documents by the number of occurrences of a keyword in a page, Google gave priority to hyperlinks pointing at the page, thereby giving a more relevant ranking. Indeed, it interpreted a link from Page A to Page B as a "vote" for Page B by page A. The importance of a website was then calculated by the number of votes it received. The PageRank algorithm also took into consideration the ranking of the voting page since the vote from a highly ranked page would have greater value. People greatly trusted this system because it was known that there was no human manipulation of the results at all, so the ranking was a source of objective information.

The software hidden behind this search engine was able to conduct a very large series of simultaneous calculations in only a fraction of second. After using PageRank to analyze the structure of the Web and to determine the ranking of a page, it determined the relevance of the page to the specific request of the user. After that, Google was able to put the most relevant result first due to the hypertext-matching analysis.

The search engine analyzed the content of a page, but instead of scanning the text, it analyzed the full content, the fonts, the subdivisions, and the location of each word. It also analyzed the content of neighboring pages in order to give the most relevant pages as the result for the user's request.

3 Business Model

The heart of the Google strategy was AdWords. This was the advertisement system of the Google search engine that displayed specific ads related to a specific request. These ads were paid for by advertisers who wanted their ads to appear when the user sent a request related to their activities. Online advertisers appreciated this system because they did not pay if the user did not click on the advertiser's link. The position of the ad on the page depended on the price paid for a click and on the popularity of the ad. The popularity was calculated by taking into account the number of appearances of an ad and the number of clicks on this ad. In order to keep the system highly relevant, Google decided to remove ads that received less than five clicks per 1,000 appearances.

Google Adsense was a good way to make a Website profitable, especially if this Website generated a large number of visitors. The partners of Google Adsense were webmasters of sites that diffused AdWords advertisements and were, as a consequence, paid by the click. In other words, the content owner would call on Google Adsense to generate ads (via AdWords) on the fly that were tailored to the user. When the user clicked on the ad, Google would share the advertiser's revenue with the Website.

After Adsense, there were many affiliated or derivative services, such as Image ads. Image ads gave the opportunity to make "visual" ads instead of text-based ones. In addition, Google tried to launch the "call per click." The user could choose to be in direct communication with an announcer. It was thought to be a nice way for Google to have contracts with announcers who did not really want to create a Website.

Even though the financial results of Google were very good (the company announced net revenues of around US$600 million for the beginning of 2006), people at the Mountain View headquarters were not as happy as they should have been.

The reason was that the business model emerging from the company's history might not have been the best one.

Google's main problem was the acknowledged fact that its revenues strongly depended on advertisement. Indeed, more than 97% of Google's revenues was generated by Adsense and AdWords, so essentially the company had not diversified its sources of revenues at all.

In addition, there were rumors of technological problems at the company (the limits of storage capacity had been reached, for example), and technical issues were often seen in the most recently launched services. Supposedly, this was why access to Gmail (the free email service) was limited: to subscribe to this service, you had to be invited by someone who already had subscribed.

To avoid declining as fast as they had risen, the management of Google tried to diversify their revenues. To do that, they diversified their targets. For example, the company bought advertisement space in real-world paper magazines. This was a great opportunity for the partners of Google to reach both webusers and magazine readers of a specific geographical location.

However, Google also became a victim of its own exploding growth. The image of the company in the world at large was not as good as it had been in the beginning now that they wanted to be everywhere on the Internet (see google-watch.org for an example of negative reactions to Google's omnipresence). The changes made for their Chinese subsidiary (blocking – some would say censoring – content that was displayed in other countries; see below) did not help Google's positive reputation that it had had since its launch in a small garage.

4 Exploding Growth

After only five years of existence, Google handled around 200 million requests a day, which represented more than 50% of the requests in the world, and in 2004 Google finally became the leading search engine, receiving around 85% of global requests. The partnerships with Yahoo, AOL, and CNN were the main reasons for this success. Indeed, these companies agreed to use Google as the search engine of their websites. Then Yahoo decided to end their special relationship with Google, creating their own search engine. That year, Gmail and Google Desktop Search were also created.

In 2006, like the other search engines, Google agreed to bridle its engine in order to reach the Chinese market. Today, when you search "Tiananmen" in the image search engine of google.fr, you discover the famous picture of a student in front of tanks, while if you make the same request on google.cn, you find pictures of happy families and nice memorials (on the Chinese Website, the images of the tanks appear only on the fourth page of the results).

Google the company was not just a "search engine." Indeed, its engineers were in the process of developing other technologies and services. More and more, the company hired the best and brightest, not only to improve existing services but to diversify into new ones. *Exhibit 3* shows the different services and software packages developed by the Google teams.

A great symbol of the exploding growth of Google was their last purchase, YouTube, in October 2006 for more than $1.5 billion. After less than ten years of existence, Google was a very powerful company trying to diversify many of its activities. After becoming the leading search engine, beating Alta Vista on its own field, they then attacked the trust of Microsoft by trying to launch new software packages and services, such as one that would try to compete with Microsoft Office.

5 Market/Competition

As the Internet was experiencing a boom in the early 1990s, many people believed they could build an efficient search engine devoted to helping users find information on the World Wide Web. When Google was launched in 1998, about ten engines were already sharing the market (*Exhibit 1*). Until then, Google had been hosted by Stanford University. The domain name google.com was gained at the end of 1998. When the domain name was acquired, Google was already famous for its link popularity ranking system, but it would have a very difficult job ahead to conquer the marketplace.

5.1 Facing Giants: Yahoo Is in the House

The main competitor of Google, at the time it came on the market, was the "giant" Yahoo. Founded in 1995 by two students, also from Stanford University, the startup had an amazing success story and quickly became a leader in the new market of search engines. After a partnership with Open Text, Yahoo signed a contract with Alta Vista in mid-1996. At that time, Yahoo held about one third of the global search engine activity. In 1997 Yahoo became the first search engine to make a profit. In mid-1998, as it saw Alta Vista as a competitor in the portal space, Yahoo preferred to end their contract and signed with Inktomi, which was powerful in the search field but which did not offer a portal. Although Yahoo was the leading search engine, the company considered itself not a "search engine," but a "media company," as its large range of services was not only focused on searching the web. After a long period of growth, Yahoo shares started to drop in May 1998, but the company stood in a comfortable position with respect to its competitors.

5.2 Google Joins the Race

At the end of the same year, google.com was created, but Yahoo did not see a threat in the new company devoted to searching the Web. Indeed Google was conquering the search field but refused to build its own portal. Larry Page, co-founder and CEO of Google in 1999, said Google would stay focused mainly on searching. "That reduces the competition we have with people we might work with," he added [5]. However, there was a need for Google to be on top in its prime activity, and month by month, Google's popularity increased, whereas Yahoo's search engine share continued to decrease.

5.3 A New Threat: When the Redmond Colossus Comes to Play

In the fall of 1998, Microsoft signed an agreement with Inktomi and created its own search engine, MSN Search, based on it. "This is an evolution of what our search service is going to be," said MSN product manager Nichole Hardy. "Our ultimate vision is to make MSN.com a place for people to get things done on the web." [6]. But at that moment, Microsoft's search engine was not optimized, as MSN Search was just one of over five services that were rotated as default choices for Internet Explorer's "search" box. On February 9, 1999, Microsoft licensed, for a period of five years, the LookSmart directory to integrate into its MSN Search service. Microsoft continued to enhance its search engine, and on March 18 the fifth version of Internet Explorer was released. With that new version came a major novelty: there were five search engines available: AltaVista, GoTo, Infoseek, Lycos, and Microsoft's own MSN Search. The very first time a user launched Explorer, one of the services had to be selected and remained the default choice from that point on, unless settings were customized.

As Danny Sullivan, editor at searchenginewatch.com, said:

> It's a good move. I think many users of both Explorer and Netscape Communicator have no idea why the search partner changes when they click on the respective search buttons – indeed, some don't even realize the partner is changing. The IE5 change allows Microsoft to give its partners a shot at its users without confusing those users in the process [7].

In reality it was a long-term strategy to clarify the origin of the results only once Microsoft's own engine was reliable. Later, in 1999, MSN added a "Top 10 Most Popular Sites" link below the search box, where the most visited websites related to the search were ranked.

A few months later, Yahoo introduced the same concept in its page to counter its recent decline. The "Most Popular Sites" ranking, only available for some categories, was born in April 2000. Due to that automatic system, the pages were listed by pertinence.

Srinija Srinivasan, Yahoo's editor in chief, said:

> We realize that a long alphabetical list can be unwieldy to navigate through, and while alphabetical order is extremely functional and democratic, it is also an arbitrary ordering as far as content goes [8].

5.4 A Big Step for Google

Google's race to become the leading reference search engine took a big step in July 2000, when it was retained by Yahoo to become their fourth search partner. That did not mean that Yahoo renounced its own search engine. In February 2001, the conservative Yahoo launched a "Sponsored Sites" program. The concept was the same as the one used by Google for a time: Commercial links related to the search topic now appeared in a separate box easily distinguishable from the search results. In October of the same year, Yahoo search performed its first major change since its creation. Previously the results had been classified under the CSP model, standing for Category, sitelistings, webpages. That classification proved boring, since users sometimes had to look at several pages of categories before viewing a single site. Still there were links to Google search results, which were available by clicking on the "Web Page Matches" button. At the same time, Yahoo released the "Yahoo Essential" tool to counter MSN. Installing it modified the behavior of Internet Explorer, so the searches realized from the address bar were actually resolved by Yahoo rather than by MSN. The Internet Explorer search button was also changed to query Yahoo.

In a new version of MSN Search released in October 2000, "Popular Search Topics" links were added so that the user could specify his search. It avoided finding results that were off topic, especially for ambiguous or incorrectly written terms. In June 2001, Microsoft had to give up its plan to add "Smart Tags," since that project was facing hostility, especially from an article published in the Wall Street Journal. The project aimed to turn words on webpages into hyperlinks that Microsoft could control and was not really welcomed.

At the end of 2001, Yahoo announced that $7.1 million was paid to Google for the search queries it had handled in the year.

5.5 Would Customers Pay for Obtaining Results?

A major novelty came on January 23, 2002 when Yahoo announced a deal with Northern Lights to use their Special Collection documents. Those documents would be available in a new service, Yahoo Premium Documents Search. Scott Gatz, vice president of search and directory for Yahoo, explained: "We're going to be the first portal to offer consumers this premium research content." [9]. Individual articles would be sold for $1 to $4, with a possibility to download up to 50 documents a month for a fixed price of $4.95. Certain documents would have an extra cost, however.

In May, Yahoo extended the contract it had signed with Overture for its paid listings, from the initial five-month trial period to three years. It was the longest deal Yahoo had ever signed, the previous record being with Google for two years. As Elizabeth Blair, senior vice president of listings at Yahoo said, "[Overture] are the only provider of paid listing within Yahoo Search." [10]. This presented a problem for Google. Yahoo was, at that time, paying Google for each search request it handled when no answer was found in Yahoo's own database. But the trend since Yahoo and Google had signed the deal two years before had moved to portals, which were getting more money, to carry a search provider's results. If Google could not run its own ads any more because of the new deal Yahoo had signed with Overture, then it would be stuck asking Yahoo to continue paying for its results. At the same time, Inktomi claimed it was looking to take over the contract Yahoo was holding with Google for search. "Maximizing and monetizing search is one of the top three priorities for Terry Semel [Yahoo chairman and CEO] this year," added Blair [11]. Yahoo had to choose between a better contract with Inktomi and resigning with Google, thus keeping the intangible but not negligible advantage of being associated with Google's brand. Finally, an official announcement on the status between Google and Yahoo was made in August. Google would continue to provide backend results to Yahoo until September.

At the same time, Nielsen//NetRatings refreshed their statistics. Google was still the most used search engine in terms of search hours, with a total of 20 million search hours of usage during the month, against five million for the second-ranked Yahoo. For that concerned audience, Yahoo remained the most popular, but MSN Search was practically reaching its level, and Google was closely approaching both of them. MSN was ranked first in the United States.

6 Google and Yahoo's Complex Relationship

In October 2002, after a period of uncertainty, Yahoo announced a renewal of its contract with Google, even though Inktomi and FAST were also interested. Both Inktomi and FAST had the advantage that a contract on their searches would not violate the paid listings exclusive deal Yahoo had with Overture. In fact the deal they agreed on was a "nonexclusive" contract. In the end, Google won because of its reputation and the fact that dropping Google could have been seen by users as a willingness to gain more cash while neglecting the relevance of searches. As a consequence of the new partnership, Yahoo and Google results were no longer split into two separate listings but instead were mixed together. However Yahoo results were still distinguishable due to a link appearing below their description.

In March 2003, Yahoo bought the search engine company Inktomi. The reason for that acquisition was clear. Even if Yahoo had survived the growth of Google so far, the reference search engine on people's mind was now Google, not Yahoo. The objective of Yahoo was to bring back the crown of search engines. Jeff Weiner, Yahoo's senior vice president of search and marketplace, explained the acquisition:

> When licensing technology, there are limitations in what you can do and in the ability to differentiate and be flexible. By owning your own technology, you are able to eliminate some of those issues and concerns [12].

He did not rule out the possibility that Google would become a competitor in the future:

> Google's one part of our current solution, and we've been pleased with them as a partner [...]. We really needed to control our own destiny in this space and not be dependent on any one third-party provider.

That is why, one year later, in July 2003, Yahoo also acquired Overture Services, Inc., and by doing so gained control over Alta Vista and AlltheWeb as well.

At the same time, Microsoft pretended not to be in a rush to renew its search engine. "We're looking at all of our strategic options right now. What's the best thing to do long term, not short term?" [13] wondered Microsoft's general manager John Krass. Some new features were added in MSN Search, but there was no great change. "We're at the initial stages of significantly expanding our investment in search. At the heart of that investment is the websearch business," said a source from Microsoft [14]. The Redmond company obviously had interests in investing in the search business. It had already shown that in the past by making its own engine the default choice in its browser Internet Explorer. That simple modification in the browser had been sufficient to boost MSN's market share, even if it could not prevent Google's use from spreading in parallel thanks to its reputation.

In Europe, MSN traffic was lagging almost 30% behind Yahoo, especially in Britain. After investing $30 million in marketing for its European activities, MSN almost reached Yahoo's level. It was estimated that about $1.2 billion would be devoted to Microsoft advertising in Europe, with up to 20% for online services.

7 Microsoft's Intentions Revealed

In May 2003, the willingness of Microsoft to enter the market became clear, since MSN Search posted a large list of jobs in its career webpages. The search spider MSNBOT was also revealed to the public. "We view it as a three horse race between ourselves, Yahoo and Google, with Google in the lead," said Lisa Gurry, a group product manager with MSN. MSN Search was already one of the most popular search sites, but the search engine Google was still first. "[MSN] doesn't have the brand recognition that Yahoo and Google have, and that's an area we hope to address over the next year or so, to raise the visibility," added Gurry. But things would not be easy for Microsoft, since a boycott page was set up, warning the reader:

Microsoft is building a websearch engine, and they intend for it to become the industry standard. Given Microsoft's track record during the browser wars, there is every reason to believe the company will again use its monopoly power to eliminate competition by building a websearch service into the next version of Windows [15].

MSN split into two divisions in November 2003. One would be focused on communications, like the portal and MSN Messenger. The other one focused on information, including searching. Paul Ryan, a former Chief Technology Officer at Overture, was hired to lead this second division.

8 Current Actors in the Market

In November 2004, Microsoft launched its own search engine in beta program. The final version was released in February 2005, two years after the first announcement. "Now we have our platform in place. We think it's super competitive to what's out there," said MSN Search and Shopping corporate vice president Christopher Payne [16]. Since then, the searches were actually performed by a Microsoft-designed engine. It also meant the end of the partnership with the former search partner, Yahoo. However, despite the search engine performing fairly well, it did not provide as much accuracy as Google or Yahoo engines.

After it had enhanced its engine, Yahoo dropped Google on February 18, 2004 and introduced its own search engine, Yahoo Search Technology, whose interface was very similar to Google's. In September 2005, however, several functions were added, distinguishing Yahoo from its competitors. Notably, there was the new "Social Search," which allowed users to put annotations with their opinions on websites provided by Yahoo search. That novelty made the results become more pertinent, as the human factor permitted better ranking of the sites.

On September 12, 2006, MSN Search was replaced by Windows Live Search. The aim of the new tool was to unify the various platforms into a single service. *"Live.com is now first and foremost a search destination,"* said Christopher Payne, Microsoft's corporate vice president [17]. Another feature was the integration of social search, which had become a key area in which Microsoft felt it could aggressively compete with Google and Yahoo.

Google was still ranked first in the most recent studies, but Microsoft was preparing its latest and strongest weapon against it. Announced on May 1, 2006, Microsoft's latest browser, Internet Explorer version 7, would have its MSN search engine installed by default. On October 18, 2006, the final IE7 version was released and would soon be proposed by Windows Update as a "recommended update" to every Microsoft Windows XP user connected to the Internet.

9 Google's Investments

Google did not acquire many search-related companies. Instead, it bought several companies doing auxiliary business to spread their activity.

9.1 Firefox

Google signed a partnership with the Mozilla Foundation in November 2004. The Google toolbar and homepage were installed by default on Firefox webbrowsers. The Firefox homepage was also hosted by Google. In exchange, Google paid for any search launched from the Firefox toolbar or homepage.

According to Mitchell Baker, Mozilla's chairwoman:

> The bulk of this [Mozilla Foundation's $52.9 million 2005] revenue was related to our search engine relationships, with the remainder coming from a combination of contributions, sales from the Mozilla store, interest income, and other sources. These figures compare with 2003 and 2004 revenues of $2.4M and $5.8M respectively, and reflect the tremendous growth in the popularity of Firefox after its launch in November 2004.

The Foundation collects money from Google for every search query. "It's not dependent on whether the searcher clicks on ads." [18]. In exchange for this, millions of queries were sent from Firefox's top Google search box, improving the search engine's popularity. There are also other signs of partnership between the two organizations. Google has hired several Firefox engineers, including lead developer Ben Goodger, and routinely provides resources to help with development issues.

9.2 Applied Semantics

This California-based company, famous for its semantic text-processing technology, was bought by Google in April 2003. The terms of the transaction were not disclosed, but the 45 employees of Applied Semantics continued working at their California center. The aim of that acquisition was to bring home their work on contextual advertising, as well as breaking the partnership Applied Semantics had with its rival of the time, Overture. "Applied Semantics is a proven innovator in semantic text processing and online advertising," said Sergey Brin. "This acquisition will enable Google to create new technologies that make online advertising more useful to users, publishers, and advertisers alike." [19].

9.3 *YouTube*

Google tried to catch the new market of video sharing by launching its Google Video service on July 12, 2006. But it was late to market and YouTube, founded in February 2005, already held a large part of it. On October 2006, Google bought YouTube for $1.65 billion.

10 Back to the Dilemma

Google CEO Eric Schmidt pondered his options with respect to Microsoft. The corporate graveyard was filled to overflowing with the remains of companies that had underestimated Microsoft in the past.

Exhibit 1
Popularity of Search Engines

Most Used Search Engines When Google Entered the Market

NDO Online Research Survey (1997)	
29%	Yahoo
17%	Altavista
14%	Web Crawler
11%	Infoseek
9%	Excite
7%	Lycos
13%	Others

Ranking by Unique Visitors (March 1998)

Ranking	Site	Unique Visitors
1	Yahoo	26,726,000
4	Excite	12,502,000
5	Infoseek	11,696,000
8	Lycos	6,787,000
9	Altavista	6,764,000
13	Web Crawler	4,477,000
21	Hotbot	2,703,000

2000 to 2002 Ranking

	N°1	N°2	N°3
October 2000	Fast-Alltheweb	Google	Northern Light
April 2001	Google	Fast-Alltheweb	Inktomi
August 2001	Google	Fast-Alltheweb	Wisenut
March 2002	Google	Wisenut	Fast-Alltheweb
December 2002	Google	Fast-Alltheweb	Altavista

Index Size Estimations

Search Engine	Estimation (millions)	Announced (millions)
Google	3.033	3.083
AlltheWeb	2.106	2.112
Altavista	1.689	1.000
Hotbot	1.453	1.500
Wisenut	1.147	3.000
MSN Search	1.018	3.000
Teoma	1.015	500

Exhibit 2
How Google Works

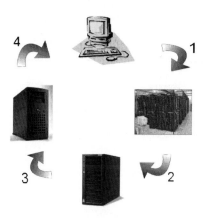

Google Process Principle

1. Directly from a computer connected to the Internet, the user sends a request to the Google webserver. It is possible to search for just about any word in any domain.
2. The server sends this request to the index servers. These are servers working like an index of a book. In other words, they tell which page contains the words that match the request.
3. The request travels to the doc servers, which actually retrieve the stored documents. A short description of the documents is generated for the user in order to show whether the results correspond exactly to the request.
4. These search results are sent to the user in a fraction of a second. On the page displayed on the user's computer, the results are ranked thanks to the PageRank system developed by Google. Users know that the first result often is the most relevant one.

Exhibit 3
Google Services and Software

Services

2001	Google news	Aggregation of pieces of information from various sources
	Google images	Image search engine
	Google catalog search	Consultation of sales catalogs
	Google groups	Brings together researchers in newsgroups
	Google movies	Movie critic search engines
2002	Google Adwords	Paid service that provides advertisements focused on the request sent in the search engine
	Google Answers	Paid service for users who can ask questions to Google teams (closed in 2006)
	Google Web API's	For webmasters
	Froogle	Search engine for products available for purchase online
2003	AdSense	Allows webmaster to sell some ad space on their Website
	Blogger	Google blog service
2004	Gmail	Webmail offering large storage space and allowing sending of large attached files
	Google Local	Local information calendar
	Google Book Search	Book digitization project
	Google Scholar	Search engine dedicated to scientific publications
	Google Suggest	Provides suggestions from Google when one uses the search engine
2005	Google Base	Online sales service
	Google Maps	Geographical search engine
	Google Transit	Itinerary service
	Google Personalized Search	Saving of previous requests and storage of favorite links
	Google Video Search	Animated feature search engine
	Personalized Google	Personalizes Google homepage
	Google Sitemaps	Register a Website on Google
	Google Reader	RSS flux search engine
	Google Analytics	Website statistics
	Google Moon	Images from the Moon
	Google SMS	SMS service
	Google Music Search	Music search engine
2006	Google Payment	Online payment solution
	Google Page Creator	Webpage creation
	Google Finance	Real time stock-option portfolio management
	Google Mars	Martian Google maps
	Google calendar	Online calendar (interconnected with Gmail)
	Google Trends	Shows popularity of keywords
	Google Notebook	Online notebook
	Google Checkout	Online payment solution
	Google code Hosting	Open source tool
	Google Base	Personal space to sell personal stuff
	Google News Archive Search	Online digitized newspapers search engine
	The Literacy project	Aims to eliminate illiteracy
	Searchmash	Experiment for new interface
	YouTube	Storage of videos
	Google Code Search	Computer code search engine
	Google Docs	Online and free competitor of Microsoft Office
	Google Gapminder	Comparison of country development
	Google Patent Search	Patent search engine

Source: http://www.google.com

Exhibit 3 *(continued)*

Software

Name	Description
Google Earth	Geographical search engine
Google Talk	Instant message service
Google Toolbar	Aid to navigation
Google Desktop	Search engine for user's own PC
Google Deskbar	Google in taskbar
Gmail Notifier	Gmail mail alerts
Picasa	To see and share images
Google Web Accelerator	Boost loading of high-ranked webpages
Google Pack	Suit of softwares
Google SketchUp	3D image creation
Google Browser Sync	Extension for Firefox

References

1   http://infolab.stanford.edu/~page/, viewed March 8, 2007.
2   http://online.wsj.com/article/SB108575924921724042.html, viewed March 7, 2007.
3   We remove the exclamation point after Yahoo! from this point forward.
4   Google Corporate Information: Our Philosophy,
    http://www.google.com/corporate/tenthings.html, viewed March 8, 2007.
5   The Search Engine Report, June 29, 1999.
    http://searchenginewatch.com/showPage.html?page=2167311.
6   The Search Engine Report, October 5, 1998.
    http://searchenginewatch.com/showPage.html?page=2166591.
7   The Search Engine Report, April 5, 1999.
    http://searchenginewatch.com/showPage.html?page=2167121.
8   The Search Engine Report, May 3, 2000.
    http://searchenginewatch.com/showPage.html?page=2162621.
9   The Search Engine Report, January 23, 2002.
    http://searchenginewatch.com/showPage.html?page=2159111.
10 http://searchenginewatch.com/showPage.html?page=2164811.
11 http://searchenginewatch.com/showPage.html?page=2164811.
12 InVancouver.com, January 21, 2003.
    http://www.invancouver.com/news_item.php?record_num=45.
13 clickz.com, March 5, 2003.
    http://www.clickz.com/showPage.html?page=clickz_print&id=2026641.
14 news.com.com, April 2, 2003. http://news.com.com/2100-1032-995086.html.

15   The Search Engine Report, July 1, 2003.
     http://searchenginewatch.com/showPage.html?page=2230291.
16   The Search Engine Report, February 1, 2005.
     http://searchenginewatch.com/showPage.html?page=3466721.
17   The Search Engine Report, September 12, 2006.
     http://tengoldenrulesblog.blogspot.com/2006/09/msn-search-is-now-windows-live.html.
18   clickz.com, January 4, 2007. http://clickz.com/showPage.html?page=3624399.
19   Google Press Center, April 23, 2003. http://www.google.com/press/pressrel/applied.html.

Chapter 5
Harvesting

Entrepreneurs invest much time, money, and personal energy into building their ventures; at some point in the future, adequate rewards are expected. The process of turning some of the value created into hard cash for the entrepreneur is referred to as harvesting. It is in effect the realization of one of the key objectives of venturing – the increase in personal wealth and financial well-being. Harvesting should never be thought of as a terminal activity, or the end of the process. Rather, it is a necessary step toward recycling the entrepreneurial talent and capital, offering the entrepreneur the opportunity to take his or her money out, partly or entirely, to pursue other plans and ventures. It is a starting point rather than an end point.

Harvesting can be realized through a number of channels, some of which do not involve a sale or even dilution of ownership by the entrepreneur. Often, though, it will involve selective disinvestment by the founder, with some portion of the equity sold to third parties. In many circumstances, the entrepreneur does not see harvesting as a key consideration: When the business is truly a lifestyle matter, for example, and generates sufficient cash to guarantee the financial security of the founder and sole investor, the notion of exit may be totally foreign. Conversely, when outside investors are involved, conflicts of interest can arise between entrepreneurs who are happy to continue running the company "as is" and investors who may be looking for a return of their principal and the value created with it.

Defining harvesting goals and crafting a strategy to achieve them, however, goes well beyond satisfying stakeholder interests. It also keeps up motivation and helps broaden the focus beyond the next quarterly results [1]. Because the topic is so "loaded," it is often ignored until it is too late to manage properly. Too many exits just happen, instead of being properly timed and planned. Not planning an exit almost always guarantees poor results. By contrast, having a clear, ambitious exit target (such as an IPO) can keep all parties focused and driven to a mutually satisfactory outcome.

Part of the reason for a lack of preparation of the harvesting phase is the profound emotional component of that step for the founder-entrepreneur. A trade sale, for example, is often compared to the experience of seeing a child leave home. It is clearly one of the most significant events in the entrepreneur's life, and frequently it is the first time that he or she has the luxury to decide about the long-term future of the business and his or her own level of engagement.

Harvesting should thus be seen as one of the most fundamental steps in the entrepreneurial process, requiring full preparation and management. First, since luck comes to the prepared, being ready means being able to seize favorable windows of opportunity for an exit. Second, decisions early in the life of a company, such as the choice of financial partners, will determine the harvest options at a later stage. Third, considering exit scenarios early on provides guidance in many decision making steps during the life of the company. Finally, the harvesting strategy is a critical piece of information for initial investors: They evaluate the investment, in part, based on the expectation of the timing and form of the liquidity event [2].

5.1 Harvesting Options

Harvesting is best understood as the owners' and investors' strategies for (at least partially) realizing the value created through an entrepreneurial venture. Several options exist for capturing this value: (1) milking the cash cow, *i.e.*, maximizing the cash-flow-generating ability of a mature company by, for example, stopping new investments, slowly extracting the value that has been built into the company; (2) a trade sale, *i.e.*, selling the whole company or part of it to new owners, be they corporate, investment funds, or individuals; (3) a merger or being acquired by a larger firm; (4) a buyout, *i.e.*, selling to some company managers, family members, or employees; (5) recapitalizing the company, for example by raising a large amount of debt to be covered by the company cash flows and paying out a large dividend now to owners (also called a leveraged recapitalization); and (6) taking the company public, *i.e.*, selling existing shares to investors in listed markets.

Each harvesting option has different requirements and consequences. The choice ultimately depends on the entrepreneur's personal goals and on the availability and "richness" of the valuation offered. Each option also leaves the entrepreneur in a very different situation – in many cases still in full control of his company (such as a leveraged recapitalization). Some key questions for the entrepreneur are: (1) How much of the company ownership does he or she want to retain and how much control? (2) How important is it to maintain some form of association with the company in the future? (3) What form of payment is most appropriate? (4) What is the desirable exit horizon?

5.1.1 Milking the Cash Cow

The simplest form of harvesting for a mature firm consists of curtailing investments and generating the largest cash flows from the entity over the longest time period. This form of planned obsolescence is based on increasing current free cash flows, reinvesting only the amounts necessary to maintain operations for as long as possible. The cash flows freed each period can then be reinvested in other activities. There are two main advantages to this approach to an exit: Ownership and control are retained, which can be important when the owner is not ready psychologically to sell, and the process does not involve the time-consuming and energy-draining process of finding a buyer and negotiating a sale. The major downsides are the inevitable decline in competitiveness of the firm over time, and the subsequent erosion of the long-term earning potential. This is really the "slow death" exit mode. Second, siphoning out the free cash flows requires patience, which not every investor has [3].

5.1.2 Trade Sale

The trade sale consists of selling the equity of the company, partly or totally, to new owners, be they corporate, institutional (such as a private equity investment fund), or individual. A trade sale usually involves the transfer of control to the new owners, but it is not uncommon for the seller to retain a small stake in the entity. Trade sales can combine a number of advantages: (1) they provide a clean exit, for cash or other considerations; (2) they are manageable, providing direct contact between buyer and seller and hence the opportunity for customized terms, which often lead to better valuations; (3) earn-out clauses, in which part of the proceeds are paid over time to the seller, can reassure buyers, and, again, support higher valuations; (4) they are quite predictable, being available in most markets, even difficult ones where other exit scenarios, such as IPOs, are not available; and (5) if well organized and advertised, they can lead to profitable bidding contests among buyers. However, a trade sale also takes time to arrange, first to find potential buyers, then to engineer the transaction. Surfing the Internet these days reveals many platforms on which small and medium-sized businesses are advertised for sale. Besides the advantage of immediate cash, the parting owners sometimes prefer this option over others as the firm is likely to survive relatively intact, a critical psychological factor.

5.1.3 Merging or Being Acquired by a Larger Firm

Technology- and science-based startups create products and services that are not only appealing to their customers but may also arouse the interest of larger firms. In fact, as the next chapter shows, most of the large, market-leading firms actively and continuously screen their environment for new products, technologies, and ideas to buy or license, rather than just relying on the prowess of their own R&D departments. Entrepreneurial ventures are a major source of new technologies for these corporations and thus are often the target of merger and acquisition activities.

From the entrepreneur's viewpoint, such transactions potentially combine a number of benefits, including large payouts, which can be increased further if a competitive bidding process can be used to prop up the bid prices. If the deals are negotiated, instead of being put to a competitive bidding process, they can be customized for the benefit of the buyer, increasing the willingness to pay. When large corporate investors come to depend on outside innovation, they are also willing and able to pay the price for it, since it becomes a critical factor for their very survival. For example, in 2005 Sheer Networks, a developer of intelligent network and service management solutions, was acquired by Cisco for almost US$100 million [4]. MySpace.com, the social networking and marketing pioneer, was acquired in July 2005 by Rupert Murdoch's News Corporation for over half a billion dollars – $580 million [5] – and Skype, the leading Voice over Internet Protocol (VoIP) provider, was sold to eBay for over $1.5 billion dollars in the same year.

On the negative side, the bargaining power of the seller in such deals can be limited and the number of potential buyers not very large. Often, the buyer brings to the table much-needed distribution channels, visibility, and the like, which give it leverage in the negotiations. Furthermore, sellers are often ill prepared to face the armies of experienced negotiators and lawyers fielded by large corporate buyers. In the case of mergers, the entrepreneur's equity can end up being tied to the assets of the merged entity, resulting in an incomplete exit, with the entrepreneur's ultimate fortune linked to the future performance of the acquiring or new company [6].

5.1.4 Buyouts

5.1.4.1 Family Sales

In family successions, the founding generation may decide to give the next generation the option to acquire the business they built up. These family successions can become quite contentious because of the complex levels of relationship between buyers and sellers. In a sense, proximity comes in the way of engineering an arm's-length transaction. Sentimental elements will play a large role, sometimes sabotaging the best intentions. Another often overlooked source of conflict is the parting owner's willingness (or not) to hand over the business and to accept that other family members will lead the company with a different style and different goals. These "ghosts from the past" have spoiled many such transactions and subsequently led to a sale to outsiders.

5.1.4.2 Management Buyouts (MBO)

A much more common plan for succession in the types of startups we look at is the management buyout. Here, the founders sell the company to existing partners, managers, or employees. In the process, the owners usually give up the majority or all of their equity. An MBO can be attractive for a number of reasons [7]:

- Speed: An MBO can be relatively quick, involving a limited number of parties in a negotiated transaction.
- Control: The selling party might not wish to surrender control to competitors for a number of reasons, psychological and/or strategic.
- Confidentiality: The seller may not wish to let competitors have access to sensitive information that would be disclosed during the trade sale process.
- Familiarity: With an MBO, the seller can continue to deal with a management team with whom he or she has an established relationship.
- Pricing: The potential for value maximization in the deal, in particular through the aggressive use of leverage, can allow buyers to up the price.

Initiators of such transactions focus on a sound and balanced management team, the standalone viability of the business, a willing seller, a realistic price, and the possibility of using a high level of leverage. Since the initiators often lack the financial resources and expertise to organize such deals, MBOs often involve specialized investment vehicles, such as private equity investors. The specialists bring to the table their industrial expertise and contacts, financial engineering skills, and bank relationships. To a large extent, leverage is used not only as a means to improve the return on investors' capital but also as a strong disciplining device, keeping everyone on the management team focused on generating the necessary cash flows to cover the debt obligations. Leveraged buyouts often lead to divestment of noncore businesses, right-sizing, cost cutting, or investing in technological upgrades that might otherwise have been postponed or rejected outright [8]. The high level of debt carries its own risks for sellers: If not able to meet the conditions, a company can be forced into bankruptcy by its creditors, with the destruction of the founder's company and sometimes the failure to make the last payments to the seller if earnouts were put in place [9].

5.1.4.3 Internal Sales to Employees

A variation on the buyout scheme is a sale to employees, who can gain ownership in several ways: They can buy stock directly, through stock options or through a profit-sharing plan. The most common form of employee ownership is the ESOP, or employee stock ownership plan. These tax-incentivized programs are viewed as potent incentive mechanisms for rewarding employees and aligning their financial interests with those of the company. In some cases ESOPs are designed to provide a retirement plan for the employees, especially if the company has serious expectations of going public in the future. These incentive programs can be extended and turned into a proper exit mechanism for the owners if employees are given the opportunity, over many years, to acquire additional shares in their company and progressively gain control. This "soft" exit retains the strong incentive component of option plans and can lead to a highly motivated workforce and smooth transition.

5.1.5 Leveraged Recapitalization of the Company

The leveraged recapitalization is, in effect, a form of "do-it-yourself" buyout. In a buyout, the acquirer normally leverages the company aggressively to boost the potential equity returns. In the case of a leveraged recapitalization, a popular technique during periods of high banking liquidity, the owners would leverage their company by borrowing from their banks and immediately pay out the fresh capital in the form of extraordinary dividends to themselves. This generates cash for the owners and does not change equity ownership at all. In other words, owners still own 100% of the shares in what is now a more leveraged, higher-risk company, but they have paid themselves significant amounts of cash. Considering the unfavorable tax treatment of such extraordinary dividend payments in many countries, such a transaction requires a well-thought-out legal and tax structure.

5.1.6 Initial Public Offering (IPO)

The most glamorous, and often profitable, harvest option is the initial public offering (IPO), through which the company's shares are floated to the public and listed on a recognized, liquid stock exchange. The IPO is often seen as the golden route to exits because it provides for both high valuation and significant liquidity for the shares, offering the possibility to calibrate exposure and ownership over time. It is also the most demanding and the most subject to the vagaries of the markets. Besides cashing out, the IPO can potentially fulfill a number of other critical functions for the company:

- It makes it possible to raise significant amounts of money to finance rapid growth [10].
- It provides liquidity to all company shares. This can be particularly important in the case of an ESOP, for example, where the incentive value of stock options is most effective with a visible stock price.
- It is recognized as a potent public relations mechanism, with listed companies often seen as the most solid entities and hence often receiving preferential treatment from parties such as banks and advisors. The disclosures and road shows accompanying the IPO are also superb opportunities to advertise the company.
- The information disclosures associated with the IPO and the ongoing public status provide reassurance to certain classes of investors, in particular institutionals with very deep pockets.
- Being publicly traded provides a boost for the venture's credibility, in particular with customers and suppliers.

Being listed also carries significant costs, starting with those associated with the repeated and extensive information disclosure. New regulations in the USA, such as Sarbanes-Oxley, also impose large governance, control, and certification expenses on listed companies. Public scrutiny is often intense, requiring permanent supervision by senior management. It is not uncommon for CEOs of listed companies to report that more than half of their time is devoted to managing the public investors. Public disclosures often extend to areas that the company would have preferred to keep confidential for competitive purposes, such as key clients and their size in the client portfolio. The IPO process itself is convoluted, expensive, and fraught with uncertainties. For small IPOs, it can consume as much as 20% of the entire value of the offering [11]. Choosing the underwriter, conducting due diligence, preparing the prospectus, setting up adequate reporting systems if needed, conducting the road show, negotiating the offer price, and the like can take between 6 and 18 months [12] and will require the active involvement of the entire top management team. The process is also deeply disturbing for entrepreneurial owners: They can find themselves totally left out and without any form of control. An IPO can be many things at once: exhilarating, frustrating, stimulating, and exhausting. It is most definitely one of the most important steps in a company's life, a rite of passage to the highest form of corporate life – life as a public company. We now briefly review some of the steps in the IPO process [13].

5.1.6.1 Lining up the Ducks

Years before an intended IPO, it is vital for the management to take a critical look at the company and start preparing it for the possibility of an IPO. In other words, it should start to organize itself as a possible listed company, with all the appropriate systems in place, both for management and reporting. Public markets have clear and stringent expectations about listed companies: They expect rigorous management systems to be in place, complete and professional management teams in all functional areas, transparent ownership and corporate structures, clear lines of responsibility, and perfectly readable business strategies. They also expect "strong" numbers, *i.e.*, growth and profitability, and a definitely stronger "story" as to why the company would present a great investment opportunity. Visibility and momentum are absolutely key to an IPO.

When all the elements line up promisingly, the IPO team can be assembled, consisting of the underwriter(s), the investment bank(s), the auditor, and the legal firm. The choice of specialist in itself can be a strong signal to the market about the quality of the IPO, with top-tier underwriters only interested in listing the most promising companies. In this regard, the firm's venture capitalists are often the critical factor between top-tier and second-tier IPO underwriters.

5.1.6.2 Preparing the Prospectus

The prospectus is a legal document outlining all the information about the offering. It is often dry and convoluted, written in arcane legalese and littered with pages of disclaimers. It includes all the financial data for the company since its inception (or at the very least the previous five years), detailed information on the management team, descriptions of a company's target markets and products, competitors, clients, and growth strategies. It is the most extensive information disclosure exercise the company will ever go through and as such needs to be organized with the greatest care since any improper information could be construed as voluntarily misleading and thus fraudulent. This is why prospectuses never include forward-looking statements (such as financial projections) since they could ultimately prove to be wrong and thus misleading. It is also important to keep in mind that securities laws in most countries are treacherous territory, allowing no "good faith" errors, *i.e.*, there is zero allowance for mistakes, and honest errors are not protected from automatic penalties. It is absolutely critical to understand, though, that the prospectus is *not* a marketing instrument for the company: It is truly an offering document, in the strictest legal sense. The marketing component, where excitement and momentum are built, is the road show.

5.1.6.3 The Road Show

The term used to describe the process is an adequate recognition of its purpose: This is where the marketing and PR spirit of the company can really express itself. Typically members of the top management team will present the company and its business plan to prospective investors – usually institutional investors. At these meetings, the underwriter will also attempt to gauge the level of interest in the IPO, which will in the next step help to decide on the price and size of the stock offering. Thus the performance of the management team is one of the most crucial factors in the IPO process. Many entrepreneurs lack the necessary experience and skills to excel in these tasks. It is thus not uncommon for the board of directors or other investors to initiate a change of management well ahead of the intended IPO, bringing in an experienced team to make the business case to public investors.

5.1.6.4 Pricing the IPO

The interest potential investors show in the "book-building" phase ultimately determines the price set for the stock in the offering. Book building can be either contractual, with interested buyers essentially "preordering" shares from the underwriters, or simply "expressing an interest," as in the USA. The issuer's investment bank plays a key advisory role in this pricing dialogue, providing the issuer with unadulterated opinions and data to fix the price with the underwriters. Underwriters in this regard are often seen as serving two masters, *i.e.*, the issuer and the institutional buyers of the shares. With the latter being more critical to the underwriter's long-term placement ability, suspicion has always run high that underwriters tend to drive prices lower to "leave a sweet taste" in investors' mouths when there is a strong early price increase after listing. The evidence indeed supports an average "first day price movement" of about +16% in the USA, which is often interpreted as the offer price having been deliberately set below the expected market value, leaving a gap between what the issuing company received initially and the investment's value at the end of the first trading day. This is often referred to as "underpricing," even though many alternative explanations exist for the phenomenon. Frequently, IPOs tend to be relatively small affairs, often representing less than 25% of the shares outstanding of a company. This "limited supply" is meant to provide better means of controlling the offer and the early after-market.

5.1.6.5 First Days on the Market

Once the offering price has been agreed on, an IPO is declared effective, and trading of the new stock starts the next day. The lead underwriter usually ensures smooth trading in a company's stock during those first few crucial days, meaning it is allowed to actively trade in the stock it just floated on the market. Similarly, underwriters have the means to penalize any of their institutional investors who are too eager to "flip" the shares in the market, *i.e.*, reselling them for a profit. Often this takes the form of a threat not to allow them to participate in future (and hopefully profitable) IPOs underwritten by the firm. Threats are usually sufficient to keep key players behaving properly.

5.2 Specific Challenges in the Harvesting Process

Besides understanding the differences between the many harvesting options, the successful exit is also a question of timing and price that can be negotiated with the purchaser – *i.e.*, valuation of the company. Anecdotal evidence suggests that the exit, partial or complete, is the most mismanaged process in the entrepreneurial life cycle, often left to look after itself and addressed when it is already too late to do much about it.

5.2.1 Timing

Only a minority of entrepreneurs have a clear vision about the harvesting phase. Many founders do not spend a minute on the issue until they face the threat of losing the entire company or receive an offer for it. Such a situation might arise quite unexpectedly because new technologies and competitors gain a foothold in the market, or a major account is lost. Unprepared, or under-prepared, companies appear at their worst, with the expected consequences of leading to the worst possible valuation. It is absolutely key for entrepreneurs to manage the exit scenarios as carefully and professionally as any other phase in the company's development since it is the one phase that will most determine the ultimate value created in the venture [14]. A harvesting strategy should be shaped early, ideally as early as the founding process: Only with this objective in mind can the entrepreneur and managers keep the company heading in the right direction and continuously monitor and scan the environment for the best exit window. Harvesting also requires immense patience and is utterly unpredictable: Most ventures will not find valuable exits in under 7 to 10 years. This should not serve as an excuse to be sloppy with the process or, even worse, ignore it entirely. Harvests and exits are the most critical elements of the entrepreneurial process; in a sense, they provide the fuel for new entrepreneurial endeavors and the proper recycling of both financial and human capital.

5.2.2 Valuation

A recurring issue in the life of a company, valuation is indispensable for the harvesting process as one cannot structure a deal without an understanding of the harvest value [15]. It determines what the entrepreneur receives for his or her equity and thus determines the success of the harvesting process. The topic of company valuation has already been discussed at length in *Chapter 3*.

5.3 Cases in This Chapter

In Chapter 4 we followed the history of EndoArt and left the company at a turning point. Here, in the case of *EndoArt: Creating and Funding a Medical Technology Startup (B)*, we will learn about the preliminary end of the story. Another two years have passed and, thanks to the new CEO, Philippe Dro, the medical technology venture survived. Almost eight years into the business, which has been mainly VC financed, the question of exit and cashing out becomes more urgent for the investors. Management recognizes the need, and the case reveals Dro's view on it. The case is an excellent vehicle to facilitate discussion among students about the tasks and role of management in preparing a company for the best value realization.

To provide an update on EndoArt's development, we have included a recent press release, in the form of a handout. The news announces EndoArt's acquisition by Allergan, Inc., a California-based provider of medical products with broad experience in the obesity market.

Much more concrete are the harvesting possibilities in the next case, *Sentron at the Crossroads (A)*. The twist for the founder here is to decide between growth and cashing out. The case describes the development of the engineering startup between 1993 and early 2004. Key factors determining the venture's business model are explored. Among them are topics such as R&D collaborations, intellectual property, and the venture's manufacturing and product delivery strategy. They all allow an assessment of the future and the respective requirements to realize their potential.

A new market looms, offering potential for an exponential growth in sales. Four alternative options are laid out with regard to how the venture could proceed from there.

Sentron at the Crossroads (B) portrays the final decision taken by the venture's management with respect to the four options outlined in the (A) case. Insights are shared about the key factors that eventually influenced the decision.

In the next case, *4M Technologies and the Optical Disk Revolutions*, we investigate the financial and strategic dilemmas facing an extremely fast-growing company striving to become the leading producer of optical disk integrated manufacturing systems in the world. Built from the ground up in Switzerland in the 1990s, the company was still smaller than some of its key global competitors and heavily burdened by the debt it had taken on to establish its position in this high-technology market. To finance the aggressive growth, the company relied mostly on bank financing, supplier credit, and customer prepayments, made possible only through the quality of its business relationships with those parties. But these arrangements had run their course and it was time for the company to put its house in order financially and to tackle the operational challenges ahead. 4M Technologies was thus contemplating an IPO, but a number of questions remained unresolved, such as where to list – or even whether to list at all. The case is a great introduction to the new growth-equity markets in Europe, which appeared in the latter part of the 1990s, offering new channels for the funding of technology companies.

An IPO decision is also part of the *Generics* case. Founded in 1986, the company has been following a hybrid business model: on the one hand acting as a contract research and technical advisory service company in close touch with the market; on the other hand playing the role of incubator aimed at turning selected home-grown business ideas into independent companies, eventually to be sold out in trade sales or IPOs. It thus functioned as a "startup factory." Since the firm's founding, well over 20 such firms had been "spun out" in this way. In addition, Generics made early-stage investments in startups active in areas with which it was familiar in order to reinforce its knowledge base and network in the markets. Overall, Generics focused on one thing: creating value from technical innovation, drawing on various vehicles to do so. Stressing that the crucial mission of business creation and growth is increasingly entrusted to the R&D function of technology companies, the case discusses the issues raised by this unusual business model, particularly as the company gets ready to go public, late in 2001. What were the pros and cons of such a move? How did the investment community see this business model? What were the implications of such a drastic change?

In contrast to the previous cases, the last venture in this chapter faces a rather unfavorable situation. The case *GigaTera Inc: Pulling the Plug?* discusses the option of exiting more to prevent further losses than to cash out. The venture GigaTera, a spinoff of the Ultrafast Laser Physics Group at ETH Zurich, commercialized laser technology for high-speed optical data transmission. Shortly after the venture's inception the entire optoelectronics industry faltered, shaking GigaTera to its foundations. The case portrays GigaTera's development from its inception in 2000 up to

2003. Besides addressing opportunity recognition and exploitation against a withering market environment, the case invites students to evaluate GigaTera's specific setup within its network of resources and financiers. By 2003, the Swiss venture faced a significant lack of financial resources and unclear market prospects. Students will be asked to indicate viable alternatives for proceeding.

References

1   Timmons, Jeffry A. *New Venture Creation: Entrepreneurship for the 21st Century*. Boston: Irwin, 1994, p. 656.
2   Smith, Richard L. and Janet Kiholm Smith. *Entrepreneurial Finance*. New York: John Wiley & Sons, 2000, p. 566.
3   Petty, William and John W. Kensinger. "Harvesting Value from Entrepreneurial Success." *Bank of America Journal of Applied Corporate Finance*, Winter 2000, 12 (4), pp. 8–19.
4   "Cisco Systems to Acquire Sheer Networks." Press Release, July 26, 2005. http://newsroom.cisco.com/dlls/2005/corp_072605.html (viewed August 2, 2006).
5   Reiss, Spencer. "His Space." *Wired Magazine*, July 14, 2006. http://www.wired.com/wired/archive/14.07/murdoch_pr.html (viewed August 2, 2006).
6   Petty, "Harvesting," pp. 434–437.
7   "Leveraged Buy-out – Company Acquisition Method." 2006. http://www.valuebasedmanagement.net/methods_leveraged_buy-out.html (viewed on August 3, 2006).
8   "Leveraged Buy-out – Company Acquisition Method."
9   Petty, "Harvesting," p. 430.
10   Timmons, *New Venture Creation*, p. 660.
11   Leleux, Benoît. "Riding on the wave of IPOs." *Financial Times*, February 3, 1997, p. 6.
12   Petty, "Harvesting," p. 439.
13   Chervitz, Darren. "IPO Basics." *CBS MarketWatch*, 2006. http://moneycentral.hoovers.com/global/msn/index.xhtml?pageid=1954 (viewed August 4, 2006).
14   Timmons, *New Venture Creation*, pp. 656–658.
15   Petty, "Harvesting Firm Value: Process and Results."

CASE 5-1

EndoArt SA: Creating and Funding a Medical Technology Startup (B)

Ralf W. Seifert and Jana Thiel

LAUSANNE, SWITZERLAND, APRIL 10, 2006. Philippe Dro felt a big sense of release. He had just closed C-round financing of CHF 13.5 million, taking the total amount raised by EndoArt to date to CHF 37 million. The new investment would enable the company to launch the gastric banding application in the European market in the third quarter of 2006. It would also support clinical developments in the USA, also scheduled to start in the third quarter.

The negotiations had, of course, taken up more time and resources than expected. But that had not been the only issue consuming Dro's energy over the past 18 months:

> At the end of November 2004, EndoArt was technically bankrupt. I had to fire all the employees. However, all of them kept on working during their resignation period. We then went back to our investors and negotiated for almost CHF 5 million of additional money. By February 2005, we could finally rehire all employees, every single one of whom had stayed through these four very difficult months! At that point, we had about CHF 1.5 million left to "bet" all on the obesity application.

The development of the gastric banding solution – named easyband® – received full attention. There was zero margin for delays. The team made a remarkable effort, and by September 2005, after successful tests (clinical trials run in Germany), EndoArt obtained the CE marking. The cardiac banding was meanwhile put on the back burner as a future niche product in the company's portfolio, and all incontinence applications were discarded.

EndoArt had come a long way from its early days as a stent developer, as Dro summarized:

> The challenge EndoArt and other startups alike face is that of a journey leading you to places where you never imagined going at the outset. What eventually determines success is the ability to clearly analyze the changes, including the external environment, and adapt the business plan accordingly. When I joined the company I tried to make everybody face the issues and not walk around them.

Copyright © 2006 by the alliance for technology-based enterprise (IMD, EPFL and ETH Zurich). Copyright permissions are handled by IMD, Lausanne, Switzerland.

On its particular journey, EndoArt had burned some CHF 25 million to date, for mere technologies, as Stergiopulos pointed out:

> Until 2005 EndoArt was basically a research company. Only then did we truly start transitioning into a product company. To a large extent it is pure luck that we are still around today, and we owe a lot to our investors, who kept supporting us through all the ups and downs.

At the time, EndoArt still had to enter the larger competitive arena, and profits were yet to be made. Being eight years into the business, however, and predominantly venture capital backed, the management team had to discuss possible exits and their timing. Dro noted three options:

- Negotiate a trade sale or strategic partnership with a larger company.
- Grow EndoArt aggressively and go for an IPO.
- Make EndoArt a successful standalone player without going public.

None of these options would happen over night. Dro's immediate issues were to keep the momentum they had built over the last year, get the additional talent in marketing on board, and, finally, start selling the product. After the launch of easyband® in Europe, the startup would tackle the Australian and Canadian markets next since these two countries fully accepted the CE marking. Nonetheless, the US market remained key and EndoArt had yet to develop, test, and register its solution there:

> It will take another CHF 20 million to launch the product in the US! Hence, our challenge at the moment is not to fancy any exit option but to work hard on making the company as successful as possible. Also think of that: we have a lot of social responsibility here. By today, we employ 20 people. All of them have family! When I had to fire people for those few months back in 2004 – that also affected their families. They depend on the management team to make the right decisions. Misjudgments, like the ones in EndoArt's past, can be very costly – not only for the owners of the firm.

The EndoArt team had gone through quite an ordeal in the past few years and only recently had a silver lining reappeared on the horizon. Looking ahead, a great deal of uncertainty still remained for this small venture.

CASE 5-1 Handout
Allergan Announces its Acquisition of EndoArt

PRESS RELEASE, IRVINE, CALIF., FEBRUARY 22, 2007

Allergan, Inc. (NYSE: AGN) today announced the completion of its acquisition of Swiss medical technology developer EndoArt S.A., a leader in the field of telemetrically controlled (or remote-controlled) implants used in the treatment of morbid obesity and other conditions. The acquisition builds upon the strength of Allergan's existing obesity intervention product portfolio, which includes the LAP-BAND® Adjustable Gastric Banding System, currently the only adjustable implant device for individualized weight loss approved in the United States and a leading bariatric procedure worldwide; and the BIBTM Intragastric Balloon System, a nonsurgical alternative for the treatment of obesity approved in many countries although not currently available in the United States.

Allergan paid $97 million, net of excess cash, for the EndoArt shares in an all cash transaction. Allergan will not alter financial guidance for 2007 as a result of the transaction. Estimates of any costs that will be excluded from Allergan's adjusted earnings per share will be provided at the time of Allergan's first quarter 2007 Earnings Release. It is anticipated that a substantial portion of the acquisition purchase price will be expensed by Allergan as in-process research and development, with the balance of the purchase price being allocated to other identifiable tangible and intangible assets acquired, including developed and core technologies, liabilities assumed and goodwill. An independent third-party valuation firm has been engaged to assist Allergan in determining the estimated fair values of the acquired intangible assets, including in-process research and development.

The acquisition gives Allergan ownership of EndoArt's proprietary technology platform, including FloWatch® technology, which powers the EASYBAND® Remote Adjustable Gastric Band System, a next-generation, telemetrically adjustable gastric banding device for the treatment of morbid obesity. The EASYBAND® device is surgically implanted around the upper stomach and can be adjusted when necessary according to each patient's individual weight loss needs and results using a simple control unit placed over the device. The control unit transmits power and commands to, and receives information from, an implanted antenna connected to the EASYBAND®, which can then be adjusted precisely and in just minutes.

"This acquisition is representative of our commitment to actively pursue the development and commercialization of next-generation products and technologies capable of providing high-quality, healthier and less traumatic weight-loss treatment solutions to patients, physicians, governments, employers and health care payers," said David E.I. Pyott, Allergan's Chairman of the Board and Chief Executive Officer. "We also believe that this proprietary technology has exciting potential across a broad range of other medical device applications and disease categories, such as urology and gastroenterology."

"We are very pleased to be joining Allergan in its effort to address the serious immediate and long-term consequences of the worldwide obesity epidemic – an effort that will benefit from the depth and breadth of Allergan's investment in the research, development and commercialization of next-generation products and technologies," said Philippe Dro, Chief Executive Officer and Chairman of EndoArt.

EASYBAND® Gastric Banding System was approved by the European Commission for use in Europe in mid-2006. Allergan anticipates seeking U.S. Food and Drug Administration (FDA) approval of the device following completion of clinical studies that will be conducted in the United States. Allergan also is establishing EndoArt's facility in Lausanne, Switzerland, as an international center of excellence for research and development in obesity-related disorders.

1 Forward-looking Statements

This press release contains "forward-looking statements," including, among other statements, the statements by Messrs. Pyott and Dro, and statements regarding the business combination between Allergan and EndoArt. Statements made in the future tense, and words such as "expect," "believe," "will," "may," "anticipate" and similar expressions are intended to identify forward-looking statements. These statements are based on current expectations of future events. If underlying assumptions prove inaccurate or unknown risks or uncertainties materialize, actual results could vary materially from Allergan's expectations and projections. Risks and uncertainties include, among other things, general industry and market conditions, technological advances and patents attained by competitors, challenges inherent in the research and development and regulatory processes, challenges related to product marketing such as the unpredictability of market acceptance for new products and/or the acceptance of new indications for such products, inconsistency of treatment results among patients and the potential for product failures, unknown risks associated with the investigational devices that are the subject of Allergan's clinical trials, potential difficulties in manufacturing new products, and governmental laws and regulations affecting domestic and foreign operations. Risks and uncertainties relating to the EndoArt acquisition include that the anticipated benefits and synergies of the transaction will not be realized and that the integration of EndoArt's operations with Allergan will be materially delayed or will be more costly or difficult than expected. These risks and uncertainties could cause actual results to differ materially from those expressed in or implied by the forward-looking statements and therefore should be carefully considered. Allergan expressly disclaims any intent or obligation to update these forward-looking statements except as required to do so by law.

Additional information concerning these and other risk factors can be found in press releases issued by Allergan, as well as Allergan's public periodic filings with the Securities and Exchange Commission, including the discussion under the heading "Risk Factors" in Allergan's 2005 Form 10-K, Allergan's Form 10-Q for the quarter ended March 31, 2006, Allergan's Form 10-Q for the quarter ended June 30, 2006, and Allergan's Form 10-Q for the quarter ended September 29, 2006. Copies of Allergan's press releases and additional information about Allergan is available on the World Wide Web at www.allergan.com or you can contact the Allergan Investor Relations Department by calling 1-714-246-4636.

2 About Allergan, Inc.

With more than 55 years of experience providing high-quality, science-based products, Allergan, Inc., with headquarters in Irvine, CA, discovers, develops, and commercializes products in the ophthalmology, neurosciences, medical dermatology, medical aesthetics, obesity intervention, and other specialty markets that deliver value to its customers, satisfy unmet medical needs, and improve patients' lives.

3 About EndoArt S.A.

Endoart S.A. is a privately owned Swiss-based medical technology company founded in 1998 in Lausanne, Switzerland, and is specialized in the research and development of telemetrically controlled (or remote controlled) implants. Historical funding for the company was provided by leading European venture firms such as Sofinnova Partners (France), Trans Atlantic Technology (Switzerland), or local Banque Cantonale Vaudoise. The last financing round was led by VI Partners (Switzerland) and Rennaissance (Switzerland), with EMBL Venture (Germany) and Genevest (Switzerland) as coinvestors.

CASE 5-2
Sentron at the Crossroads (A)

Ralf W. Seifert, Christopher L. Tucci, and Jana Thiel

Dr. Christian Schott, head of development, was reflecting on the outcome of his recent talk with Professor Popovic, managing director of the Swiss-based, privately held Sentron AG:

> In terms of financial success as well as market maturity and credibility, the automotive industry could represent a quantum leap forward for our company. Even though it's a brutal, competitive business, the market prospects are extremely promising.

It was January 2004 and Sentron had just closed the books for its tenth business year. The company had established an excellent reputation for the production of magnetic sensors for scientific and industrial applications. Although recognized as a promising target, the automotive market had yet to be captured.

> There are sustainable growth options for Sentron without automotives, but if we actively refuse, our decision will always continue to hover as "opportunity threat" above us. It is simply common business sense to enter into and prosper in that market.

With an eye on the future, Sentron's main shareholders felt that the company needed to revise its strategy. Four alternatives had been identified for consideration:

1. Maintain Sentron's well-established startup management style, aiming predominantly for organic growth.
2. Introduce a more aggressive management style and seek venture capital financing to expand into a medium-sized company more rapidly.
3. Look for a strategic investment from a large partner already established in the automotive industry.
4. Make a full trade sale, *i.e.*, sell the entire company to a strategic investor, at its current stage.

The decision called for a critical reflection on the company's history to date and a comprehensive look at future perspectives.

Copyright © 2005 by the alliance for technology-based enterprise (IMD, EPFL and ETH Zurich). Copyright permissions are handled by IMD, Lausanne, Switzerland.

1 Company History

Sentron started business in Zug, Switzerland, in 1993. It was founded by Radivoje S. Popovic, a well-known expert in the field of Hall-effect-based semiconductor devices [1] and former vice president of Landis and Gyr Central Laboratory, Zug. The company's name was derived from its target business: magnetic SENsors and interface elecTRONics. Sentron's slogan was "Magnetic measurement is our business!"

In 1994 Popovic was appointed full professor of microtechnology systems at the Ecole Polytechnique Fédérale de Lausanne (EPFL). Subsequently, parts of Sentron's research activities were directed from Lausanne, utilizing EPFL's excellent research facilities while the original office in Zug remained operational.

At that time Sentron's product portfolio included both optical and Hall-Effect-based sensors (*see Exhibit 1 for more information on the Hall effect and Sentron's technology*). But soon Sentron focused exclusively on Hall sensors. Its early business activities included the customized manufacture of magnetic sensors, probes, and magnetic measurement instruments (transducers and teslameters) for a scientific niche market (*see Figure 1 – Scientific*). The main customers were scientists and engineers from research labs such as CERN, Fermilab, and Brookhaven. Building on Professor Popovic's reputation, the company expanded by primarily providing R&D services for scientific magnetic sensor applications.

This expertise was soon recognized in the broader market. The Swiss-based Swatch Group approached Sentron to investigate a compass application for one of its prestigious wristwatch projects – the Tissot "T-Touch" (*see Figure 1 – Industrial*). Jointly, Sentron and ASULAB S.A., Swatch's R&D laboratory, adapted Sentron's technology. The "T-Touch" was finally launched in 2000, fueling sales of Sentron's two-axis Hall magnetic sensor. Whereas scientific measurement in-struments had consumed about 100 of these sensors per year, the annual demand now rose from 50,000 in 2000 to 200,000 in 2003 [2].

> The challenging issue was to convince a large company to develop a new product based on our sensor without any guarantee of delivery in the long term. Sentron could not afford a penalty agreement to ensure against nondelivery. This led to difficult negotiations.

Having ventured into the industrial market, Sentron soon identified a broader range of industrial applications [3]. ASULAB served as a lead contact, which had a catalytic effect on potential customers and research partners. Shortly after, Sentron began a strategic collaboration with the highly diversified Japanese Asahi Kasei Corporation. Asahi Kasei obtained exclusive exploitation rights for Sentron's IMC[1] technology for the Asian market and in return financed some of Sentron's R&D activities [4]. This allowed Sentron to extend its development

---

[1] IMC: integrated magnetic concentrator; see also *Exhibit 1*.

capabilities into CMOS-based[2] sensor applications, which was vital for targeting the original equipment manufacturer (OEM) business on a larger scale.

By 2003 Sentron employed eight people. Its total revenues from R&D services and scientific and industrial applications amounted to CHF 2.2 million.

2 Sentron's Business Model and Operations

2.1 Sentron's Business Pillars

Sentron's product portfolio rested on three different business pillars: scientific instruments and industrial and automotive applications. Examples of typical products that use the company's magnetic field sensors are illustrated in *Figure 1*.

While the scientific and industrial pillars actually contributed to revenues, the automotive sector was only addressed by means of prototypes or product ideas. No marketable product had been launched for this specific sector yet (*Table 1*).

In terms of future revenue growth, the automotive pillar seemed to be the most promising. Potential demand for several million sensor units per year had been identified. (*For an overview of potential sensor applications in the automotive industry, see Exhibit 2.*)

However, automotives is known as a very demanding business. Special requirements had to be met in respect of a supplier's financial health, market presence, and quality standards. Although Sentron was well established in the scientific arena and was able to provide excellent references from the Swatch Group, automotive customers had not yet started buying from Sentron.

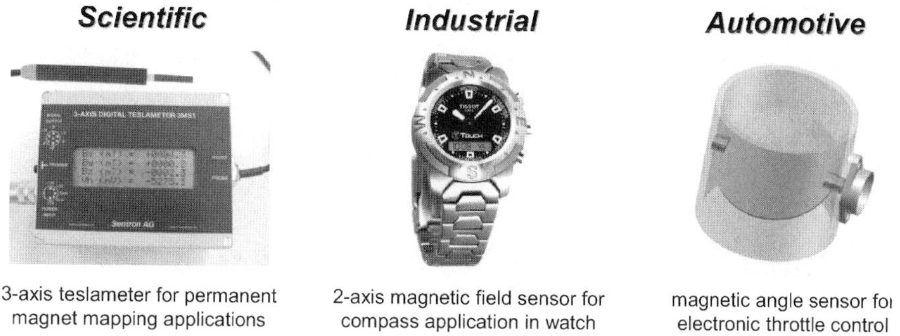

Scientific	**Industrial**	**Automotive**
3-axis teslameter for permanent magnet mapping applications	2-axis magnetic field sensor for compass application in watch	magnetic angle sensor for electronic throttle control

Figure 1 Product Examples
Source: Schott, Christian. Sentron: Magnetic Measurement is our Business. Company Presentation, IMD, January 2003.

---

[2] CMOS: complementary metal-oxide semiconductor, a major class of integrated circuits.

Table 1 Market Readiness Assessment of Sentron's Product Portfolio in 2003 [5]

	Scientific Instruments (<1 K units/year)	Industrial Sensors (1 K – 100 K units/year)	Automotive Sensors (>100 K units/year)
Marketable product	• Particle accelerators • Superconductor magnets • Fusion reactors	• Control knobs • Motor drives • Watch compass	
Evaluation prototypes		• Tooling machinery	• Wiper system • Electronic gearshift • Steer-by-wire • Power windows • Headlight control
Potential identified		• Joysticks • Power converters	• Seat adjustment • Exhaust gas recombustion

2.2 Technology and Intellectual Property

Sentron's IMC-Hall® technology was innovative in the way it measured magnetic fields. Whereas conventional Hall sensors only responded to magnetic fields perpendicular to the chip surface, Sentron's IMC-Hall® chips were able to respond to fields parallel to their surface. Combined with a conventional Hall element, this allowed for the first time both perpendicular and parallel components of the magnetic field to be measured in a single chip [6] (*see also Exhibit 1*). In addition, the sensor's small size, low cost, and great robustness against electromagnetic interference constituted a clear competitive edge.

Sentron followed the strategy of protecting its key knowledge for large markets through simple and broadly described patents, which covered the structural makeup of its sensors. These patents secured its business and prevented infringement by competitors. They also served as an effective means of building credibility and maintaining Sentron's technological leadership position (*see Exhibit 3 for Sentron's patent portfolio*).

2.3 Product Delivery

In order to address the high-volume OEM market, Sentron's delivery process was increasingly built around standardized products.

The delivery was organized in two steps. First, standard products were offered, largely serving industrial needs. Custom-tailored modifications were provided by Sentron's application engineering services if required. The second step aimed at fulfilling nonstandard requests like designing entirely new products. These products usually needed to pass a feasibility stage before entering the actual development phase. In such cases customers were invoiced for both the proof of concept and the subsequent development expenses. Sentron followed a strategy of building on its existing product base as far as possible. Thus, getting paid for each proof of concept allowed it to reap additional profits from previous R&D projects.

> Being paid was the best way to gauge our customers' commitment to the projects. Besides, we could not have survived if the feasibility studies had been carried out free of charge. The most gratifying part was when developed know-how could be reused for another client.

2.4 Manufacturing

Chip manufacturing has become a highly standardized process. Thus, parts of it could be outsourced to readily available foundries. But to benefit from outsourcing, Sentron had to apply a "design for manufacturing" strategy. For Sentron's IMC-Hall ASICs[3] this meant that almost 95% of the conventionally integrated circuit technology could be applied. The remaining specialized 5% was carried out in one step – the Sentron-specific IMC® postprocess. This step used the existing interface between the first two phases of the conventional manufacturing process [7]. Consequently, the Sentron-adapted manufacturing process consisted of four steps: (1) CMOS process, (2) IMC postprocess, (3) assembling, and (4) testing. The standard processes were outsourced to different external contractors. Sentron, in turn, covered the customized design and programming. It also coordinated the whole chain in terms of logistics and quality assurance.[4]

2.5 Staffing

Sentron's proximity to university-based research facilities had always allowed it to attract a creative and flexible staff with strong technical and scientific capabilities.

---

[3] ASIC: application-specific integrated circuit, a chip specifically designed for a particular application.
[4] For a more comprehensive description of the manufacturing process of Sentron's chips refer to Popovic/Fahrni/Stuck 2003.

Our skill resulted in fast and very effective problem solving. I remember a case in point when the technology we used became obsolete and the Sentron subcontractor decided to discontinue the component's production. We not only risked losing one of our main clients, but also credibility. In less than a month, a totally new solution was redeveloped for our client.

In 2003 eight engineers and technicians were employed, two of them located at the Lausanne office. Several factors, such as fairly unrestricted working conditions and the employees' common desire to work on appealing products (as ascertained from trade fair feedback), created a sense of community and motivation to stay with the company.

3 The Sensor Technology Market

The sensor technology market had progressed in recent years. It was now composed of niche markets such as scientific research and large-volume OEM markets such as consumer electronics and automotives. Automotives stood out from the other industries because its market size was estimated to be five times greater (*Exhibit 4*).

In terms of sensor production, the market was driven by the cost of sensors as well as their robustness, reliability, manufacturability, and ease of integration into the OEM product. To adapt sensors to specific product needs, a number of companies provided additional engineering services.

3.1 Sentron's Revenue Buildup

Sentron's early revenues came almost exclusively from engineering services. This changed between 1996 and 1998 when the company started to commercialize its IMC-Hall technology, leading to increased sales of scientific instruments. From 1999 industrial sensor sales caught up, as the Swatch Group became a client and the cooperation with Asahi Kasei began. Sales of industrial sensors grew at nearly 100% per year, to reach CHF 1.1 million by 2003, representing 50% of Sentron's total revenues (*Figure 2*).[5]

---

[5] For a more detailed analysis of Sentron's revenue buildup see Popovic/Fahrni 2003.

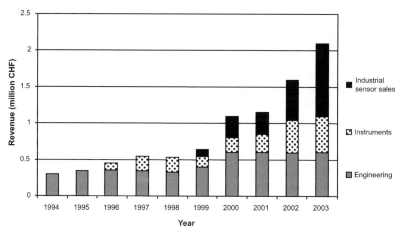

Figure 2 Sales by Market Segment from 1994 to 2003
Source: Schott, Christian. Sentron: Magnetic Measurement is our Business. Company Presentation, IMD, January 2003.

However, eventually Sentron covered only 1.0% of the scientific and 0.5% of the industrial applications market. Neither the company's market share nor the overall market size of either segment was expected to increase significantly.

The automotive sector, with a projected total market size of more than CHF 500 million, clearly represented a more appealing target. Even though a small company like Sentron could not expect to loom large, a market share comparable to the industrial or scientific one would provide a jump in sales. If a successful market entry could be achieved, total revenues were expected to climb as high as CHF 11 million by 2007 (*Figure 3 and Exhibit 4*).

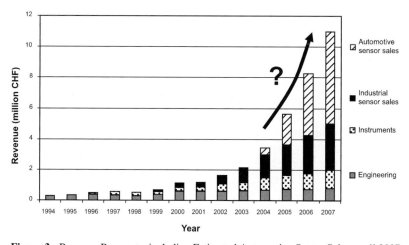

Figure 3 Revenue Prospects, including Estimated Automotive Sector Sales, until 2007
Source: Schott, 2003.

4 Competitors' Technologies and Products

Sentron was not the only player in the magnetic sensor market, which was scattered in terms of competing technologies.

Technologies that competed with the IMC-Hall® sensor were conventional Hall sensors, magnetoresistive sensors such as AMR (anisotropic magnetoresistive) and GMR (giant magnetoresistive), inductive sensors, Fluxgate sensors (compass) and MEMS (microelectrical mechanical systems). Even though some of them could not match Sentron's technology in terms of maturity or cost effectiveness, they were historically well established.

Several direct competitors of varying size and market power were identified, namely Philips, Infineon-Siemens, Melexis, Honeywell, Allegro Micro, Micronas, and Asahi Kasei. The chief threat posed by competitors of this size was their financial strength. A large player could drop its prices by cross-subsidizing any losses for some time. Thus, despite its superior and affordable solutions, Sentron was in potential danger of running dry.

> It is very difficult to penetrate a market in which there are such large competitors. Even if we do have a product that is an order-of-magnitude better with respect to the quality-price ratio, competitor technologies and products will still have the main market share in many applications. The sensors' interface standards and compatibility with established production leads to strong inertia.

5 Alternative Future Strategies

By and large, Sentron was a typical high-tech startup with excellent scientific and technical skills. Like most startups, its main strengths were based on know-how and intellectual property, but its market penetration and distribution capabilities were still limited. For Sentron to expand into a larger company, these would now require more rigorous attention.

Four alternative directions were identified through which expansion could occur:

1. *Maintain Sentron's well-established startup management style, aiming predominantly for organic growth.*
   The company's niche markets were consolidating and would most likely support moderate growth within known limits. In addition, companies in the automotive industry could be granted licenses and thus provide added profits at low risk. Sentron would thus stay profitable without having to make major changes. However, if licensing turned out to be a less viable solution than projected, and if the company wanted to target mass production markets, several new challenges would have to be faced. Mass production in a highly competitive environment with low margins such as the automotive industry would, for one, require new manufacturing mechanisms. High-volume production would also need further investment in various equipment – meanwhile more time might go by while this option was explored.

2. *Introduce a more aggressive management style and seek venture capital financing to expand into a medium-sized company more rapidly.*
   This option could advance Sentron's manufacturing capabilities and improve competitiveness more immediately. However, penetration of the automotive market would still be a challenge. Several licenses as well as ISO certifications might be required for various technologies. In the short term, the automotive market might still not be directly accessible. On the other hand, success would allow the company to achieve a main player position with very large profits, and once automotive took off, it would need to deliver swiftly.
3. *Look for a strategic investment from a large partner already established in the automotive industry.*
   An investment from a larger partner would lead to new manufacturing solutions and Sentron would be able to capitalize on the partner's established marketing channels. Cost effectiveness could be improved through internal service exchanges. However, negotiations could be difficult and time consuming – possibly requiring the disclosure of sensitive information in the process. Even if the company were successful, it would have to surrender some of its equity and thus its decision-making power.
4. *Make a full trade sale, i.e., sell the entire company to a strategic investor, at its current stage.*
   From the shareholders' perspective, this option would allow them to cash in on resources that had been acquired and maintained over the last ten years. Depending on the investors' interest and the final contract details, it might even be possible to start a new business. On the other hand, Sentron's valuation could be higher in the future. Moreover, selling the intellectual property might close quite a few avenues that could be profitably investigated.

6 Decision Looming

Despite careful consideration of all the facts, there was still no obvious direction.

It was never just a walk in the park. There have been many risky decisions along our path before.

Exhibit 1
The Hall Effect and the Hall Sensor Principle

The effect was named after physician Edwin Hall who discovered it in 1879: if an electric current flows through a conductor or semiconductor that is placed within a magnetic field with a magnetic flow at a right angle, this field exerts a force that results in a measurable voltage (Hall voltage) perpendicular to both directions of the electric current and the magnetic flow (*see figure below*).

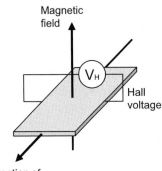

Because the output voltage VH responds proportionally to the magnetic flux density through the Hall plate, it is an ideal sensor for magnetic fields.

The concept can also be generalized for other parameters such as current, temperature, pressure, and the like. In such cases, an input interface transforms the corresponding physical quantities into a magnetic system to which the Hall sensor responds. The output interface can convert the outgoing voltage into any applicable signal required by the final application.

The first practical application of Hall sensors was in the 1950s when they were used as microwave power sensors. With the increasing technical ability that allowed the Hall effect device and associated electronics to be combined within a single integrated circuit (mass production of semiconductors), it became feasible to use the Hall effect in high volume products.

Sentron's IMC-Hall® Sensor[6]

Sentron's Hall ASICs[7] were a combination of Hall elements and appropriate interface electronics integrated with a magnetic flux concentrator (thus IMC – integrated magnetic concentrator) on a chip. Consequently the chips were sensitive to

---

[6] For a more comprehensive overview of the IMC-Hall technology refer to Popovic/Fahrni 2003 and Popovic/Fahrni/Stuck 2003.

[7] ASIC: application-specific integrated circuit.

Exhibit 1 *(continued)*

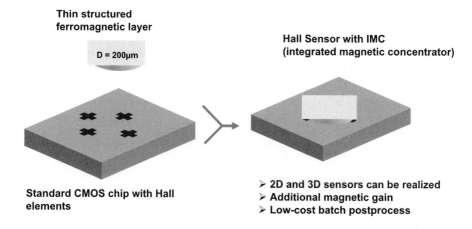

Thin structured ferromagnetic layer

D = 200μm

Hall Sensor with IMC (integrated magnetic concentrator)

Standard CMOS chip with Hall elements

➢ 2D and 3D sensors can be realized
➢ Additional magnetic gain
➢ Low-cost batch postprocess

magnetic field components parallel to the chip surface and not, like the conventional Hall magnetic sensors, perpendicular to the device surface. This allowed unique high-performances sensors to be manufactured for applications that required very high sensitivity (position/current/compass) or for angular sensors (contactless potentiometer/brushless motor sensor).

Advantages of Sentron's IMC-Hall® Sensor

The success of the IMC-Hall® technology is mainly related to the following three factors [8]:

a. IMC technology allows for a dramatic improvement of the performance of integrated Hall magnetic sensors: an increase in sensitivity up to a factor of 10 and the unique possibility to design low-cost angular position sensors.
b. The concept of IMC technology corresponds entirely to the paradigm "design for manufacturability": it allows the use of 95% conventional processing and only about 5% new technology. Moreover, this new technology block fits very well into a natural interface between the conventional processes.
c. The manufacturing process of IMC-Hall® ASICs is consequently organized through cooperation with other specialized companies.

Factor (a) makes the new product technologically very attractive for many applications, factor (b) makes the product affordable, and factor (c) is the key to avoiding big initial investments, which makes production of this magnetic sensor microsystem affordable even for small companies.

Exhibit 2
Typical Automotive Applications that Could Use IMC-Hall Sensors

Typical applications in the automotive industry that could build on IMC-Hall®
sensors are throttle position sensors, gas/brake pedal positioning, and window/mirror positioning. Additional possibilities are shown below.

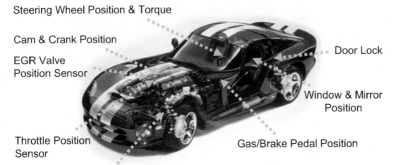

Steering Wheel Position & Torque

Cam & Crank Position

EGR Valve
Position Sensor

Door Lock

Window & Mirror
Position

Throttle Position
Sensor

Gas/Brake Pedal Position

Ride-Height Position

Source: Schott, Christian. Sentron: Magnetic Measurement is our Business. Company Presentation, IMD, January 2003.

Exhibit 3
Sentron's Patent Portfolio

Year	Patent number Additional patent #s	Patent title
1994	CH683577	Integrated resistance of semiconductor material
1996	EP0746100	Circuit for operating a resistive, capacitive, or inductive sensor
1997	EP0772046	Magnetic field probe and current and/or energy probe
1999	US5942895	Magnetic field sensor and current and/or energy sensor
2000	EP1045461	Method of making Hall-effect devices
2001	US6184679	Magnetic field sensor comprising two Hall elements
	EP0947846/US6278271	Magnetic field sensor
2002	EP1182461/US6545462/ US2002021124/JP2002071381/ CA2355682	Sensor for the detection of a magnetic field
	EP1243897/US6731108/ JP20023228046/US2002167306	Device with a magnetic position encoder
	EP1260825/WO02097463/ US2004232913	Magnetic field probe
	US2002167306	Device with a magnetic position encoder
	EP1243898	Magnetic position sensor
2003	JP2003329749	Magnetic sensor and current sensor
	JP2003302428	Substrate mounting type current sensor and current measuring method
	JP2003294818	Magnetometric sensor and method of manufacturing the same
	JP2003262650	Current sensor
	EP1498697 WO03081182/AU2003221189	Angle determining apparatus and angle determining system
2004	EP1395844	Magnetic field sensor
	WO2004025742	Magnetic field sensor comprising a hall element
	WO2004013645	Magnetic field sensor and method for operating said magnetic filed sensor
	EP1443332/WO03038452/ US2005030018	Current sensor and current sensor manufacturing method

Exhibit 4
Revenue Prospects and Targets Through 2007

Market segments		Target market share (%)			Target revenue by segment (mil CHF)		
Business	**size CHF**	**2002**	**2004**	**2007**	**2002**	**2004**	**2007**
Scientific	50M	1.0%	1.6%	2.4%	550 K	800 K	1.2 M
Industrial	100M	0.5%	1.5%	3.0%	550 K	1.5 M	3 M
Automotive	500M+	0%	0.3%	1.2%	0	1.5 M	6 M
Total	**650M+**	–	–	–	**1.1 M**	**3.8 M**	**>10 M**

Source: Company literature

References

1   Prof. R. Popovic is the author of the widely acclaimed book Hall Effect Devices, 2nd ed., Bristol: Institute of Physics, 2004.
2   Popovic, Dragana R. and Fritz Fahrni. "Launching the first mass product of a high-tech start-up company," IEEE International Engineering Management Conference, Proceedings of IEMC 04, Singapore, October 2004.
3   Popovic, Dragana R. and Robert Racz. "Advanced Magnetic Sensors," invited paper, Research Day Nav, Proceedings of the conference, Lausanne, Switzerland, March 2004.
4   Popovic/Fahrni 2004.
5   Schott, Christian. Sentron: Magnetic Measurement is our Business. Company Presentation, IMD, January 2003.
6   Popovic, Dragana R. and Fritz Fahrni. "Modelling the development of a high-tech start-up company," International Conference on Management of Engineering and Technology, PICMET 03, Proceedings of PICMET 03, Portland, USA, July 2003.
7   Popovic, Dragana R., Fritz Fahrni and Alexander Stuck. "Minimizing investments in production of a sensor micro system," The 8th International Conference on the Commercialisation of Micro and Nano Systems, Proceedings of COMS 03, Amsterdam, September 2003.
8   Popovic/Fahrni/Stuck 2003.

CASE 5-3
Sentron at the Crossroads (B)

Ralf W. Seifert, Christopher L. Tucci, and Jana Thiel

In the next couple of weeks, events evolved rapidly. Melexis – a leading integrated semiconductor device manufacturer for the automotive industry, based in Belgium – approached Sentron with an attractive offer to purchase the company for an undisclosed amount. Sentron's management needed to evaluate the company's position and decide on its next steps.

Prior to the offer, the companies had partnered with each other in R&D projects. Furthermore, both of them relied on the same ASIC [1] manufacturer, Xfab – a fact that would ease the integration of Sentron's delivery process into Melexis' operations. Sentron's progress in the field of magnetic sensing technology was an excellent complement to Melexis' manufacturing and automotive expertise.

Melexis was a fully qualified automotive supplier, well established in applications such as position and speed sensors, engine timing management sensors, and electrical DC motor drivers. It served a wide customer base of original equipment manufacturers (OEMs) in Europe, North America, and Japan. Sentron's IMC technology offered many opportunities to extend Melexis' product line regarding both contactless position and current sensors for the automotive environment and a wide range of industrial applications.

Copyright © 2005 by the alliance for technology-based enterprise (IMD, EPFL and ETH Zurich). Copyright permissions are handled by IMD, Lausanne, Switzerland.

1 The Proposed Merger

Having considered all the available information, Sentron's management eventually decided to sell. As of February 2004, Sentron became a fully owned subsidiary of Melexis, whose CEO Rudi De Winter commented:

> This acquisition will allow Melexis to offer the automotive industry major advantages in using Melexis Hall sensor technology in their products. It will give Melexis a considerable leading edge over its competitors [2].

Whereas Melexis covered the automotive application market, Sentron maintained its magnetic sensor development and the associated activities and engineering services in respect of the industrial market. The industrial products and services continued to be available directly from Sentron AG or through its exclusive distributors: GMW Associates in North America and CEFRA Spa in Italy.

Prior to this development, Sentron had decided to discontinue the manufacture of magnetic measurement instruments in order to concentrate on Hall sensor components and modules. The analog teslameters (transducers) and digital teslameters business line was taken over by the spinoff SENIS, which became operational in Zug, Switzerland, at the end of March 2004.

2 Afterthoughts

Within a few months, Melexis successfully launched a group of new automotive products, building on Sentron's Hall technology. To accomplish this task on its own, Sentron would have needed, among other things, to pass a time- and resource-consuming supplier qualification process.

> We never worried about the fact that our products did not conform to the automotive market's requirements. However, to qualify for this market takes a long time. We just saw it as a resource problem, not a political one.

Due to its size, Melexis had more flexibility in responding to price pressure. It was able to source much greater volumes than Sentron could have and thus benefited from discounts which, in turn, made lower product prices possible.

> Building up all the capabilities – from qualification to distribution – to successfully compete in the automotive market might not have ultimately increased Sentron's value in a trade sale down the road, but might have consumed far more time and investment.

By April 2005, Sentron's office at EPFL had been merged with the Melexis branch office located in Bevaix, Switzerland, while the office in Zug remained.

References

1   ASIC is short for application-specific integrated circuit, a chip specifically designed for
     a particular application.
2   Melexis press release, February 4, 2004:
     http://www.melexis.com/NewsDetail.aspx?nID=301.

CASE 5-4
4M Technologies and the Optical Disk Revolutions

Benoît Leleux and Lisa (Mwezi) Schüpbach

In early October 1999 Multi Media Masters and Machinery (also known as 4M Technologies) and its CEO and founder Adel Michael faced a life-defining decision. Despite strong growth and a pristine reputation as one of the leading producers of optical disk integrated manufacturing systems in the world, 4M Technologies was still smaller than some of its key global competitors and heavily burdened by the debt it had taken on to establish its position in this high-technology market.

Effectively launched in 1991, the company had experienced massive growth over the years, going from no product to an expected turnover of CHF 230 million in 1999 (around US$ 140 million, a 158% increase over 1998 figures) and a net benefit of some CHF 10 million, following on a loss of 9.7 million in 1998 (*see Exhibit 1 to view the company's consolidated income statements*). To finance such aggressive growth, the company had relied mostly on bank financing and an equally aggressive use of supplier credit and customer prepayments, permitted only by the quality of its business relationships with those parties. But these arrangements had run their course and it was now time for the company to put its house in order financially to tackle the operational challenges ahead. The demand for its products, in particular for CD-recordable and DVD manufacturing systems, remained extremely strong, but keeping up with such fast growth required new financial resources. And it was clearly not the time for 4M Technologies to let competitors steal a march on its path to the future. Founder Adel Michael was not about to let that happen. He was actively considering an IPO (initial public offering) for his company to provide the strong equity capital base on which to continue growing the business and challenge the more established players.

Copyright © 2000 by IMD – International Institute for Management Development, Lausanne, Switzerland. Not to be used or reproduced without written permission directly from IMD.

1 Company Background

Adel Michael, a Swiss national born in Cairo, was only 35 years old when he founded the company in 1988 in Yverdon-les-Bains, Switzerland, on the shores of the Lake of Neuchâtel. With an engineering background, Michael had worked as a technical consultant after having spent several years in the machinery tools sector in southern Switzerland (Ticino). Impassioned by the evolution in the world of electronics and data storage, he quickly understood that the CD and other optical devices were the wave of the future and the machines needed to produce them would represent a niche that had to be occupied quickly. The demand for optical disk manufacturing solutions was growing faster than the big firms could meet. 4M Technologies was launched to capitalize on that window of opportunity and hopefully gain enough of a foothold to become a significant player in this high-end market.

After a comprehensive study of the market and contacts with firms like Sony and Philips to evaluate the market requirements, Michael diagnosed the most pressing shortcomings of the first-generation CD production lines – too much expensive labor involvement. His first distinctive contribution to the industry, and claim to fame, was the introduction of automated systems of production.

In 1991, 4M Technologies delivered its first inline solution for prerecorded media and followed this in 1994 with the first of what has become a success-ful CD-R (CD-Recordable) product range. The year 1998 saw 4M Technologies acquire the CD/DVD unit of Leybold Systems. The acquisition transformed 4M Technologies into one of the top players in the optical media equipment market by sales revenue (*see Exhibit 2 for various applications in the digital world*). Michael explained:

> This acquisition doubled the size of 4M Technologies' R&D staff, which dramatically reduced the company's product development cycle. 4M Technologies' equipment pro-duction capacity has been substantially increased, with a faster reaction time to the mar-ket's capacity growth requirements, and the company's sales and service network has been extended to a truly worldwide reach.

With over 200 million CD-ROM desktop drives and 600 million CD-audio players installed worldwide and a continuous growth, optical media was the univer-sal conduit for information exchanges. Industrywide sales of CD/DVD-Recordable media had grown by 150% over the past year, with continued strong growth pre-dicted by industry analysts.[1]

To cover the entire optical disk market and meet the demands and needs of an ever-growing global customer base, 4M Technologies had acted to aggressively expand its operations globally by creating subsidiaries and liaison offices and appointing a comprehensive network of agents and representatives. Specifically, newly expanded sales and service centers were located in Santa Clara (USA), Taipei (Taiwan), Hong Kong, and Aprila (Italy). These markets, particularly active

---

[1] 4M Technologies. Press Release. May 1999.

and important for 4M Technologies, were served by local service engineers, extensive local spare parts suppliers, and professional sales staff.

Locating the 4M Technologies headquarters in Yverdon-les-Bains had not been an obvious choice. A number of factors intervened. For one, the development and manufacture of high-technology optical disk manufacturing systems required superior R&D and microengineering skills – skills that were in relatively large supply in the region because of the reputed engineering schools in Neuchâtel and Lausanne. Next, the apprenticeship system was strongly imbedded in the Swiss educational system, and it offered the ability to efficiently train specialist workforces. Finally, Swiss tax regulations were particularly supportive of such endeavors, providing a combination of relatively loose rules for carrying losses forward and low tax rates for years.[2]

2 Optical Storage Media: A Short Introduction

Everything is going digital! Just think of it! Your work, your home, your leisure time are all becoming more and more digital. Multimedia has blurred the borders between radio, television, computers, pictures, movies, games, books, and the Internet. The quality of digital data increases every second, and with the unprecedented access to information provided by the Internet, there is a growing need for digital data storage.[3]

Digital data could essentially be stored on three different storage media: optical, magnetic, and semiconductor-based. But optical storage media possessed a number of advantages over the other two technologies: (1) they offered very high storage capacities of up to 17 gigabytes (Gb) per disk; (2) they were very handy, easily portable and user-friendly; and (3) the medium was relatively durable. They unfortunately also tended to exhibit relatively slow access times, which made them optimally suited for applications such as storing music, photos, films, and software (which did not require speedy access to the data).

Optical disks were often classified into three distinct categories: (1) written (prerecorded); (2) write once (recordable); and (3) write many times (rewritable).[4] The prerecorded category included the ubiquitous CD, or compact disk, which was launched on the market as a recording medium for music as early as 1982 (CD-audio). Despite superior sound quality, smaller footprint, and greater tolerance to wear and tear in comparison with vinyl records and audio cassettes, it was only from 1985 onwards that the CD-audio began to assert itself as the standard medium for recording data in the music industry. In 2000, over 80% of all music recordings worldwide were on CD; the rest was divided up between singles,

---

[2] Investext Broker Reports. 1 March 2000.

[3] 4M Technologies Annual Report 1999.

[4] The digital storage study is from the Swiss Research Report by Bank Julius Bär on 4M Technologies Holding, September 24, 1999.

cassettes, and minidisks (MDs). Since 1987 the CD had also been used as a storage medium for software on CD-ROM, which took off commercially around 1993, again about seven years after the product introduction. CD-video, to store video films, was also introduced but never materially took off and represented less than 5% of total CD production in 1998. It was constrained by its 74-minute recording capacity and relatively poor resolution specifications.

To tackle the capacity needs of high-definition video storage, the DVD (digital versatile disk) format was launched in 1996. Very similar to CD technology, DVD relied on greater compression ratios and the use of both sides of the disk to squeeze more information (up to 17 Gb) onto the medium. DVD recording density was up to 26 times higher than that of a CD. If the DVD format was initially developed for films (DVD-video), rapidly increasing capacity demands in the software industry led to a flowback of the technology into that field as DVD-ROM, a substitute for the saturated CD-ROM.

As of 2000, there were five major DVD versions, each with its own characteristics. Seventy-six percent of all DVDs sold in 1998 belonged to the DVD-5 standard, with data stored on only one side of the disk and a capacity of 4.7 Gb [compared to 650 megabytes (Mb) on a CD]. DVD-10, which used both sides of the disk for storage, doubled the capacity to 9.4 Gb per disk, but had not caught on massively yet, representing less than 10% of the market. DVD-9, despite its lower numbering, was a markedly superior technology, which also used the two sides of the disk but managed to read it with only one laser beam, using a semireflective layer on one side to achieve the feat. The capacity reached a respectable 8.5 Gb, enough to store a 266-minute film in eight languages and 32 subtitles. It represented 13% of the DVD production in 1998. In the wings of the future were DVD-14 and DVD-18 formats, with capacities to reach 17.2 Gb and 13.2 Gb, respectively.

Recordable disks, on the other hand, left the factory without information stored on them, apart from a spiral groove that later served the laser as a guide during recording. The recordable disk was coated with a chemical layer (dye) into which the data were later transferred (burned) during recording. Again, as with prerecorded material, two different formats needed to be distinguished: CD-R (CD-Recordable) and DVD-R (DVD-Recordable). The CD-R was a blank, 650- to 700-Mbcapacity CD that could be written only once. Four hundred forty-five million CD-Rs were sold in 1998, while the DVD-R was still in its early introduction stage. It was based on the DVD-5 standard and would represent the legitimate successor of CD-R, with a capacity of around 4 Gb.

A further development of the recordable disks was the rewritable format. CD-RW (CD-Rewritable) and DVD-RW (DVD-Rewritable) were considerably more expensive to produce than their write-once brethren since a number of metal layers had to be applied to the disk. The actual information layer was an alloy occurring in two phases (crystalline and amorphous). The laser in the player initiated the phase change and thus generated storage or deletion of the information on the disk. Consumer acceptance of these products had been hampered by the lack of a uniform standard for rewritable disks. In 2000, three formats were vying for attention (DVD-RW; DVD+RW; DVD-RAM), none of which was compatible with any of the others. DVD-RAM seemed to be ahead due to the availability of drives.

2.1 The Market for Optical Storage Media

The different formats of optical storage media demonstrated very different growth rates, depending on the stage they had reached in their product life-cycle.[5] (*See Exhibit 3 for expected market growth rates in the various digital media formats.*) Two formats were clearly the growth leaders in the market: DVD and CD-R. The original forecasts for sales of CD-Rs of around 600 million units in 1999 had been achieved by July of 1999, with production expected to exceed 2.5 billion units for the entire year. DVD sales were also exploding, with around 120 million units expected to be sold in 1999.

After exceptional growth in the early 1990s due to the replacement of vinyl records, there had been a gradual reduction in recent years of the growth rates in the market for prerecorded disks to around 5% to 8% per annum. Seventeen years after its launch, the CD-audio market was best described as mature. Growth in the sector was driven primarily by the number of new hit albums, compilations of old titles, and higher penetration of the media in new markets like India, Latin America, and China. According to market estimates by IRMA, CD-audio would reach its maximum in terms of units sold in 2001/2002; thereafter, sales figures would decline slightly.

Competition for CD-audio was expected to come from different angles. After the industry had agreed on common DVD-audio standards in April 1999, all major producers of playback devices and the music publishers announced that they would start production toward the end of the year. DVD-audio would take time to establish itself, though, as most consumers appeared to be quite pleased with the audio quality of CDs and saw little need to upgrade rapidly. A more potent threat to the whole prerecorded music industry was the growth in recordable formats. The advent of the Internet, combined with the increased facility to burn one's own disks, had the music industry scrambling for effective copy-protection mechanisms. The likelihood of this succeeding, however, was dubious. A final development was the success of the MP3 format, a competing technology that relied on massive compression algorithms to make music files suitable for storage on chips or CD-Rs.

In 2000, CD-ROMs were still growing at double-digit rates: 2.8 billion copies sold in 1998, up 28% from the year before. Over the previous five years, the average annual growth rate in CD-ROMs had topped at 95%. It was expected, though,

---

[5] Swiss Research Report by Bank Julius Bär on 4M Technologies Holding, September 24, 1999.

that over the coming years, CD-ROMs would be cannibalized and slowly replaced by DVD formats due to the natural growing need for higher storage capacity.

The 1999 Julius Bär Research Report counted some 470 producers of CD-audio disks around the world. In addition to the major recording studios (Bertelsmann, Polygram, EMI, Warner, Sony), the market was characterized by a number of smaller independent labels. Another 220 CD-ROM producers were highlighted. To a large extent, most of the major software manufacturers, such as Microsoft and Lotus, had outsourced their production to a number of independent suppliers such as Nimbus and MPO that undertook actual disk production.

The market for CD replication machines was estimated at around 750 units per year, with a growth rate of around 10% per year. The gap between the growth rate in machine sales and actual number of CDs and DVDs produced was due to the increasing productivity of the replication machines. Whereas the cycle time (the time required to produce one disk) five years before had been above six seconds, it had fallen to less than three seconds for the latest generation of machines. The number of manufacturers of CD replication machines was still around 20, but a rapid consolidation was expected.

3 4M Technologies: Integrated Manufacturing Solutions for the Digital Age

4M Technologies led the market in the production of integrated manufacturing solutions for all CD and DVD optical disk formats: prerecorded (audio, video, and ROM), recordable (CD-R and DVD-R), and rewritable (CD-RW, DVD-RAM, DVD+RW, and DVD-RW). 4M Technologies' customers then produced the optical disks that served as platforms of choice for the development, storage, and distribution of much of the digital content that originated in existing and emerging professional, commercial, and consumer applications.

3.1 A Strong Performance-based Culture...

Highly unusual for the Swiss market, 4M Technologies had built a very strong performance-based culture, a critical achievement for its founder. This performance orientation was exemplified by a number of attributes. For example, employee compensation, incentives, and bonuses were based on attaining preset personal and group goals. Similarly, management compensation was directly and materially tied to corporate and divisional results. Stock option plans for all employees strengthened their sense of ownership. These were employee-owners, with a strong sense of pride in their jobs and achievements and direct incentives to deliver performance for the company as a whole.

3.2 ...and Some Strong Results

With such a strong incentive base, it was not surprising that the company had shown some astounding numbers of its own. As presented in *Exhibit 4*, sales growth had been spectacular. The financials had also been stellar, with the exception of the 1998 numbers, which had been mainly lowered by the crisis in Asia, where a majority of 4M Technologies' clients were located. Going forward, the vision was summarized by Michael in 4M Technologies' annual report:

> To be one of the world's top three leading manufacturing solution providers in all profitable segments of the data storage media industry where our core competencies apply.

3.3 Global Presence and Customer Focus

Most of 4M Technologies' business was conducted in the Asia-Pacific region (65% of sales in 1998), but it capitalized on the escalating demand and acceptance of optical disk technology in high-growth markets such as China, India, and South America to diversify its customer base and more evenly distribute its business. Recent product announcements from leading Japanese consumer electronic companies regarding DVD-RW-based video recorders had generated much interest in the possible replacement of video tape by new high-performance versatile systems that could seamlessly interface with PCs and the Internet. Their sales and marketing network provided a global coverage with an installed base of over 650 machines in 40 countries, thus providing the company the infrastructure for its internationalization and the risk-lowering distribution of its primary markets.

At the same time, 4M Technologies was increasing its market share in the European (27% of 1998 sales) and American markets (8% of sales in 1998) by allocating increased resources, regional marketing, and advertising campaigns, as well as through market development strategies. This not only included a form of technology brokering, but also the provision of valuable, consolidated market information and research to the various parties involved. For 4M Technologies this would be the preferred business model of the future, since the company had already realized significant returns from its implementation.

Customer satisfaction was at the top of 4M Technologies' priority list. To support that key objective, a rationalized spare parts and service network was in place. With strong local capabilities in these areas and a track record for superb customer responsiveness, 4M Technologies believed that it would be more effective than its competitors – Singulus and Steag (based in Germany), and the Dutch company Toolex (*see Exhibit 5 for a quick overview of 4M Technologies' major competitors*). To this end, 4M Technologies had initiated an active program of technology transfer and brokering to ensure that its customers got access to the latest technology and manufacturing process know-how. Michael explained:

> Our approach is to engage the customer in a much more broad-based approach. It begins with an in-depth facility and resource planning, comprehensive on- and off-site training, a supply of extensive documentation and guides, 24-hour-a-day technical support, and in many instances, specialized assistance with marketing and business issues. We go the extra mile to support their businesses and ensure their long-term viability because we realize that we can only be successful when our customers are successful.[6]

4M Technologies was involved in the sharing of strategic market information between key players, thereby creating a synergistic process that the company believed better equipped all parties to proceed with greater speed, efficiency, and security. In other words, it saw itself as an intermediary among all involved parties, with the aim of bringing new technology to the market in the field of optical storage media. In addition to disk producers and manufacturers of replication lines, this also included software producers from the PC, games, music, film, and photo-editing industries, producers of computer hardware and related consumer goods, as well as providers in the telecommunications and Internet sectors.

3.4 4M Technologies' Business

4M Technologies' product philosophy was to deliver complete turnkey manufacturing solutions instead of being a system integrator or equipment supplier. The company mastered all the relevant technologies (including the manufacturing process and media design technology) to produce all CD/DVD formats (prerecorded, recordable, and rewritable).

In 1998, 4M Technologies acquired the CD/DVD activities of Leybold. The former subsidiary of the Swiss Oerlikon-Bührle Group had a strong expertise in thin-film vacuum deposition and metallizing technology as well as a long-standing tradition in the optical disk business that was vital to enhancing and strengthening 4M Technologies' leadership. By 2000, 4M Technologies had positioned itself as one of the four largest suppliers in this market.

> We, at 4M, do not consider ourselves system integrators or simply equipment manufacturers. On the contrary, we view ourselves as solutions providers to the global multimedia industry. After all, we provide the machines that make the disks used for the global distribution of music, software, games and video.[7]

In all segments, machines were designed to switch from CD to DVD production in the shortest possible time with a minimum of moving parts, which

---

[6] 4M Technologies. Press Release. May 1999.
[7] Michael, Adel. 4M Technologies.

made them reliable and easy to install, run, and service. The critical key components were manufactured in house, with the spin coating systems in Switzerland and the sputtering systems being made in Germany and assembly modules and subsystems manufactured by subcontractors. (*See Exhibits 6 and 7 for an overview of 4M Technologies' range of products.*)

4M Technologies' business and strategy were intimately intertwined and dependent on the direction and developments in the global multimedia industry. To this end, the company had created strong partnerships and relationships with key players in all major segments and believed that, in doing so, it would maintain a leadership position in the creation of innovative solutions for its customers.

4 The Future

Information traditionally available in voluminous books like timetables, phone books, catalogs, and encyclopedias was increasingly distributed on optical disks or accessible through the Internet. First popularized as audio-CDs, optical disks were the ideal, inexpensive, reliable, and universally standard medium for storing not only music, but also information of any kind.

Optical disks had initially replaced traditional vinyl LPs and singles and were now rapidly supplanting audio and video cassettes, floppy disks, and, to a certain extent, computer hard disks. For 4M Technologies the reasons were obvious:

> Optical disks clearly outperform any other data storage media in some regards. For example, at a price of around US$1 apiece and a data capacity of 650 MB, CDs have a clear cost advantage over floppy or super floppy disks. Optical drives are being rapidly used in new applications categories: by the end of 1998, roughly 250 million CD-ROM drives were installed worldwide. Virtually every computer sold today features a CD-ROM or DVD-ROM drive. The cost of CD–R/RW (rewritable) drives has fallen considerably and has recently led to their incorporation as a standard option in many PCs. This has obviously triggered an enormous demand for CD-R (recordable) disks. Most industry analysts have predicted sustained robust growth for many optical disk formats, and 4M Technologies expects this to translate into strong sales of its core products for many years to come.

The high demand for CD-R manufacturing systems in 1999 due to the aggressive entry of new participants and strategic expansion by key media manufacturers had been expected to lead to a slight overcapacity in installed CD-R manufacturing systems by the end of 1999. 4M Technologies believed that this trend might depress order flow and sales into the early part of 2000. However, Michael remained confident:

> Our most probable scenario for 2000 has been constructed on the basis of an expected 10% to 15% price erosion in the CD-R segment. Despite these reservations we intend to strengthen and even expand our market share in the CD-R segment in 2000 and remain one of the leaders in the industry. The target for 4M Technologies in the new millennium is to continue developing Multi Media Masters' worldwide image as 'The Quality Reference' through outstanding product and service quality, as well as its recognition as a benchmark in the optical media industry.

4.1 Tapping the Stock Markets to Sustain the Growth

To support these ambitious plans, 4M Technologies needed more money than ever, not only to fund the working capital expansions that go hand in hand with growth but also to keep investing in the research and development that would keep the company ahead of the learning curve. A strong equity base would also equip the company with the necessary flexibility to face the inevitable changes and technological discontinuities in the leading-edge markets it served. As media markets evolved, so too would the optical support systems and the manufacturing lines to produce them.

Floating part of the company to the public was seen as the way to move forward and guarantee the future fundraising ability of the business. But many questions remained to be solved, such as where to list, how many shares to sell to the public, and what underwriter to rely on for this initial foray into the public domain. None of these questions was simple, even considering the heady markets for technology IPOs in the fall of 1999.

In particular, Michael had been approached by one of 4M Technologies' bankers, who suggested a listing on the soon-to-open New Market (NM) of the Swiss Stock Exchange in Zurich. This clearly offered some exciting potential but also some serious challenges. First of all, the New Markets for growth equity were still relatively new creatures in Europe (AIM had opened in 1995, followed in the next three years by the Nouveau Marché in Paris, the Neuer Markt in Frankfurt, the NMAX in Amsterdam, the Easdaq in Brussels, and the Nuovo Mercato in Milan, the latest market to join the pan-European Euro-NM trading platform). They were untested, and even though their performance over the period 1997 to 1999 had been astounding, there was no clear indication of their ability to withstand serious market turbulence. Second, the Swiss New Market was, for all practical intent, nonexistent, since it had never listed a Swiss company. It was not clear how investors, analysts, and regulators would react to it. Third, it was not clear either what level of valuations this market would be able to provide, compared to the lofty numbers attached to high-tech companies floated on the Nasdaq. Getting a higher valuation was not only a question of prestige; it also helped build a strong equity base for future growth and development. Liquidity was an issue as well, especially considering how aggressively the company had been using stock option programs. Finally, the company also had to consider where its competitors were listed (two on Neuer Markt and one on the Nasdaq) since it would offer some comfort as to the analysts' level of understanding and familiarity with the industry in general, and specifically where its most important partners were, since the IPO was, after all, a strong public relations signal directed to them.

Michael started to line up both the positive and negative issues with the IPO and the various venues on which he could float the company.

Exhibit 1
4M Technologies' Consolidated Income Statements
(in CHF Thousand)

	Expected year ending December 31, 1999	Year ended December 31, 1998
Gross sales	**238,163**	**92,166**
Sales deductions	(9,023)	(3,018)
Total net sales	**229,140**	**89,148**
Cost of goods sold	(195,663)	(81,922)
Gross profit	**33,477**	**7,226**
Research and development expenses	(8,387)	(1,709)
Marketing and sales expenses	(9,685)	(5,796)
General and administrative expenses	(6,402)	(5,093)
Amortization of goodwill	(533)	(133)
Operating profit/(loss)	**8,470**	**(5,505)**
Financial income/(expense), net	(209)	(4,064)
Profit/(loss)	**8,261**	**(9,569)**
Income tax credit/(charge)	2,259	(97)
Net profit/(loss)	**10,520**	**(9,666)**
Basic and diluted earnings/(loss) per share (in CHF)	**30**	**(806)**

Source: 4M Technologies

Exhibit 2
Applications in the Digital World

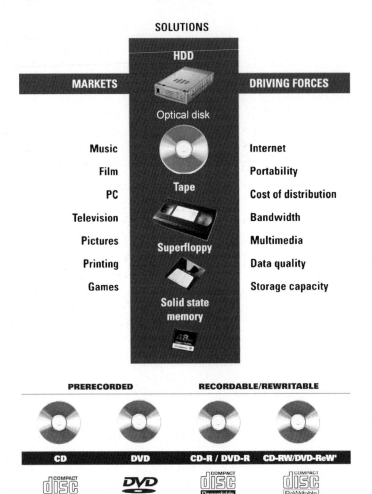

Source: 4M Technologies

Exhibit 3a
The Market for Digital Disks (in Billion Disks)

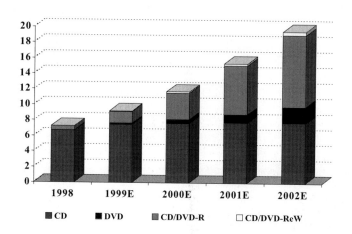

Exhibit 3b
CD-R/CD-RW Media Demand Forecasts (in Billion Disks)

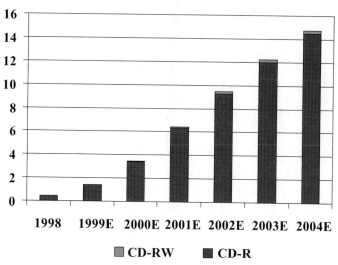

Source: 4M Technologies estimates

Exhibit 4a
4M Technologies Sales Performance: Past and Future Sales (CHF Million)

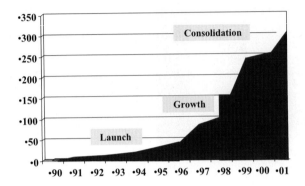

Exhibit 4b
4M Technologies Earnings and Cash Flows (1999 Figures Expected)

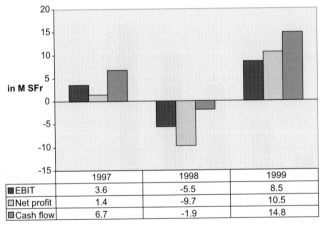

	1997	1998	1999
■ EBIT	3.6	-5.5	8.5
□ Net profit	1.4	-9.7	10.5
■ Cash flow	6.7	-1.9	14.8

Source: 4M Technologies estimates

Exhibit 5
4M Technologies Competitor Overview

Company	Founded	Offices	Employment	Specialization	Financials
Singulus	1996	*Headquarters:* Singulus Technologies AG, Alzenau, Germany *Other offices:* Singulus Technologies Inc. Windsor, CT – Jan. 1st 1996 West Coast office in Pleasanton, CA – May 1996 *Subsidiaries:* Singapore for the Asia-Pacific region, Watchfield, Swindon, UK, Sao Paulo, SP, Brazil for Latin America *Representatives:* Australia, Benelux, France, Hong Kong, India, Japan, Korea, Malaysia, Spain, Portugal, Taiwan, Thailand www.singulus.com	Employs more than 140 people – 25% are located in own subsidiaries	CD/CD-ROM and DVD metallization and replication	Forecasted consolidated sales for 1999 have been announced at DM 340 million. Revenues for Q1/2000 are expected to again exhibit strong growth over the previous year by at least 50%. This growth can be attributed primarily to the strong demand for DVD replication lines, expected to more than double in 2000. 1999 distribution by region: Europe 22.7%; North and South America 10.9%; Asia 66.4%. Singulus trades on the New Market (Neuer Markt), a market segment of the Frankfurter Stock Exchange (Deutsche Börse Frankfurt).

Source: Singulus annual report

Exhibit 5 (continued)

Company	Founded	Offices	Employment	Specialization	Financials
Steag	1958	*Headquarters:* Steag Electronic Systems AG, Essen, Germany *Subsidiary:* Steag HamaTech AG, Sternenfels Germany (production plant for CDs and DVDs) www.steag.de	30% of group's 4,000 employees	Combined in divisions Power Generation in Germany, IPP International (power generation abroad), and Electronic Systems (manufacture of equipment for the production of semiconductors and optical data carriers, and microsystems technology).	20% of the total group sales of DM 2.5 billion. Subsidiary floated on stock exchange in May 1999. In October, turnover for 1999 was forecast at DM 480 million, while net profits were targeted at DM 37.7 million. Multiples comparison Hamatech/Singulus: Share price 23.10 43.25 P/E 2000E 25.4 25.6 P/E 2001E 19.4 21.1 CAGR EPS '99 – 01E* 24.3% 29.0% EV/EBIT (2000E) 11.7 12.3 (*'99 figure adjusted for IPO)

Source: Investext Broker Reports

Exhibit 5 *(continued)*

Company	Founded	Offices	Employment	Specialization	Financials
Toolex		*Headquarters:* Toolex International N.V., The Netherlands Manufacturing operations in the Netherlands, Sweden, Germany, and the USA *www.toolex.com*	1999 employees: 453 1-year employee growth: 381.9%	Optical disk manufacturing equipment	Nasdaq symbol: TLXAF Mastering systems and systems for duplicating prerecorded disks account for more than two thirds of sales, and most sales are in Europe and Asia. 1999 sales (expected) ($ million): 232.5 1-year sales growth 18.3% 1999 net inc. (expected) (million): 20.6 1-year net inc. growth 57.3% For the first quarter of 2000, Toolex expected gross revenues of close to €50 million, a 50+% increase over the same period in 1999.

Source: Toolex Annual Report and Dow Jones International News

Exhibit 5 (continued)

Expected earnings data by company (1999)

	Singulus	Steag	Toolex	4M
Revenues (in millions of €)	166.3	264.8	231.8	141.3
Gross margin (%)	39.0	28.5	36.0	15.9
EBITDA margin (%)	24.4	20.1	14.7	9.2
P/E (1999) (x)	69.3	55.0	21.8	55.1
EV/EBITDA (x)	33.7	23.5	13.8	34.8

Source: Investext broker reports

Exhibit 6
4M Technologies Sample Product Range

MCL 50 range	MCL 200 range	MCL 800 DVD-ReW	MCL 600 DVD-R	MCL 500 CD-R
4M Technologies' latest solution for the manufacture of both prerecorded CD and DVD formats and based on 4M Technologies' new proprietary PHOENIX Sputtering platform and its EXPERT® cathode. It has been designed to maximize productivity and minimize total cost of ownership. It combines a small footprint of a just 1.3 m² with high reliability and a down-stream cycle time of 2.8 seconds and is one of the fastest systems on the market. The MCL 50 D is based on the same platform but is capable of replicating either CD or DVD5 formats.	This family product comes in a range of flexible configurations for all prerecorded CD and DVD formats. Since injection molding has traditionally been the "rate-determining" step in the disk production process, this high-speed twin CD replication line with two injection molders is designed to offer customers the highest productivity and can easily be converted between DVD-5, 10, and 9. This product is configured with a fully integrated bonding unit.	Designed initially according to the most advanced manufacturing process requirements of DVD-RAM, this system assures smooth and efficient production of any DVD re-writable format such as DVD-RAM, DVD+RW, and DVD-RW.	A flexible architecture suitable for CD-R and DVD-R production, with many advanced features, including modular design, and centralized control. It assures maximum flexibility for future manufacturing process improvements and has sophisticated and advanced disk handling and conditioning capabilities. Specific care has been taken to ensure identical thermal histories for all parts. This is critically important for the higher capacity DVD-R format due to much more restrictive tolerances on finished disk quality and shape.	This new manufacturing system is specifically designed for cost-effective, dedicated CD-R manufacturing. Based on 4M Technologies' many years of experience in the CD-R industry and with the latest technology employed in all manufacturing modules, the system is designed to save space and capital, guaranteeing low total cost of ownership. The MCL 500 is equipped with 4M proprietary TORNADO® high precision spin coater.

Source: 4M Technologies

Exhibit 7
Digital Media Formats: Technological Bases

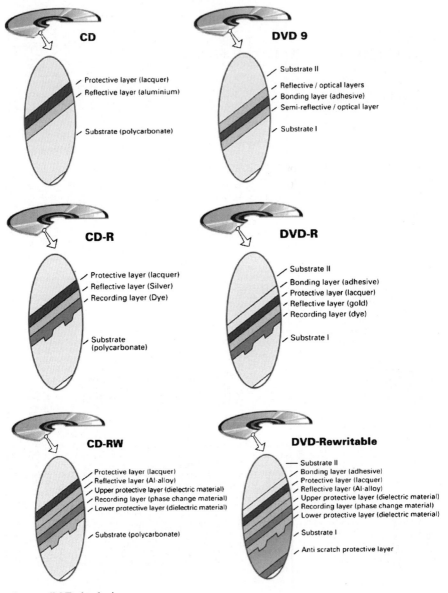

Source: 4M Technologies

CASE 5-5
Generics

Benoît Leleux, Georges Haour, and Laurent Piguet

Just outside Cambridge, UK, in the lush countryside, Gordon Edge (*Exhibit 1*) entered the glass-fronted building of The Generics Group.[1] Moving swiftly past the greenery framing the door, he waved hello to the people in the main reception area and climbed the stairs into the uniformly furnished open-plan office area. There, as he surveyed the busy staff, he realized what progress had been made.

With a staff of nearly 250, the company was entering a new phase, which in terms of the demands on its management would not be trivial. In particular, the company required a much-needed cash infusion that would allow it to foster and grow its own business, as well as to continue financing the spinoffs it had created over the years and keep generating innovations to fill its intellectual property portfolio. But the terms of this cash infusion threatened the specific cultural conditions for generating innovation – the core of the company's business – that Edge had deliberately put in place when he created the company.

Figure 1 A View of "The Mill" This renovated old mill forms part of Generics' facilities in Cambridge, UK, and houses the incubation activities.

Copyright © 2002 by IMD – International Institute for Management Development, Lausanne, Switzerland. Not to be used or reproduced without written permission directly from IMD.

[1] Generics changed its name to Sagentia in February 2007.

1 The Genesis of Generics

Edge shaped Generics following the model of a few other companies he had started: Cambridge Consultants in 1964 and PA Technology in 1969. PA Technology, a trust with a fee-driven business model, had a compound annual growth rate (CAGR) of 28%, always turned a profit in the 16 years of its existence, and attained revenue of £28 million in 1986. Like Cambridge Consultants, it was a technology consultancy and advisory company offering interdisciplinary scientific resources in a wide range of domains, from life sciences to electronics, communications, advanced materials, and optics. Its customers were Fortune 500 companies looking for very innovative research ideas to address specific problems. The resources provided by both consulting companies made it worthwhile for customers to outsource some of the research under contract rather than developing it internally.

To provide a high quality of service, Cambridge Consultants and PA Technology both relied on integrating a broad base of technology skills with business savvy – in essence, helping people make money with technology. Both were also highly entrepreneurial in the way they conducted their operations, focusing early on innovation supported by specific processes. This innovation soon gave rise to emulation, and over the years, hundreds of people left PA Technology and Cambridge Consultants to set up their own businesses, creating a long history of innovation and entrepreneurship in the region that came to be known as "the Cambridge Phenomenon."

But Edge did not stop there. He believed there should be a better way to capture the value of innovation in these companies. In 1983 he proposed creating an equity vehicle within PA Technology to fund spinoffs and improve the return on "innovation investments." After repeated rejections by the board, Gordon Edge left and established Generics in 1986.

The original business model was remarkably simple. (1) Set up a consulting business first, populated with great scientific minds from many disciplines, and bring it to profitability. (2) Use the cash flow from this now profitable entity to invest in intellectual property. (3) When enough cash becomes available, start creating spinoffs in which Generics maintains an equity stake. (4) Dispose of those investments when they start turning a profit a few years down the road by selling to an eager buyer or floating on the stock exchange. Gordon likened this process to a goose laying eggs:

> We are not a strategic business: we don't try to create a strategy in life sciences and say that's where we are going. Our strategy is to create a process, and that process I call the "goose." Our job is to feed the goose, keep the blood supply going, keep it healthy – and it lays eggs. And sometimes it lays goose eggs, duck eggs, whatever eggs, golden eggs now and again. It's the best way to think about it. It's the nature of this business.

Simon Davey, sales and marketing director and previously director of the innovation consulting division at Generics, added:

> The underlying business is about the combination of technologies to create opportunity. So, the strategic context is in the process that generates this combination, not the actual components of the combination. It doesn't matter which sector the innovation is in, what matters is the quality of the innovation.

2 A Culture of Innovation

To provide revolutionary innovation (and – it hoped – value) to its customers, Generics tried to "systemtize" innovation generation. Unlike most strategic goals, innovation and the capacity to find new solutions to problems require something else besides organizational structures. For Edge, all the factors required to drive innovation were cultural.

2.1 Interdisciplinary Working

Edge noted:

> Interdisciplinary working is one of the most fundamental drivers of innovation. It means you don't recognize any barrier between physics, chemistry, biology, life sciences, whatever. You encourage people to work across boundaries, to read each other's discipline, to talk, to discuss, to meet and think about new opportunities. And a vast amount of innovation comes from that space. It's just that most companies don't occupy it.

Davey commented that this was not observed in most companies, where a great divide remained between the centers of expertise in technology and business.

> Often you'll have a research organization here, and over there you'll have a business unit that will be taking care of the day-to-day business. And the linkage between those two is essentially some business process. Typically, the business unit will define, through a strategic process, a set of business objectives for the firm, which will be translated into a technology strategy, some sort of R&D portfolio definition and, finally, some kind of project definition. That's the usual "connecting" process between technology and business. Generics operates differently: the technology and business expertise are integrated at the level of the individual. So, actually, even the least business-aware person within Generics is going to be an order of magnitude more business-aware – almost – than the most business-aware person in a typical corporate technology organization.

Julian Fox, a Generics employee, liked to use a combination of metaphors to describe the way interdisciplinary work creates innovation:

> It is almost like a Brownian motion phenomenon. The molecules collide and there is a statistical uncertainty as to which directions they go next, because what you are looking for

is innovation from the intersection of different fields, experiences and skills. And this is what your pressure-cooker environment gives you. For that model to work, you have to have loosish controls, because you can't actually tell when you are going to get something good, because it is a fairly random process. If you look at some of the new and interesting technologies that are around now, the whole life sciences field is a great mixture of bio and electronics and all sorts of things all mixed in together that have come about through the sort of intersections that we build here.

To demonstrate the effectiveness of this environment, Edge used a historical parallel:

If you look at Florence in the 12th, 13th, 14th century, you will find all the elements of Generics there: you've got rich backers, you've got sculptors, painters, writers, *Section 1* all working in the same sort of ferment and cross-fertilizing each other. This was called the Renaissance, and we try to build that same Renaissance in here. It's not that much different actually in terms of the cultural concepts that make it happen, so there is nothing new really about our model: we are just applying to the scientific community what worked with the artistic community. After all, it's always about human beings being creative in an environment.

2.2 Removing All Barriers to Communication

To promote this work culture, all physical barriers to communication were removed. Everybody at Generics, from the CEO to the secretary, sat in a common, open-plan configuration, in an effort to force everyone to communicate. Edge recalled:

The whole space was designed so that there were no verticals in it. Where safety was critical, like a biology lab, then you made the walls glass, so that people could see through, and this creates a familiarity with the unfamiliar.

Conference rooms were used for team meetings and times when people needed privacy or required a lively discussion without disturbing their colleagues.

2.3 Removing Status Factors

According to Edge:

You can't have status factors interfering with anything that has to do with science and innovation, because status is a completely artificial term and it inhibits people's willingness to talk to each other.

Much like in young Silicon Valley startups, the management structure at Generics was always designed to be flat, in order to be as lean as possible and do away with any rigid framework that could jeopardize the spirit of innovation.

A management hierarchy was present, but it was thought of as a sort of meritocracy of skills – people took and left positions of influence based on peers' recognition of their ability to contribute significantly to different processes and on their specific area of expertise. This process stimulated the emergence of recognized leaders in each area of business – centered on the charismatic personality of Edge – and promoted the conception and evolution of a diverse set of innovative ideas being championed by people with very different backgrounds and skill sets.

2.4 Taking Risks

Edge noted:

> The top management of the business has to be risk taking, people have got to see that I take risks, not just financial risks, but risk with what I say.

This was a way to generate new ideas, revolutionary concepts, to show employees the way and stimulate them to speak out, make challenging statements, and not worry about their image. In the end, Edge concluded, the purpose was to create

> … a shared set of beliefs, so everybody believes this is the right way to behave, by example and by doing. So, if you add all these things together, you get a more innovative organization than if you don't put them all together. And once you start to lose any element of this, the organization slowly slips away into a reduced innovative output.

3 Consulting Activities: Creating Value Through Intellectual Property

Edge liked to call Generics a "technical consulting and advisory" company because of the proactive role it took in bringing opportunities, or ideas for solutions, to customers they thought could benefit:

> For example, we had [a company] here today, a very diversified family business. Some of their top people came to Generics to discuss very specific things we had taken to them, which will convert into consulting jobs plus relationship-type projects. One of the things we are trying to get them to look at is a new way to extract titanium. This is something that gets me very excited because it's a very traditional business that has been around for 50 or 100 years, but when you get a technology that allows you to extract titanium and manufacture it for the same price as aluminum, it means you can introduce structural changes in an industry. The car industry could start using titanium. If the car industry uses titanium, it means you get lighter cars, stronger cars, environmentally more friendly [cars]. Now that's innovation!

The most difficult part of the "sales" process was to get the customer sufficiently attracted to an opportunity that he would pay for the development of the solution for the particular problem while allowing Generics to retain the rights of exploitation of the solution in other areas. Generating intellectual property (IP) from customer contracts was no trivial matter. Typically, when a customer came with a problem to solve, he would expect all of the resulting IP rights, along with their exploitation, to belong to him. The strategy for Generics then was: (1) to get the customer company to be interested only in applications relating to its core business, in a narrowly defined market segment or technology application area; and (2) to agree to transfer the rights to other areas back to Generics. For Edge, this made perfect sense.

> We talked to an elevator company; they had a problem about getting elevators to stop within a tenth of a millimeter on a floor. So we looked at the problem and came up with an innovation to solve it. But this was a very fundamental innovation, of which the elevator was just a minor application, but it was sufficiently important for the company to pay us a fee to develop it. They got their solution, and we kept the overall intellectual property. Now, I argue very strongly that we have a right to do that, and the reason we have a right to do that is because the cost of arriving at a point where we can actually solve that problem is huge. The fact is, we've had to invest in labs and people, we've had to take risk in building those facilities and you can't come in and say I'll pay you a thousand pounds a day for your time and the innovation. It just doesn't cover the cost. It covers the day-to-day cost, but it doesn't do anything else. So, we've been quite strong from the beginning in saying: [...] our ability to create intellectual property, of which your application is a subset, is a function of our own investment, our historical investment and our process. That is our asset, and that's what we are selling, and if you want to buy innovation, that's the condition. If they don't like it then we have to either compromise or go our separate ways.

In all cases, a very clear understanding was necessary to avoid any potential conflict with customers over IP rights. Generics could not afford to have any of its customers feel they had been deprived of fundamental IP as a result of awarding a research contract to Generics.

However, the company could certainly justify why customers should have every confidence in it. According to Edge, Generics' specific model for innovation – as well as its ability to expose scientists constantly to challenging problems defined by customers in real-world conditions – was the key to a rate of innovation about three times as high as that obtained at more traditional organizations.[2]

4 Optimizing the Creation of Value

In 1991, five years after it began, Generics spun off its first company. Diomed Inc. was established to develop a new optical technology enabling substantially

---

[2] Using the number of patents filed each year as a metric to measure innovative output.

increased power output from solid-state laser arrays. The technology's main application areas were in surgery for angioplasty[3] and photodynamic therapies.[4]

In addition to facilitating the creation of new companies through spinoffs, the technology expertise of Generics' staff gave it the unique ability to assess the value of technologies developed outside its own walls. To harness this value, in 1987 the company created an investment fund, Generics Asset Management Ltd. (GAML), to invest in external companies, using its internal staff to do the evaluations. The focus of the investments was seed-stage companies, very like the ones being spun off.

4.1 Formalizing the Model

By 1998 the level of activity related to the exploitation of IP had grown significantly. Initially, this process had been managed within the consulting organization, but the increased workload soon led to strain and conflicts. In particular, managers in the consulting organization faced a basic conflict when deciding whether to increase the level of effort on a particular piece of technology: any time spent on the project for later exploitation resulted in a decrease in billable hours and revenue. Edge commented:

> We realized in 1998 that this model really did look as though it could be used to generate substantial value and it did have a net positive impact on our ability to manage the company. Up until that point, most of us would have thought it had a net negative impact. Things like conflict on the marketplace over intellectual property or internal strains over the management of HR would outweigh the benefits. I think by 1998, we were pretty convinced it wasn't the case. So, we took the decision then to formalize the management of the IP exploitation side of the business. Prior to that, there hadn't been a specific management structure addressing that investment arm.

The strategic review formalized a representation of Generics' business model, taking into account the dual nature of its activities, to help present the company to future customers, business partners, and analysts (*Figure 2*). The decision was also taken to create GenTech, a specific structure to help manage innovation exploitation.

---

[3] Repair of the heart's blood vessels.

[4] Selective targeting of tissue using laser light. A substance sensitive to the specific laser frequency is injected into the targeted tissue. Applying laser light then activates the substance to kill the specific tissue.

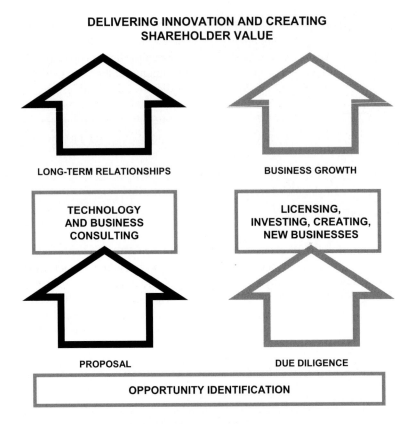

**DELIVERING INNOVATION AND CREATING
SHAREHOLDER VALUE**

Figure 2 Generics' Business Model
On one side, value is created by the constant interaction with clients bringing challenging problems to solve. On the innovation exploitation side, the stimulation and identification of inventions creates an intellectual property portfolio that is then converted to value through equity in spinoffs, licenses, or joint ventures.

4.2 GenTech: Early Internal Funding for Projects

GenTech (GenTech Investment Group AG) was created to provide a framework in which managers could focus on the investment decisions linked to the exploitation of technology specifically. It functioned as an internal investment fund with over $3 million accumulated from consulting activities and was available to fund opportunities arising from the IP stimulation and identification process that required consultants' time and attention. A group of four or five people was created to take over these responsibilities, using resources provided by the consulting organization.

4.3 The IEB: Stimulating Innovation

Core to the incubation model was the Innovation Exploitation Board (IEB) comprising a wide range of Generics' professionals. The IEB provided funding, time, and business support for the best ideas from employees until they were ready to be exploited commercially, whether via a spinoff company or other means. It met monthly to provide peer review of innovative ideas presented by Generics' staff and external sources; when necessary, it also encouraged Generics' people to innovate in specific fields.

Once an idea was approved, the IEB recommended financial backing and staff time, enabling patents to be filed and prototypes to be developed.

4.4 The Generics Value Creation Process

Figure 3 outlines Generics' value creation process in more detail. There were seven major steps:

1. Consultancy and advisory services were sold to clients in a wide range of technology fields. The resulting contracts generated cash and IP for Generics.
2. While part of the IP remained with the client, Generics aggressively pursued other fields of applications for the innovative solutions it generated. The resulting IP was added to the IP portfolio. In addition, some of the proceeds from the consulting activities were used to fund an internal funding mechanism, GenTech.
3. The IEB reviewed ideas proposed by consultants, employees, and external people. It also stimulated people to come up with innovations. Once a concept was deemed potentially interesting, resources were allocated to pursue it. Funds from

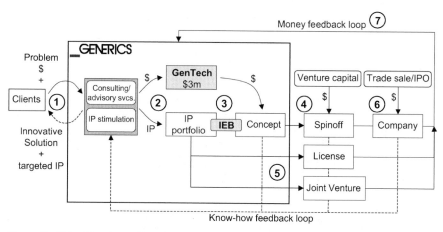

Figure 3 Value Creation at Generics

GenTech were used to write a business plan, develop the technology, build prototypes, evaluate markets, Section 1 This phase was considered the "incubation" phase
for Generics, a period during which a concept was matured within the walls (both
physical and legal) of the parent company.

4. Once the concept had matured enough and the value of the underlying business
   had been established, Generics created a spinoff company, initially retaining
   100% of the equity. Generics employees who had championed the project internally would leave the Group to become the first employees of the company.
   Group and venture capital money was then used to fund the company initially,
   with about 20% of the equity distributed to founding employees.

5. The IEB could also identify alternatives to creating a full spinoff company.
   Licenses or joint ventures could be pursued, requiring investments and generating cash flow either immediately or later.

6. As the spinoff company developed its markets, it would become interesting for
   a buyer or would go to the public markets for an IPO. Typical Generics spinoffs were more likely to be acquisition targets because of their high-technology
   and science focus.

7. Generics realized the value of its equity participation, and the resulting cash
   was returned to the Group to fund further consultancy services and innovation
   exploitation.

(An overview of Generics' finances is given in Exhibit 2.)

4.5 Examples

4.5.1 Absolute Sensors: Exploiting IP Generated by a Consulting Assignment

In 1994 an elevator company approached Generics to solve the problem of precise
positioning of an elevator. To ensure it stopped exactly level with each floor, precise computer control systems requiring precision sensing were used. The challenge arose because of the dirty operating environment, where precise alignments
and positioning could not be guaranteed. In addition, the sensor would be subject
to changing temperatures and, to guarantee a longer life span, had to have no contact with the moving elevator.

Generics developed a technology that became known as Spiral. Fundamentally,
this new sensor design functioned using magnetic induction via an electronic
board that generated alternating current. The current was converted to a magnetic
field that could be transmitted to the sensor electronics on the other side of the
gap, causing it to resonate. This signal was then picked up by the processing electronics and converted to a precise location and angle measurement. The Spiral
technology was very attractive: not only was it noncontact and able to function in
a dirty environment, but it could also be produced cheaply, was highly precise,
and could be used for measurements in two and three dimensions.

After discovering the range of potential applications of this sensing technology and successfully obtaining two licene deals for it, Generics decided to form Absolute Sensors Limited on July 1, 1998. Ian Collins, whose group had developed the technology, David Ely, an engineer specialist in resonance systems, and Malcolm Burwell, an experienced technology consultant, left Generics to become Absolute Sensors' first employees. With equity initially entirely held by Generics, the company eventually obtained £ 487,000 in seed funding from the Group and made its capital available to Group employees, with a portion of shares for the three founding partners.

With applications in diverse markets, such as the automotive, game, and pen input devices – used for example for palm assistants – the company was soon able to raise additional funding from Generics (£ 460,000) and other investors. Finally, Generics sold its 80.3% interest in the company to Synaptics Incorporated for cash and equity having an aggregate value of £ 3.3 million in 1999. By the end of that year, Generics retained a 2.7% interest in Synaptics Inc.

The cycle of creating value from innovation was completed when Synaptics Inc. floated on Nasdaq in January 2002

4.5.2 quantumBEAM: Exploiting Internal Ideas

quantumBEAM started in 1999 as a result of a question from Edge. Generics had been working in broadband network technologies for a number of years. It had done some of the original DSL development for an American company on cable and wireless radio-based broadband. Gordon then realized that a common problem with all broadband technologies was their reliance on the heavily regulated radio spectrum. Consequently, a good solution would be to find a technology that would allow the use of the optical spectrum (which is not regulated). Edge remembered:

> I said to our guys in the optics group: "Look, we need a broadband local loop communications system based on optics, which can compete in cost terms with the other access technologies. So, the terminal in people's houses will have to cost less than a couple hundred dollars. Go away and think about it." And they came back with one of the most fundamental innovations you could have hoped for.

It was a new materials technology involving gallium arsenide, which turned the entire concept of transmission and reception – everything known about conventional networks – on its head. With this technology, the more subscribers that were added, the more bandwidth was available, as opposed to a normal system where more subscribers resulted in less bandwidth. This allowed the systems to become more and more cost effective as they scaled:

> There was a lot of work to do in basic material science to develop the gallium arsenide; it's a vertical, classical vertical integration with the roots bounded in sciences and the top end in networks, and quantumBEAM occupies the entire range from new materials technology, new components into products, systems and networks.

After some initial laboratory development, quantumBEAM was demonstrating high-capacity information transmission by optical means. By the end of 1999 the company had a local loop version on trial in Cambridge and was expected to attract a great deal of interest from venture capitalists (VCs) and customers. The company was established as a spinoff company in June 2000 and received subsequent investments in cash and IP worth £2.4 million.

Applications included broadband communications to and from moving vehicles, local area networks for laptops, PCs, and workstations, and embedded computing such as in domestic appliances as well as local loop broadband applications. (*See Exhibit 3 for more examples.*)

5 Managing a Complex Process

Generics was a complex organization. Providing consulting services, which is cash flow and profit driven, was a different business from creating and growing companies, which is capital gains driven. Managing the two within the same organization created organizational issues management had to deal with. Davey remembered:

> If you look at the way the performance of consulting operations is traditionally measured, essentially it's profits and cash flow. No one is interested in anything else. Clearly, profit and cash flow [are] completely inadequate to measure the performance of the innovation exploitation side of the business, and if you attempt to do a straightforward revenue profit/cash flow calculation on that side of the model, what you get is a situation that is noisy: it doesn't mean anything anymore. Gordon has had a long-standing view that the only way you can measure performance on the exploitation side [is] on a valuation basis. And [a valuation perspective was adopted] for the whole group, including the consulting side.

In addition, once a concept had been identified and resources allocated to develop it, the organization had to be able to follow on with significant commitments in cash and resources, which tended to push toward an early exit, rather than a longer value-building period. Davey continued:

> You've got a privately financed company, with a finite cash flow, essentially generated by the consulting organization, which means whenever you create a piece of intellectual property and you invest in it, you are immediately under pretty serious management constraints, because the thing is burning cash, either as the result of the development requirements, or as the result of being an incubated company. So immediately, the management of the parent company is looking at the cash burn versus the amount of headroom they've got in terms of cash flow and the almost inevitable result of that is a push to exit too early. That certainly happened with Absolute Sensors, there is no question. I was on the board when that exit decision was taken, and part of the thinking behind that was that the burn rate was too hard to sustain, and we needed a deal.

Finally, the Generics model was a mix of two very different decision-making styles: (1) the "charismatic," or "nonlinear," style, which was well adapted to innovation and intuitive research; and (2) the "systematic" style, which was required for management and consulting activities. However, Davey thought the IEB was an interesting instrument to mitigate the two styles and get the best out of everyone:

> The IEB is a genuine peer group structure which allows the charismatic and systematic decision styles to co-exist within a single platform. So the IEB is quite capable of taking an entrepreneurial decision, but it will also quite often apply a systematic style, and it will move between those two on different occasions. And the individuals within the IEB will experience those, so the charismatic guys will be moderated by the systematic guys, and the systematic guys are stimulated by the charismatic guys. And that's actually the principal form in the company in which that takes place.[...] I think that lent a degree of stability to what can otherwise be a quite tense situation and I think it's quite widely recognized in the company.

6 Growth Challenges

With the ever-increasing financing requirements of Generics' portfolio companies, it was necessary to complement the financing capability of the company. In 1996 Catella Holding acquired a controlling interest of 79.5% in Generics. This recapitalization, as well as a loan made available in January 2000, enabled Generics to better exploit the projects in which it owned IP rights. However, with the loan coming due for repayment at the end of 2000, the company would be put under financial pressure and a solution had to be found.

Even though the issue had arisen before, it was now time to consider seriously the possibility of the IPO route for Generics. The business plan had mentioned it, but it had been put aside several times over the past 14 years. It was the end of 1999, and the market for technology IPOs was starting to decline. Many high-profile companies reported very bad results, and all tech markets were showing signs of a slowdown.

The situation was even worse for companies linked to incubation and high technology, which were starting to see their share prices fall heavily after investors realized how difficult it would be for these companies to realize any profit from their mostly Internet-related portfolios in the recent dot-com meltdown. But Generics seemed ready, as Edge recalled:

> There was an accumulation of evidence that we were ready, in terms of the company being able to establish a very long term profit record. In the entire history of the company, 15 years, only one year had we not made a profit, so we could demonstrate profit growth. We could also demonstrate international operations, and we'd also been able to demonstrate that we had been through the entire cycle of identifying a technology, building a business around it, growing that business and selling it. We'd been through that cycle three times already, so we had a unique track record.

But becoming a publicly traded company required a clear focus that would allow analysts and investors to understand what they were buying. Edge believed that the interdisciplinary nature of the company made the Generics model difficult to explain:

> There is no focus. The spread of investments covers the whole spectrum of the business. And that's the nature of the whole model. The fact is that an interdisciplinary organization will by nature produce interdisciplinary concepts, which don't sit comfortably in any

particular focus area. So when we have a business like one of our most recent spin-outs [spinoffs] called Adaptive Screening, what is that? It's life sciences, it's optics, it's electronics, it's computing, it contains IP from Imperial College, from the University of Glasgow and from Generics, all wrapped up together to make up an exciting business. It's a completely new approach to Genomics. And that's the kind of thing we do best. If we try to pretend we are a highly focused organization going into the life sciences sector, we'll trip flat on our face, because it's not how the business works.

Fox put it in a slightly different way:

You have to take this relatively amorphous organization, which you may have sort of packaged up in some form or another for the sake of convenience, and construct something that will enable the investing public to have confidence in it, and want to invest in it, and be able to understand it. Which is difficult.

Also, for a number of years, Edge and the principal shareholder, Catella, had likened Generics to a goose laying eggs. The internal processes were generating innovation and building long-term value out of it, financed by the consultancy and advisory services. Conversely, the founders – reaching retirement age – and the investors – like all investors – needed an exit strategy. Was it time to take a short-term profit and bring the goose to market?

Finally, considering an IPO meant that the company had to find a way to position itself. It was not a pure consulting company, but it was not a pure IP company, or even an investment fund, either. Jeremy Klein thought this was an issue that could prevent it from attaining a good valuation:

We have always found it difficult to represent Generics as a value generating proposition into our market in a way that makes sense. Because we actually do lots of different things. I use business skills, I do reports, presentations, work with senior management, go to committee meetings, Section 1 But there is another division where people sit there and design circuit boards, doing the aesthetics of the casing of a product, and there are other people designing communications networks, Section 1 The difficulty is to find some single strand that somehow connects what I am doing with the person doing a communications network and the person designing a circuit board. What we always say is that we are interested in managing technology, or making money out of technology: engineering people do it through making circuit boards and we sit and talk about the strategic side of doing that. I am just not sure that it really comes together as a powerful selling proposition in the marketplace.

Being quoted on the market would also push the company to focus on whatever aspect of the business the city would find most attractive. In an age of inflated valuation expectations, the innovation exploitation side would be sure to attract much attention, and management would be required to focus efforts on it. In bear markets, the steady revenues and profits from consulting would likely be favored, with a risk of neglecting the IP pipeline and its development. Trying to be both at once would be challenging, noted Klein:

Large worldwide service organizations only come to exist if they fundamentally give priority to client service. You look at the best organizations, the McKinseys, or some of the advertising agencies, you have to be fundamentally committed to your clients, and I would say the culture here and the mission statement has a little way to go in achieving that. So, I think there are some very very big issues around the business model that we now have, the priority that is now given to the IP exploitation side.

7 Moving On

The questions were challenging. Should Edge push for a Generics IPO in less than ideal market conditions? Was this move going to jeopardize the very culture that seemed so critical to the company's success and established credibility? If he proceeded with the IPO, what were the risks associated with failure?

Would it be possible to sell this culture of innovation – this interdisciplinary process that requires care and feeding to produce long-term results – to a public market that was known to be brutally analytical and focused on the short term?

The company had been developing well so far. With 250 people, it now featured impressive facilities complete with fully equipped laboratories in wide-ranging technology sectors. The staff of scientists had demonstrated their ability to generate innovation for customers and new companies alike. But the requirements of the newly created companies it spun off and the focus of the Group's valuation strategy were starting to prove impossible to meet with the current limited resources. An IPO would bring a well-needed cash infusion that could be used to finance new projects and expansion plans.

Exhibit 1
Profile of Gordon Edge

Gordon Edge, founder and executive chairman of The Generics Group plc. Edge is an electronics engineer (Dip. Tech), a chartered engineer (C. Eng), a member of the Institution of Electrical Engineers (MIEE), and a fellow of the Royal Swedish Academy of Engineering Sciences (IVA). He has a doctorate (D. Tech) from Brunel University.

Edge holds several professorships and fellowships, sits on the board of several international groups, joint ventures, and colleges, and is a member of the Foresight Panel on Information Technology. He has spoken and published widely on a variety of business- and technology-related subjects. During his career, he has contributed to several hundred patents in science and technology.

He received the CBE (Commander of the British Empire) in the 1992 Queen's Birthday Honours.

Exhibit 2
Financial Statements

Balance Sheet (thousands of £)

	1997	1998	1999
Fixed assets			
Goodwill		88	129
Tangible assets			
Land and buildings	2,245	2,232	2,403
Furniture and fittings	384	363	399
Equipment	756	827	906
Investments			
Associates	535	319	106
Other investments	520	1,069	3,276
Total fixed assets	**4,440**	**4,898**	**7,219**
Current assets			
Debtors			
Due within one year	3,898	4,480	4,480
Due after one year	68	89	
Cash at bank and in hand	2,497	1,413	751
Total current assets	**6,463**	**5,982**	**5,231**
Creditors: amounts falling due within one year	(5,093)	(4,822)	(4,357)
Net current assets (liabilities)	**1,370**	**1,160**	**874**
Total assets less current liabilities	5,810	6,058	8,093
Provisions for liabilities and charges	(568)	(390)	(440)
Net assets	**5,242**	**5,668**	**7,653**
Capital and reserves			
Called-up share capital	3,465	3,465	3,465
Merger reserve	92	92	92
Own shares			(38)
Profit and loss account	1,635	1,888	3,767
Shareholders' fund - all equity	**5,192**	**5,445**	**7,286**
Minority interests - all equity	**50**	**223**	**367**
Total capital employed	**5,242**	**5,668**	**7,653**

Exhibit 2 (continued)

Profit and Loss Accounts (thousands of £)

	1997	1998	1999
Income			
Fee income	13,582	14,900	14,125
License/royalty income	444	1,942	92
Total income	**14,026**	**16,842**	**14,217**
Operating expenses			
Selling and marketing costs	(1,349)	(1,476)	(1,831)
Administrative expenses	(12,526)	(14,320)	(16,012)
Total operating expenses	**(13,875)**	**(15,796)**	**(17,843)**
Amounts written off investments	(147)	37	
Operating profit (loss)	**151**	**1,046**	**(3,626)**
Share of associates' operating losses	(440)	(283)	(235)
Profit on sale of discontinued operations			3,286
Profit on disposal of fixed asset investments	1,184		1,913
Profit on issue of shares by subsidiary to minorities	747		421
Profit (loss) on ordinary activities before finance income (charges)	**1,642**	**763**	**1,759**
Finance income (charges)	81	128	35
Tax on profit (loss) on ordinary activities	(554)	(363)	(219)
Minority interests	29	100	306
Profit (loss) for the financial period	**1,198**	**628**	**1,881**
(Loss) gain on foreign currency translation	(3)	(17)	(2)
Total recognized gains and losses relating to the period	**1,195**	**611**	**1,879**

Exhibit 2 (continued)

Cash Flow Statement (thousands of £)

	1997	1998	1999
Net cash inflow (outflow) from operating activities	**1,504**	**952**	**(2,858)**
Returns on investment and servicing of finance			
Interest received	128	111	84
Interest paid	(49)	(15)	(60)
Dividends received	2	2	
Total returns on investment and servicing of finance	81	98	24
Taxation			
Corporation tax (paid) received	(308)	(507)	(466)
Total taxation	(308)	(507)	(466)
Capital expenditure and financial investment			
Purchase of tangible fixed assets	(532)	(482)	(1,023)
Capitalization of development costs			
Purchase of investments	(955)	(574)	(467)
Sale of investments	1,743	25	2,400
Total capital expenditure and financial investment	256	(1,031)	910
Acquisitions and disposals			
Purchase of subsidiary undertakings		(437)	(256)
Purchase of minority interest in subsidiary under-taking			(14)
Net (overdraft) cash acquired with subsidiary undertaking		(67)	267
Sale of subsidiary undertaking			224
Net overdraft (cash) disposed of with subsidiary undertaking			124
Investment in associates	(300)		(11)
Total acquisitions and disposals	(300)	(504)	334
Equity dividends paid	(372)	(272)	(358)
Cash inflow (outflow) before management of liquid resources and financing	**861**	**(1,264)**	**(2,414)**
Management of liquid resources			
(Increase) decrease in cash deposits	(1,249)	1,531	
Total management of liquid resources	(1,249)	1,531	0
Financing			
Issue of shares by subsidiary undertaking to minor-ity interests	25	171	800
Repayment of unsecured loan	(826)		
Increase in unsecured loan			952
Total financing	(801)	171	1,752
Total net cash inflow (outflow) from operating activities	**(1,189)**	**438**	**(662)**

Source: Company information

Exhibit 3
List of Spinoffs at End 2000

Year	Name	Description
1991	Diomed, Inc.	Optical technology for increased power output for lasers used in surgical applications (angioplasty and photodynamic therapy)
1996	Flying Null, Ltd.	Magnetic tagging technology permitting a gap between tag and reader head; used to track goods, art
1997	Imerge, Ltd.	Hard-disk-based audio and video software and devices for consumer, retail, and corporate sectors
1998	Absolute Sensors, Ltd.	Noncontact positioning technology; applications in automotive, game, and pen input markets; sold to Synaptics Inc. in 1999
2000	quantumBEAM, Ltd.	Free space optical communication technology, transmitting bit rates up to 5 Gb/s; fixed wireless technology
2000	RaceTrace, Inc.	Radio frequency location technology, used for accurate representation of sports items and/or players (equipped with device)

CASE 5-6
GigaTera Inc: Pulling the Plug?

Christopher L. Tucci, Ralf W. Seifert, and Jana Thiel

ZURICH, SWITZERLAND, AUGUST 31, 2000. Today, the Ultrafast Laser Physics Group at the Swiss Federal Institute of Technology in Zurich (ETH Zurich) has spun off a new venture – GigaTera Inc. The new firm will focus on the development and commercialization of high-speed laser components for the optical networks industry.

FEBRUARY 21, 2001. Six months after its inception, GigaTera raises US$7 million in its first-round venture financing led by Broadband Capital of Zug, Switzerland, and JK&B Capital of Chicago, Illinois.

NOVEMBER 27, 2001. GigaTera receives a follow-on investment of $2 million from Cypress Semiconductor and its wholly owned subsidiary Silicon Light Machines, both based in Silicon Valley, California.

NOVEMBER 29, 2002. Following a price reduction in its ERGO laser in October 2002, GigaTera Inc. has now released a wavelength tunable version of the laser device, hoping to capitalize on recent market developments.

JANUARY 15, 2003. GigaTera demonstrates a new single laser source that is able to generate 32 different wavelengths from only one device. The firm hopes to be able to market the new device quickly by using the same chassis as for its ERGO laser.

***

Despite its technical success, GigaTera was running low on cash. And the overall market prospects and the firm's still marginal revenue streams left investors hesitating. In early 2003 GigaTera's future was under threat. Would the venture be able to succeed? What could it do to improve its chances of survival?

Copyright © 2006 by the alliance for technology-based enterprise (IMD, EPFL and ETH Zurich). Copyright permissions are handled by IMD, Lausanne, Switzerland.

1 Optoelectronics: the Soaring Opportunity

If you are an entrepreneur and you mention the word "optical," people stand in line to write you a check.

These words by Rob Chaplinsky, a California-based VC, gave a fairly accurate description of the situation in 2000 – the year GigaTera was founded [1]. Back then, just about any optical communications company was judged to be a winner [2], promising profits that exceeded those of the Internet companies, which were slowly declining.

One reason for the success of the optical industry was its ability to solve a problem that had been triggered by the Internet wave and the accompanying advances in digital technology – a huge hunger for bandwidth. This meant that ever-increasing amounts of data had to be transmitted per unit of time. At this point, fiber optics made an appearance. The use of glass fiber to transmit data through light instead of electrical signals via metal wire had decisive advantages [3]: Optical systems were immune to electrical interference, they did not radiate, and the signals could travel much longer distances. Most importantly, however, optical networks made it possible to send data with higher frequencies – terahertz instead of megahertz and gigahertz – which ultimately allowed a much higher data throughput.

Optical transmission had in fact been used in long-haul transmission [4] (*i.e.*), distances over 40 km) since the late 1970s. However, the underlying technology was out of date and increasingly restricted speed. It needed to be replaced to meet new performance requirements. Moreover, optics had also started to invade the realm of metro-area networks [5], where bandwidth needed to be added to the "the last mile." [6]. This was the result of more and more end users demanding Internet connection at reasonable speeds to keep up with the growing data traffic and streaming content.

These technical developments had a significant economic impact. Optical transmission called for a large variety of enabling parts and components besides fiber – most of them highly integrated with electrical elements. The increasing demand for these components led to the development of a whole new industry. All over the globe, optoelectronics startups sprang up like mushrooms, supported by the upbeat forecasts of industry experts. Growth rates of up to 80% annually until 2005 were promised [7]. In Europe alone, over a hundred new optoelectronic ventures were founded [8]. Most of them focused on the development and manufacture of components such as optical transmitters, connectors, or receivers.

Switzerland, in particular, boasted a large number of startups in 2000 and 2001. With their strong tradition in nanotechnology, Swiss research centers continued to spawn new ventures, which meant the country was soon ranked second in Europe after Scotland [9]. Often these new firms were generously funded by VCs.

2 Company Background

Amidst all the hype in the optoelectronics industry, scientists at the Ultrafast Laser Physics Group at ETH Zurich were working on a cutting-edge all-optics laser principle. It was designed for high-frequency transmission in long-haul optical networks. To develop and commercialize the invention, GigaTera was founded in 2000. As an official spinoff of ETH Zurich, it received tremendous support from the universitiy's technology transfer office in finding the right location for its business and setting up the patents to protect its proprietary technology [10].

2.1 The GigaTera Team

To be successful, optoelectronic firms needed to assemble a team of scientists who were highly specialized and competent in various fields such as physics, optics, electronics, and mechanics, among others. These competencies were typically found only at the top universities and research institutes. GigaTera was a case in point, as it was built to a large extent using the resources and capabilities of its founding lab. It was the research group's second spinoff, following in the footsteps of Time-Bandwidth Products (TBP), also an optoelectronics company, founded in 1994. Like GigaTera, TBP dealt with ultrafast laser systems, but using a different technology. With its wide range of pico- and femtosecond laser systems, TBP enjoyed steady success at the time.

Professor Ursula Keller, head of the Ultrafast Laser Physics Group and a renowned expert in laser technology, was a founding member of GigaTera. She had also cofounded TBP but did not play an active management role in either firm, although she was on GigaTera's board [11].

The actual founding father of GigaTera, as well as of TBP, was Dr. Kurt Weingarten – Keller's husband [12]. He was GigaTera's chief technology officer (CTO) and also president of TBP's board. Weingarten, who held a degree from Stanford University, had previously worked with laser and optoelectronics startups in Silicon Valley for a number of years.

Another Stanford graduate was GigaTera's CEO, Andros Payne, who had worked with the Hughes Space and Information Group (now Boeing) and with Mercer Management Consulting in Switzerland before joining GigaTera [13]. He did not have a particular knowledge of the optoelectronics domain but brought significant expertise in business strategy and operations. Thus, while Weingarten managed the technology issues, Payne managed public relations and engaged heavily in the search for funding.

Payne also dealt with team-building issues, holding together a pretty diverse group of high-profile scientists. GigaTera had quickly built up its resource base to some 20 people. It had over a dozen researchers, all from Switzerland and the

USA, with impressive track records and practical experience on its payroll [14]. In addition, it profited from the excellent profile of Keller's research lab, with which the spin-off maintained close ties.

2.2 GigaTera's Technological Edge

GigaTera's laser aimed at the next-generation standard in optical data transmission, transmitting data at 40 GHz (40 billion pulses per second). In 2000 the existing norm was still 10 GHz, but it was moving inexorably toward 40 GHz networks [15], since higher frequency typically meant higher bandwidth in data transmission. According to Payne:

> Generating light pulses at rates as high as 40 billion pulses per second is one of the key remaining challenges to increase the bandwidth of current communication networks. We believe our approach has the advantage of simplicity and compactness [16].

GigaTera's laser utilized the so-called return-to-zero (RZ) pulse approach [17], where the light level falls back to zero after every data bit. RZ pulses were placed at the core of all future systems because, at frequencies above 10 GHz, the previously prevailing and easier to achieve non-RZ technology failed to deliver reliable and stable transmission. Traditionally, however, RZ lasers needed an electronic modulator with a corresponding electronic driver to cut the otherwise continuous stream of light into regular RZ pulses [18]. GigaTera's laser featured a novel way of achieving this. Its all-optics approach rendered the two expensive electronic parts unnecessary. Thus, the new laser came in at lower cost and smaller size. Moreover, it consumed less electricity while at the same time achieving a higher output power, which improved the signal intensity and longevity. The lower number of necessary parts also reduced potential sources of failure, making the transmission all the more robust. So far, no other firm besides GigaTera had been able to transmit such pulses without additional electronics that would solve the issue of laser instability at the high frequencies of 40 GHz [19].

This was the theory; in practice; however, GigaTera's device could not yet generate true 40-GHz pulses. Above 10 GHz, its device sent out light at wavelengths outside the industry norm. Hence, the fundamental frequency output of GigaTera's device was still 10 GHz, which was then converted to 40 GHz with the help of a special multiplexer. This additional converter had been developed in collaboration with the Zurich Research Laboratory [20]. Although the Swiss startup continued to work on true 40-GHz transmission, a solution was not expected before 2003 [21].

2.3 Application Areas

GigaTera's optical laser was intended to be used as the source of light in a variety of devices like transmitters and amplifiers. Its characteristics mapped the technology specifically to applications in long-haul transmission.

On the other hand, the jitter [22] and footprint [23] parameters of GigaTera's device – in which the firm claimed to be the industry leader [24] – made it compatible with test, measurement, and systems development applications. As optical networks had become increasingly complex, test and measurement technology had started playing a key role. Both equipment manufacturers and network operators needed the most advanced tools available to examine and verify system and component performance [25]. (*See Exhibit 1 for a tabular overview of fields of application.*)

2.4 Funding GigaTera

GigaTera closed its first-round venture financing in February 2001. The startup had then raised $7 million from two institutional investors: the Swiss-based Broadband Capital and the US-based JK&B Capital.

Only nine months later, in November 2001, GigaTera complemented its funds with another $2 million. This time, corporate venture capital was raised from Silicon Light Machine, a subsidiary of Cypress Semiconductor. Both Keller and her husband Weingarten had been students of David Bloom, the founder of Silicon Light Machine [26]. Robert Corrigan, senior vice president of the Cypress subsidiary, who joined the GigaTera board as an observer, commented on the strategic investment:

> The investment relationship between Cypress/Silicon Light Machines and GigaTera is based on the recognition of a strong set of complementary skills. There is significant potential for broader cooperation between our organizations in the future [27].

Both financing rounds provided sufficient resources to push research and product development ahead for about a year. After that the firm would have to look for more financing. In general, optoelectronics startups were a costly endeavor because of their sophisticated R&D processes. Typically firms burned more than $100 million from three financing rounds before achieving any revenue stream. Hence, GigaTera had made plans to raise another $15 million in the following year and between $45 and $60 million sometime down the road [28].

3 Brakes on the Development of the Optoelectronics Market

With the collapse of the Internet bubble at the end of 2000, the demand for bandwidth did not increase as much as had been expected. Hence, more and more telecom providers had to scale back operations or even close their businesses. This cut the demand for advanced optical components dramatically, putting the entire optoelectronics industry under severe economic pressure.

3.1 Updated Market Forecast

Originally, optical components and networking applications were forecast to reach a market volume of $24 billion by 2004 [29]. But then business slumped by 85% from Q4 2000 to Q4 2001. The forecasts had to be amended, with the optical components market expected to approach only $3 billion by 2003 [30]. Although there was hope of regaining some growth after 2004, the overall optical business was unlikely to expand above $10 billion by 2006. This number also included network services. Hence, the actual market for a component provider such as GigaTera was even smaller.

The forecast resumption of growth after 2004 was induced by the renewed request for broadband access at the end-user level. However, only firms that had a stake in the metro-area technology – which relied on different components than long-haul networks – were likely to benefit from this development. For long-haul networks, by contrast, vast overcapacities had been created, which would prevent additional investments – even in the unlikely event of an upturn in the market. New approaches and cutting-edge technology in long-haul transmission were much harder to sell than those in metro-area applications, or, as Ron Kline of RHK Inc., a market analyst company, put it:

> It's evolution, not revolution, and that is because there are billions and billions of dollars of equipment out there that already works. Anything new has just to work as well or better [31].

Facing considerable uncertainty, firms narrowly specializing in a particular field looked to have a less rosy future than suppliers with a broader base.

3.2 Venture Financing

VCs were hesitant about throwing money into the depressed market. By 2001 only about half of what had been invested the year before was available for venture funding [32]. Several industry experts stressed that startups without enough funds to last the next two to three years would most probably go under [33].

3.3 Impediments to Achieving Technological Maturity

Due to the economic environment and the drying up of the venture financing market, the technological conversion from 10-GHz to 40-GHz networks came to a halt. A wider range of products supporting the higher frequency was only likely to emerge by 2004 [34]. And although 40 GHz was promoted as up-and-coming technology, 10-GHz infrastructures were still working well and not up for immediate replacement, as Rick Berry, CTO of Sycamore Networks, a provider of intelligent switching solutions in optical networks, summarized:

> Although prototypes have been field-tested, real 40-gig products will not be carrying customer traffic until 2002, with volume in 2003 and 2004. It will be many years before 40-gig replaces 10-gig, similar to the evolution of 2.5-gig and 10-gig systems [35].

Impediments to the deployment of 40-GHz technology were manifold. For one, 40 GHz pushed the limits of technology in both electronics and optics. The supporting technologies were far from mature – and first had to catch up with some of the imposed technological challenges [36]. A more serious problem, however, seemed to be that, compared to its predecessor, the 40-GHz technology did not yet compete on price. There was a significant shortage of low-cost components manufactured on a large scale [37]. The reason for this was that, for a long time, the optoelectronics sector had been nurtured exclusively in the academic environment, which had led to a typical lack of standards and to a highly scattered manufacturing landscape a long way from being able to produce in large batches [38].

Standardization was also an issue, with a growing number of competitive carriers raising the importance of interoperability. Networks needed to carry traffic from many different sources, which meant that compliance with telecommunications standards was essential. In a market that lacked standards, speed could bring success. Early design wins, by being first to market, were especially important for smaller companies to develop into second-tier providers after such well-entrenched players as JDS Uniphase or Corning [39].

The manufacturability of optoelectronic components was another concern. Automation was the area where future profits were to be made [40]. In a period of recession, cost savings through high-scale production were of utmost importance. A single laser, for example, cost up to $500 to manufacture and was sold for three times as much. But the industry needed lasers that cost tens of dollars, not hundreds. One of the most pressing goals of the market was hence to reduce the cost of parts – in some cases by 90% [41]. Generally, it was believed that only those companies that were able to focus on cost advantages over technology would be able to survive [42]. And most often it was only established vendors that were able to bring prices down. Small startups had trouble competing in this area, which led to a surge in merger and acquisition activities and a consolidation of the market.

3.4 Supplier Landscape

This market consolidation reinforced itself as most customers preferred to turn to larger, established suppliers, which were typically better positioned and capitalized than startups [43]. Money was no longer readily available, and when it was spent on next-generation optical communications products, the business went to component industry leaders such as JDS Uniphase, Alcatel, or Nortel. As a result, many startups vanished from the scene, having neither enough cash nor streams of revenue to allow them to survive. The optoelectronics market was henceforth increasingly dominated by incumbents [44].

4 How GigaTera Dealt with the Market Changes

The question was how the small Swiss startup GigaTera would be able to cope with the dramatic reduction in overall business volume and the shifts in market trends. Payne was well aware of GigaTera's unfashionable approach:

It's an excellent time to develop new technologies but a difficult time to be selling them [45].

Still, 2001 had been comparatively better for the small startup than for the rest of the market. The funds raised helped the Swiss venture face the economically tough year to come. One of GigaTera's first actions in early 2002 was to add another industry expert to the board – Harry L. Deffebach. He had over 30 years' experience in the North American electronics and optoelectronics market. Specifically, he had headed the JDS Uniphase Transmission Group, where he was responsible for five product divisions in the USA [46].

4.1 Bringing a Product to Market

Especially at first, GigaTera was clearly a research-dominated company as it had yet to prove the concept of the all-optics RZ laser before a marketable product could be developed. By September 2001 GigaTera had succeeded in demonstrating a first working prototype. The market required precision and high reliability, which called for diligent testing to guarantee the necessary performance. Only once all checks had been done could production on a large scale begin. Nevertheless, standard batch sizes were not expected to be reached before late 2002 or even early 2003 [47].

The startup finally released its first laser device, ERGO, in March 2002 [48]. Over the following months, GigaTera presented its technology at different technical conferences in the USA and Europe, each time at the booth of Agilent Technologies, a well-known provider of test and measurement equipment [49]. In June Gigatera demonstrated the functioning of ERGO in a live, real-world application in the Photonics Laboratory at Sweden's Chalmers University, Gothenburg. According to the Swedish researchers, both the performance and form factor of GigaTera's device made it an interesting candidate for the next generation of fiber-optic transmission systems [50].

However, GigaTera's continued proof of concept after it had released the product to market made it clear that the development of a saleable product posed several challenges. GigaTera was still far from steady revenue streams. It hoped for a time when data traffic would increase significantly, which most probably meant holding out for three or four more years before demand for network capacity rose again [51].

4.2 Selling the Product

GigaTera openly announced several collaborations, for example with Agilent, Chalmers, and ETH Zurich, but it never reported significant sales or any particular reference project. In fact, two years into business the firm did not seem to have much of a track record. By contrast, the more successful companies in the industry did not necessarily suffer from a lack of customers but from an inability to serve incoming requests on time and in the quality and quantity demanded. Of course GigaTera faced similar problems, but more obviously it had to grapple with finding buyers for its technology.

In October 2002, in a move that might have been partially motivated by the lack of customer base, the company reduced the price of ERGO to $49,000. Payne justified the mark-down as follows:

> ERGO has been priced aggressively with the intention of passing on the economic and performance benefits of our unique design to our development customers [52].

GigaTera also claimed to have reduced the price to further enable emerging RZ applications, which would not have been either attractive or feasible at the previously existing market prices. The company also announced that it would continue development efforts to reduce the size of its laser. That way, the ERGO technology would eventually scale to meet the size and cost requirements for network line card applications [53] to be installed in high-speed routers – one potential growth sector [54].

4.3 Defining a Flexible R&D Portfolio

With uncertain prospects in the market, firms had to be flexible in their portfolio – and flexibility was what GigaTera was working toward. In November 2002 the startup announced new results from its research – a wavelength tunable version of the ERGO laser. Tunable lasers had been hailed as the driving force in the rejuvenating optoelectronics market [55]. They permitted frequency regulation after installation, which it was believed would help service providers cut down their laser inventories, as they would no longer need to carry a broad portfolio of lasers with different frequencies. Now, one source could replace many.

Two months later, in January 2003, GigaTera hit the news again with another invention – a new laser technology emitting 32 channels of light simultaneously. An integral part of the device was a so-called dynamic gain equalizer (DGE) – a piece of technology from the labs of GigaTera's strategic partner Silicon Light Machines. GigaTera claimed to be able to bring the new technology to market quickly by putting it on the same chassis as its existing ERGO laser [56]. This time, the application specifically addressed business in metro-area and regional networks. CTO Weingarten commented:

> Network operators want to add capacity and flexibility in their metro and regional optical networks, but they are also under tremendous pressure to reduce the space [requirements] and operating expenses of DWDM [dense wavelength division multiplexing] systems. Our technology will enable what's currently done with racks of separate lasers to be condensed onto a single linecard [57].

The advantage of the new device was that telecom providers that traditionally employed 32 fixed-wavelength sources with 32 lasers and 32 sets of driver electronics, *etc.* would use a single device to attain the same 32 channels. However, the basic idea was not new. Bell Labs had worked on a similar principle as early as 1997 [58] as Tom Hausken, director of optical components research at market forecasting firm Strategies Unlimited, pointed out when commenting on GigaTera's approach:

> Yes, it is a very interesting technology. However, I think, the fact that Bell Labs didn't pursue it, says a lot [59].

4.4 Competitive Landscape

Just a few days after GigaTera's announcement, another startup, the Israel-based KiloLambda, came out with its own version of a multiwavelength laser module. Unlike GigaTera, though, the Israeli firm was backed by public grants and corporate venture financing "scheduled" for the next three years. During this time KiloLambda's business roadmap did not speculate on any sales [60]. By contrast, US-based Opticalis, another startup developing multiwavelength sources, had closed in 2002 because it forecast a long-term lack of commercial activity for its technology.

On the tunable-laser front, GigaTera was also not alone. Wide research into tunable lasers had started as early as 2000, when leading analysts predicted a rise in market volume from $5 million to $1.2 billion by 2004. Even though this forecast had since been revised downward, the area was well exploited by established players such as Nortel, Lucent, and Fujitsu Networks. It had also seen a lot of innovative startups appear [61]. And the wide variety of approaches in tunable laser technology, with no clear winner in terms of design, made a market entry all the more challenging.

5 Deciding on the Future

All in all, GigaTera performed excellent research but still lacked significant revenue streams. If this continued, the Swiss venture would soon need to secure new funding to continue operations. By 2003, although GigaTera had CHF 1.3 million assets on its balance sheet – against CHF 350,000 in liabilities [62] – investors were generally reluctant to put more money in, particularly as the firm did not have any large revenue prospects in the forseeable future. The chances of convincing investors to put more money into the company were slim.

This was a fate the startup shared with many others in the optoelectronics industry. Struggling firms employed different tactics to handle the situation. Some chose self-liquidation, sometimes even returned money that was left to their shareholders. This, however, was only when management came to believe that their product was not working at all, which evidently did not apply in GigaTera's case. Other ventures pursued mergers mostly with larger and more broadly positioned companies to prevent going officially bankrupt. In any case, firms passively waiting or hesitating typically went under [63].

GigaTera had to decide on a course of action. But what were its alternatives? How much more time did it have to act? What could it do to improve its chances of survival?

Exhibit 1
Fields of Application of GigaTera's Laser Source

Telecom	Test and Measurement	Other
R&D	• Reference RZ pulse source	• Optical docking
• 10- to 160-GHz RZ transmission	• Impulse response testing	• Quantum cryptography
• 3.125- to 50-GHz DWDM transmission	• Dispersion measurement	• Optical A/D
• Optical 2R, 3R	• Reference multiwavelength source	• Metrology
• Wavelength conversion	• WDM testing	• Sensing
• OTDM Demux	• OTDR	• Defense/ high-speed processing
• Optical signal processing		
Network		
• 10- to 40-GHz RZ transmission		
• Metro, access, WDM transmission		

Source: Company website (http://www.gigatera.com, accessed December 15, 2005)

References

1   Shinal, J. and T. Mullaney. "At the Speed of Light." Business Week Online, October 9, 2000.
    http://www.businessweek.com/2000/00_41/b3702185.htm, accessed December 15, 2005.

2   Jaffe, J. "Optical Delusion." *The Deal.com*, October 16, 2003: 1.

3   The idea of transmitting data via light has in fact existed since the late 19th century. A brief
    history of fiber optics can be found at http://www.fiber-optics.info/fiber-history.htm.

4   "Dense Wavelength Division Multiplexing (DWDM) Performance and Conformance
    Testing." *The International Engineering Consortium*,
    http://www.iec.org/online/tutorials/dwdm_perf/, accessed December 15, 2005.

5   A loosely defined term generally understood to describe a data network covering an area
    larger than a local-area network (LAN) but less than a wide-area network (WAN).

6   Payne, A. "Optoelektronik – eine vielversprechende Industrie; Dank Know-how ist die
    Schweiz in einer günstigen Startposition." Neue Züricher Zeitung, December 12, 2000: 81.

7   Payne, A. "Optoelektronik – eine vielversprechende Industrie; Dank Know-how ist die
    Schweiz in einer günstigen Startposition." Neue Züricher Zeitung, December 12, 2000: 81.

8   "Downturn Puts Brake on Component Start-Ups." *Opto & Laser Europe*, September 2001.
    http://optics.org/articles/ole/6/9/8/1, accessed December 15, 2005.

9   Thompson, V. "Swiss on a Roll." *Light Reading*, November 24, 2000.
    http://www.lightreading.com/document.asp?doc_id=2625, accessed December 15, 2005.

10 *Thompson, V. "Swiss on a Roll."*
    http://www.lightreading.com/document.asp?doc_id=2625, accessed December 15, 2005.

11 Company Profile GigaTera. *VentureXperts*, October 1, 2003.

12 "Thriving in a Man's World." *Opto & Laser Europe*, October 2001.
    http://optics.org/articles/ole/6/10/5/1, accessed December 15, 2005.

13 Heywood, P. and P. Rigby. "GigaTera AG." *Light Reading*, May 2, 2002.
    http://www.lightreading.com/document.asp?doc_id=14709, accessed December 15, 2005.

14 "Die Schweiz tut GigaTera gut." *Neue Züricher Zeitung*, September 25, 2001: 124.

15 Clavenna, S. "40-Gig Forecast." *Light Reading*, May 1, 2001.
    http://www.lightreading.com/document.asp?doc_id=4849, accessed December 15, 2005.

16 "GigaTera Demonstrates All-Optical 40 GHz Pulse Source at the Optical Fiber Conference
    in Anaheim, California." *GigaTera Press Release*, March 18, 2002.
    http://www.gigatera.com/cont/5_press_rel_2002_03_18e.html, accessed December 15, 2005.

17 A pulsing signal is required to generate "0s" and "1s," a principle on which all data trans-
    mission is based.

18 Clavenna, S. "40-Gig Forecast." *Light Reading*, May 1, 2001.
    http://www.lightreading.com/document.asp?doc_id=4849, accessed December 15, 2005.

19 Heywood, P. and P. Rigby. "GigaTera AG." *Light Reading*, May 2, 2002.
    http://www.lightreading.com/document.asp?doc_id=14709, accessed December 15, 2005.

20 "GigaTera Demos 40 GHz Pulse Source." *Light Reading*, March 18, 2002.
    http://www.lightreading.com/document.asp?doc_id=13091&site=ofc, accessed December 15, 2005.

21 Freeman, T. "Gigatera's RZ Laser Cuts Costs, Complexity." *Fibers.org*, December 2, 2002.
    http://fibers.org/articles/news/4/12/1/1, accessed December 15, 2005.

22 Jitter is the deviation in or displacement of some aspects of the pulses in a high-frequency
    digital signal.

23 Footprint in this sense means the amount of space occupied by a device.

24 Freeman, T. "Gigatera's RZ Laser Cuts Costs, Complexity." *Fibers.org*, December 2, 2002.
    http://fibers.org/articles/news/4/12/1/1, accessed December 15, 2005.

25 "Dense Wavelength Division Multiplexing (DWDM) Performance and Conformance
    Testing." *The International Engineering Consortium*,
    http://www.iec.org/online/tutorials/dwdm_perf/, accessed December 15, 2005.

26 Heywood, P. and P. Rigby. "GigaTera AG." *Light Reading*, May 2, 2002.
    http://www.lightreading.com/document.asp?doc_id=14709, accessed December 15, 2005.

27 "GigaTera Gets Cypress Cash." *Light Reading*, November 28, 2001.
http://www.lightreading.com/document.asp?doc_id=9955, accessed December 15, 2005.

28 "Die Schweiz tut GigaTera gut." *Neue Züricher Zeitung*, September 25, 2001: 124.

29 Jaffe, J. "Optical Delusion." *The Deal.com*, October 16, 2003: 1.

30 "Survive Now, Flourish Later." *Newswire NewsInc.*, Vol. 22, No. 17, April 29, 2002.

31 Lewotsky, K. "Long Haul Looks to the Future." *spie's oemagazine*, May 2002.
http://oemagazine.com/fromTheMagazine/may02/specialfocus.html, accessed December 15, 2005.

32 "Downturn Puts Brake on Component Start-Ups." *Opto & Laser Europe*, September 2001.
http://optics.org/articles/ole/6/9/8/1, accessed December 15, 2005.

33 "Survive Now, Flourish Later." *Newswire NewsInc.*, Vol. 22, No. 17, April 29, 2002.

34 "Survive Now, Flourish Later." *Newswire NewsInc.*, Vol. 22, No. 17, April 29, 2002.

35 Clavenna, S. "40-Gig Forecast." *Light Reading*, May 1, 2001.
http://www.lightreading.com/document.asp?doc_id=4849, accessed December 15, 2005.

36 Clavenna, S. "40-Gig Forecast." *Light Reading*, May 1, 2001.
http://www.lightreading.com/document.asp?doc_id=4849, accessed December 15, 2005.

37 "Optical Component Market Opportunities, Market Forecasts, and Market Strategies, 2004–2009." *Global Information Inc.*, May 2004.
http://www.gii.co.jp/english/wg19976_optical_component.html, accessed December 15, 2005.

38 Payne, A. "Optoelektronik – eine vielversprechende Industrie; Dank Know-how ist die Schweiz in einer günstigen Startposition." Neue Züricher Zeitung, December 12, 2000: 81.

39 "Survive Now, Flourish Later." *Newswire NewsInc.*, Vol. 22, No. 17, April 29, 2002.

40 Payne, A. "Optoelektronik – eine vielversprechende Industrie; Dank Know-how ist die Schweiz in einer günstigen Startposition." Neue Züricher Zeitung, December 12, 2000: 81.

41 Reinhardt, A. and John G. Shinal. "The World's Most Glamorous Cottage Industry." *Business Week Online*, October 9, 2000.
http://www.businessweek.com/2000/00_41/b3702185.htm, accessed December 15, 2005.

42 "Survive Now, Flourish Later." *Newswire NewsInc.*, Vol. 22, No. 17, April 29, 2002.

43 "Survive Now, Flourish Later." *Newswire NewsInc.*, Vol. 22, No. 17, April 29, 2002.

44 "Worldwide Next-Generation Optical Networking 2004 Vendor Shares." June, 2005.
http://www.gii.co.jp/english/id30382-optical-networking.html, accessed December 15, 2005.

45 Jaffe, J. "Switzerland VCs Have Hand in Optical Startups." The Daily Deal. November 29, 2001.

46 "GigaTera Adds to Board." *Light Reading*, January 9, 2002.
http://www.lightreading.com/document.asp?doc_id=10745, accessed December 15, 2005.

47 "Die Schweiz tut GigaTera gut." *Neue Züricher Zeitung*, September 25, 2001: 124.

48 Freeman, T. "Gigatera's RZ Laser Cuts Costs, Complexity." *Fibers.org*, December 2, 2002.
http://fibers.org/articles/news/4/12/1/1, accessed December 15, 2005.

49 Heywood, P. and P. Rigby. "GigaTera AG." *Light Reading*, May 2, 2002.
http://www.lightreading.com/document.asp?doc_id=14709, accessed December 15, 2005;
refer also to Walko 2005.

50 "GigaTera Transmits Pulses." *LightReading*, June 4, 2002.
http://www.lightreading.com/document.asp?site=lightreading&doc_id=16967, accessed December 15, 2005.

51 "Die Schweiz tut GigaTera gut." *Neue Züricher Zeitung*, September 25, 2001: 124.

52 GigaTera Cuts Laser Pricetag." *Light Reading*, October 3, 2002.
http://www.lightreading.com/document.asp?site=lightreading&doc_id=22083, accessed December 15, 2005.

53 A line card is a circuit board that provides a transmitting and receiving port for a particular protocol. Line cards plug into a telco switch, network switch, router or other communications device.

54 Thompson, V. "New Zurich Nanotech Lab Will Help Advance Industry." *Small Times*, July 23, 2002.
http://www.smalltimes.com/document_display.cfm?document_id=4217, accessed December 15, 2005.

55   "The Market for Integrated and Discrete Optoelectronic Components: 2002–2011." Global
     Information, Inc., October 2002.
     http://www.gii.co.jp/english/cf12008_optoelectronic_components.html, accessed December
     15, 2005.
56   "GigaTera Goes Multiwavelength." *Light Reading*, January 16, 2003.
     http://www.lightreading.com/document.asp?site=lightreading&doc_id=26970, accessed December
     15, 2005.
57   "GigaTera Demonstrates 32 Optical Channels Suitable for WDM Transmission from a
     Single Laser." *GigaTera*, Press Release, January 15, 2003.
     http://www.gigatera.com/cont/5_press_rel_2003_01_15.html, accessed December 15, 2005.
58   "Lucent's Bell Labs Scientists Use an Ultra-Fast Laser to Transmit Data over 206 Wavelength
     Information Rainbow." Lucent Technologies, Press Release, February 17. 1997.
     http://www.lucent.com/press/0297/970217.blb.html, accessed December 15, 2005.
59   "GigaTera Goes Multiwavelength." *Light Reading*, January 16, 2003.
     http://www.lightreading.com/document.asp?site=lightreading&doc_id=26970, accessed December
     15, 2005.
60   "Molex Backs Multiwavelength Modules." *Light Reading*, January 31, 2003.
     http://www.lightreading.com/document.asp?doc_id=27738, accessed December 15, 2005.
61   "Optical Components Markets 2000–2004." Global Information Inc., November, 2000.
     http://www.gii.co.jp/english/ci5651_optical_components_markets.html, accessed December
     15, 2005.
62   Time-Bandwidth Products AG. Commercial register (Handelsregister), Vol. 122, No. 126,
     Friday, July 2, 2004.
63   Jaffe, J. "Optical Delusion." The Deal.com, October 16, 2003: 1.

Chapter 6
Corporate Entrepreneurship

In the preceding chapters, we have studied the creation of new businesses as independent startups. However, we recognize that many well-established and large companies do, in fact, engage in entrepreneurial activity [1]. Corporate entrepreneurship (CE) may be broadly viewed as (1) the creation of new businesses within existing organizations, either through internal innovation or joint ventures and alliances, or (2) the transformation of organizations through strategic renewal [2]. Large firms embark on CE for various reasons (*why*) and using various approaches (*how*), which we will now explore in more detail.

6.1 Why Do Large Corporations Pursue Entrepreneurship?

Entrepreneurial pursuits are often considered strategically imperative to rejuvenate a firm, create value, and sustain competitive advantage [3]. As such, CE is intentional in nature [4]. By deliberately introducing what is perceived as an entrepreneurial attitude, large corporations aim to reduce bureaucracy, increase flexibility, and speed up their product development. As a result, they hope to create more innovative products and services that will enable them to stay ahead of the competition. Specific goals stretch from making a new product introduction more successful to reinventing the entire firm.

All the case studies included in this chapter feature technology-based businesses that are trying to shape or redefine their markets through CE. This throws up issues not unlike those encountered by technology startup entrepreneurs. In addition, however, particularities due to firm size and structure emerge. We find that CE is not only seen as a cure for the staleness and lack of breakthrough innovations big companies sometimes experience, but also as a curse that pushes the boundaries of existing structures and hierarchies [5]. A sizable number of companies have difficulty dealing with the changes and conflicts their entrepreneurial approaches entail.

6.2 How Can Corporations Promote Entrepreneurship?

Corporate entrepreneurship has been described as manifesting mainly in the form of sustained regeneration, organizational rejuvenation, strategic renewal, and domain redefinition [6]. All four can exist concurrently. It should be noted, however, that their emergence is not necessarily a result of a deliberate managerial strategy. Rather, they seem to be the most common firm-level manifestations of an experimentation-and-selection process [7].

In *sustained regeneration* – the most widely accepted form of CE – firms seek to continuously introduce new products or enter new markets. Companies that are successful at this tend to align corporate culture, values, and procedures, all to foster entrepreneurial traits among their employees and to support innovation. These companies view the capacity for innovation as an essential core competence that must be protected, and leveraged. *Organizational rejuvenation* aims to improve a firm's competitive standing by altering its value chain activities and internal resource allocation. It often leads to organizational transformation, *i.e.*, a rearrangement of resources and processes, which changes the way business has been done till now. *Strategic renewal* includes the redesign of internal processes but goes even further, as it leads to fundamentally rethinking how the firm ought to compete. Hence, the focal point is not the firm *per se* but the firm in its environmental context. Finally, *domain redefinition* focuses on exploring possible new markets and products to create first-mover advantages.

The inherent challenge in all the cases presented in this chapter is that firms redefine, renew, or replace their innovative and competitive capabilities. At the same time, they must ensure that the resulting changes in policies, priorities, and processes will be accepted throughout the existing organization [8]. This requires them to act on at least two interrelated levels – adapting the organizational design and empowering the individual employee.

6.2.1 Encouraging Corporate Entrepreneurs

The individual entrepreneur can be viewed as the motor of organizational innovation. However, the entrepreneur in a large business creates new products, services, or processes alongside the company's existing business. Consequently, he or she is embedded in an established set of culture, values, and procedures, whereas the small-business entrepreneur still has the freedom to define them [9]. Our cases focus on large companies as they deliberately create a corporate culture that fosters the entrepreneurial mindset, in general, and encourages enterprising individuals, in particular, to recognize and exploit opportunities.

Typically, large firms are focused on selling existing products and services rather than developing new ones, which naturally inhibits the recognition of opportunities [10]. And even if opportunities are recognized, their subsequent exploitation requires the addition of resources or changes in resource deployment; it also requires learning and the development of new capabilities. This all entails increased risk for the corporation. Managers in large firms, however, operate within fundamentally different decision-making structures. Typically, they are rewarded to sustain performance at the lowest possible risk rather than placing big bets. Hence, the corporate reward system turns out to be a major brake on CE. Furthermore, CE requires top management to actively promote entrepreneurial decision making and risk taking throughout all hierarchical levels. It also requires flexibility and tolerance of ambiguity and eventually – without being exhaustive in this list – the willingness to grant managers the time required to pursue new opportunities [11].

Nonetheless, not all employees in large firms will or indeed even need to engage in entrepreneurial activities. Employees responsible for day-to-day operations will work alongside those driving entrepreneurial projects. Naturally, we often find resistance to change among the former and resistance to following a well-trodden path among the latter. Managing the resulting conflicts and balancing both groups is just another challenge of CE. Typically, traditional organizations have difficulty accommodating this intricacy.

6.2.2 Designing the Entrepreneurial Organization

Different choices of organizational design have led to successful CE. The right approach to CE naturally interrelates with the four types of CE discussed above.

Two main determinants of possible organizational design have been suggested: (1) the envisaged strategic importance of the project and (2) the degree to which it relates to the firm's current capabilities and skills [12]. Depending on these two dimensions, CE initiatives are anchored either inside or outside the firm. Fully integrated innovating systems or separate new venture divisions are examples of internal CE, whereas spinoffs or joint ventures exemplify external attempts (*Table 6.1*).

Table 6.1 Organizational Designs for Corporate Entrepreneurship [13]

	Unrelated	Special business unit	Independent business units	Complete spinoff
Operational Relatedness	*Partly related*	New product/ business department	New venture division	Contracting
	Strongly related	Direct integration	Micro new ventures department	Nurturing and contracting
		Very important	*Uncertain*	*Not important*
		Strategic Importance		

Our cases exhibit different designs; and they show that these designs are subject to change over time. Typically, the organizational form is defined in the early development of a new product. However, the challenge continues when the product matures and processes need to be institutionalized, *e. g.*, a separate venture division needs to be merged into another division or turned into a full business unit [14]. Hence, the organizational frame leading to initial success in the innovation, *i. e.*, the core entrepreneurial part, is not necessarily the same needed to yield longer-term success. These and all the other dynamics discussed so far make CE an intricate task that, nonetheless, promises high returns, as the cases in this chapter illustrate.

6.3 Cases in This Chapter

This chapter consists of seven cases presenting five different firms. All develop and market technology-based products and, like small businesses, have to deal with the inherent uncertainty in both new technologies and new markets. In addition, they have to deal with their inertia and their size, both impediments to recognizing and fully exploiting opportunities.

The first two cases, entitled *Rebuilding a Passion Brand: The Turnaround of Ducati (A)* and *(B)*, highlight why companies seek out CE. A famous Italian motorcycle manufacturer is facing serious cash issues and declining customer enthusiasm in the mid-1990s. Ducati had rested on its laurels for too long and had fallen behind the competition. Problems pervaded all of its operations, starting from inefficient production and marketing capabilities to the cultural setup that was totally risk averse and not passionate about the company or product. The (A) case introduces Federico Minoli, appointed CEO to turn the company around after it had been taken over by Texas Pacific. The student will be challenged to step into Minoli's shoes to identify and prioritize the steps necessary to redefine Ducati's competitive position. The (B) case describes the path Minoli chose to reestablish the Ducati brand and streamline its operations. It ends in 2003 with Ducati facing its next challenge – the rejuvenation of its R&D engine. Was Minoli successful in creating the entrepreneurial spirit to take the firm to the next level?

The third case, *Innovation Leadership at Logitech*, takes the matter of entrepreneurial culture one step further. Contrary to Ducati, Logitech has consistently flourished up to this point in 2003 when the case opens. Logitech's senior management had understood early to put innovation at the core of business. The case discusses the role of Logitech's leadership in fostering the entrepreneurial culture that drove the company to become a champion of innovation. It also discusses elements of the innovation process at Logitech, highlighting how the company tries to ensure that business opportunities are identified both internally and externally. Students will observe which corporate values are enforced to permit risk taking and failure while, at the same time, having the necessary executive rigor to push projects through. It will be up to students to assess Logitech's setup in terms of its viability, especially how this setup would need to be adapted to further grow the company.

With the knowledge gathered from the innovation leadership case, students then have the opportunity to look behind the scenes of a particular project and see how Logitech's innovation process works in practice. The case *Logitech: Getting the io™ Digital Pen to Market* discusses a product launch in a new product category. Students are asked to recommend target markets for introducing a (potentially) disruptive technology but will also be asked to choose the right organizational anchoring for the innovative endeavor – either as a project within an existing division or as a standalone unit.

Organizational anchoring is likewise of concern in the internal venturing approach described in *The "mi adidas" Mass Customization Initiative*. The case discusses the very practical implications associated with expanding a radical innovation in manufacturing and service from a small pilot to a wider operation within adidas-Salomon AG. The internally nurtured project gets challenged by the internal structures and procedures that are not prepared for the new business model and its fast success. Along the way, the case outlines an interlinked set of issues, from marketing, retailer selection, and information management, through production and distribution, to project management and strategic fit. The case offers three alternative routes for moving forward and challenges students to decide the future direction of mi adidas.

The sixth case – *New Business Creation at Tetra Pak: Reinventing the Food Can* – is already past the decision point presented in the previous two cases. The case provides historical information on the development of a retortable carton within Tetra Pak, a project managed internally for eight years before it was spun off to try to turn it into a real, profitable business. We get to know Jan Juul Larsen, who has taken over the newly founded Tetra Recart AB and faces choices familiar to every entrepreneur. Students are invited to analyze the development process to date and to assess the product and market strategy, as well as to provide suggestions about how this new business could reduce the time to profit realization. The case also provides a vehicle to discuss the various pitfalls in this particular R&D process.

The case *Innovation and Renovation: The Nespresso Story* encompasses many of the issues highlighted earlier. The case traces the development of the Nespresso System in a 100%-owned affiliate deliberately placed outside of Nestlé's main organizational structure. Like the earlier cases, the Nespresso story touches on opportunity recognition, the characteristics of the corporate entrepreneur, and the necessary organizational structures. In addition, however, it covers the delicate transformation of an initiative from the entrepreneurial stage to institutionalization for sustained growth.

References

1   Shane, Scott and S. Venkataraman, "The Promise of Entrepreneurship as a Field of Research." *Academy of Management Review*, 2000, *25* (1), pp. 217–226.

2   Dess, Gregory G., G. T. Lumpkin, and Jeffrey E. McGee. "Linking Corporate Entrepreneur-
     ship to Strategy, Structure, and Process: Suggested Research Directions." *Entrepreneurship:*
     Theory & Practice, 1999, *23* (3), pp. 85–102.
3   Covin, Jeffrey G. and Morgan P. Miles. "Corporate Entrepreneurship and the Pursuit of
     Competitive Advantage." *Entrepreneurship: Theory & Practice*, 1999, *23* (4), pp. 47–63.
4   Dess, Gregory G., R. Duane Ireland, Shaker A. Zahra, Steven W. Floyd, Jay J. Janney, and
     Peter J. Lane. "Emerging Issues in Corporate Entrepreneurship." *Journal of Management*,
     2003, *29* (3), pp. 351–378.
5   Thornberry, Neal. "Corporate Entrepreneurship: Antidote or Oxymoron." *European Man-*
     agement Journal, 2001, *19* (5), pp. 526–533.
6   Covin and Miles. "Corporate Entrepreneurship and Competitive Advantage," pp. 47–63;
     Dess *et al.*, "Emerging Issues," pp. 351–378.
7   Covin and Miles, "Corporate Entrepreneurship and Competitive Advantage," pp. 47–63;
     Burgelman, Robert A. "Corporate Entrepreneurship and Strategic Management: Insights
     from a Process Study." *Management Science*, 1983, *29* (12), pp. 1349–1364.
8   Dess *et al.*, "Emerging Issues," pp. 351–378.
9   Kimberly, John R. "Issues in the Creation of Organizations: Initiation, Innovation, and
     Institutionalization." *The Academy of Management Journal*, 1979, *22* (3), pp. 437–457.
10   Burgelman, "Corporate Entrepreneurship and Strategic Management," pp. 1349–1364.
11   Thornberry, "Corporate Entrepreneurship: Antidote or Oxymoron," pp. 526–533; Burgel-
     man, "Corporate Entrepreneurship and Strategic Management," pp. 1349–1364.
12   Burgelman, "Corporate Entrepreneurship and Strategic Management," pp. 1349–1364
13   Burgelman, Robert A. "Designs for Corporate Entrepreneurship in Established Firms."
     California Management Review, 1984, *26* (3), pp. 154–166.
14   Kimberly, "Issues in the Creation of Organizations," pp. 437–457.

CASE 6-1
Rebuilding a Passion Brand:
The Turnaround of Ducati (A)

Dominique Turpin and Rebecca Chung

In mid-1996 the Italian motorcycle manufacturer Ducati Motor SpA was losing money and experiencing cash flow problems (*Exhibit 1*). American venture capitalists Texas Pacific Group (TPG) acquired Ducati and appointed Federico Minoli as the new CEO. As a management consultant from Bain and Co., Minoli had been involved in the due diligence for the acquisition, and his mandate now was to turn Ducati around and define a new strategy for the long-term viability of the company.

Minoli believed that this was an attractive opportunity for TPG since Ducati had several important assets: top-notch engineers and sophisticated engine technology; stylish bike design; and a racing team that had achieved numerous victories and much media attention. An earlier survey had shown that the main reason customers chose Ducati was for its racing achievements. In addition, in major European markets, Ducati's customer loyalty was solid. For example, in the key markets of Italy, Germany, and France, about 64%, 54%, and 52% of customers, respectively, expressed their intention to purchase Ducati again.[1]

However, Ducati faced many challenges. The company did not have money to pay its suppliers, so they refused to deliver any more, leaving the factory idle, employees frustrated, and thousands of orders unfulfilled. Even worse, the quality of Ducati motorbikes had deteriorated and warranty costs were soaring. As a result, even though some Ducati fans originally described its Superbike 916 model as "the bike of the century" or "the sexiest machine in the world," they considered turning to another brand because they did not want to pay a price premium and have to wait for at least a year to receive a motorbike that would have poor craftsmanship, break down easily, or even catch fire in some instances.

By fall 1996 Minoli had moved to Ducati's headquarters in Bologna. He now had to set his priorities for the coming few years to rebuild the troubled company.

Copyright © 2004 by IMD – International Institute for Management Development, Lausanne, Switzerland. Not to be used or reproduced without written permission directly from IMD.
[1] Source: CSM International 1995; 3,000 motorcycle customers were sampled for Italy, 2,000 for France, and 2,500 for Germany.

1 Federico Minoli

Minoli, aged 54, was born in Gallarate, Italy, originally the home of the famous motorcycle-racing icon Meccanica Verghera. He began his career at Procter and Gamble, Italy, in 1974. A few years later, he joined McKinsey and became involved in strategic change in the fashion and entertainment industries. He then became CEO of the US subsidiary of the Italian apparel company Benetton and turned it around in less than four years. After that, Minoli returned to his role as a management consultant, at Bain and Co. in Boston, and specialized in turn-around management.

2 Situation at Ducati in 1996

2.1 Evolution of the Company

During the due diligence, Minoli was surprised to find very little internal information about Ducati's history. The company seemed to suffer from corporate amnesia. The key information that Minoli managed to uncover, although patchy, showed Ducati's strong Italian heritage and glorious racing victories.

In 1926 the Ducati family had founded the company, a radio and electrical components business, just outside Bologna. At one point, it was even the second largest manufacturing company in Italy. During World War II, the company produced materials for the Italian army, but the Allies bombed the factory in 1943. After the war, the founders had no money left, so the Italian government took over Ducati through the state holding company, Institute for Industrial Reconstruction (IRI). In the late 1940s the company moved into the motorcycle industry when the Ducati brothers bolted Il Cucciolo (literally, the puppy), a four-stroke engine of less than 48 cc, to a bicycle frame to create the first motorcycle in Italy (*Figure 1*). As the motorcycle was fast but inexpensive, it became a blockbuster in Italy.

In the 1950s Ducati started to develop engines for motorbike racing. In 1955 Ducati's legendary engineer Fabio Taglioni made his first design for Ducati – a single-cylinder 100 cc engine that drove the Marianna model (*Figure 2*). Within two years, Marianna had won three victories in the Motogiro d'Italy race and two in the Milan-Taranto race. In 1957 Taglioni made a breakthrough, creating the Desmodromic engine. Minoli commented:

> The Desmodromic engine makes Ducati unique! It had a valve operating system, which allowed the engine to make more revolutions per minute and generate greater power.

Figure 1 Il Cucciolo
Source: Company documents

For many years, Ducati and its Desmodromic engine dominated the World Super-bike Championships, the race devoted to motorcycles derived from production rather than prototype. The desmo engine was fitted on every bike, and its deep pulsing resonance soon came to be identified as the signature sound of a Ducati.[2]

In the early 1980s, IRI decided to shift its investment from the motorcycle business to other businesses, thus reducing Ducati's ability to invest in engine innovation and racing. In 1983 an Italian conglomerate and manufacturer of small road motorcycles, the Cagiva Group, bought IRI's shareholding in Ducati. The company quickly regained its reputation in engine innovation and racing. But due to poor management, it started to suffer liquidity and quality problems in the mid-1990s. In 1996 Ducati was close to bankruptcy, and TPG bought a 51% stake in the company for € 36.8 million.

Figure 2 On November 30, 1956 Marianna broke 44 speed records at Monza, with an average speed of 160 km/h
Source: Company documents

---

[2] To hear the sound of the Desmodromic engine, click http://www.ducati.com/bikes/my2004/superbike.jhtml, choose a model, and then click "sounds." In the 1990s some Ducati fans in Sweden even recorded and globally released a CD of the engine sounds of about 20 different Ducati models.

2.2 The Motorcycle Industry

2.2.1 Market Segmentation

The motorcycle market could be divided into the following segments, sub-segments, and categories (*Figure 3*). Each category was different in its technical and design features as well as its target customers.

Road motorcycles with an engine capacity of more than 500 cc were mainly for recreational purposes such as weekend leisure riding, offroad adventure, long-distance touring, and racing. Recreational motorbikes were generally more expensive than those that were used essentially as a means of transport. More than 95% of riders were male. In 1996 the global market size of this segment was approximately 1.2 million units, with Europe, Japan, and the USA constituting 45.6%, 27.2%, and 25.4%, respectively. The European, Japanese, and US markets were projected to expand at a compound annual growth rate – in units – of 1.3%, 0%, and 4.3%, respectively, until 1998.

The cruiser subsegment targeted easy riders, emphasizing style over comfort and speed. Dual bikes, as the name suggests, were multipurpose, designed for both onroad and offroad uses. Touring bikes emphasized comfort for long-distance travel. All these subsegments targeted middle-aged customers.

Ducati focused on the sports subsegment, which targeted younger, fast riders, who dreamed of "knee sliding" on the curves. In 1996 the global market size of

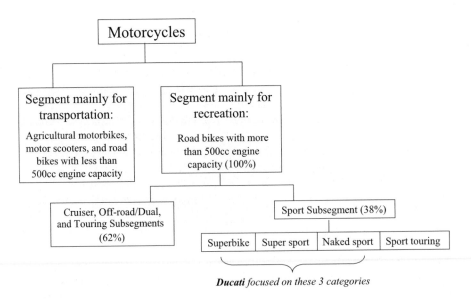

Figure 3 The Motorcycle Market

Figure 4 Superbike World
Championship: Carl Fogarty
Source: Company documents

this subsegment was about 38% of the over 500 cc road motorbike market. The design of sportbikes was inspired by racing and emphasized speed, so the frame was lighter and the rider sat further forward than on other bikes.

There were four categories of bikes in this subsegment:

- *Superbikes*, the closest to racing bikes, offered top performance for racing professionals and enthusiasts; age group: 18 to 35 (*Figure 4*).
- *Super sportbikes* offered high performance and great handling; age group: 18 to 35.
- *Naked sportbikes* offered good performance, lightweight and sensational design, making them perfect for racing beginners; age group: 18 to 35.
- *Touring sportbikes* offered speed as well as comfort to long-distance travelers; age group: over 30.

2.2.2 Competition

Japanese manufacturers such as Honda, Suzuki, Yamaha, and Kawasaki emerged in the 1970s as efficient mass producers of motorcycles. Japanese motorbikes were reliable and competed on price, gaining significant market share from smaller European manufacturers like BSA, Enfield, Norton, Agusta, Laverda, Moto Guzzi, and Triumph, and even putting some of them out of business. In the mid-1990s, these Japanese manufacturing giants made both large and small motorbikes. They dominated the offroad/dual, touring, and sport subsegments for over 500 cc road bikes (*see Exhibit 2 for market presence and share*). Honda was the largest motorbike manufacturer. In 1995 it sold five million large and small motorcycles and had more models than all its competitors.

BMW, a leading German automaker, sold 50,000 motorcycles in 1995 and was present in the sport, offroad/dual, and touring subsegments. The company focused on technical innovations for the whole bike, pioneering, for example, antilock brakes. Its motorbikes were renowned for excellent quality, safety, reliability, and comfort.

The American Harley-Davidson was perceived as a heritage brand. When the company was founded in 1903, it developed a culture associated with freedom and rebellion and, hence, had a strong following in the USA. Harley-Davidson was also renowned for its emphasis on lifestyle. Although it had not entered the sport subsegment, it dominated the cruiser subsegment, especially in the US market, and was also present in the touring subsegment. In 1995 it sold 105,000 motorcycles, more than 70% of them in the USA. The median age of its owners was 46. Of all the manufacturers, Harley-Davidson paid the most attention to apparel and accessories, which accounted for 16% of its sales. It was also successful in organizing the world's largest club of motorbike owners, with 600,000 members.

In 1995 Ducati sold 20,000 motorbikes. Europe accounted for 77% of Ducati's sales, with key markets in Italy, Germany, France, and the UK. Only 12% of Ducati's sales were in the USA, 8% in Japan, and 3% in the rest of the world.

Although superbikes were Ducati's flagship products, the company's most popular line was the lower-priced naked Monster sportbike. The lower seat of the Monster also appealed to female riders. The revenue contribution of the superbike, super sport, and sport naked lines was 42.8%, 30.1%, and 27.1%, respectively. Like most of its competitors – except for Harley-Davidson – less than 1% of Ducati's revenue came from accessories and none from apparel.

Ducati set its prices at a significant premium over comparable models. For example, the price premiums of the Superbike 916 in the USA, Germany, and Italy were 48.6%, 42.2%, and 21.3%, respectively. In Japan, the price premium of the Monster M900 was as high as 47.2%. The average price of a Ducati bike in 1996 was about € 7,650. In terms of sensation and image, some fans regarded Ducati as the "Ferrari" of the motorcycle industry.

Because of the sporty nature of the bikes, most Ducati customers were aged between 25 and 35. At least 30% of Ducati owners had more than one bike. Minoli added:

> Ducati owners are more into the mechanics of the bikes. They enjoy fixing the bikes and changing spare parts. As well, although we are "red" like Ferrari, all our fans are our customers while fans of Ferrari are not always their customers.

2.3 Financial Issues

Back in the early 1990s the Cagiva Group had a great deal of control over Ducati's finance and administration (*see Exhibit 3 for the original organizational chart*). Ducati relied heavily on the Cagiva Group for product design, information system, human resources, advertising, and distribution. Minoli found out during the due diligence that Ducati's accounting was largely mixed with Cagiva's. He recalled:

> It was very difficult to have a clear picture of where the money went. For example, the transfer prices between Cagiva and Ducati were not clear.

TPG's due diligence team cleaned up the data and discovered that if the transactions were at arm's length, Ducati could save € 3.5 million a year.

In addition, although Ducati's growth in market share and profitability had been strong until 1995 (*see Exhibit 1 for financial performance*), the company had a problem in cash-flow management. It held inventory too long and was not able to collect receivables fast enough (*Table 1*). Therefore, it did not have enough cash to pay suppliers on time.

Table 1 Inventory Calculations

	Ducati in 1996 (days of sales)	**Expected by TPG** (days of sales)
Inventory	120	40
Accounts receivable	90	60
Accounts payable	100	60

Source: Company literature

By the end of 1995, Ducati's accounts payable and debt amounted to € 20.7 million and € 93.8 million, respectively. Late accounts payable accumulated and, by June 1996, was estimated to be as high as € 17 million. Payment delays made Ducati's suppliers – especially those that relied heavily on the company – refuse to deliver any more. Missing components led to a large number of "work-in-progress" bikes and an average output of only ten bikes a day (maximum production was 140 bikes per day). The factory workers were almost idle. By mid-1996 Ducati had accumulated 6,000 unfulfilled orders. Customers had to wait for more than a year to receive a Ducati bike that they had ordered (the average waiting time for a Harley-Davidson was six months.)

2.4 Marketing Issues

When Minoli came on board at Ducati, there was no real marketing department. He commented:

> The existing management just naturally focused on the niche sport subsegment but was unclear about Ducati's product positioning against its competitors. There was no marketing strategy. Again, Ducati relied heavily on the trading arm of the Cagiva Group – Cagiva Trading – for advertising and distribution. Marketing efforts were insufficient. For example, there was minimal advertising in specialist magazines such as Motorcycle Consumer News and Sport Rider. The distribution services Cagiva Trading provided were expensive but neither exclusive nor customer-focused. Cagiva Trading charged 70% of its costs, € 10.5 million, to Ducati. However, activity-based analysis by the due diligence team showed that only 54% of these costs were attributable to Ducati.

Like its competitors (except for Harley-Davidson), Ducati did not have its own retail stores. It used dealers as retail outlets. Cagiva Trading had a sales force to manage 154 dealers for Ducati in Italy. These dealers sold 3,700 Ducati motorbikes in 1995. For the USA, Cagiva Trading had a subsidiary to manage about 200 dealers, which sold about 2,000 Ducati bikes in 1995. For other European countries and the rest of the world, Cagiva Trading used a group of importers – most of them personal friends of the head of Cagiva Trading – to deal with about 1,000 dealers, which sold about 15,000 Ducati bikes in 1995.

The dealers could be divided into two types: Cagiva's network, which might sell Cagiva's own brand and another brand – Husqvarna – in addition to Ducati, and multifranchises selling other manufacturers' brands as well. In Italy, for example, 107 dealers were multifranchises and 47 were Cagiva's network members. Of the latter, 28 dealers carried Ducati as well as Cagiva and Husqvarna, 17 carried Ducati and Cagiva, and only 2 carried Ducati exclusively. Ducati was not a significant brand for these Cagiva network dealers – 60% of them generated less than 50% of their business from Ducati.

Because of nonexclusivity, the average number of bikes sold in Ducati's multibrand dealers was much lower than the number sold in an exclusive dealership for the competition, such as Honda (*Table 2*). Additionally, dealers of Japanese brands could sell products other than >500 cc bikes, such as smaller bikes and scooters of the same brand, thus lowering their break-even points. Moreover, the dealers were often unable to provide customers with satisfying sales assistance and after-sales service.

Table 2 Bikes Sold by Competitors

	Honda	Yamaha	Suzuki	Kawasaki	BMW	Ducati
No. of dealers in Italy	200	190	130	110	100	154
Average no. of bikes (>500 cc) sold per dealer	60	37	30	27	35	24

Source: Company literature

However, it was difficult to turn multibrand dealers into exclusive ones because each dealer selling exclusively >500 cc bikes needed a sales volume of at least 100 bikes per year to break even. In the USA and France, for example, the average number of bikes sold per year was only 11 (*see table below*). Therefore, dealers were unlikely to carry Ducati exclusively. Even worse, these dealers had territorial exclusivity, which made it difficult for Ducati to shift to better dealers that could sell more bikes.

Table 3 Bikes Sold per Country

	Germany	France	UK	USA	Japan
Units sold in major markets	2,500	2,008	1,870	2,367	1,596
No. of dealers	105	180	47	213	51
Average no. of bikes sold per dealer	24	11	24	11	31

Source: Company literature

Figure 5 Motorcycle
assembly at Ducati's factory
Source: Company documents

Minoli added:

> Another headache was the distressed relationships between Cagiva Trading and the importers. For example, Cagiva asked importers to pay cash advances and buy a few Cagiva products if they wanted to buy a Ducati 916 bike. Therefore, if we fired an importer, it would be a double hit to us. First, the importer would not sell any bikes for us during the three-month notice for termination. Second, we needed to pay them back the cash advances received by Cagiva to get back the inventory. It was a lot of money!

2.5 Production Issues

Using the spare parts supplied by its vendors, Ducati carried out assembly and machining in its sole factory in Bologna (*Figure 5*). Minoli remembered that when he arrived the factory was worn out – for example, the roof leaked. Production flows were inefficient. It was difficult to implement process control because standards for worker productivity measurements were obsolete. In addition, the factory was still using the old-fashioned block order system, whereby orders were placed with suppliers only every four months. By contrast, Japanese manufacturers had implemented the just-in-time process long before, and most of Ducati's suppliers had started planning on a rolling basis.

Minoli also saw some potential problems with suppliers. In 1995 material costs of € 81 million constituted close to 90% of the total variable expenses. Ducati dealt with 420 vendors of these materials; 65 of them were major suppliers, representing 80% of total purchases. A quarter of these major suppliers had quality problems and were not financially strong. Some did not have enough capacity to satisfy Ducati's requirements when the company resumed maximum production.

Minoli was also alarmed by the escalation of the quality problems, which had resulted from poor production process control and below-standard suppliers. In 1995, 31% of final test time was spent on repairing defects. But in January, February, and March of 1996, the figures shot up to 72%, 100%, and 90%, respectively. Even worse, 50% of product technical refinements were made after the bikes were already on the market – and they were often triggered by customer complaints.

2.6 Organizational Issues

Minoli felt that, historically, Cagiva management had been paternalistic and rigid. For example, managers wore a coat and tie. The organization structure was R&D and production focused. Gianfranco Giorgini, operations director recalled, "Ducati was an engineering company for racing professionals."

Most of the staff had started working for Ducati immediately after graduation and were in their 40s. They shared a *status-quo* mentality. When Minoli took over, these employees were frustrated by the poor state of the company.

3 Quick Fixes

The first thing Minoli had to do was form his new management team. He hired David Gross, an international corporate attorney specialized in mergers and acquisitions, from the USA to be responsible for strategy development. Before going to law school, Gross had been a journalist for both Time and the New York Times Magazine, where he had covered popular culture and business. Minoli also hired Pierre Terblanche, who had worked at an external design studio for Ducati and was obsessed with motorbikes, to be responsible for design and engineering, and Christiano Silei, Minoli's former colleague at Bain and Co. who was involved in the due diligence for the Ducati acquisition as well, to be responsible for product development.

Minoli also brought with him other high achievers with diverse backgrounds from the USA. Some were from Ford Motor Company. Minoli commented:

> Although none of these young talents spoke Italian or had worked in the motorcycle industry, I preferred them because they were visionary, driving, passionate, and not afraid of change and risks. However, I knew that bringing people from Detroit with thinking and habits different from Italians was a potential recipe for disaster.

From the outset, Minoli and his "foreign" management team wore business casual wear to work to signal change. The team had a long "to do" list, including letting managers and office employees who were unreliable or incompetent go. Minoli commented:

> I had the luxury from the due diligence of finding out exactly whom I needed to fire. Basically, only the R&D department stayed.

In addition to taking back full control of finance and administration from the Cagiva Group, Minoli added a sales and distribution department to the organization structure to take care of marketing management (*Exhibit 4*).

Another step was to pay all the suppliers. Within a week, Minoli had written 4,000 checks. He recalled:

> Three weeks after sending out the checks, the local priest thanked me because he was able to collect more money during the masses.

For the first eight months of 1996, the factory had produced only 4,000 bikes. Minoli aimed to ramp up production to maximum capacity – 140 bikes a day – in the last four months to produce 7,000 bikes to clear the backlog of orders. He therefore needed to hire more workers, but it was difficult because of the company's bad reputation. He was left with two options: women from a faltering intimate apparel factory or new immigrants. Minoli remarked:

> Ducati had never hired women to work in the factory before. We did not have separate bathrooms for women.

4 Priorities for the Next Few Years

After solving these operational problems, Minoli had other immediate decisions to make. For example, the workers' union protested that the new "American" management of Ducati did not care about the working conditions of the workers and urged them to fix the leaking roof of the factory. It would cost € 1 million. The new director of the sales and distribution department had recommended building a museum like Ferrari's to regain the pride of Ducati in the early 1920s. It would cost another € 1 million. The R&D department was working hard to persuade Minoli to refocus on engine innovation and enter the Grand Prix Championship, a prestigious racing competition from which Ducati was still absent. R&D expenses and racing costs would amount to at least € 4 million a year.

Minoli knew that he would not have enough money to fund all these new projects and he had to set priorities quickly. Although some of his financial problems had been solved by restoring the suppliers' confidence, his list of problems to be fixed was still long: how to restore staff morale, what should be the new marketing strategy for Ducati, and how to reinject passion into the brand.

Exhibit 1
Ducati Financial Performance 1993–1996

	1993	1994	1995	1996 estimates
Motorcycles sold (units)	16,507	18,163	20,017	13,480
- year-over-year growth		+10.0%	+10.2%	-32.7%
Average unit price per motorcycle (€)	4,287	5,165	6,611	7,644
- year-over-year growth		+20.5%	+28.0%	+15.6%
Motorcycle revenue	70.8	93.8	132.3	103.0
- year-over-year growth		+32.6%	+41.1%	-22.1%
Total revenue (motorcycle + others)	84.5	105.5	150.8	103.0
- year-over-year growth		+24.9%	+42.9%	-31.7%
Total variable expenses	54.3	68.5	91.5	66.2
Variable contribution	30.2	37.0	59.3	36.8
Selling, general, and administrative expenses	15.6	17.8	28.1	23.8
EBITDA*	14.6	19.2	31.2	13.0
- year-over-year growth		+32.0%	+62.8%	-58.1%
EBITDA margin	17.2%	18.2%	20.7%	12.7%
EBIT**	11.4	16.1	28.1	(0.7)
- year-over-year growth		+40.7%	+74.9%	-102.6%
EBIT margin	13.5%	15.2%	18.6%	(0.7%)
Market share	3.7%	4.0%	4.5%	3.0%

Notes: All figures are in € million[3] except where specified. Some figures are approximate because Ducati's accounting was mixed with the Cagiva Group. Figures for 1996 are estimates based on preliminary figures for January through June.
*EBITDA = earnings before interest, taxes, depreciation, and amortization
** EBIT = earnings before interest and taxes
Source: Company literature

---

[3] €1 = 1,936.27 lire.

Exhibit 2
Competitor Presence in Different Subsegments and Market Share Information in Sport Subsegment for 1995

	Cruiser	Offroad/Dual	Touring	Sportbikes				Total
				Superbike	Super Sport	Naked Sport	Touring Sport	
Honda	✓	✓	✓	23.4%	42.2%	26.1%	16.4%	
Yamaha	✓	✓	✓	23.9%	4.6%	9.0%	29.2%	
Kawasaki	✓	✓	✓	21.9%	18.7%	7.0%	20.5%	
Suzuki	✓	✓	✓	19.3%	17.8%	28.2%	10.1%	
Subtotal (Japanese brands)		Dominating	Dominating	88.5% Dominating	83.3% Dominating	70.3% Dominating	76.2% Dominating	
BMW	✗	✓	✓	✗	✗	14.8%	19.0%	
Ducati	✗	✗	✗	11.3%	15.8%	9.2%	✗	
Others, e.g., Triumph	✓	✓	✓	0.2%	0.9%	5.7%	4.8%	
Harley-Davidson	✓ Dominating	✗	✓	✗	✗	✗	✗	
Total				100%	100%	100%	100%	

Notes: ✗ = absent in subsegment or category ✓ = present in subsegment
Source: Company literature

Exhibit 3
Original Organization Structure before Takeover

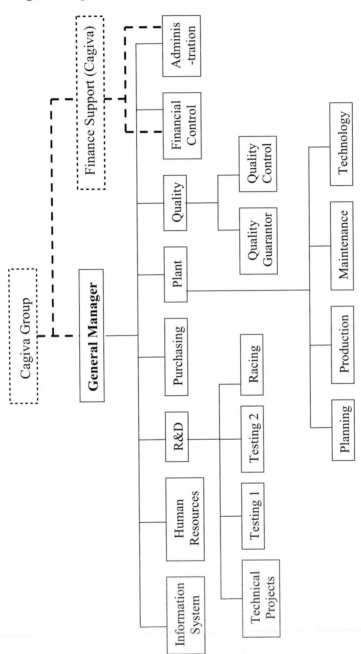

Source: Company literature

Exhibit 4
New Organization Structure after TPG Takeover

Source: Company literature

CASE 6-2
Rebuilding a Passion Brand:
The Turnaround of Ducati (B)

Dominique Turpin and Rebecca Chung

Minoli and his management team decided to turn Ducati from "a metal-mechanical company" to "an entertainment company based on a passionate brand." They believed that building the brand should be their top priority and that rationalizing the factory and refocusing on R&D could come later (*Figure 1*):

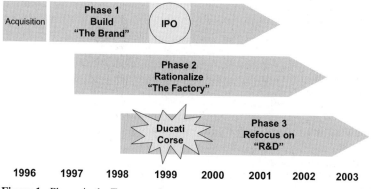

Figure 1   Phases in the Turnaround

Copyright © 2004 by IMD – International Institute for Management Development, Lausanne, Switzerland. Not to be used or reproduced without written permission directly from IMD.

1 Phase 1: Build the Brand

They identified the five components for building the brand (*Figure 2*).

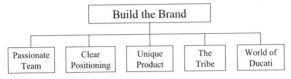

Figure 2   Brand Building Components

1.1 Passionate Team

Minoli described the importance of forming a passionate team to turn Ducati around:

> It was difficult to build a new culture at the beginning. I tried to find the common ground between my American and Italian staff, and then, I realized that the passion for Ducati was the glue. The turnaround will be successful only if everybody wants to be part of it…

Minoli explained how he nourished his employees' passion for Ducati:

> When I first came here, I saw a parking lot reserved for top managers in front of the main building. Everybody else parked far away. Then I granted the employees who own Ducati bikes the privilege of parking their bikes there, while the parking spaces for top managers were moved further away…I made every employee go to motorbike school. We organized buses to take us to watch races together. I also allowed employees to use our bikes for going to Ducati events over weekends. And I offered discounts to staff who wanted to buy Ducati bikes.

1.2 Clear Positioning

In 1996 Ducati's product positioning was in the "performance and function" quadrant (*Figure 3*). While maintaining this positioning, Ducati decided to push the boundaries of its niche toward comfort and lifestyle in order to broaden its fan base and increase repeat purchases. Ducati thus entered the last category of the sport subsegment – sport touring – to target racing fans who were over 30 and looking for comfort in long-distance traveling. It also introduced special models of the Monster line to satisfy racing fans with a taste for lifestyle.

Ducati also aggressively targeted female customers, establishing a riding school for women, introducing apparel for them, and including them in its marketing communication.

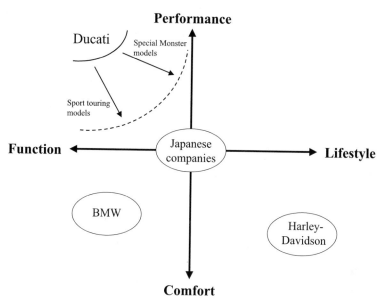

Figure 3   Product Positioning

1.3 Unique Product

The company clearly communicated the following unique brand characteristics of Ducati to its fans: Desmodromic valves, 90° L-shaped twin engine, tubular trellis frame, Italian design, and its signature sound.

1.4 The Tribe

Ducati also focused on building up its fan base. Minoli commented:

> I also realized that we don't actually have customers. All we have are passionate fans – Ducatisti! When I look at our fans, I cannot find any good ways of segmentation. Maybe the only thing they have in common is Ducati. I can think of them as a tribe in a village: it does not matter who they are, the link between them is the object they love and are passionate about. We used every opportunity to engage our fans in racing. In the Ducati tribe, what is important is not winning, but fighting together. We make them feel as if they are fighting and winning with us.

1.5 The "World of Ducati" Branding Strategy

Minoli and his management team developed the "World of Ducati" branding strategy and executed the components in parallel (*Figure 4*).

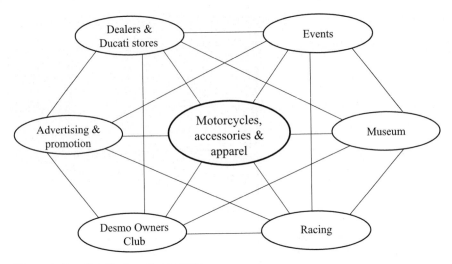

Figure 4 Branding Strategy Components

1.5.1 Motorcycles, Accessories, and Apparel

Ducati added accessories and apparel to its product portfolio. In 1997 it acquired a 50% stake in Gio. Ca. Moto, a producer of Ducati accessories, and in 2000 increased its stake to 99.9%. Ducati also cobranded with Dainese, a leading Italian manufacturer of technical riding gear, to sell apparel. By the end of 2003, the accessories and apparel business had grown from nothing in 1996 to € 28.3 million, 7.3% of Ducati's total sales.

1.5.2 Advertising and Promotion

To establish a consistent new image, Ducati immediately started an identity campaign. It commissioned a new logo and used it in every possible communication medium (*Figure 5*). In 1997 Ducati launched its first global ad campaign, "Ducati

Figure 5 Ducati Logo before (left) and after (right) the TPG Takeover
Source: Company documents

Figure 6   "Ducati People"
Ad Campaign 1997
Source: Company documents

People," featuring black-and-white retro photos of its employees with Ducati bikes at various places in Bologna (*Figure 6*). It aimed to allow Ducati fans to identify with young and sporty riders who were passionate about the company's Italian heritage and racing glory.

Minoli added:

> The people who make Ducati are the best ambassadors of the brand. I saw a great picture of Ducati in the 1920s in which all employees were in front of a building with a big Ducati sign. I commissioned a designer from New York to replicate this old picture with the new Ducati board members and employees, and then put both pictures on the Christmas card for the 1997 season (*Figures 7 and 8*). This was a message to the world that we were back to rebuild Ducati.

Figure 7   Ducati Employees in front of the Factory in the 1920s
Source: Company documents

Figure 8 Picture Replicated in 1997
Source: Company documents

In 2002 Ducati launched another "Ducati People" global ad campaign, using Ducatisti from all over the world to communicate the fans' passion for racing performance.

Over the years, Ducati bikes were featured in fashion and design magazines, department store showcases, exhibitions, television series, and movies. For example, Warner Brothers bought eight motorcycles from Ducati for some of the racing scenes in the blockbuster movie Matrix Reloaded.

1.5.3 Dealers and Ducati Stores

Minoli rationalized distribution in three stages:

- Ducati established wholly owned subsidiaries – replacing Cagiva's importers – to manage overseas dealers and handle promotions and after-sales service in principal markets like Germany, France, the UK, Benelux (Belgium, Netherlands, and Luxembourg), the USA, and Japan. The company hired 250 professionals, who had a passion for motorcycle racing and respect for customers, to join these subsidiaries.
- The company changed the dealer structure, forming a new network with greater emphasis on exclusivity and quality in order to reinforce Ducati's brand image and improve sales assistance and after-sales service. For example, in Italy, the number of dealers decreased from 154 in 1995 to 58 in 2003, and as a result, the average number of bikes sold per dealer per year improved from 24 in 1995 to 207 in 2003.

Figure 9   First Ducati Store in Manhattan
Source: Company documents

- Ducati also took the big step of creating of its own flagship stores to strengthen its brand image. The first Ducati store was opened in Manhattan in 1998 (*Figure 9*). By 2003, Ducati had opened 50 stores in Italy, 19 in Germany, 11 in France, 9 in the UK, 8 in Benelux, 3 in the USA, and 13 in Japan.

To reinforce its cutting-edge image, the company launched its virtual store www.ducati.com, becoming the first in the industry to sell bikes on the Internet. On January 1, 2000, within 31 minutes, it sold 1,000 units of the limited edition MH900e online. In July 2000 Ducati included apparel and accessories in the eStore for online sale. Ducati also used this channel to communicate with Ducatisti about racing news and product updates and to form a global Ducati tribe by enabling them to chat about mutual interests such as motorbike racing and maintenance. By 2003, about eight million people had visited the Web site and about 140,000 racing fans were registered as users.

1.5.4 Events

In June 1998 Ducati held the first World Ducati Weekend (WDW) at the Misano racetrack near Bologna to allow Ducatisti from all over the world to celebrate the company's successful turnaround. About 10,000 people attended the big party with their Ducati bikes. Ducati even brought its legendary engineer Taglioni (aged 77 and referred to as Dr. T by Ducatisti) from his hospital bed to join this milestone event. Minoli recalled:

> When Dr. T arrived, all the Ducatisti welcomed him with the roar of 10,000 engines. It was an intensely emotional moment.

The company organized the WDW again in 2000 and 2002, attracting even more Ducatisti each time – 23,000 and 25,000 respectively.

1.5.5 Museum

Minoli decided to build a museum before fixing the factory roof. He explained:

> We could have done both. But I like to dramatize. I wanted to tell the employees that we are a marketing-oriented company. So I chose to build a museum to send the message out. And eventually, we fixed the factory.

Minoli hired Marco Montemaggi to build the Ducati museum to restore its pride and showcase its new developments. Montemaggi talked to Ducati employees from the early days to collect interesting stories and traveled around the world to convince collectors to lend the museum their racing bikes and unique Ducati memorabilia. Montemaggi recalled:

> It was like a competition. After I got the first collector bike, the other collectors were keen for their bikes to be in the museum too! With their enthusiastic support, we were able to open the Ducati Museum within three months, in June 1998.

Minoli realized that Ducati's workers were becoming increasingly proud of their work and some even wanted to show their working environment to their families. Hence, he turned the factory into a showroom and opened it to the general public for visits seven days a week. He commented, "I want the public to see what goes into building the bikes." Every year, more than 10,000 people visited the museum, together with the factory downstairs.

1.5.6 Racing

Since racing was the root of the brand, in 1999 Ducati created a subsidiary, Ducati Corse, and appointed Claudio Domenicali as CEO. Domenicali's mandate was to improve racing knowledge and performance and to generate profits from racing. By the end of 2003, Ducati Corse's staff had grown to 90 – 52% of them graduate engineers – with an average age of 32. In 2003 it spent € 9.8 million on R&D and was able to generate sponsorship revenue of € 22.1 to cover almost 85% of the racing and research costs. (In 1996 Ducati had spent € 3.9 million on racing with no sponsorship revenue.)

1.5.7 Desmo Owners Club

In 1999 Ducati set up the Desmo Owners Clubs (DOC) to intensify the passion among Ducati fans. By 2003, there were more than 400 clubs worldwide.

2 Phase 2: Rationalize the Factory

In 1997 Ducati started implementing KAIZEN[1] to streamline the assembly process and also introduced just-in-time production (*Figure 10*). Productivity increased from 140 bikes a day in 1995 to 220 bikes a day in 2003. In addition, the time taken to build a bike decreased from seven days in 1995 to just ten hours in 2003.

Ducati also chose some high-quality vendors of key components to be its principal suppliers and had them deal with the subsuppliers. Thus the number of suppliers Ducati dealt with directly decreased from 420 in 1995 to 185 in 2003.

Figure 10　Factory before and after rationalization
Source: Company documents

3 Phase 3: Refocus on R&D

With the branding strategy on track and production and supply chain improvements introduced, Ducati began investing in engine innovation again in 1998.

---

[1] KAIZEN: A Japanese approach in which improvement is gained in small steps, with the help and collaboration of every worker involved and without any significant investment.

Although the company bought some top-notch design software, the engineers needed two years to learn how to use it and product introduction was delayed. However, once the technology was adopted, product development time decreased from 36 months in 1995 to 15 months.

The engineering team was then able to refocus its efforts on innovation. For example, the Superbike 999 and the new Dual product line – Multistrada – were well received. The R&D team also developed the Desmosedici MotoGP, the company's fastest engine to date. With this bike, it made a debut in the 2003 MotoGP, another popular and influential motorbike race. Domenicali revealed their plan:

> We've been very strong at developing engines. Now, we want to be strong at developing both the engine and the bike as a whole. We are developing a proprietary dynamic simulation software for bikes to learn how to improve different aspects, such as suspension.

4 Results

Ducati successfully went public on the Italian Stock Exchange in 1999. From 1996 to 2003 the company's net sales and EBITDA[2] increased at a compound annual growth rate of 20.4% and 21.1%, respectively (*Exhibit 1*). Its market share increased from 3.0% in 1996 to 5.6% in 2003. It successfully expanded in the US market: The number of motorbikes sold in the USA was up from 2,000 in 1995 to 4,575 in 2003. The company had grown from 500 employees in March 1996 to 1,100 employees – including 980 factory workers – in 2003. By 2003, Ducati had won 13 of the last 14 World Superbike Championship titles and ridden to victory in its debut MotoGP race.

Exhibit 1
Ducati Financial Performance 1996 and 2003

	1996	2003	CAGR**
Net sales (€ million)	105.8	388.2	20.4%
Value of production (€ million)	130.6	428.5	
EBITDA* (€ million)	11.8	45.2	21.1%
EBITDA as % of net sales	11.2%	11.6%	
Net profit (loss) (€ million)	(21.9)	0.04	
Motorcycles registered (units)	12,117	38,128	17.8%
Market share	3.0%	5.6%	

* EBITDA = earnings before interest, taxes, depreciation, and amortization
** CAGR = compound annual growth rate
Source: Company literature

---

[2] EBITDA = earnings before interest, taxes, depreciation, and amortization.

CASE 6-3
Innovation Leadership at Logitech

Jean-Philippe Deschamps and Atul Pahwa

> Even though we might soon be a $1 billion company, I still look at Logitech as being a small company, made up of business units, each of them having to survive in this world. Each of them acts or should act like a startup. In a startup, the reason to be is only to innovate; so, for us, innovation – and I would even extrapolate for many companies nowadays – is the way to survive.
>
> *Daniel Borel, cofounder and chairman of the board, 2002*

Logitech passed the one billion mark in sales in fiscal year 2002/2003. During the preceding five years, its stock price had risen over 580% while the NASDAQ stock exchange, on which it was quoted, lost ground. With a discipline for consistent product innovations in PC peripherals, the company had received several industry accolades and had realized considerable success with word-of-mouth advertising.

As the company's senior management prepared for their next annual general meeting, to be held on June 26, 2003, they could not help but ponder over several of the challenges the company would face during the course of the coming year:

- Can we keep meeting the market's expectations quarter by quarter? What will happen if/when our markets saturate? How do we find new growth opportunities in a maturing market?
- Will we be able to leverage our key technical, operational, and marketing strengths to enter new markets that may be further away from our familiar PC peripheral world, for example in the mobile telecommunications area?
- Will our current organizational culture and leadership resources bring us to the next stage in our growth? How do we expand without losing our startup mentality?

Copyright © 2003 by IMD – International Institute for Management Development, Lausanne, Switzerland. Not to be used or reproduced without written permission directly from IMD.

1 Company Background

Logitech started as a software development company before becoming the well-known specialist of peripheral devices that people use to work, play, and communicate with and through their computer.

In its early days Logitech installed a manufacturing facility with a capacity of 25,000 mice, even though the worldwide market was far below this capacity. But its founders saw the potential.

> We were a small group of people with a great dream. From the beginning, we dreamt of a day when Logitech would be established in the world market with a recognized name, providing fun and innovative products.
>
> *Daniel Borel*

Logitech's vision was to be the interface between technology and people, and the mouse was intended to be the tip of the iceberg of this vision. The idea was to make the personal computer personal and utilize the optimal combination of design, technology, and manufacturing processes to introduce appealing, user-friendly, and affordable products in the market.

> Maybe our products are special because the computer is a technology that often is under the table. The products we make, they are products you hold; they are products that must carry emotions, like your watch, which is your own, your keyboard, your mouse; you touch it, you feel it!
>
> *Daniel Borel*

Logitech's growth was linked with the development of the computer mouse. Its origins lay in the OEM sector, which remained an important part of its business. To its OEM customers, who included most of the world's largest PC manufacturers, Logitech stood for high volume, low cost, and quality in design and manufacturing. Its strengths lay in its responsiveness to changes in technology or market conditions and in its efficient worldwide distribution.

Over the years, the company built a strong presence in the retail sector as consumers enhanced their basic systems with more fully featured interface devices. Logitech had, indeed, dramatically broadened its product offering with a range of mice and track balls, keyboards, Web cams, audio speakers, gaming accessories, *etc.* These products added functionality and cordless freedom to desktops and catered to new applications such as gaming, multimedia, or visual communication on the Internet. Its latest product offerings included peripherals and accessories for the digital photography, PDA, mobile note taking, and wireless phone environments.

The company enjoyed strong brand recognition in retail outlets in over 100 countries, as well as on several Web-based retail sites. To provide the market with best of category in a broadening array of products, Logitech supplemented its internal engineering and manufacturing strength with additional products and technologies through strategic acquisitions and industry partnerships.

Logitech barely survived tough times between 1993 and 1995, when fierce competition from Asian imports, falling prices in the computer industry, and a high cost structure forced the company to restructure. Manufacturing was consolidated in China, management was reshuffled, and the company moved from being a functional organization to one built around business units. After two quarters in the red, Logitech resurfaced.

(For details on Logitech and a timeline of events see Exhibit 1. For selected financial data, including stock price, revenues, and profits, see Exhibit 2.)

2 Company Organization

In 2003 the company operated under a fairly classical matrix structure. Sales and marketing was divided into four zones: EMEA (Europe, Middle East, and Africa), Americas (including Australia), Asia-Pacific (excluding Japan), and Japan. It also operated three worldwide business units: control devices, video and audio, and interactive entertainment.

Business unit senior vice presidents, who also played the role of marketing directors for their respective businesses, reported directly to the CEO. In "Control Devices" – the largest business unit – there was also a vice president of engineering. Business units were comprised of a number of product units, focusing either on the OEM or retail sales channels (*Exhibit 3*). Some of them, as exemplified by the Retail Pointing Devices unit, which focused on mice sold in retail channels under the Logitech brand, could be very large. But unlike business units, a single person did not head product units. A team consisting of an engineering head and a product-marketing head, working in partnership, ran them jointly. As one manager explained:

> By forming these pairs, we made sure that we would remain a strong, engineering-driven company while not losing touch with the market. Business ownership is shared by both functions. At the end, they have to come to the business unit head and show a roadmap that is agreeable to both!

Making this "management duo" work as a team was one of the personal tasks of the business unit head:

> How do we make the cooperation between Marketing and Engineering work? Well, sometimes we don't! It requires a tight synchronization! But at the end of the day, they must come to an agreement.

Business Unit Senior Vice President

This pairing of marketing and engineering was extended to the operating level. Product managers from marketing regularly teamed up with project leaders from engineering. Both were sharing the responsibility for a given project while reporting to different bosses.

Besides the usual corporate functions, such as HR, MIS, finance, or legal affairs, Logitech had established a corporate business development function. Reporting directly to the CEO, the vice president of corporate business development had two roles pertaining to growth and innovation: (1) act as the implementer of the business development initiatives launched or approved by Logitech's executive team and (2) be a source of stimulation, ideas, and challenge to help business units find promising new business opportunities and potential partners to grow and expand in their markets and beyond.

3 Innovation Philosophy and Drivers

3.1 A Compelling Vision

Logitech's claim to fame was its ability to carve a market space for its brand next to the powerful PC manufacturers. Mice were standard peripherals with each PC sold for over a decade. Yet, with its determination to reshape the computing environment to one that enhanced the user experience, Logitech created products with emotion and color and captured the theme of "user-friendly innovation" as its official mission (*Exhibit 4*).

From its early days when he envisioned Logitech having "fun and innovative products," Borel felt that providing employees with such a vision gave them something to aspire to.

> With more than 50 industry "firsts" to its name, Logitech is dedicated to innovation. But more importantly, the company does not innovate for innovation's sake. We create new products that people want to buy and love to use. If people say, "Intel is inside," then I would say, "Logitech is outside." Outside meaning it is the visible part of the iceberg. This vision captures the minds of people working for us, and in some way continuously provokes them to say, "What could be the other way for us to bridge the gap between the technology itself and the end user?"

> *Daniel Borel*

Within Logitech, innovation went beyond the realm of product development. It also extended to systems and processes within the organization.

> Innovation is not just in product development. We try to be innovative in how we implement our systems so that they can grow and scale up.

> *Product Unit Engineering Vice President*

3.2 A Strong, Noncomplacent Push for Growth

Most product unit managers were used to working under a very strong top management pressure for continuous product renewal and growth. The pressure from the stock market – the tyranny of the quarterly performance outlook – was felt throughout, but it was accepted. Competition was an especially strong driver of managers' energy.

> Unless you are paranoid, you will sit back, be comfortable, and rest on your successes and say, "I know what is happening out there." We listen, we are humble when it comes to competition, and part of that humbleness comes from experience. Logitech has had its share of cycles, good and bad, and that reinforces that this success could all go away tomorrow!

> We have a great competitor in Microsoft. Microsoft is a wonderful brand, credible! They do great things. We challenge each other very nicely. We may do a new product and get them going. They may create a new product and they get us going. It is not that we are obsessed about Microsoft at all, but it creates points of energy. A periodic infusion of energy is the best way I can say it.

> *Business Unit Vice President*

3.3 A Focus on New Products

Logitech was continuously renewing and expanding its product lines with new introductions (*Exhibit 5*). In its 2002/2003 fiscal year, it introduced 91 new products. With the average market lifecycle of its products between 12 and 18 months, more than 50% of its annual revenues came from new products. To deliver consistently on such a demanding product launch schedule, the company had put a very strong emphasis on design.

> We don't talk about design anymore…it is a given! Logitech needs to create products that have emotion in their shape or color. The PC is now a mainstream consumer market. We moved from boring beige to more exciting colors.

> *Product Unit Marketing Director*

The company also kept a watchful eye on burgeoning industry trends. Realizing that the industry ecosystem was shaped by a multitude of players, Logitech worked closely with other companies to improve the overall user experience.

> We watch players who influence the hardware and software infrastructure. Microsoft introduced significant improvements into managing digital media with Windows XP. When we look at the media evolution, we can leverage some of these trends that are cool, useful, and fashionable. We then ask ourselves, "what can I make the user discover which is hidden in the operating system today? How can I take that and bring it visually into my product and market?"

> *Product Unit Marketing Director*

3.4 Openness to External Technologies and Ideas

The company had always been open to looking at both internal and external sources for new product innovations. Despite having deep engineering-based roots, Logitech had learned not to rely solely on its own technological resources. Even its engineers recognized that the important thing was not where the technology was sourced, but what ultimately delivered greatest value to the customer.

> A company like ours was started by engineers, and you can imagine that we tend to value and reward only engineered products made in-house. So, what we have had to develop over the years are people who are part of the organization, whose only job is to source technology or compare what is being developed inside with what is being developed outside.

> One of the characteristics of a leader of this innovation is to be open-minded, but I would say, open-minded in a way that goes much beyond accepting any ideas. It is to be able to get out of the box! This is the ability to take input from inside, from outside, to take technology here and there and do the equation that brings a product that will be unique for the end user at the end of the day. Whether we develop our own cordless technology or we buy it from outside, the consumer doesn't care, because he/she buys a Logitech product which will work no matter what, which has a cool design, an affordable price. He doesn't buy a cordless module made by Logitech!

> *Daniel Borel*

3.5 An Obsession with Costs

But the most striking aspect of Logitech's innovation philosophy was, perhaps, its obsession for developing very affordable, yet quality, products. It needed low costs to reach its ambitiously low, targeted retail price points (95% of its products retailed for less than $99 and most of them for under $49). Logitech senior management made a point of recognizing those engineers whose main task was to extract a penny out of a design to make the product affordable and profitable. Such a focus on costs actually helped make technology sourcing more attractive, as explained by the corporate business development head:

> Innovation is linked to costs! We have been drilled so much by the OEM market to try to analyze, reduce and factor in the ultimate cost, that this forces you to agree that the time to get there is a cost as well; so it is an incentive to use outside technologies to save time, thus costs. We had an even higher culture of costs than the Taiwanese! A lot of engineers don't like that, but it is a challenge. Ultimately, they can see it as rewarding. Today, we have a consumer brand valued by the market with the inherent cost structure of an OEM supplier!

But even when it sourced technology outside, Logitech found new ways to work on it and optimize it in terms of performance and/or features so as to adapt it for mass production and cost reduction.

> Given our strong OEM background, we cannot buy, retag and sell low-cost stuff. We are forced to think of buying technology, then bringing it inside and working on it.
>
> *Corporate Business Development Vice President*

4 Corporate Culture and Values

Corporate culture and values were very strongly felt at Logitech, despite the fact that many of them were more tacit or implicit than formally expressed[1] (*Exhibit 6*). Nevertheless, managers frequently referred to them in their discussion. It was not difficult to notice that conformance to these values was essential for thriving within the company, as stated explicitly by a senior manager:

> Logitech regroups people that share the values and accept to learn. People who come from larger companies or believe that they have a lot of experience may have a hard time to embrace our values. We make sure that we have the right people in the bus...generally young people, in the stage of their careers where they are very malleable. We look for the ability of people to jell, to team up, to see the big picture. People that are not paid by the hour but are paid for results and contribution. There are no 9 to 5 jobs at Logitech.
>
> *Product Unit Engineering Vice President*

4.1 A Sense of Passion, Dedication, and Commitment, from Top to Bottom

When Borel was looking to hand over the reins of the company to a new CEO, he too pointed to the importance of the cultural fit between the existing organization and the person he would ultimately choose. It was important to him that his successor should have a strong global experience, be a charismatic leader, have known business success and failure, and most importantly be passionate.

> I felt that I would want my child – Logitech – to be happy and successful, and I wanted to get a replacement for myself who was passionate!
>
> *Daniel Borel*

---

[1] In the absence of a written creed on values at the corporate headquarters – at least until recently – a product unit (Retail Pointing Devices) had taken it upon itself to reiterate its guiding principles in a set of values. The unit ensured that its functional managers transmitted these values to their teams.

The new CEO he selected, Guerrino De Luca, had been Apple's executive vice president for worldwide marketing. He brought with him the experience of a company that Logitech had long admired for its passion to create exciting new products with revolutionary designs.

> The key success to Logitech over the last five years has been the power and energy that Daniel and Guerrino can create together. Guerrino is an excellent businessman, excellent marketer. He is a tough guy when things are not working well, and he can be very demanding. He is brilliant on the operating side. Then you take Daniel who is more of an entrepreneur and a visionary, a little bit more of a dreamer, excellent for PR. He is a very charismatic person. The team is a very good team. They work well together.
>
> *Regional Sales and Marketing Vice President*

At all levels, Logitech managers made sure they hired the "right kind" of people for key positions. When asked what they looked for in a candidate, apart from basic intelligence and technical fit for the position, several senior managers replied that they looked for a sense of passion, emotion, and connection with the product.

> I would like someone to give me a sense of love of products, not necessarily from Logitech. Someone who is passionate about the product is also going to be passionate about the user.
>
> *Business Unit Vice President*

Another senior manager echoed the same sentiments, suggesting that one could generally spot the enthusiasm at the job interview. One reason for the success of the organization, according to him, was that the people believed in what the organization stood for and in the products that they were selling.

> We have people that are driven, self-motivated. I would rather have a "Smart Alec," fast-moving guy that irritates me on occasion, that I have to hold back, rather than a guy that's fast asleep, that I have to kick in the backside to get going. Experience is that the guy who's running fast has the intelligence that I am looking for. You can always hold him back and put him in the right direction. If you've got somebody that doesn't move fast, he's never going to move fast.
>
> *Regional Sales and Marketing Vice President*

Of course, that passion was expressed differently by different functions. Engineers were passionate about technologies and products; marketing and sales people about beating competitors.

One of the attributes of successful organizations is their attention and dedication to their core business, as Logitech realized during the crisis the company faced in the early 1990s where it lost sight of its bread-and-butter product – the mouse – as it turned to other products.

> Dedication is key. You need people living, breathing, dreaming of these products all the time. The worst we did, ten years ago, was having the mature business almost die. The new business is so much more "sexy" than the core business – you need to keep attention on the core business.
>
> *Product Unit Engineering Vice President*

While financial incentives were obviously a strong driver at Logitech, they were never the main driver.[2] What got employees enthusiastic was the job content. The fact that they saw the fruits of their labor in products sold worldwide was one of the most tangible positive attributes of working there.

> Some engineers doing the soldering have a million in their bank account! And those people still continue working! They didn't change anything in their life...they just continue to have this passion for their job.
>
> *Product Unit Engineering Vice President*

4.2 A Global Mindset

Logitech prided itself for its truly global culture, taking the best of each of the cultures of its various locations: Swiss engineering rigor and sense of perfection, US marketing and innovation finesse, Irish design excellence, Taiwanese pragmatic engineering talent, and Chinese cost-consciousness. Interestingly, in each location, the company was perceived as a local manufacturer. Top management made sure that local managers staffed each operation. Yet, staff exchanges across labs and offices were frequent.

Besides facilitating interoffice cooperation, this global mindset provided a healthy sanity check on competition and removed any tendency to be complacent, as expressed by a Swiss project leader after a three-year posting in Taiwan:

> Being in Taiwan gives you a strong sense of competition. The competition is just behind you. Copycat products generally appear six months after you launched an innovation. You have that in the US as well! The dangerous place is Switzerland because you do not see competition. It removes your sense of urgency. If Logitech had remained Swiss, it would have gone bust years ago!

4.3 A Willingness to Empower and Trust People Based on Openness and Sharing

From its startup time, Logitech had kept a strong culture of encouraging personal initiatives and empowering its managers. In the technical area, it meant giving project responsibilities to young engineers fairly early in their careers, but with coaching assistance. In all cases, it meant encouraging staff to call for help in case of problems.

> As a manager, if you have an issue, if you have a problem, you typically go to your boss or other managers and ask for advice on how to resolve it. At another Swiss multinational company I worked for, if you had a problem, you tried your best to hide it because, if anyone else found out about it, they'd use it to stab you in the back.
>
> *Regional Sales and Marketing Vice President*

---

[2] The stock boom of the late 1990s made a significant number of millionaires among the "old-timers" with stock options – over 100 at last count.

As part of his/her induction process, each engineer would typically spend an hour or two with every functional manager within the engineering team, not just his/her own manager. This allowed newcomers to understand what the issues, technologies, and work processes were across the organization. It also encouraged a culture of cross-collaboration among various product groups, promoted a family atmosphere, and reduced tensions between product groups.

> One of the key influencers is personal coaching. The engineering family coaches you. What we have is a strong learning process – it's not a process, it's more about information sharing. Part of the culture is transmitted in this one-to-one, or in the small team sessions. It is a very informal learning, through coaching and observing.
>
> *Engineering Program Manager*

In the market area, empowerment meant letting executives run their business without too much interference.

> Take our General Manager-France. He's like an entrepreneur. He's got a team of 12 people, he's got a lot of pressure on him, but he runs Logitech France as if it were his own company. And I think it's hard to work in a more centralized organization after you are used to that concept. If you empower people, it's certainly a motivating factor. Our philosophy is to find the right people in the local markets and empower them. A long time ago, we realized that you couldn't fully understand and manage it from a distance.
>
> *Regional Sales and Marketing Vice-President*

4.4 A Willingness to Take Risks and Accept Failures

Taking risks was not only tolerated but also encouraged, and that suggested an acceptance of failures. Logitech products tended to be near perfect, technically and functionally, but some had fizzled out on the market. Borel was the first to recognize that the Logitech museum, at the Swiss headquarters, contained a number of great products that did not make it. Many of those products, like the world's first digital camera, had simply been introduced too early.

> In our business, you would rather be right six times out of ten than two out of two. Because you need to take those risks to be able to capture the unique moment, the unique opportunity out there, which is going to make you different, which is going to make you successful, and as such you accept failure. You should not punish failure! Failure is part of the game.

> The innovative leader has to accept that, in part of his activity, there will be things that will not work, and people who have worked on it might be even more valuable afterwards because they will learn from their failure next time and adjust their gut feeling and their judgment for the future. There is a fine line between success and failure! What makes the difference is the experience. As people usually say: "good judgment comes from experience!" But they forget to say that experience comes from bad judgment!
>
> *Daniel Borel*

This sentiment was seconded by a product unit engineering vice president who reflected:

> There is a "no-blame" culture at Logitech, although it depends on the size of the mistake and how many times you repeat it. I have made a certain number of serious mistakes in the company in the almost fourteen years I have been there – one where I should have been in much trouble. But my boss dismissed it saying "You know, you have several successes, so that is good."

4.5 An Informal, Hands-on Management Style

To the visiting outsider, several Logitech managers – from Borel and De Luca down to the more junior operating level – shared common traits:

1. A strong sense of congeniality and informality. People dealt across hierarchical levels on a first-name basis and dressed casually.
2. A pragmatic, hands-on management style. This style was described vividly by the business unit head of control devices:

> I am a fairly involved executive. I am involved in product updates, and through these updates, the questions that I ask and the discussions we end up in are critical to them feeling what is important to me; it helps reinforce some of the priorities and values I want to convey. And secondly, it helps people feel a better sense that management cares, understands, appreciates and is sensitive to what the challenges are.

4.6 A Healthy Respect for Realism and Rigor in Execution

Through its past experience as an OEM supplier and its financial crisis in the early 1990s, Logitech learned the hard way the importance of rigor in execution, the nitty-gritty part of innovation.

> Our people have to have emotion at the level of design, at the level of technology. They have to be passionate in what they are doing. And yet – which is a very difficult thing in managing innovation – you have to deliver the product in a way where execution is perfect. Otherwise you might end up just with innovation for the sake of innovation, leaving you nowhere! Execution is down-to-earth! Execution, sometimes, is not something that makes people dream. The real leader is the one who appreciates the passion aspect, the emotion, but yet is able to put it in a framework where execution is going to deliver to the end-user, to the customer, a product which eventually will be profitable for the company.

> *Daniel Borel*

A senior manager summed up the need for such implementation rigor. Given the size of the business unit and the kinds of volume he was dealing with on an annual basis, he said:

> If I have a problem, a major issue – if I need to recall 500,000 mice – I'm dead!

4.7 An Acceptance of Constructive Confrontation

Logitech managers were expected to speak their mind, and challenging each other was accepted, as explained by a young program manager:

> There are tons of conflicts but we have to solve them. I often disagreed with my boss. If the conflict is irrational, based on gut feel, then you are in trouble. But, in most cases, the conflicts are about facts, not persons."

Senior managers made a point to live by the tacit rule, as stressed by one of them:

> We are probably about as democratic as a company can get. People question my decisions. I'll sit down. I won't throw them out of my office and I'll talk to them, no matter who it is, and if the person comes back to me with some good valid points, then I'll change my mind.

But some managers saw things a bit differently and feared that the culture was becoming too tolerant. When asked whether managers challenged each other as peers, *i.e.*, across business or product units, he declared:

> We are a very gentle company…too much sometimes! We are very friendly with each other. Sometimes, we are not challenging enough. Many times I complained that we are too gentle. So, Guerrino is pushing to have the business units working more and more together. He forces them to challenge each other on how we can differentiate ourselves from our competitors.

Senior Technologist

4.8 A Quest for 24-hour Efficiency

Time, not just cost-efficiency, was a real concern at Logitech, and the company was an eager adopter of new information and communication technologies. Engineering and manufacturing databases were centralized and accessible by all engineers, no matter where they were located worldwide. E-mail and Web-based messenger services were used extensively, and the people in the company were avid users of the latest innovation, the digital pen.

More surprisingly perhaps, the company promoted absolute transparency to ensure that everyone worked with the same information; all documents – even the more strategic ones – were available online to all people potentially concerned. And certain managers went so far as to encourage people to work on e-mail during meetings, recognizing that not all issues were relevant to all attendees.

The location of Logitech's engineering offices and plants in all corners of the world helped make optimal use of a 24-hour day, as a senior technologist explained:

> When I have a problem one evening, I can explain it by e-mail to my colleagues in California. They will work on it while I'm asleep, then possibly pass it on to our colleagues in China or Taiwan to work on it. When I get back to the office, the next morning, more often than not, I find an answer to my questions.

5 Innovation Process

5.1 Strategy and Planning

Given its dynamic industry environment and demanding investors, Logitech felt the need to be on top of its sales forecasts and budgets on a very regular basis. Three-day operational meetings were held every quarter with the full top management team (the CEO and his 25 top officers.) Their main focus was to review quarterly business unit results and plans and discuss product and technology roadmaps. Parallel to this, budgets were redone each quarter based on prevailing market conditions. For example, projects were reprioritized if there was more or less money than originally budgeted. Products, markets, sales, and competition took up the majority of these meetings.

In addition, each year in January a more restricted group of top managers would meet for a one-week strategic planning retreat offsite. During that week, participants typically reviewed a three-year roadmap of each business unit, new business ventures, and the upcoming annual budget. Time was also spent on discussing core competences of the company, its vision, mission, and values.

5.2 Market, Competitor, and Customer Intelligence

Logitech had two marketing functions that provided insights on the market. The regional sales and marketing organizations were in day-to-day contact with customers and retailers, thus indirectly with competitors. Their focus was clearly on selling, not creating new products. Nevertheless, they provided invaluable insights on market developments, channel strategies, competitors' perceived tactical moves, and customer reactions, albeit within a rather short time horizon. The product marketing organizations within each business and product unit were responsible for integrating all market inputs, developing product strategies and roadmaps, and specifying the new products to be developed.

Engineering departments were also deeply involved in discussions on market trends and customer preferences and in decisions on product design and features. At Logitech, senior engineering heads were expected to turn "market savvy" and to think about more than just their function. The company culture demanded that everyone wear a broad corporate hat. At the end of the day, their bonus was structured not just on engineering metrics but also on how well their unit performed and how Logitech performed.

Logitech senior management recognized that the company was far from having developed a good grasp of motivations and unarticulated needs of its customers.

> I would honestly say, and maybe this is part of our humbleness, that we are not yet a very good customer company today. We can get a lot better. We can raise the bar in lots of different areas. How close are we to the customer? We don't know as much as we think!

Talking about that, challenging that, forces us to raise the bar, dig deeper, do things to get closer. I think one of our big challenges is to find out why someone buys a cordless mouse versus a cordless desktop with mouse included? What is their behavior? What is their decision process? Where do we go from here? We are not going for the early adopters anymore.

Business Unit Vice President

5.3 Idea Generation and Concept Development

Logitech's top management left a great amount of freedom to its business units as to how they should develop ideas into concepts and turn concepts into commercially viable products. The company did not try to enforce a common, stereotyped process. The degree of formalization varied greatly from unit to unit. Some product unit heads were perceived as relatively structured, and they used systematic approaches and tools for generating and screening ideas. Others were perceived more as "mavericks," relying much more on gut feel or intuition. Despite these differences in work style, they all shared the same rigor when it came to addressing their target of delivering affordable quality products and planning a seamless and smooth execution.

5.4 Product Creation and Development

Logitech had implemented, companywide, a simple but rigorous process to steer its product-creation projects that, typically, lasted from six (for product extensions) to eighteen months (for totally new concepts). This process left a lot of day-to-day freedom to the project teams, but it required them to prepare for, and pass, three tough management reviews, or "toll gates," before commercial launch. These gates were passed in the course of animated meetings attended by the business and product unit heads as well as senior engineering and marketing managers.

The first one was called *Gate 0 or Project Authorization Gate*. This first review gave a group within Logitech the license to work on the project. The debates involved at this level included whether the new product concept was interesting and promising enough and whether the company should spend money on it.

The most important gate was the next one, *Gate 1 or Go Gate*. Here, the team was supposed to articulate the full product concept convincingly to management and present prototypes that validated its technical feasibility. It was also asked to present the key elements of its business case, including market and sales volume estimates, detailed price and cost assumptions, and the resulting margins after deduction of expected marketing and distribution costs. Finally, the team was expected to commit to the broad project deliverables in terms of development cost, schedule, and performance. A key "deliverable" was the product availability date – the date when shipments to retailers could start.

Each element of the plan was scrutinized, and not many proposals made it through the *Go Gate* on their first attempt. Management got involved in the details, probed financial, operational, and marketing aspects of the plan, and asked as many questions as they needed to feel comfortable with the project. Management's tough stance was intended to minimize potential issues down the road, silence the strongest of critics, and appeal to the most discerning of customers once the product was launched in the market.

The last gate, *Gate 2*, was held just before mass production. All plans were carefully scrutinized, once again, before committing with suppliers and building millions of dollars of inventory. For totally new products, a final market check was conducted prior to that decision point.

In between these gates, the overall project responsibility lay in the hands of the program manager, reflecting a decentralized management style. Each week, project leaders for specific modules of the overall project would report on the status of their individual modules to the program manager, and an entry into a Notes® database would ensure speedy project status updates within the rest of the organization.

> At Logitech, a program manager does not have to do detailed reporting on a project unless there is a problem. He/she is the only person who ultimately makes the decision of what to communicate to management. There is a lot of trust from management. But of course, project leaders are coached in the beginning, either by their functional managers, or by their program managers.
>
> *Engineering Program Manager*

The program manager summarized the status of his team's progress through a system of green, yellow, or red flags posted on the project tracking system. If, during the course of the project, a problem or delay occurred, a "yellow flag" would be raised, triggering notifications to all involved. The project manager was still responsible for tackling the problem. But if it got out of control, a "red flag" alerted senior management. The Business unit head was the only person empowered to alter the project schedule and reset the product availability date.

5.5 New Business Development

The corporate business development group was responsible for scoping out potential partnerships and acquisitions outside the current business areas of the three existing business units. It interacted with external proposals (both solicited and not) and directed potentially interesting opportunities toward the appropriate individuals within the organization. The group did not look for new technologies but for new business opportunities. Scanning the horizon for disruptive technologies was left to the individual business units, which were aided in the process by a senior technologist acting as an advisor to the process.

Early on in the process, we realized that some of the assets and strengths that the company had were a very strong brand and access to channels, plus an ability to define what people might buy. One way to leverage this is not necessarily to reinvent all products and product lines, but to accept proposals from the outside world.

Corporate Business Development Vice President

When they received a proposal that met the necessary criteria, they put a team of people together from different business units – product marketing, operations, and engineering – and worked aggressively on turning the opportunity into something tangible. There was no formal, dedicated incubator at Logitech, only a virtual one.

The io™ digital pen was a typical outcome of this process. Anoto, a Swedish high-tech company, had commercialized its core technology within the mobile phone market. Logitech's business development team steered Anoto into the PC environment, created a product strategy around it, and successfully sold the concept to both Anoto's and Logitech's executive management.

To be accepted into the organization, new ventures brought to the table had to be not too far away from existing areas of business to benefit from potential synergy in technology, operations, and sales and marketing. And this usually led to one of the business units ultimately taking sponsorship/ownership of the new venture.

5.6 Spinning in and Spinning off Projects

It enriches people to have some participation in the future as well as the core.

Product Unit Engineering Vice President

This is what happened to the io™ pen, which was entrusted to the Retail Pointing Devices product unit. Leveraging existing resources was the key reason for doing so; but as importantly, it acted as a motivator for the unit to embark on a new venture.

When the io™ had grown into a product ready for launch, senior executives recognized that it required a full-time, dedicated management team, beyond the current resources of the product unit. Consequently, it was spun off as a separate unit of its own, within the control devices business unit. Some of the earlier project team members stayed on with the new venture, including the project leader on loan from his original engineering unit. When asked how one felt about grooming the project leader and the new business up to the exciting moment, only to give it up, the engineering head of the product unit replied:

It is a pride to see how people develop and go from subordinates to peers. I think it is a pride to see that, somewhere, we added value to the company by moving some people who are now able to fly by themselves. We believe it was a natural promotion.

If we did all this work and grooming, went through the pain of delivery, and went through the process of setting up the new unit, I have an interest in seeing that this project is flying and bringing revenues…We put the best people in this new unit to increase the chances of being successful. The worst thing that could happen is that the whole thing would collapse. Besides, some part of my compensation is tied to Logitech and its success as a company. We need to deliver growth. Unless we have new units, it is extremely difficult to deliver this kind of growth just with our core business.

5.7 Product Launch Decisions

The decision of whether or not to launch a new product was sometimes based on pragmatic gut feel, intuition, and first impressions.

> The user may react on a product design – it may be a conscious or an unconscious reaction – but the fact that we dealt with such questions for over twenty years makes us see what the user feels but will not express. Sometimes, when the mouse is too big or some functions are not well positioned, the user is not going to say it because he cannot compare one device to another. He will buy it, use it and get used to it. But at the end of the day, by having the experience of the full range, you can predict what the user is going to say and what he is going to feel but not going to be able to express.
>
> *Business Unit Engineering Vice President*

To feed this intuitive approach of predicting the success (or failure) of a new product, Logitech conducted usability testing through focus groups. Users were carefully monitored on various activities, from taking the sealed package and opening it to getting accustomed to using the product. It was all videotaped. Logitech immediately began to recognize patterns.

> You need no more than ten users to get very good feedback. 80% of what you need to know, you can get from ten users picked from among random buyers.
>
> *Business Unit Engineering Vice President*

5.8 Sales and Marketing

Once the products were engineered and manufactured, it was time for sales and marketing to do their magic and translate the engineered new concept into revenues and satisfied customers. Products, however, did not sell themselves, and it took a great deal of salesmanship to get them on store shelves.

> We have excellent products, but there are a lot of beautiful products that do not sell. If you want to succeed in a company like ours, you have to be a super salesman. You need to have a very strong customer focus. You need to be able to generate revenue for the company. But you also need to be a general manager. We are one of the few companies that are growing. A lot of other companies could be growing if they had our attitude as well.
>
> *Regional Sales and Marketing Vice President*

Logitech benefited from a strong brand franchise, slowly built through clever word-of-mouth, selective advertising, clever packaging design, and aggressive merchandizing. Yet, there were limitations to what the brand could do.

> In a relatively complicated consumer electronics environment where people don't always understand what they are buying, I believe you tend to put more faith on brands. A brand has to bring value. But in this day and age, you cannot overcharge based on brand. If you are competitively priced, if the value proposition you are offering is equally good as a Taiwanese product, then our brand might bring us to a 15–20 percent premium, but you've got limits on that.
>
> *Regional Marketing and Sales Vice President*

The company was aware that over 50% of revenues in the coming year were typically coming from products that had not yet been manufactured, and people realized how critical the product launch process was to the company.

> We are betting the whole company on an annual basis. If the product platform does not work, we are going to have a 12-month negative cycle.
>
> *Product Unit Engineering Vice President*

6 Future Challenges

The challenges for the future were many. In some respects, getting to the first billion-dollar mark in sales was the most difficult. However, as the annual general meeting in 2003 came closer, its executive management could not help but think of the rocky road ahead. Some questions on everyone's mind were:

- How could Logitech continue to rejuvenate, redefine, and reinvent its core businesses? Could they continuously top the last successful product? What unarticulated customer needs (if any) should they go after?
- How would they continue to find new businesses that could leverage their core design, manufacturing, and branding strengths and that would fit with their comfort zone in terms of price points, volumes, and channels?
- If they came up with more new concepts like the digital pen that required novel marketing approaches, would they be able to expand their channel, brand, and market coverage? If so, at what cost?
- How would they grow outside the PC environment toward the living room or the mobile space? How would they find the next disruptive technology, nurture it, and attract the necessary skills to manage it?
- How would they maintain their innovative informal culture and keep growing at the same time? Would their current organization be able to cope with more complexity, in terms of countries, channels, segments, and products?
- What new leadership talents were required to steer the company in all its new promising directions? What should top management do, specifically, to define, detect, and groom such talent?

Exhibit 1
Company Details

- Company Headquarters   Fremont, California
- Publicly Traded   Swiss Exchange and NASDAQ
- Regional headquarters   Romanel-sur-Morges, Switzerland, Hsinshu, Taiwan
- Employees   4,800 worldwide
- Revenues $1.1 billion (85% from retail business, 15% from OEM business)

Company History

1981	Daniel Borel and Pierluigi Zappacosta set up Logitech as a software development company.
1983	Logitech builds a manufacturing plant in Switzerland with a capacity of 25,000 mice.
1985	Logitech's OEM customers include HP, ATandT, and Olivetti. Production capacity is increased to 300,000 units.
1986	Logitech enters the US retail market with a direct sales approach. It sets up manufacturing in Taiwan. Production capacity reaches 10 million units.
1988	With more business originating in Europe, Logitech adds Cork (Ireland) to its manufacturing sites.
1990	Worldwide mouse sales top $500 million. Logitech reaches a 35% share of the OEM market and a 27% share of the retail segments.
1997	In March the company goes public and is quoted both on the NASDAQ and the Swiss Stock Exchange.
1998	Logitech acquires Connectix and introduces PC video cameras. Daniel Borel hires Guerrino De Luca as CEO and stays as active Chairman.
2001	Logitech acquires LabTec (PC audio speakers) and begins inroads into the audio peripheral market.
2002	Logitech launches the io™, a digital pen that captures handwritten notes and drawings on the PC.

Exhibit 2
Select Financial Data (1993–2003)

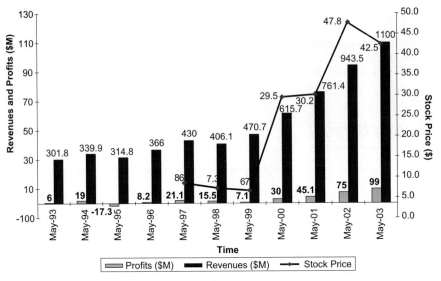

Source: Public information

Exhibit 3a
Corporate Structure at Logitech

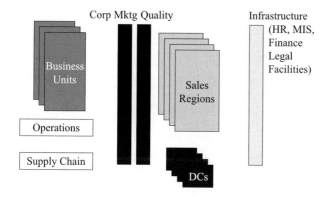

Exhibit 3b
Control Devices Business Unit

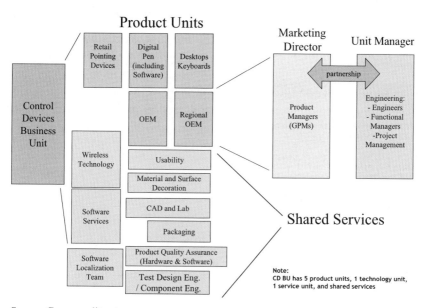

Source: Company literature

Exhibit 4
Logitech's Vision/Mission over the Years

In the beginning ...

The main driving force behind the creation of Logitech was simply the very strong desire to participate in and belong to an industry which had a huge potential to impact society – in the same way that the train, the airplane, or the TV did. It was a time when so many new developments were going on... when nothing seemed impossible. In fact two posters from the time sum up the very early days of the company:

- A small blondish 6-year old boy walking in the middle of a high ripe golden wheat field (with "golden" hair like the field itself) barely able to see where he is going, but saying: "*I don't know where I am going but I am on way!*»

- Seagulls flying, with the caption: "*They can* (meaning flying) *because they didn't know it was impossible!*»

Such early statements showed a clear strategic direction for the new startup...

Beginning in late 1985, the mouse, beyond its role as a pure product, helped to define the first vision for Logitech, i.e., the icon of the interface between technology and people – an icon which continues to express the company's mission today – to provide this human interface not only on the PC, but beyond the traditional desktop as well.

Source: Company literature

10 years later...

As Logitech gained a stronger understanding of the business challenges and rewards that the future held, the company refined the key elements that it believed would support growth and success:

Vision: a humanized computer in the hands
           of every individual
Mission: to provide every computer user with
           the very best computer senses

20 years later...

Beginning with its initial listing on the Nasdaq in 1997, Logitech reinforced its vision: "The Interface is the Computer," illustrated by the slogan "Intel Inside, Logitech Outside," and later on, by "It's what you touch." With the emergence of the Internet and the fast growth of the home market since 1996, Logitech brought to market numerous innovative products, which helped users to naturally and intuitively access the ever-richer digital world. Whether working, learning, communicating or playing (including non-PC platforms), Logitech offers the ideal Interface to bridge the gap between People & Technology.

As the company entered the year 2000, Logitech's vision & mission evolved once again:

Vision: The interface that links people and
            information will transform the way
            they work, learn, communicate and play

Mission: Logitech brings to market tools that enrich
            the interface between people
            and information

Exhibit 5
Product Releases (1981–2003)

Year	1981	1987	1988	1989	1991	1992	1993	1995	1997	1998	2000	2001	2002	2003
Mouse	First Mouse	Logitech Branded Mouse		Trackball	Radio Based Cordless				Wheel Mouse		Optical	Cordless Optical	Bluetooth	Rechargeable Cordless Optical
Monitor			Monitor											
Scanner			Handheld Scanner											
Camera						Digital Camera		Color Digital Camera					Pocket Digital Camera	
PC Camera								First PC Camera - VideoMan		QuickCam Series				Detachable ClickSmart Series
Audio							Handheld mic & speaker for PC		SoundMan Series			LabTec Acquisition	Cordless Headsets	
Keyboards										Cordless		Optical Cordless	Folding PDA Keyboard	Bluetooth
Gaming								WingMan Joystick	Force Feedback Joystick			PlayStation Products	Microsoft Xbox & Nintendo GameCube Products, Cordless Joystick for PC	
Digital Pen													io launch	
Year	1981	1987	1988	1989	1991	1992	1993	1995	1997	1998	2000	2001	2002	2003

Source: Company literature

Exhibit 6
Key Values – Control Devices Business Unit

- Innovate anytime, anywhere, anything (not only products)
- One global culture (that encompasses the best of each site culture)
    - Fight the National Institutes of Health
- Transparency and visibility (plans, issues, design, decisions, *etc.*) due to geo dispersion
- Partnership and mutual respect (rather than internal customer/suppliers)
    - Challenging is encouraged
- Strong project management practices
    - Risk management – proper usage of postmortem reports
- Spirit and acknowledgement of contribution
    - Focus energy on tasks that add value to the company
- Process orientation
    - Flawless execution, attention to details
    - Issues translate into process improvements and training
- Efficiency
    - Judicious usage of IT tools available: Notes DB, E-mail, telephone, Net-meeting, IM
    - Meetings have agenda, preparation, and minutes. Laptops encouraged in meeting rooms to get info and fire action items.
    - Documentation of proceedings, decisions, action items
- Humility (success is never final)

Source: Company literature

CASE 6-4
Logitech: Getting the io™ Digital Pen to Market[1]

Jean-Philippe Deschamps and Atul Pahwa

In November 2002 Logitech launched its io™ digital pen in the USA and Germany with a considerable PR campaign. Encouraged by an enthusiastic response from the press, many early adopters tried the product and had generally favorable comments about their experience. A clear and simple marketing message had enabled Logitech to convey the basic features and value proposition of its pen, both at consumer and press levels.

Still, sales were nowhere close to the numbers originally hoped for. By April 2003 the company had started to appreciate the challenge of marketing a totally new product concept in a retail environment. The pen required considerable end-user education on both the product and its application. While not insurmountable, these challenges had slowed down its adoption. Management was wondering how it could further capitalize on the early-adopter market and generate enough buzz about the product to extend the initial sales momentum.

Logitech's management was committed to the io™ for the long haul. Still, with a limited marketing budget, the io™ team needed to evaluate future market opportunities carefully. They would have to redirect efforts for increasing both short-term revenues and long-term product adoption by various user groups. There were now decisions to be made on how to increase sales:

- What should Logitech do to spur retail sales activity among segments other than early adopters? Which specific segments should it target and how?
- How should Logitech address industry-specific (vertical) market opportunities and professional user segments?
- Should Logitech focus instead on the enterprise software market and work toward developing applications (in house or with third-party developers) specifically for this environment?

Copyright © 2003 by IMD – International Institute for Management Development, Lausanne, Switzerland. Not to be used or reproduced without written permission directly from IMD.

[1] This case follows and builds upon the case "Logitech: Launching a Digital Pen" prepared by Ken Mark under the supervision of Professor Mark Vandenbosch (Richard Ivey School of Business ref. 9BA03A002), which is strongly recommended as background reading.

1 Logitech's Entry into the Digital Pen Environment

The 1990s saw the emergence of handheld computing with Apple's Newton™ and Palm's PDA.[2] Pen computing emerged shortly thereafter with the advent of the IBM TransNote™ and the Sony Pen Tablet™. In 2001 Microsoft announced its Tablet PC™ initiative, promoting it as a new platform for notes organization and management.

Anoto, a Swedish firm founded in 1999, had developed a digital pen and paper technology that attracted Logitech's interest. Slightly thicker than a ballpoint pen, the digital pen included a camera that captured one's writing on a specially designed paper and saved this as strokes in its memory. The paper featured a proprietary pattern of small dots that were not easily visible to the eye. This grid allowed the pen to recognize its absolute positioning with respect to a particular area of the page, and one page among several pages. The user could thus go back and forth from one page to another as with regular pen and paper.

Anoto's founding team came from the mobile telecommunications world and thus found it natural to work with SonyEricsson to develop the ChatPen™. Launched in Q2 2002, it allowed users to write a message on Anoto digital paper and send it via a mobile phone. Logitech convinced Anoto of an even greater opportunity in the computing environment, and in March 2002 the two companies announced an alliance. Logitech introduced an initiative of "extending the power of the PC beyond the desktop." [1] (That alliance was reinforced in June 2003 when Logitech made a 10% equity investment in Anoto). Subsequently, Logitech decided to manufacture a digital pen as well as develop software to connect the pen to the PC and several popular applications including Microsoft Office.

At the end of 2002 there was competition brewing on the horizon, with Nokia releasing a product similar to the SonyEricsson ChatPen™, Microsoft committed to the Tablet PC™ environment, and Seiko aggressively promoting its InkLink™ system. (*See Exhibit 1 for more details on the competitive environment.*)

2 Logitech's Digital Pen Project

After signing a licensing deal with Anoto, Logitech's management assigned the development of a digital pen and related software to the Retail Pointing Devices unit.[3] The project followed the company's established product development process with a few significant differences. One in particular was the high level of top management attention that it received throughout, given the novel nature of the product concept.

---

[2] Personal Digital Assistant, the ubiquitous "Pilot™".
[3] Retail Pointing Devices was one of four product units within Logitech's Control Devices Business unit. It designed and developed a broad range of mice and trackballs.

Logitech traditionally managed the marketing and engineering sides of its projects closely together and in parallel. As a consequence, product launch issues were not pushed toward the end of the development process but handled from the very beginning concurrently with product design, development, and engineering.

Logitech's development process was built around two critical gates or management decision points: *Go Gate* (or *Gate 1*) and *Gate 2*. Both required senior management approval, and the io™ launch issues straddled these two gates.

2.1 Preparing for and Passing the Go Gate: Authorization to Start a Project

Although the engineering and marketing groups typically worked hand in hand on new project ideas for months before that gate (in what was called the "strategic front-end"), the *Go Gate* was the official start of a development project. At this first important decision point, which the pen project passed in August 2001, the core project team was supposed to:

1. Define the product concept convincingly to management.
2. Validate its technical feasibility.
3. Commit to the broad project deliverables in terms of a CSP package (cost, schedule, and performance). If and when senior management was satisfied with the project team proposal, it gave the engineering group authorization to design and develop the concept.
4. Present elements of a business case. The marketing group was expected to present and defend its market and sales volume estimates, its detailed price and cost assumptions, the resulting margins, and expected marketing and distribution costs.

2.1.1 Product Concept

Right from the start, and almost intuitively, the digital pen was conceived as a branded product for the retail channel. An OEM launch would have required the buy-in from several partners and a significant time and resource commitment. Besides, the retail channel was more profitable for Logitech.

The pen could leverage Logitech's strong and growing brand franchise, good retail relationships, and efficient distribution system. OEM, enterprise, or professional market segments were not excluded but considered only as additional business opportunities for a later stage. However, the detailed nature and size of those markets were not investigated.

For the io™ team, the product concept was clear and compelling (*see Exhibit 2 for a description of the concept as seen by the press at launch.*)

2.1.2 Cost

Since its early days as a supplier to the OEM market, Logitech had developed an obsession for targeting the lowest possible product costs. This allowed Logitech's products to be sold generally below $99, a psychological threshold for impulse purchases (95% of Logitech products retailed for $99 or less). However, it soon became clear that the digital pen, with its sophisticated technology (sensor/camera, memory, and battery), could not fit with that pattern, something that management accepted.

The business case also included specific financial information, including contribution margins that reflected the CEO's obsession for costs.

2.1.3 Schedule

The CSP package at the Go Gate traditionally required the team to specify a self-imposed product availability date (PAD). This was defined as the date when the product was expected to meet its specified quality targets and become available for shipment to the distribution channel. Anticipating the PAD – an analytical project scheduling exercise – meant putting together a detailed project path. At the Go Gate, management approved the team's proposal for a PAD of September 2002.

2.1.4 Performance

Industrial design and some software features were still not set, but the key performance attributes, from memory to battery life, were conceived, defined, presented to, and approved by management at the *Go Gate*. Service was a nonissue since the battery lasted the life of the pen. However, considerable attention would have to be given to technical and user/application support (hotline).

The product that would be launched 15 months later ended up pretty much in line with that *Go Gate* concept in terms of its basic functionality and product features.

2.2 *Formalizing the Project and Advisory Teams*

In the Logitech tradition, the project team was composed of members from Europe (engineering), the USA (marketing), Taiwan (engineering and pilot manufacturing), and China (manufacturing). On the engineering side, the team was initially composed of senior members of the Retail Pointing Devices unit, but it also borrowed manpower resources from other units. Full-time personnel would be hired as sales developed (therefore justifying the additional overhead).

In parallel, the core product development team set up several task forces to handle specific challenges, particularly on software development and user documentation. As the project progressed, everyone realized that these areas would need more attention than originally anticipated.

Given the particularly risky nature of the project, Logitech set up an advisory team to steer and approve the work of the project team. The advisory team consisted of (among others) three senior executives, including the head of the Control Devices business unit who had initiated and closely monitored every aspect of the project (*see Exhibit 3 for a description of the team organization*). They met in person every 8 to 12 weeks. Given the tight PAD deadlines and the difficulty of holding frequent meetings across different locations, the advisory team was forced to make on-the-spot decisions that eventually speeded up the launch process.

2.3 Refining the User Concept After Go Gate

The project team quickly realized that a number of product decisions, particularly on software, would condition important user applications. Such decisions would have to be made in a vacuum unless the team had a better understanding of what customers would want to do with the pen. All involved in the project had been excited about the proposed product from their own perspectives, but they were concerned about introducing their personal bias into the decision-making process. It was decided that the team should develop a hypothetical persona through whom decisions could be channeled. In this way, they could imagine how the customer would react to alternative product features or applications.[4]

Mike, a 21st-century knowledge worker, was imagined. He would be in his mid-30s, work at a mid-sized construction company in middle management, have a significant level of computer literacy (though he would not be defined as a geek), travel, and attend meetings frequently. As the team built the business case for the io™, it weighed benefits as they specifically related to Mike. However, as the marketers in the team built the business case for their digital pen, they obviously went beyond Mike to create a more general customer value proposition.

---

[4] The team followed the example of Palm as described in the book "Piloting Palm."

2.4 Shaping a Launch Strategy

After the *Go Gate*, a number of launch strategy decisions were taken, particularly regarding the geographic scope of the launch, pricing strategy, channel strategy, signing up of stationery partners, and sales targets.

2.4.1 Geographic Scope

Logitech was good at launching new iterations of existing products (mice, keyboards, Web cams, *etc.*) on a global scale. Introducing an entirely new product category required a more experimental approach, and this is why a selective launch strategy seemed more appropriate. Logitech would launch the pen in the USA and Germany first before rolling it out globally.[5] The USA, Logitech's natural home market, was an obvious choice. Germany was selected because it was Europe's largest electronic market. It was also Logitech's biggest market in Europe, thanks to a strong brand and channel position. Japan, a key market for new electronic devices, was briefly considered but rejected due to its complexity.

This selective launch strategy was strongly influenced, if not determined, by Guerrino De Luca, Logitech's CEO,[6] who had been following the project throughout and was acutely aware of the risks. His message to the team was:

> Get the product right and make sure that we have the right paper stationery channels in place. Let's be humble and learn from the market. When you will have proved that we master all needed attributes, then we can extend our reach!

In the USA and Germany, the target would be key retailers rather than a massive full-blown launch. Press and consumer feedback during the gradual launch process would allow management to closely monitor progress and take swift corrective action if necessary.

Once the launch countries were selected, regional marketing teams composed of experienced marketers well versed in various retail channels were mobilized. Their involvement ramped up rapidly as the project moved toward launch.

---

[5] In contrast, SonyEricsson, which was dependent upon telecom operators for its service, launched its digital pen first in Sweden, then in Denmark, and much later in Italy.

[6] Several Logitech managers commented that De Luca's plea for a prudent launch was strongly influenced by his past experience as Executive VP Marketing at Apple, where he had lived through the trauma of the Newton[TM] launch failure (*see Exhibit 4 for a story of the Newton[TM] flop*).

2.4.2 Pricing Strategy

According to the product unit head, the rationale behind the pricing strategy was "part art, part science." Normally, Logitech would find out what price points brought in the most margin dollars. However, the pen was in an entirely new product category with no antecedents to extrapolate information from, and it was clear that the product cost structure would not allow a retail price of $99 or less. Based on preliminary product cost estimates, the team had proposed a US$149 retail price. At the CEO's objection that it would be foolish to leave money behind in an early adopter market, the team settled for $199.

All agreed that there might be no perceptible difference between $149 and $199, but they feared that exceeding the $200 mark (for example $219) would no longer qualify the io™ for an impulse purchase. Pricing was tested in an online survey where the price point was found to be acceptable. At $199, the team felt confident that the product benefits and features, which were to be well described on the packaging, would be sufficient to attract a broad range of customers. That introduction price of $199 was maintained at launch in the USA, while the price in Germany was set at € 249.[7]

2.4.3 Channel Strategy

Retail was known to be a very unforgiving channel if one did not meet anticipated sales volumes. Besides, committing to too many retailers at the same time would create inventory issues. A gradual release and selective retail distribution would limit the risk while allowing for learning and product enhancements before reaching the mass market. As a consequence, Logitech decided to focus initially on a few selected retail channels:

- Technology-oriented stores, where early adopters would shop. These chains, like Best Buy in the USA and Media Markt in Europe, were the traditional outlets for Logitech products.
- Office superstores, where people cared less about the technology but more about the organization.
- In the USA, Franklin Covey stores. While a subset of the second category, these were stores dedicated to the personal organizer product category.

A number of well-established webtailers were also selected. In the long run, the project team speculated that the io™ could open up new sales channels for Logitech, from campus bookstores (targeting students) to office product catalogues (targeting corporate users) and duty-free shops.

---

[7] The original retail price implemented at launch included the pen, two notebooks (of 80 pages in the US and 96 pages in Germany); one Post-it® pad, and a set of ball pen refills.

2.4.4 Stationery Partners

Securing stationery supply partners and organizing the paper distribution was one of the key tasks of the team between the two gates. The concept and basic layout design of the Anoto pad existed. For its mobile phone Chatpen™, SonyEricsson had established a supply partnership with Esselte, a European supplier. Logitech felt that it needed to sign up strong and credible paper and stationery brands to secure the retail success of its pen. As a consequence, it gave an exclusive contract to Mead for the USA and to Groupe Hamelin for Europe.[8] The company also enlisted 3M for the provision of Post-it® notepads with Anoto functionality.

In the USA, retailers like CompUSA or Staples had strong distribution networks. They had centralized purchasing for all their stores nationwide and managed the distribution process down to the individual retail store level. Mead worked directly with the retailers without Logitech's involvement.

In Germany, managing retailers proved to be an arduous task. Media Markt, in the interest of minimizing the number of its suppliers, required that an existing supplier provide the digital paper. Logitech was thereby forced to get into the distribution business for digital paper. Also, since the retail stores placed individual orders from their central warehouse, Logitech had to make sales calls at the individual store level to make sure that the pen and paper were stocked at all times.

Logitech began to realize that it was reluctantly getting involved in "consumables," something management had never considered as part of its vision. Not having envisioned a consumables business model, Logitech could not see itself as a paper supplier or see stationery as a viable business opportunity.

Although the production cost of the stationery was competitive, retail price points were set high by both paper suppliers and retailers (Logitech was not involved in stationery price setting).[9] Implicit in this decision were the anticipated low sales volumes and the hypothesis that customers would not be very price sensitive.

2.4.5 Sales and Financial Targets

Since the pen was a new product category, much of the quantitative aspects of the io™ business plan were hypothetical. Given the lack of industry and market data, the goal was to sell 100,000 pens in the first year. Target sales figures were based on an assumed sales volume per month per store. For example, the only initial metrics discussed were a target of two pens per retail store per month and reaching

---

[8] Groupe Hamelin was a French office stationery specialist with several known brands such as Oxford™, Super Conquérant™, and 001 International™.
[9] The thick (160-page) Anoto notebook would sell for $9.95 in the USA (3 packs for $24.95) and the Post-it® notes (50 sheets) would sell for $4.95 (3 packs for $14.99.) In Germany, a thinner (96-page) notebook would sell for €6.95.

a certain percentage of the customer base in Franklin Covey stores. The German forecast was established as a percentage of the expected US volume. There were no defined metrics for e-commerce sales either.

However, senior management strongly believed in the long-term potential of the io™ and the CEO had explicitly stated that the team should not worry about first-year sales or profitability. The goal was to get the product out in the market and by the second year build a strong base of repeat customers. Logitech had high hopes for sales of about $100 million for the third year.

2.5 Launch Planning and Preparation

Before *Gate 2*, a number of critical launch issues were handled, such as deciding on the product name, finalizing the packaging, selecting the launch retailers, testing the product for usability, and preannouncing the product.

2.5.1 The io™ Name and Advertising Copy

In parallel with engineering efforts, the marketing team worked diligently on a launch strategy starting in April 2002 with finding a name for the pen. It was decided that the name should not be too descriptive, and after several brainstorming sessions with an ad agency options were narrowed down to "e-pen" and "i-pen." The team first agreed on "io pen" (with a pun on the similar sounding *eye open-ing* experience), but it decided to remove "pen" from the name and leave it as "io." The name was well received and perceived as being "cool." An advertising agency came up with several copy concept proposals including an idea to highlight the io in words like evolution, revolution, innovation, and information (*see Exhibit 5 for io™ advertising*).

2.5.2 Packaging Design and Documentation

The novelty of the io™ concept and applications had to be clearly communicated to the consumer at the point of sales, particularly in the sparsely staffed, mass-retail environment in which Logitech operated. The company had always given a lot of attention to designing compelling and informative packages. For the io™, packaging and user documentation were even more critical. Clear pictures of the product, as well as steps outlining how it should be used (including pictures to illustrate these steps) were carefully conceived and designed. The box also detailed novel usage applications. In short, the packaging had to convey the essence of the io™ value proposition. The project team also spent a lot of time explaining the product to their colleagues responsible for preparing technical/software documentation and the user manual.

2.5.3 Launch Retailers

In the USA, CompUSA (a technology retailer with 225 locations) and Staples (an office superstore with over 1,100 locations) were chosen as prime retail outlets. Not all stores were to carry the product, only those that drew in considerable traffic in markets with strong technology product sales. Franklin Covey, an exclusive retail outlet chain, was also selected. In addition, several e-tailers including Amazon.com, Buy.com,[10] and CDW[11] were to carry the product. In Germany, Media Markt (a tech retailer with 160 locations in Germany) and its sister chain Saturn (another tech retailer with 70 German stores) were selected. Karstadt[12] and Misco, a medium-sized mail-order firm, were also chosen.

2.5.4 Testing

While Logitech routinely conducted extensive prelaunch tests of its traditional products (mice were extensively consumer-tested for design, color, and functionality), the testing of the io™ was limited. Basic usability testing was done on the product.[13] Pen designs were tested online, and the feedback was almost evenly split between a more conservative pen-oriented shape and a radical design. Logitech decided to go with the latter. Implicit in the choice of not having more detailed testing was the perception that testing totally new product concepts might prove misleading as customers lacked reference points.

2.5.5 Preannouncement

Creating a new product category required educating the customer. Logitech decided to focus on industry opinion leaders to sell them the digital pen concept as a new, legitimate tool that filled a real user need (as opposed to being perceived as a new "tech toy.") Between February and March 2002, Logitech's CEO and Senior VP Control Devices participated in several road shows with analysts, thought leaders, and journalists. The objective was to ensure that the io™ would not be lumped, in the minds of these opinion leaders, with the slew of earlier products that had failed in that market. Logitech also wanted to differentiate the

---

[10] Buy.com was named the "Best e-Commerce Site" by *PC World*, "Best Overall Place To Buy" by *Computer Shopper*, a "Best of the Web" in the computer and electronics category by *Forbes*, and the No. 1 electronics e-tailer by Forrester Research, Inc.

[11] CDW's business model focuses on small- and medium-sized businesses with 97% of sales derived from commercial accounts. Its website attracts 69,000 unique visitors each day.

[12] Karstadt had 189 department stores in Germany, a 36% share of the German department store market, 92% brand awareness, and 2.5 million customers daily.

[13] Usability testing encompasses methods for identifying how users interact with a product. In a typical approach, users use the product to perform tasks, while observers watch and take notes.

io™ from competing products by positioning it in a new "personal digital pen and paper" category.[14] It also made an effort to explain the Anoto partnership.

CeBIT in Hanover, Germany (March 2002), provided a venue for the pre-announcement of the io™. To provide for a single worldwide announcement, a Web-based conference call was set up for investors to learn details of the announcement. Communication objectives were clear: to convey the product positioning and its value proposition. It was also intended as a platform to generate industrywide interest (including news, discussion forums, and reviews) and secure ongoing media coverage.

2.6 Gate 2 (Authorization to Go for Mass Production)

Over the first half of 2002, the product was prepared for tooling and mass pro-duction, and packaging design and user documentation were done and translated. So, by July 2002, the io™ was ready to pass the *Gate 2* milestone. This gate was perceived as a sanity check before spending huge amounts in buying components and committing firmly to partners and suppliers. The io™ project passed that gate according to schedule and the September 2002 PAD was reconfirmed.

3 Logitech's io™ Launch

At project inception, it was decided that the product would be released to the retail channel in September 2002. Along the way, due to several delays in mechanical design, software, and hardware, the team came out eight weeks late and the PAD had to be postponed to the third week of November 2002. By itself, that delay would not have been critical, except it meant not getting the io™ to retailers in time for the crucial Christmas season in the USA. This is why, in the USA, the io™ was launched on the Web before hitting the retail shelves. The actual public launch date in retail stores was now effectively set for January 2003.

Logitech traditionally put a lot of effort into the launch of its new products. The company knew well how to introduce established products like mice. Launching the io™ was another ball game. Management had to make assumptions and be prepared to discover and tackle issues as they arose. So, whereas Logitech's launches were generally handled without much consumer media advertising, an exception was made for the io™. It was to be selectively advertised in US consumer media and on technology focused websites. In Germany, however, management

---

[14] One of the key differentiating factors of the io™ pen was that it did not change the way people wrote notes, unlike SonyEricsson's Chatpen™ or Seiko's Inklink™.

hoped that the product would sell by itself, thanks to its strong brand franchise,[15] and advertising support there was therefore limited.

3.1 The Logitech Sales and Marketing Organization

A senior VP, responsible for OEM and retail, headed Logitech's sales and marketing organization worldwide.[16] Reporting to him were the VP Corporate Marketing, as well as sales and marketing VPs for the four geographic regions: the Americas (including Australia, and New Zealand); EMEA (Europe, Middle East and Africa); Asia-Pacific (excluding Japan); and Japan.

In North America and Europe, Logitech sold through wholesalers and distributors, like Ingram Micro and Tech Data. These channels serviced systems integrators and value-added resellers who built integrated IT solutions, as well as smaller retailers not directly served by Logitech. While Logitech preferred to maintain a direct relationship with its larger retailers such as Best Buy in the USA and Media Markt in Europe, these larger retailers would sometimes buy from the wholesalers (who would provide the logistics function) or in some instances directly from Logitech. The company also sold to online e-tailers and select catalogue firms.

The sales and marketing organization was complemented with account management teams responsible for maintaining relationships and managing promotions at each level of the sales process, both with retail and OEM customers.

3.2 Mobilizing the Sales Force and Technical Support Staff

Management decided early on that the io™ launch would follow the same pattern as its other products. Mobilizing the sales force on the pen, which was not an easy product to explain to retailers, was therefore a critical challenge.[17] Consequently, the launch team did a lot of internal PR over sales meetings to position the pen as a strategic new product and thereby created enthusiasm despite the fact that no special incentives were available to sell the io™. In parallel, the project team spent a lot of time training the user support staff on software and pen functionalities. While a new product category provided future and long-term sales potential, Logitech's difficulties of introducing new categories in the past made the sales team skeptical of management's long-term commitment to the io™.

---

[15] In Germany, many stores had set up a dedicated "Logitech shelf." In the USA, however, while reasonably well known, Logitech was not perceived as a "destination brand" *per se*.

[16] OEM sales accounted for 15% of company sales, retail sales under the Logitech brand accounted for the remainder.

[17] On average, a Logitech sales person had 200 products in his/her portfolio and was accountable for reaching a challenging sales target each quarter.

3.3 The Launch Venue

It was now September 2002 and Logitech decided to launch the io™ (and release the io™ name) at DEMOMobile in La Jolla (California), the most influential venue in the USA for introducing significant new products in the mobile and wireless markets. Palm had launched its revolutionary PDA in 1996 at that same event, creating a new product category and spurring a slew of imitators. Many of the journalists who had covered the Palm introduction were present once again.

Of the 200 companies launching their products at DEMOMobile in 2002, 35 were short-listed and given six minutes on stage to communicate their product message and strategy. Logitech was not only on the shortlist, but it also walked away with a DEMOGod award for taking the energy on the stage to new heights! The team had succeeded in its goal: create buzz among an audience of industry movers and shakers. Logitech hoped that people would not only remember the show but also talk about it, thus generating word-of-mouth advertising.

3.4 Postlaunch PR and Media Promotion

The press coverage that followed, between September and December 2002, went far beyond typically tech-focused publications (*see Exhibit 6 for an example of press coverage*). ABC, USA Today, Teen People, Sales and Marketing Management, Esquire, Time, Forbes, Popular Science, and Laptop Magazine all had write-ups on the io™. The idea was to generate press activity through as many channels as possible to enable potential users to learn about the product and buy it.

Around this time, to connect the pen with the Logitech brand, the company signed on leading search engines to ensure that the io™ pen would pop up in front of all other search results when triggered by specific keywords or product categories.

In April 2003 the io™ won the Best Product and Best New Technology awards at RetailVision, a leading technology event for the consumer channel, organized by Gartner. Since several large retailers visited the show, this provided significant visibility for Logitech as well as a reassurance for retail channels.

The company also tried an innovative marketing idea. It advertised on TiVo,[18] hoping that the same early adopters who chose TiVo would potentially be interested in the io™. TiVo users would see a brief spot on the io™ whenever they switched on their device. They were then directed to the Logitech io™ website to learn more about the product and offered discount coupons plus participation in sweepstakes for winning a free pen. Within four days of the spot being aired, over 50,000 people (over 6% of all TiVo users) registered on the Logitech io™ website to learn more about the product.

---

[18] TiVo is a service that automatically records your favorite TV shows every time they air.

In addition, Logitech produced a TV show on PBS with Morley Safer[19] to introduce the io™. The PBS production cost under $10,000.

Logitech also benefited from having Microsoft launch its Tablet PC™ in the same time frame. Conceptually, this was a more complex product than the io™, but it generated significant press coverage as journalists tended to compare the Tablet PC™ with other newly released products in the same market arena.

3.5 Communicating the Value Proposition

Logitech's pen was not a product that customers knew they wanted before its existence. As a consequence, the company needed vehicles to tell people about it.

The first means of communication was online advertising, which was pursued extensively in the USA on technology-oriented sites. The spirit of the campaign was to intrigue and drive people to the io™ website (www.logitechio.com). Once on the website, prospective customers were able to view a Macromedia Flash® demo of the pen. The strategic aspects of the advertising were developed in-house while the company contracted a creative agency to handle the actual copy writing and ad layouts. Later research showed that 75% of people who had bought an io™ had visited the website to learn more about the product before buying it.

The second means of communicating with potential customers was through point-of-sales promotion. In the USA, stores were equipped with point-of-purchase displays, some with video-looping sequences of typical applications of the io™. This approach required dedicated and elaborate real estate within the stores, the cost of which – Logitech felt – would pay off in the long run. Close to 1,000 stores received this type of display. Those stores that did not have the video-looping display either had io™ experts on hand to discuss customers' queries or in-store material that customers could take with them.

Stores in Germany had only a static display – about an A3 size – with messaging and education on the product. The bet was that Logitech was known well enough as a brand in Europe for a new Logitech product to attract attention as soon as it was put on the shelf.

These displays won both praise and criticism from store and/or department managers. Initially, the retail chains loved the product and the displays and were keen to get them into their stores. Later, however, the perception changed. The io™ was seen more as a showcase product that was not bringing in the revenues expected, and some stores even removed Logitech's in-store displays.

This drove home two points. First, that the product was definitely not an impulse purchase, and second, that the consumer education process was more complex than

---

[19] PBS is a not-for-profit public broadcasting channel in the USA, available to 99% of US households. News correspondent Morley Safer has been honored with numerous awards, including 11 Emmys, 3 Overseas Press Club Awards, and the Robert F. Kennedy Journalism award's first prize for domestic television.

initially anticipated. Management felt justified in having recommended a gradual retail launch strategy and handling of retailer issues at the individual store level.

3.6 Managing the Retail Channel

3.6.1 Retail Sales Calls

Getting individual stores to carry the product in Germany proved to be a painstaking process. Just because the central purchasing department at the retailer had agreed to carry the product did not mean that the product would automatically show up on the shelves. Individual stores had completely decentralized operations and individually decided what did or did not get ordered. Logitech had to sell at the individual store level to make sure that the product was going to be available on the shelf. In the USA, this was not the case as the store distribution process was more centralized. Once the retailer had placed an order for all its stores, one could be confident that the product would show up at all the stores it was intended for.

If the io™ did not have the required run rate at an individual store, Logitech did not get reorders. At some point it became a mind-share problem. When the department manager at a retail store had 150 different products to sell, how much time was he/she going to spend on a problem child?

3.6.2 Shelf Placement

Getting the product into the right store aisle turned out to be trickier than expected. There were several options regarding product placement:

- The io™ could be placed in the same area of the store that had other Logitech products, like keyboards and mice. This option was followed in Germany to leverage Logitech's strong brand awareness.
- It could also be placed next to the paper and pen products (e.g., Mont Blanc, Franklin Covey organizers), which would allow targeting the affluent organized person. Even though io™ was a technology product, the second option would bring the "pen and paper" world into the technology age.
- Finally, it could be sold on the PDA shelf space. This third option was frequently chosen in the USA, and it seemed to work best.

3.6.3 Stationery Supply and Sales

Each major retailer had different buyers for technology and for paper products. Paper buyers did not buy technology products, and vice versa. It was a strange situation in which both wanted to stock the product but neither wanted to handle the part they were unfamiliar with. Logitech had to make sure that there was an abundant supply of paper pads in the stores. Customers would be hesitant to buy the pen if they were concerned about paper availability. One of the ways to reduce the risk of running out of paper stock was to bundle a bunch of paper pads with the pen and give those pads to retailers for free, who could then sell them at full retail price. Paper pads were also easily available on the Logitech website.

4 Evaluation Phase

In the spring of 2003 Logitech's management hived off the io™ from the Retail Pointing Devices unit, which had housed it so far, and set it up as an independent product unit, reporting directly to the head of the Control Devices Business unit.

By July 2003 the feedback from the retail front, both from the USA and Germany, was starting to flow in. The two market launch teams, who had run their show individually since January, were also planning a formal get-together for a first exchange of experience. The senior management advisory team – still officially in existence – was also expected to meet shortly to review the results so far. These varied significantly between the USA and Germany.

In the USA, despite the fact that sales volumes were not what management or the team had hoped, the launch was judged "reasonably successful." In Germany, however, results were disappointing. In both cases, eight months into the product launch, the team had not yet identified really big user segments at the retail level. While Logitech's channel contacts had generated an initial momentum and the shelf space and visibility desired for an initial launch, the io™ sales volume per store had proved disappointing.

However, several useful observations could be drawn from the market feedback:

1. Negative feedback on the io™ expressed by a few opinion leaders was that it was too thick – one compared his experience with writing with a cucumber – and that its handwritten character recognition capability was limited.
2. Advertising in the USA had apparently been positively received and resulted in strong product awareness. Germany, on the other hand, had done limited advertising, and management wondered whether this explained the difference in awareness and penetration.
3. The in-store point-of-purchase support had turned out to be a key marketing element, particularly in the USA.
4. The io™ had quickly become the best-selling product on Logitech's US website, but volumes remained much more limited than in retail channels. In Europe, Web-based sales still had to take off.[20]

---

[20] Management had noted a tight correlation between retail sales and Web sales. Products that moved well at retail generally also moved on theWeb.

5. Nearly all the users surveyed were actually using the pen to capture their notes, with about 40% using it to capture pictures as well. The latter number was higher than anticipated.

6. As identified by the end-user research conducted after launch, the largest individual user segments turned out to be senior managers, followed by sales people and lawyers.

7. The ioTM required considerably more technical and application support than products like keyboards or mice. The inquiry flow was extensive.

8. While initial PR helped generate the necessary buzz, the challenge was to keep the attention going. Consumers had a short memory.

9. Although Logitech expected that most of its purchased pens would be treated as professional expenses, it did not have much clue on the way people wanted to handle stationery refills.

10. In retrospect, management realized that branded paper appeal proved marginal at best; consumers did not care about the brand of their paper. Logitech-branded paper would have done the job equally well.

5 Looking Ahead

5.1 Internal Turmoil at Logitech

In June 2003, after 22 consecutive quarters of growth in sales and profits, Logitech did not meet analyst expectations. This announcement triggered a steep stock price decline that was immediately followed by a series of strict cost-reduction measures. Although the investment in the io™ launch would continue, its marketing budget was significantly curtailed.

De Luca had often criticized the conventional practice of supporting a product launch with an artillery of media advertising, which tended to be supported by fat budgets. He admired instead the way Palm had ingeniously used word-of-mouth advertising for the launch of its Pilot™. De Luca was therefore constantly challenging his sales and marketing organization to develop more innovative and less costly marketing approaches. The freeze on the marketing budget would, by necessity, force the launch team to rethink their strategy.

5.2 Opportunities in Vertical and Enterprise Markets

During the course of the previous twelve months, Logitech had also received several inquiries from third parties interested in partnerships for developing both vertical as well as enterprise markets for the io™. This former category included developing software for the medical environment where, for example, using the io™, prescriptions would be digitally recorded by doctors and transmitted to pharmacies over the Internet. The latter category could include the addition of the io™ into a sales force automation software or a CRM package.

The io™ team had also realized the endless possibilities for the pen to be used in forms management. Many government agencies or insurance companies tended to be forms-based in their paperwork. There was a potentially huge cost savings in going digital in these markets. A similar opportunity existed in the pharmaceutical industry where paperwork was involved in handling drug applications. Exploiting these opportunities, however, would require the development of customized software and specialized sales teams.

A dedicated person was assigned to a first-level screening of these proposals. Management wondered, however, whether these vertical and enterprise segments could help reach initial target volumes. Was the io™ going to be a hit at the retail level? Should they invest more into end-user education for the retail market? Or focus on the new possibilities in the enterprise and vertical segments?

5.3 Major Developments in the Digital Pen Category in 2003/2004

In the fall of 2003, the io™ was scheduled to be released in France, just in time for the back-to-school season. The company was also launching a second generation of io™ software that would introduce an advanced handwriting recognition capability, a feature judged critical for the future of the io™. In summer 2004 Logitech would launch the next version of the io™ hardware; this version of the pen would be approximately 20% smaller than the current product.

Nokia was about to launch its bluetooth-based Anoto pen for use with its mobile phones. While this would launch more media and end-user interest in the digital pen space, there was also the fear of early adopters embracing the digital pen in the mobile phone environment instead of the mobile computing environment.

5.4 Next Steps

The io™ project team wondered what they could do to trigger the interest of a broader group of customers and further spur end-user education without much advertising support. Realizing the danger of charting into an unknown market, they also wondered how to address new opportunities in the io™ space.

Exhibit 1
Competitive Landscape

Current products in pen computing environment

The Seiko InkLink is a tool for instantly capturing your handwriting or drawings directly to your handheld, laptop, or desktop PC. It's easy to use, works on ordinary paper up to legal size, and easily clips onto 50-page tablets. Plus, InkLink comes with its own carrying case that fits in a pocket or purse. Using Seiko's exclusive binaural technology, the InkLink Data Clip continuously listens for communications from the InkLink pen while it tracks the natural movement of your hand. As you write anywhere on the paper pad with the InkLink pen, the InkLink Data Clip reads precisely the location of the pen tip and communicates its exact position to your handheld, laptop, or desktop PC. Sounds complicated? Well, InkLink is about as difficult to use as a paper clip. With InkLink, all you have to do is think it, ink it, and link it.

Designed to address the problem of natural communication in a complex digital world, the VPen™ gives users freedom to express themselves in the most natural way – with handwritten words and ideas, instructions, and drawings. VPen™ combines two familiar tools, the PC mouse and a regular pen, into a single digital writing instrument.

VPen™ converts handwriting to ASCII text and sends it straight to your handset, PDA, PC, or iTV.

It writes e-mails, SMS notes, calendar entries and URL addresses. VPen™ supports both Latin and Asian characters. You use the VPen™ just like you would a regular pen. Backspace, punctuate, and capitalize with a gesture. Moving the VPen™ makes things move in the digital world. Just like your PC mouse.

Navigate and select items in menus and graphical interfaces through "point and click." Direct cursor positioning for location-based services and website links. Use preprogrammed shortcuts or create your own for instant access. Use the VPen™ for drawing and creating graphical messages. Add graphics to text messages on your handset, and sketch concepts and diagrams directly to your PC.

The SonyEricsson Chatpen™ works via a mobile phone to let you communicate both digitally and on paper – simultaneously. Since writing with Chatpen™ is at once physical and digital, there is no longer a need for reentering scribbled notes into a computer. Perhaps more appealing is the fact that you are no longer restricted to text-based SMS or e-mail but are able to send handwritten notes with Chatpen™ to anyone with a mobile phone, PC, or PDA. Everything from sketches to graphical e-mails can be transmitted instantly, adding a more personal dimension to messaging and simplifying the process of sending text in languages that do not use the Roman alphabet.

When paired with a compatible phone, the Nokia Digital Pen lets you send colorful handwritten MMS messages to a compatible phone or email address. You can also use it to store handwritten notes on your compatible PC.

Exhibit 1 (continued)

Key Features:
Natural use of pen and paper
Wireless connection to compatible phones using Bluetooth technology
Send your personalized note or drawing via MMS to a compatible device
Store your personal notes on a compatible PC
Supports the use of compatible third-party open services

Products that failed in pen computing space
Combining a mobile computer with a digital notepad, the ThinkPad TransNote™ helps highly mobile users integrate paper-based information into the digital workspace efficiently. It's ideal for mobile professionals, such as executives, sales reps, bankers, or attorneys. The ThinkScribe™ digital notepad, digital pen, and Ink Manager™ Pro capture handwritten ideas on paper and transfer them to the computer, where they can be organized and searched. Easily share notes and ideas by dropping them into e-mail, word processing documents, or presentations. The ThinkPad TransNote™ organizes handwritten "to-dos" and can be used to fill out customized forms.

The Sony Tablet is a new LCD display technology that allows the user to draw, sketch, erase, and navigate displayed content directly on the LCD monitor. Whether it's image editing or Web surfing, the Tablet offers a more intuitive and natural way to interact with a PC.

The Tablet's ergonomic design allows for an LCD display to fold down to a near horizontal angle, providing a position similar to that of writing with a pen and paper. The extremely high position sensing accuracy of the Tablet allows users not only to operate applications running on the Windows® operating system as precisely as a mouse, but also to use the interface for professional image editing and drawing. To allow the user to take full advantage of the Tablet, Sony has developed new software that optimizes the intuitiveness of the stylus operation.

PenPoint was the founding project of the GO Corporation, which began in 1987. It was a broad project, which included the design of hardware, systems software, and initial applications that were seen to be the foundation for a new part of the computer industry: mobile, pen-based computers. The PenPoint operating system and interface incorporated a number of innovations, as well as providing a degree of object integration that had been missing in personal computing since the days of the Xerox Star, which was a similar project in its attempt to provide unified integration of hardware, operating system, and applications. The interface was based on a notebook metaphor, and the use of the pen as the primary (for most uses the only) input device.

PenPoint never reached the commercial market, for reasons that are a complex mixture of economic, business strategy, and technology maturity issues. It nevertheless serves as a good illustration of design that is strongly guided by thinking about the fit between the tasks a user wants to do and the characteristics of the devices that are appropriate.

Source: Company information from respective websites

Exhibit 2
Concept Description

The Logitech® io™ Unites Pen, Paper, and PC
18 September 2002, PR Newswire
First PC-based Personal Digital Pen That Remembers Everything You Write!

Logitech (Nasdaq: LOGI; Switzerland: LOGN, LOGZ) today announced its entry into a new market – digital writing – with the launch of the Logitech® io™, a personal digital pen, at IDG's DEMOMobile 2002 Conference in La Jolla, Calif.

With the Logitech io pen, users can easily share, store, organize and retrieve their handwritten information by simply writing with ink on paper the way they have for thousands of years. An optical sensor embedded in the pen captures the handwritten images, storing up to 40 pages in memory. This captured digital information can then be transferred into the PC by synching the pen via a USB cradle. The Logitech® io™ solution offers total mobility, since all that is necessary to carry is the pen and a digital paper notebook.

Available in November in the U.S. and select countries in Europe, the Logitech io uses functionality developed by Anoto, a company whose pen-and-paper technology is emerging as a new standard for digital writing. The pen itself is the key component of an ecosystem that includes leading paper manufacturers – Mead Cambridge Notebooks from MeadWestvaco, Post-it® Notes from 3M and productivity tools from FranklinCovey® in the U.S., and the Groupe Hamelin in Europe – combined with Logitech's proprietary software. Suggested U.S. retail price for the complete system starter kit is $199.

"Logitech is taking a very different approach to digital writing for the PC," said David Henry, senior vice president and general manager of Logitech's Control Devices Business Unit. "While other attempts at pen input have started with the PC, with the goal of making the PC more friendly, our point of departure is pen and paper, with the goal of making these elements more effective in the digital world.

"With the Logitech io, there's no need to change the way you work, or to lug your PC to meetings," Mr. Henry continued. "We believe this product will be well received by today's mobile workforce, as well as consumers who are looking to be more effective and creative with their note taking."

The digital paper that enables the capture of handwritten notes contains patented pre-printed tiny dots, which form a light screen effect. This pattern is read by an optical sensor embedded in the pen, which then stores the information in nonvolatile memory until the pen is docked in its USB cradle.

Upon docking the Logitech io pen, users can export their handwritten information to popular applications such as Microsoft® Word, Adobe® Illustrator, and calendaring tools including Microsoft® Outlook, Lotus Notes®, or any MAPI-supported email application. They can also create Post-it® Note reminders on their

Exhibit 2 (continued)

PC desktop. Note takers can also categorize and search their handwritten documents through the support of limited handwriting recognition by means of ICR (Intelligent Character Recognition) fields.

"The myth of the paperless office has not come true," said Christer Fahraeus, founder and chief executive officer of Anoto. "People everywhere are still using pen and paper in the majority of note taking situations, struggling with how best to merge the information that they want to have available in their PC with their everyday habits."

At launch, the Logitech io retail package will include the pen, a rapid-charging USB cradle, Logitech io software, one Mead Cambridge Limited Digital Notebook, Post-it® Software Notes-Lite and one pocketbook of Post-it® Notes for Digital Pens from 3M, an AC adaptor and five ink refills. Later this year, FranklinCovey will also introduce FranklinCovey® iScribe Digital Planning Pages for use with the Logitech io pen.

System requirements for the Logitech io pen include a PC running Windows® 98 or higher, a 90 MHz or higher processor, 64 MB RAM (minimum – 128 MB recommended), Internet access, Microsoft® Internet Explorer 5.01 or later and Microsoft® .net framework (both included on CD in the retail package), an available USB port, CD ROM drive and pre-printed paper (initial supply included). Email support requires a MAPI-compatible email client such as Lotus Notes® or Microsoft® Outlook®.

Source: http://tbutton.prnewswire.com/prn/11690X02642129

Exhibit 3
Team Organization

io™ Advisory Team

David Henry, Senior Vice President - Control Devices
Collette Bunton, Vice President - Regional Sales and Marketing, Americas
Robin Selden, Vice President - Worldwide Marketing

io™ Project Team

Marketing Team

Chris Bull, Dir. of Mktg., Retail Pointing Devices
Vanessa Torres, Senior Product Marketing Manager
Brigitte Maier, Business Development Manager, Germany
Jan Edbrooke, Business Development Manager, US
Claire Jenkins, Sr. Manager, Direct Marketing and Advertising
Betty Skov, Director of Public Relations
Nathan Papadopoulos, PR Manager

Engineering Team

Jean Claude Etter, Retail Mice Program Manager

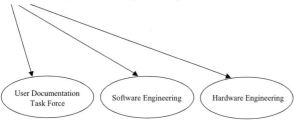

Source: Company literature

Exhibit 4
Why the Apple Newton Failed

Why the Apple Newton Failed
A former member of the Newton development group offers his thoughts on the
failure of Apple's PDA.
By Larry Tesler, CEO and chief scientist, Stagecast Software

Before there were Palm PDAs and Pocket PC devices, there was the Apple Newton. Like many of Apple's products, the Newton was ahead of its time. Unfortunately, Apple failed to implement key customer requirements uncovered during market research for the design phase of the product. Because Apple ignored its customers' demands, the Newton was a failure from the start.

I led the Newton group for more than two years, stepping down to make way for the third of a series of five top managers. I left the group a few months before the Newton shipped.

In my view, Apple prematurely launched the Newton for competitive reasons. From 1990 to 1993, Apple felt that pen computing and handwriting recognition was going to be the next big thing. The first product that satisfied the user's needs would surely dominate a huge new market.

Why Apple Rushed the Newton

Apple felt it was in a race with two competing companies: Microsoft and GO Corporation. GO's PenPoint operating system was seen as a potential threat to Apple's dominance. Microsoft's Pen Windows was a largely vaporware project from that company's cowboy days.

The Newton group redefined the project several times. Winning the race was a matter of survival, and we tried our best to meet and beat whatever specifications and ship dates our competitors were telling the press. Apple even forged a partnership with the expert group at Sharp Electronics that had developed the pioneering Wizard device. But in the end, we cut corners and ignored problems to try to meet a price range and a ship date that we had prematurely announced to gain an edge in a reckless public relations battle.

The Newton's False Marketing Claims

Before the Newton first went public, Apple marketing had written a draft brochure that boasted, "The Newton can read your handwriting." I instructed the marketing team to remove the statement and not to make such claims. Marketing initially disagreed but later promised to comply.

Exhibit 4 (continued)

When the brochures arrived, not only was that statement still there, but it was featured on the front page as the most prominent feature. I was shocked. The marketing group's desire to attract attention through dramatic claims ignored the fundamental tenet that a product must never seriously under-deliver on its promises.

As soon as I saw the final Newton brochure, I knew we were doomed.

The End of the Newton

Soon after the first Newton model bombed, Jeff Hawkins, creator of the Palm Pilot and a cofounder of Palm Computing, realized that technology had developed to a point where the features that customers really wanted in a PDA could be implemented at an affordable price. Suffice it to say the Palm line has been a roaring success. Microsoft reentered the market with Windows CE and continues to refine its strategy as it learns from each attempt it makes.

The Newton dramatically improved after its ignominious launch. Apple and its partners produced numerous prototypes and products. But it was too late, and few of these devices received any attention. Eventually, Apple ended the Newton product line. Rumors of a new Apple PDA have floated around for years, but I know of no such project myself.

Source: http://www.techtv.com/print/story/0,23102,3013675,00.html

Exhibit 5
IO^{TM} Advertising

Source: Company literature

Exhibit 6
Example of Press Coverage

Business Week
Dec 13 2002
By Stephen H. Wildstrom

Real Ink, Real Paper, Real Digital

Remember the CrossPad? Now Logitech has a slick new approach for getting handwritten notes from pad to PC.
The notion of getting handwritten text into a computer has been around for quite a while. A few years ago, A.T. Cross brought out a device called the CrossPad that let you write on a standard pad with a special pen and paper holder, then transfer your scribblings to a computer. IBM (NYSE: IBM), which was responsible for the CrossPad's technology, refined it into the ThinkPad TransNote, a laptop with a CrossPad-like tablet. Like the CrossPad, it flopped. And, of course, Microsoft (NYSE: MSFT) has invested heavily in the Tablet PC, which lets you write on screen with a special pen and saves the results as "digital ink" (see BW, 11/18/01, "The Microsoft Pen Is a Mite Clunky").

Now, Logitech (NASDAQ: LOGI) is betting that maybe the Cross-IBM idea of capturing writing with real ink on real paper was the right way to go. Its new io™ Personal Digital Pen dispenses with the CrossPad's clunky digitizing pad holder, which read radio signals from the special pen. Instead, it incorporates a tiny camera in the pen – similar to the sensor used in an optical mouse – that traces a pattern of tiny dots printed on special paper.

When you put the pen into a dock that connects to a Windows PC through a USB port, the digital ink is transferred to the computer. Once there, it can be pasted into a number of applications, from Microsoft Word to FranklinCovey's PlanPlus scheduler. (Since the pen uses ordinary ink, you can also use it to write on ordinary paper, but the results won't be captured.)

THE OLD-FASHIONED WAY.
The $200 io™, based on technology from Sweden's Anoto, requires a fairly bulky pen that people with smaller hands might find uncomfortable, though its weight of a bit under 2 ounces (53 grams) actually makes it lighter than many conventional pens. It can supposedly hold 40 pages of written material in its built-in memory, but since the battery is rated at only 25 pages, you'll probably have to dock the pen for recharging long before you would have to dock it to dump its memory to a PC.

The io's advantage is that it lets you take notes the old-fashioned way, far more unobtrusively than with a Tablet PC, while still creating an electronic record of what you write. Although Logitech chose not to include full-fledged handwriting recognition, feeling that the state of the art wasn't good enough for a satisfactory user

Exhibit 6 (continued)

experience, you can create keywords or date notations in capital letters and search for them later.

One significant drawback to this approach is the cost and limited availability of the special paper required. Three 160-page spiral-bound notebooks from Mead cost a stiff $24.99. A three-pack of large Post-it® notes from 3M costs $14.99. Oddly enough, the trusty legal pad isn't yet in the catalog. If the io pen catches on, however, the special paper should get much cheaper, and many more varieties should be available.

Source: http://www.businessweek.com/technology/content/dec2002/tc20021213_7467.htm

References

1   Logitech Press Release, March 2002.

CASE 6-5
The "mi adidas" Mass Customization Initiative

Ralf W. Seifert

Rolf Reinschmidt, head of the Forever Sport Division of adidas-Salomon AG, was reviewing adidas' mass customization (MC) initiative: "mi adidas":

> We all talk a lot about experiences these days – experiences that consumers and retailers expect to have with brands like ours [adidas]. Well, here is an experience our brand is uniquely able to offer, differentiating us significantly from the competition and building an incredible image for the Forever Sport Division.

It was October 2001. Reinschmidt sat down in his office and reflected on his experience to date. He had been sponsoring mi adidas to create a customization experience. The journey had started many years earlier, with the company providing tailor-made shoes for top athletes. Now, customized shoes had been made available on a much broader scale. Competitors also tested the market, and a trend toward MC was visible in other industries from PCs to made-to-measure jeans. The time had come to make specific recommendations on the best course of action for mi adidas.

Reinschmidt had three alternative routes to choose from:

- *Withdraw*: Celebrate the success and PR effect accomplished to date but quietly withdraw from MC in order to focus on adidas' core business.
- *Maintain*: Maintain the developed capabilities and selectively run mi adidas fairs and planned retail tours following top events such as World Cup Soccer and world marathon series.
- *Expand*: Expand mi adidas to multiple product categories and permanent retail installations; elevate it to brand concept status while further building volume and process expertise.

mi adidas had gained substantial momentum – it needed direction.

Copyright © 2002 by IMD – International Institute for Management Development, Lausanne, Switzerland. Not to be used or reproduced without written permission directly from IMD.

1 Global Footwear Market[1]

In 2000 the global footwear market was US$16.4 billion.[2] North America accounted for 48%, Europe 32%, and Asia-Pacific 12% of the market. The degree of concentration in the footwear segment was relatively high, with the three largest companies controlling roughly 60% of the market: Nike commanded 35%, adidas 15%, and Reebok 10% market share. Nike was particularly strong in the US market, with a 42% market share in footwear, but also led in Europe with 31%. adidas was significantly stronger in the European footwear market, holding a 24% market share compared with its 11% market share in US footwear. (*See Exhibit 1 for regional market share information and trends for adidas, Nike, and Reebok.*)

2 adidas-Salomon AG

For over 80 years adidas had been part of the world of sports on every level, delivering state-of-the-art sports footwear, apparel, and accessories. (*See Exhibit 2 for a history of adidas.*) In 2001 adidas-Salomon AG's total net sales reached € 6.1 billion and net income amounted to € 208 million. Its main brands were adidas with a 79% share of sales, Salomon with 12%, and TaylorMade-adidas Golf with 9%. The company employed 14,000 people and commanded an estimated 15% share of the world market for sporting goods. Headquartered in Herzogenaurach, Germany, it was a global leader in the sporting goods industry, offering its products in almost every country of the world: Europe accounted for 50%, North America for 30%, Asia for 17%, and Latin America for 3% of total sales.

2.1 *Forever Sport Division*

In 2000 adidas was reorganized into three consumer-oriented product divisions: Forever Sport, Originals, and Equipment.[3] Forever Sport was the largest division with products "engineered to perform." Technological innovation and a commitment to product leadership were the cornerstones of this division. Sales fell into a few major categories: running 32%, soccer 16%, basketball 11%, tennis 9%, and

---

[1] Sporting goods intelligence (SGI).
[2] Total market value based on wholesale prices.
[3] The reorganization officially took effect January 1, 2001. In 2002 the organization was revised again and the Forever Sport division became the Sport Performance division.

others 32% [1]. Reinschmidt was the head of the division. He reported directly to Erich Stamminger, head of global marketing for adidas-Salomon AG.

3 mi adidas

"mi adidas" was envisaged as an image tool and a center of competence for the Forever Sport division. Christoph Berger, director MC, was responsible for mi adidas and led a small but dedicated team. Berger came from an old shoemaking family and followed a traditional apprenticeship as a shoemaker himself. Having earned an Executive Master of Business Administration (EMBA), he started working for adidas in 1995. The pilot was sponsored directly by Reinschmidt and Stamminger. Without formal line authority, however, Berger had to draw implementation support from the various functions and use external contractors to complement his team. (*See Exhibit 3 for a project breakdown.*)

mi adidas was launched in April 2000 to provide consumers with the chance to create unique athletic footwear produced to their personal specifications. The idea was not entirely new, as adidas had provided tailor-made shoes to top athletes for many years. Now mi adidas could be experienced by many consumers at top sporting events and select retailers. The project initially offered only soccer boots but was to be expanded in 2001 to offer running shoes. For 2002 the plan was to further build volume and expand the offering into the customization of basketball and tennis footwear. (*See Exhibit 4 for the initial mi adidas rollout plan.*)

4 Phase I: The mi adidas Pilot

The first phase of the mi adidas project was a small pilot to evaluate the feasibility and prospects of mass customizing athletic shoes. The objectives were clear-cut: offer a customized product, test consumers' demands for customized products, and fulfill their expectations as far as possible. The pilot allowed both project team and functions to gain hands-on experience in marketing, information management, production, distribution, and after-service of customized shoes. It also provided a basis for a rough cost-benefit analysis and future budgeting.

4.1 Product Selection

The pilot mandate was soccer boots. A Predator® Precision boot already in production was offered for customization with regard to fit (size and width), performance (outsole types, materials and support), and design (color combinations and embroidery).

4.2 Marketing: Event Concept and Communication

The pilot was 100% event based. Over a two-month period in 2000, six events were held in different European cities: Newcastle, Hamburg, Madrid, Marseille, Milan, and Amsterdam. Consumer recruitment was very selective, using local market research agencies, phone calls, written invitations, and pre- and postevent questionnaires. (*See Exhibit 5 for details of the customer recruitment process for the pilot.*) The target group was 50 participants per event and country.

The MC unit was designed in a rather neutral technology-oriented style stressing the brand's tradition as the athlete's support. A white cocoon (evoking a mysterious atmosphere with its shape and color) housed the 3-D foot scanner, the heart of the unit (*Figure 1–left*). A newly developed matching matrix software supported scanning and fitting. At a separate fitting terminal, a selection of sample boots was available for testing fit preferences (*Figure 1–right*).

In addition, a design terminal (not shown) had a laptop on which the participants, assisted by adidas experts, could customize their soccer boot in terms of materials and design. The stations also displayed material and color samples to facilitate the decision-making process. (*See Exhibit 6 for details of the customization steps.*)

> Today, it is a brand new, revolutionary, and futuristic experience. Soon it will be as normal as buying individualized glasses at an optician.

Figure 1 MC Unit: 3-D Foot Scanner Station (left) and Fitting Terminal (right)

4.3 Consumer Feedback

adidas consumers greeted mi adidas with tremendous excitement:

> Consumers loved the product. One hundred percent want a customization service available in the future.

Shortly after the introduction of mi adidas, even adidas headquarters started to receive direct inquiries from interested consumers who wanted to purchase a customized shoe. Franck Denglos, marketing coordinator, reflected on his experience with the mi adidas pilot:

> The concept and its execution gave consumers a strong positive impression of the brand. They left with the perception that adidas was acting as a leader. However, we have to keep in mind that their perception was highly influenced by the impressive tool, run by highly qualified adidas experts. Plus, during the pilot, the shoe was free.

5 Competitors' Footwear Customization Initiatives

Several competitors offered a similar service.

Nike Inc.

Nike decided to bring MC to the Web in November 1999. Nike's NIKEiD program enabled online customers to choose the color of their shoe and add a personal name of up to eight characters [2]. For this service, Nike asked the regular retail price for the shoe plus a $10 custom design fee and shipping charges. Delivery of the footwear was advertised as being within three weeks for the US market. To keep fulfillment and distribution under control, however, Nike imposed an artificial ceiling and only accepted up to 400 US-based orders per day [3].

Reebok International, Ltd

As of 2001, Reebok had not launched (or announced) its own mass customization initiative. Instead, it marketed its full foot cushion for its top-of-the-line running shoe, the Fusion C DMX 10. Utilizing DMX®10 technology and 3-D ultralite sole material, Reebok provided ten air pods to help distribute air for custom cushioning and to achieve the ultimate in shock absorption [4].

New Balance Athletic Shoe, Inc.

New Balance opened its first "width center concept unit" at Harrods in London in April 2001. Coming from a long tradition of making arch supports and prescription footwear to improve shoe fit, the US-based company first manufactured a performance running shoe in 1961. By 2001 New Balance featured a range of athletic shoes and outdoor footwear. Although New Balance did not offer shoe customization, it typically offered its products in three (at times up to five) different width sizes to optimize shoe fit [5].

Custom Foot Corp.

Custom Foot Corp. was one of the leading pioneers of mass shoe customization in terms of fit and design. Featured in cover stories of the New York Times, Forbes, and Fortune, the company seemed to show a whole industry new ways of doing business: custom-made Italian shoes, delivered in about three weeks, at off-the-shelf prices. But after a glorious start in 1995, Custom Foot went out of business and closed operations in summer 1998. Its concept for blending customization and mass production had failed, as the whole system could not handle the enormous complexity of the process [6].

Creo Interactive GmbH

In 1998 Creo Interactive designed a totally new shoe based on a modular concept for the sole, the main body, and the tongue. This shoe could be produced in just 83 working steps compared with Custom Foot's 150 to 300 steps [7]. Leveraging the Internet as an interface for configuration, Creo offered pure design customization in terms of colors and patterns. By locating production in Germany, it was possible to swiftly fulfill European market needs. However, three years into the venture, Creo Interactive closed operations in 2001.

Customatix.com of Solemates, Inc.

Based in Santa Cruz, CA, Customatix.com allowed consumers to log on to its Internet site and choose from a vast array of colors, materials, graphics, and logos to create their own personalized portfolio of designs online: "150 choices you can put on the bottom of your sneakers." [8]. The blueprints were then transmitted to the company's factory in China, where the shoes were manufactured and shipped to the consumer's doorstep within two weeks. The shoes retailed for $70 to $100 per pair, including import duty and delivery charges.

> The biggest problem we have is, people don't believe we can do what we do.
>
> *Dave Ward, CEO of Customatix.com [9]*

6 Phase II: The mi adidas Retail Tours

The decision to proceed with the "Customization Experience" project was made after the successful completion and stringent evaluation of a pilot project conducted in the second half of 2000 in six European countries. During the test project some 400 pairs of the revolutionary adidas Predator® Precision soccer boots were custom built and delivered to a select group of consumers in Germany, France, England, Spain, Italy, and the Netherlands. Delivery time took two weeks on average. Consumer satisfaction was overwhelmingly positive.[4]

---

[4] adidas-Salomon AG, Press Release, April 2, 2001 (*see Exhibit 7 for the full press release*).

Figure 2 New Retail Unit Running

The pilot project was mainly seen as a first attempt to evaluate the requirements of "normal" consumers, as opposed to those of top athletes, with whom adidas had an ongoing relationship. Taking the successful concept of the pilot to the retail channel, however, meant facing different and new challenges. For the pilot, certain issues to do with back-end processes were adapted to current processes or not covered at all. Now, these would require more attention. In addition, a new retail unit (*Figure 2*) had to be created that was smaller (10 to 20 m²), easier to transport, more durable, and user friendly.

6.1 Retailers

Retailer interest in mi adidas was overwhelming. In Germany alone, almost 1,000 athletics specialty shops wanted to participate. However, only 50 German retail stores could be part of this second phase: the first retail tours in 2001. Soon retailer selection became a sensitive issue within adidas: Marketing preferred small athletics specialty shops for a maximal image effect and utmost retailer commitment.[5] Sales, however, favored big key accounts for reasons of relationship management. In addition, country selection was controversial: in some countries retailers were accustomed to paying a fee to a manufacturer for being able to host a promotion such as mi adidas. In other countries, retailers had never paid a fee for in-store promotions and might even demand a fee from the manufacturer instead. Depending on the final selection verdict, retailer feedback ranged from enormous enthusiasm to vast disappointment (even sporadic threats to withdraw business from adidas altogether).

---

[5] Some specialist stores got very excited about mi adidas and lined up local sponsors to equip entire sports teams with customized shoes while hosting the mi adidas retail unit at their outlet.

Once selected, the retailers took care of consumer recruitment. To support them in marketing mi adidas, they were given a package of communication tools: CDs, posters, invitation cards, registration cards, and folders. Some retailers felt that the material was not engaging enough and demanded more support. Subsequently, the countries modified and translated the tools to fit the needs of their consumers more directly. Yet consumer turnout (and order placement) varied greatly from one retail store to another, depending on the commitment to mi adidas.

Whereas the pilot was 100% event based, retailers played the central role in the second phase and accounted for roughly 90% of the order volume. Using multiple mi adidas retail units, well over 100 retailers participated across Europe in 2001.

6.2 Customization Process

The customization process was still run by adidas experts and emphasized the "brand experience" theme. The 3-D foot scanner, however, had been replaced by a simpler Footscan^{TM} unit, which was used in combination with a static measurement device for length and width measures. At the same time, the proprietary matching matrix software continued to evolve and directly conformed to consumer preferences in three out of four cases. The overall process had become very stable. Fifty to 80 "customization experiences" could be handled per day during an event while about 15 to 20 were possible at a retail outlet.

Recent survey results seemed to confirm European consumers' interest in customized shoes. Although a focus on design customization was much simpler from a configuration perspective, consumers rated a customized design as much less important than a customized fit. In addition, individual preferences varied significantly across different European countries (and to a lesser extent also between men and women), necessitating further research for a targeted offering [10].

6.3 Product and Pricing

The athletic footwear market was characterized by rapid product turnover. In 2001 mi adidas already featured its second product generation in soccer boots. The customized version of this soccer boot sold for a 30% to 50% price premium over the catalog price of € 150 [11]. In addition, the product offering was expanded to running shoes. After successful internal presentations of mi adidas for Running at the adidas global marketing meeting in March 2001 and Investor Day in July 2001, preparations were made to launch the project in the Running market. In September 2001 mi adidas for Running was introduced to the market during the Berlin Marathon. Consumers were either recruited or invited by PR, or they were impulse buyers who passed by and became interested. Within three weeks, they received the shoes. For 2002 the plan was for mi adidas for Running to be present at all adidas-sponsored marathons (*i.e.*, Paris, Boston, London, Madrid, Rotterdam, Prague, and Berlin) and to go on a retail tour in the relevant country after each event.

6.4 Consumers

Consumer feedback was excellent. In particular, the short delivery time and the opportunity to design their own shoes impressed the consumers. mi adidas also attracted strong interest from the press: two television stations (Bayerischer Rundfunk and Fox TV) and many articles featured adidas' MC initiative and hailed it as a major milestone.

> Although we received this good feedback, there were several technical problems that had to be tackled. These problems caused delays and in some cases wrong production.

6.5 Information Management

Information management throughout the entire process was critical: basic consumer data, product options, biometric knowledge, and product specifications had to be merged for order taking. In addition, sourcing, production, distribution, payment, and reordering required appropriate IT backing [12].

> Information is the most important conversion factor of successful mass customization.

Many challenges in terms of the scope and integration of the required IT infrastructure remained:

1. The mi adidas kiosk system for order creation led to technical problems with synchronizing information generated offline (*e.g.*, order numbers and customer records) with adidas backbone systems such as the sales system and customer master database.
2. The traditional sales system was not designed to process orders of individually customized shoes with detailed information on each article.
3. The IT systems for distribution needed to be extended for an organized distribution and return process.
4. Consumer data captured via mi adidas could not be transferred to the adidas CRM system.

There were ways around these problems, but they resulted in limited centralization and poor accessibility of data.

The initial rollout was clearly underbudgeted. For example, eRoom [*Exhibit 10*] was chosen as the web-accessible repository for the technical documents. This decision was not entirely supported by Global IT and is seen as a short-term solution until an alternative can be found.

All development, configuration, and support for mi adidas had thus far been absorbed by the business and no costs had been charged to the project budget for IT solutions, beyond the mi adidas kiosk application and scanning software. The kiosk application was developed by a contractor. However, no helpdesk was available for support and future system integration. The IT department was worried:

> The speed of implementations, the time needed to support both SAP and non-SAP countries, and the limited resources Global IT presently has to support this leads to the conclusion that we may fail to maximally deliver mi adidas globally.

mi adidas had progressed fast – calling for a completely new set of requirements.

> mi adidas, even such a small project, has forced the IT department to think about how close we are getting to our consumers and what is needed to support this development.

6.6 Production

By 1992 most sporting goods companies had outsourced the main part of their footear production to the Far East to reduce production costs. adidas followed suit and outsourced all textile production and 96% of footwear production during its turnaround in the mid-1990s. The outsourced footwear production was divided between Asia (China, Indonesia, South Korea, Taiwan, Thailand, and Vietnam), Eastern Europe, and North Africa. Depending on the quality of the shoe, between 20% and 40% of production costs was related to personnel costs, which were the main driver for cost differences between regions [13]. Contract manufacturers focused on footwear assembly and sourced input materials from local suppliers as needed. (*See Exhibit 8 for a supply chain overview.*)

adidas maintained a small footwear factory in Scheinfeld, Germany, near its headquarters. Here, models, prototypes, and made-to-measure performance products could be manufactured and tested. In addition, special shoes for Olympic sports such as fencing, wrestling, weightlifting, and bobsled were made. However, Scheinfeld was not excited at the prospect of taking mi adidas production in-house. Furthermore, material provisioning for a vast set of customization options could be more costly in Scheinfeld because it was too far away from volume production sites and suppliers.

The production processes used for the mass customization shoes were the same basic processes used in mass production. For the MC events, however, a combination of development sample room and mass production facilities was used (*see Exhibit 9 for a comparison*). This combination was chosen to allow for the highest level of control and quality while providing a minimal "disruption" to the factory's daily mass production schedule.

A program like mi adidas, without dedicated facilities, manpower, and materials resources, will always be perceived as an interruption to the overall process of creating shoes.

Yet the capacity of the sample rooms was limited[6] and its operational format was not designed for volume scale effects. The mass production facilities, by contrast, were not meant to handle a lot size of one[7], nor were they set up to allow for close linkage of individual product flow with corresponding customization information. Such a process was not in place, and the workers lacked training and language capabilities to handle production according to detailed written product specifications.

Variability is simply not in our business model!

Although the assembly of a customized shoe was theoretically straightforward, provisioning the required material proved to be time consuming. Delays were exacerbated when material was needed that was not currently available for inline production. In this case, special material provisioning resulted in significant inventory costs as materials for the top-of-the-line models in question were expensive. From a production perspective, a better understanding was needed of the value-cost tradeoff between the marketing perception of customer value added versus inventory and production costs for specific customization options. For example, design customization in terms of multiple colors was not ideal from a material provisioning perspective because different shoe sizes already meant different component sizes (*e. g.,* strip length varied with the shoe size), which would now have to be available in a range of colors. These tradeoffs and the options available for new shoes should ideally play a much more prominent role right from the start – in designing products for MC. Karl-Josef Seldmeyer, vice president, head of global supply chain management, summarized his experiences:

For today's volumes, the combined complexity of fit, performance, and design is too much.

6.7 Distribution

Timely mass customization also depended on proper execution of communications and logistics to meet the seven-day lead time from order receipt to ex-factory shipping. Starting in July 2001, the mi adidas process was changed from a pilot with deliveries direct to the final consumer to a process that involved the retailers in customization and distribution. After customization at the retailer shops, orders were no longer transferred directly to the sourcing systems. Instead, they were routed from the retailer to the respective subsidiary's sales system and from there to logistics ordering systems, using the subsidiary's regular buying process. The addressee was no longer the final consumer but the individual retailer in whose

---

[6] For the mi adidas pilot, volume was limited and production was not a problem. In general, a development sample room, however, could not handle more than 500 to 1,000 pairs per month.
[7] Production setups were often made only once per day producing large batches of footwear.

shop the customization had taken place. The individual retailer was now responsible for distribution to the end consumer. (*See Exhibit 10 for an overview of the mi adidas order and product flow.*)

6.8 Communication and Competing Initiatives

With the push into multiple product categories, communication became more difficult. In particular, the extremely technical and highly advanced customization process could not be adequately promoted as the mi adidas budget did not support targeted messaging by category. However, increasing the marketing budget was not then an option, since marketing saw MC as just one of many initiatives. After all, they already supported top athletes via a special care team and tailor-made shoes made in Scheinfeld. Since it was naturally in competition for resources and management attention with other recent initiatives, mi adidas was often seen as secondary to designated brand concepts such as "a3"[8] and "ClimaCool™".[9] Hence a³, "Football never felt better," and Clima acted as overriding messages for the upcoming marathons, World Cup Soccer, and Roland Garros (French Open Tennis Championship), respectively.

> Communication activity and spending need to be regulated, ensuring that brand concepts are not undermined by ongoing mi adidas activity.

6.9 adidas' Own Retail Activities

To further strengthen its brand, adidas had also just stepped up its own retail activities, increasing the number of its own retail outlets from 37 in 2000 to 65 in 2001. Most notable here was the opening of two concept stores in Paris and Stockholm as well as an adidas Originals store in Berlin [14].

6.10 Negotiating Continued Internal Support

By October 2001 mi adidas was an established initiative, and the generally positive brand image effect was widely accepted within the organization. Although mi

---

[8] a³ was a functional technology combining cushioning, stability, and light weight. It managed the foot's natural movement by dissipating harmful impact forces, stabilizing and guiding the foot through the entire footstrike, and retaining and redirecting energy from the rear foot to the front foot. adidas planned to introduce the concept for running shoes in 2002 as the most technical and functional design available.

[9] ClimaCool™ was a footwear technology concept offering 360 degrees of ventilation and moisture management. In scientific tests, it produced 20% dryer and 20% cooler feet. Targeting regular or serious athletes, adidas planned for a staggered market introduction across products in 2002.

adidas had become bigger, the organizational setup had not substantially evolved. To date, much of the support for the project from different functions of adidas was granted on a goodwill basis. As time progressed and volumes increased, it naturally became more and more difficult to persuade core business units to fully support this initiative, especially out of their own cost centers.

The annual budget for mi adidas had basically stayed identical during its first years.

The situation was not ideal. Although the functions continued to support mi adidas and took pride in its success to date, the ultimate responsibility for mishaps, of course, rested with the project team. Should mi adidas be elevated and play a more independent role or should it be better integrated into the existing matrix to be in sync with adidas' core business, with the functions in turn assuming more accountability? A clear evaluation was made difficult by the current practice of attributing mi adidas sales to the respective countries, hindering separate accounting.

7 The Future of mi adidas

Reinschmidt wondered if the time was right to scale mi adidas to the next level (Phases III and IV; *see Exhibit 3*). The pilot (Phase I) had been very successful and adidas had developed and refined important new capabilities. Consumer feedback was enthusiastic, and retailers fared much better during repeat offerings of mi adidas. Yet the initial retail rollout (Phase II) had been somewhat slower than projected, falling 40% to 50% short of the targets established in the original rollout plan.

7.1 Future Alternatives

Reinschmidt had come to the conclusion that mi adidas needed clearer direction. Once again he reviewed the three generic alternatives that the company could embark on:

mi adidas could be turned into a commercial tool over the course of the next years and now was the time to decide upon this.

Alternative 1: Withdraw – Celebrate the success and PR effect accomplished to date but quietly withdraw from MC in order to focus on adidas' core business.

mi adidas had been launched two years earlier and now featured a soccer and running shoe. As the product life for these model cycles ended, so would mi adidas. Current commitments would be honored, but any further investments in the MC initiative were to be avoided. New PR tools would soon take the place of mi adidas.

Alternative 2: Maintain – Maintain the developed capabilities and selectively run mi adidas fairs and planned retail tours following top events such as World Cup Soccer and world marathon series.

mi adidas would continue in its current form and scope and be allowed limited organic growth. Investment would be minimal and MC responsibilities would be more fully integrated into the existing functions. mi adidas would be part of (and governed by) adidas' annual planning cycle. As new boots were introduced to the market, a customizable derivative of those models would be created for mi adidas; the kiosk application, promotional material, and back-end processes, *etc.* would be adapted accordingly.

Alternative 3: Expand – Expand mi adidas to multiple product categories and permanent retail installations; elevate it to brand concept status while further building volume and process expertise.

mi adidas would be scaled up in terms of both volume and product categories. Increased marketing spending and revised back-end processes would support its rollout. Permanent installations at select retail stores would complement the event and retail tour concepts to foster more continuous order flow and steady volumes. Further investments would ensure a degree of independence for mi adidas and help develop MC into a potential business model in its own right for adidas.

7.2 Decision Looming

Reinschmidt had the alternatives lined up, and it was up to him to come to a sensible recommendation based on the various inputs received. He had just started to summarize a set of key issues that should determine which alternative to choose, as well as his assessment of the alternatives, when he was interrupted.

Exhibit 1
Global Footwear Market: Regional Overview

adidas footwear sales [15]

Region	2001 Net Sales	Net Change *vs.* 2000
North America	€ 818 million	−10%
Europe	€ 1,200 million	+15%
Asia	€ 371 million	+39%
Latin America	€ 122 million	+3%

Note: adidas' total net sales in 2001 were € 4.8 billion.

Nike footwear sales [16]

Region	2001 Net Sales	Net Change *vs.* 2000
USA	$3,209 million	−4%
Europe	$1,423 million	+9%
Asia-Pacific	$632 million	+14%
Americas	$360 million	+5%

Note: Nike's total net sales in 2001 were $9.5 billion.

Reebok footwear sales

Region	2001 Net Sales	Net Change *vs.* 2000
USA [17]	$931 million	+1%
UK*	$484 million	+1%
Europe*	$410 million	−4%
Row*	$256 million	−4%

Note: Reebok's total net sales in 2001 were $3.0 billion.

---

* Footwear share of net sales estimated based on Reebok International, Ltd., 2001 annual report.

Exhibit 2
A Short History of adidas: 1920 – 2001

It all began in 1920, when Adolph "Adi" Dassler and his brother Rudolph made their first shoes in Herzogenaurach, a small village in the south of Germany. Using the few basic materials available after World War I, Rudolph began making slippers with soles made from old tires. Adi converted the slippers into gymnastics shoes and soccer shoes with nail-on studs or cleats. The idea was as simple as it was brilliant: provide every athlete with the best possible equipment.

At the 1928 Amsterdam Olympics, German athletes showcased Dassler shoes to the world. In 1936, the brothers achieved a major breakthrough when Jesse Owens agreed to wear their shoes in the Berlin Olympics, where he won four gold medals. By 1937 the Dassler brothers were manufacturing shoes for more than 11 different sports. In 1948 the two brothers quarreled and Rudolph left to establish the Puma sports company, while Adi registered the name adidas and, the following year, adopted its now famous three-diagonal-stripe trademark. The first samples of adidas footwear were used at the 1952 Helsinki Olympics.

In 1954 Germany won the World Cup, wearing new screw-in studs on adidas soccer shoes. In 1963 the first adidas soccer ball was produced and clothing was added to the product range in 1967. By the Montreal Olympics in 1976, over 80% of medal winners were adidas-equipped athletes. Business was booming. adidas had become a household name in the sports arena, synonymous with athletic achievement.

In 1972 Nike entered the American market with low-quality, fashionable leisurewear targeting teenagers. Reebok followed suit in 1979. Following the death of its founder in 1978, adidas struggled through turbulent organizational and management changes and was quickly outrun by changes in the industry. Its street popularity faded as newer, more aggressive companies like Nike and Reebok stepped up the pace of competition.

In 1989 the Dassler family withdrew from the company, and the enterprise was transformed into a corporation. Bernard Tapie, a French business tycoon, took over but was soon jailed following his involvement in a soccer-fixing scandal. Subsequently, adidas was declared bankrupt and left to a number of French banks. In 1993 French-born Robert Louis-Dreyfus was appointed chairman of the executive board of adidas. Having purchased 15% of the company, Louis-Dreyfus was a majority shareholder and led a stunning turnaround for the company and initiated adidas' flotation on the stock market in November 1995.

In 1997 adidas acquired Salomon Group, and the company's name changed to adidas-Salomon AG. With the brands adidas (athletic footwear, apparel, and accessories), Salomon (skis, bindings, inline skates, adventure shoes, and accessories), Taylor Made (golf clubs, balls, and accessories), Mavic (cycling components), and Bonfire (snowboard apparel), adidas-Salomon AG substantially broadened its portfolio of sports brands, offering products for both summer and winter sports.

Exhibit 2 (continued)

In 2001 Herbert Hainer took over as chairman and CEO. adidas-Salomon AG's total net sales grew to €6.1 billion with net income of €208 million. It employed 14,000 people and commanded an estimated 15% world market share.

Source: http://www.adidas.com and IMD adidas case IMD-3-0743 (GM 743), January 26, 1999

Exhibit 3
Working Breakdown Structure Plan

Project Management	Product Development	Product configurator/ design tool	Marketing → Event → Communication	Information management	Production	Logistics/ shipment	Service/ fulfillment	Payment tracking
Project organization	Product development	Layout	*Product configuration*	General IT structure	Data reception	Customized packaging and shipment	Hotline	Clarify types of payment
Tracking	Pattern engineering	Image to transport	Color variations	Data tracking	Creation of bills of material	Define partner for shipment	Satisfaction panel	→ Online (credit card)
Budgeting/ payment	Foot bed	Configuration principle	Positioning/layout of all decorations	Handling	Planning	Speed	Returns/ replacement	→ Credit card at event location
Project kick-off	Fit test	Handling	Clarify: deco stitching or printing	Storage	Production operation	Costs	Next steps	→ Invoice
Project evaluation	Sample production	General architecture	Cosmetics	Research	on demand	Quality	New offers	→ Currencies
Meetings and presentations	Budget	Potential for future	*Event*	Speed	QC on demand	Budget	Ensure enduring customer-brand relationship	→ Accounting and billing
Handle critical situations		Other categories	Concept	Intelligence	Preparation for shipment			
		eCommerce	Message we want to bring across	Future plans	Packaging			
		Intelligence	PR-image we want to create	Budget	Interfaces to systems and projects (e.g., scanner, payment, eCommerce, content management)			
		Data flow interfaces	Communication strategy (name of program/ product!)		Production site selection			
		Budget	Preevent		Budget			
			Event organization					
			Postevent					
			Cooperation with regions					
			Conceptualization of product Configurator					
			Budget					

Source: Company information

Exhibit 4
Mass Customization Rollout Plan

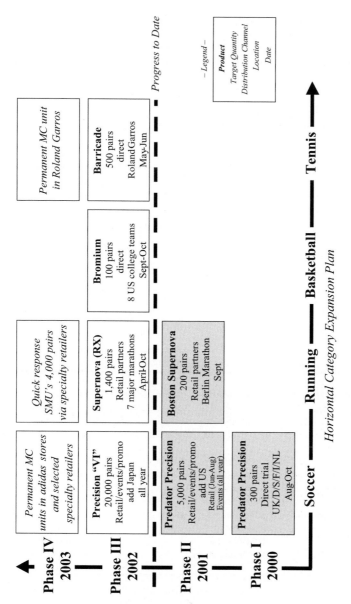

Source: Berger, C. "The Customized Revolution at adidas." Kundenindividuelle Massenproduktion: Von Businessmodellen zu erfolgreichen Anwendungen. Eds. M. Schenk, R. Seelmann-Eggebert, F.T. Piller. Die dritte deutsche Tagung zur Mass Customization, Frankfurt, November 8, 2001

Exhibit 5
One-to-one Communication with Event Participants/Recruits

–15 days	Local market research agency contacts potential participants
	Questionnaire-guided telephone interview
–10 days	Selection of recruits based on questionnaire answers
	Telephone availability-check of recruits
	Setup of substitute list
–7 days	Official written invitation accompanied by preevent questionnaire
–2 days	Ultimate check if recruits will take part in the event
	Telephone invitation of substitutes as required
Event	adidas customization experience and onsite interview
+2 weeks	Delivery of customized soccer shoes
+6 weeks	Telephone follow-up interview

Not only should the product and services be individualized and unique, but also the personal relationship and interaction with the consumer as being part of the customization experience. Are we prepared?

Christoph Berger, Director MC

Source: Company literature

Exhibit 6
Customization Process

Step 1 check in >	Step 2 scanning >	Step 3 fitting >	Step 4 testing >	Step 5 design >
Get ready for "mi adidas"! Get registered now for the chance to create your unique pair of customized shoes.	First your feet will be scanned by means of the adidas Footscan system to determine the exact length, width, and pressure distribution of each foot. This will enable you to determine which technologies your shoe will need for optimal performance.	Here you consult with an adidas fitting expert to review the results of your footscan. Then this information, combined with your personal fit preferences, is entered into a computer to determine the best-fitting shoe.	Once you have determined your personalized function and fit, you will have the opportunity to test your shoes before heading into the final design phase.	Now you can put the finishing touches on your one-of-a-kind shoes. You will be able to choose different colors, materials, and even personalized embroidery – all of which can be viewed on the computer screen as you make your selections.

All that is left now is to confirm and order your customized shoe.[10] Within two weeks your personalized footwear will be with you ready for a new level of performance.

Source: http://www.miadidas.com

---

[10] Customers were not pushed to accept a customized shoe – returns upon delivery were minimal.

Exhibit 7
Press Release: "adidas to Launch Customization Experience"

Custom-built footwear for consumers – Retail launch mid 2001 – Pilot project unit on display at ISPO

Herzogenaurach, 02/04/2001 – adidas, as the first brand in the sporting goods industry, is set to launch a pioneering "Customization Experience" project in footwear. The project starts in the Soccer category, but will be expanded into other major sports categories. With the "Customization Experience" project, adidas will give consumers the opportunity to create their own unique footwear to their exact personal specifications in terms of function, fit and looks, thus providing services that were so far only available to soccer stars like David Beckham and Zinedine Zidane.

The decision to proceed with the "Customization Experience" project was made after the successful completion and stringent evaluation of a pilot project conducted in the second half of 2000 in six European countries. During the test project some 400 pairs of the revolutionary adidas Predator® Precision soccer boots were custom built and delivered to a select group of consumers in Germany, France, England, Spain, Italy and the Netherlands. Delivery time took two weeks on average. Consumer satisfaction was overwhelmingly positive.

With the start of the "Customization Experience" project, adidas is entering the new age of the "experience economy." adidas introduces a new business model in the industry, influencing and changing the whole value chain and potentially the sporting goods marketplace, creating a new level of relationship between the consumer and the brand.

The adidas "Customization Experience" unit used during the pilot project will be on display at ISPO in Munich, February 4–6, 2001. Launch in the retail marketplace is scheduled for mid 2001.

Source: adidas-Salomon AG

Exhibit 8
Traditional Order and Product Flow

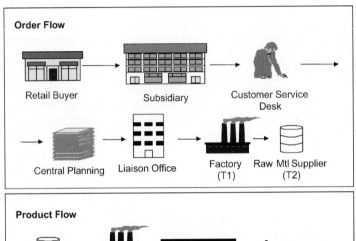

Order Flow

Retail Buyer → Subsidiary → Customer Service Desk →

→ Central Planning → Liaison Office → Factory (T1) → Raw Mtl Supplier (T2)

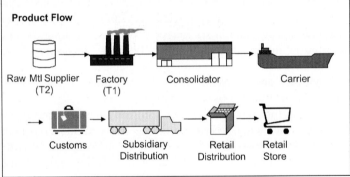

Product Flow

Raw Mtl Supplier (T2) → Factory (T1) → Consolidator → Carrier

→ Customs → Subsidiary Distribution → Retail Distribution → Retail Store

Source: Company literature

Exhibit 9
Comparison: Classical Production vs. Mass Customization
of adidas Footwear

cp (classical production)	mc (mass customization)
production	
• Third-party production • Mass/bulk production • Weekly production planning	• Third-party production • Fast moving lines • "On demand"
delivery-logistics	
• Transportation outsourced • Distribution centers (adidas owned)	• Complete outsourcing • (From factory to retailer)
lead time and production time	
70-110 days lead time orders - delivery in dc (distribution center) **10-12 days in production process**	**14 days lead time** order - delivery to customer (with material in stock) = critical component **3-4 days in production process**
delivery-logistics	
US$ 0,70 seafreight (landed DC) US$ 4,50 airfreight (landed DC) + customs and distribution freight	~ 6 $ courier service (landed retailer) per pair

Source: Berger, C. "The Customized Revolution at adidas." Kundenindividuelle Massenproduktion: Von Businessmodellen zu erfolgreichen Anwendungen. Eds. M. Schenk, R. Seelmann-Eggebert, F.T. Piller. Die dritte deutsche Tagung zur Mass Customization, Frankfurt, November 8, 2001

Exhibit 10
mi adidas Order and Product Process Flow

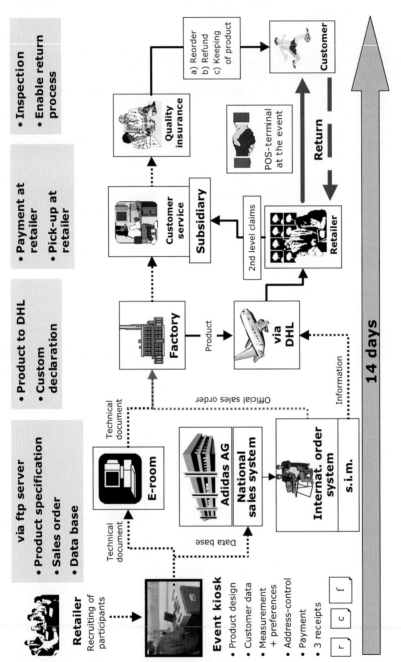

Source: Company literature

References

1   adidas-Salomon AG, 2001 Annual Report.
2   http://nikeid.nike.com.
3   "Nike offers mass customization online." Computerworld, November 23, 1999.
4   http://www.reebok.com.
5   http://www.newbalance.com and New Balance Athletic Shoe, Inc., Press Releases, April 20, 2001.
6   Piller, F.T. *Mass Customization*. Wiesbaden: Gabler Verlag, 8/2001: 397.
7   Piller, F.T. "The Present and Future of Mass Customization: Do It – Now!"
8   http://www.customatix.com.
9   K. Smith, "Fancy Feet." *Entrepreneur's Business Start-Ups*, 7/2001.
10 Jäger, S. "Market Trends: From Mass Production to Mass Customization." EURO ShoE project, March 3, 2002, Innovation in the European footwear sector conference, Milan.
11 "Individuelle Maßanfertigung von Sportschuhen." *Schuhplus/infocomma*, October 15, 2001.
12 Dulio, S. "Technology Trends: From Rigid Mechanical Manufacturing to Mass Customization." EURO ShoE project, March 3, 2002, Innovation in the European footwear sector conference, Milan.
13 IMD adidas case IMD-3-0743 (GM 743), January 26, 1999.
14 adidas-Salomon AG, 2001 Annual Report.
15 adidas-Salomon AG, 2001 Annual Report, Year Ended 12/31.
16 Nike Inc., 2001 Annual Report, Year Ended 5/31.
17 Reebok International, Ltd., 2001 Annual Report, Year Ended 12/31.

CASE 6-6
New Business Creation at Tetra Pak: Reinventing the Food Can

Jean-Philippe Deschamps and Atul Pahwa

As he sat in his office in Lund (Sweden) one sunny morning in October 2001, Jan Juul Larsen wondered how he could rapidly turn a revolutionary packaging development project into a real, profitable business. He had agreed, a few months earlier, to leave his job as managing director of Tetra Pak-Taiwan to head a new company, Tetra Recart AB. Its mission: Bring a retortable carton to market, thus offering food canners a modern alternative to the 150-year-old metal can and extending the reach of Tetra Pak into solid foods.

The retortable carton had existed as a development project within R&D for eight years. During that time, engineers had managed to perfect their laminated carton technology to make it "retortable," *i.e.*, be able to withstand high boiling temperatures for hours. They had also developed a complete system to form carton boxes from printed blanks at high speed, fill them with various kinds of food, seal them, and sterilize them at high temperature before packing them on pallets for shipment to stores. This innovative system, Tetra Recart™, was cost-competitive with new canning lines and offered significant advantages at all stages of the food canning chain, from manufacturers to retailers and consumers.

With two field test customers lined up, Nestlé and Bonduelle, and encouraging retailer and consumer reactions, Tetra Pak management believed they had developed a winning concept. Why couldn't Tetra Recart™ replicate the success it had achieved with its ubiquitous milk and juice cartons, with solid foods?

Yet Larsen was fully aware of the many challenges that remained to be addressed before Tetra Recart™ could bring a profit contribution to the company. The new business unit had to define its market and product strategy. It also had to decide how to market and sell its system. Finally, it had to reduce its "time-to-profit" while keeping its product edge, as competitors would undoubtedly soon emerge with their own systems.

Copyright © 2004 by IMD – International Institute for Management Development, Lausanne, Switzerland. Not to be used or reproduced without written permission directly from IMD.

1 Company Background

Tetra Pak AB was founded in Sweden in 1950 by entrepreneur Ruben Rausing who had a vision to develop efficient carton-based milk packaging solutions for the emerging supermarket channels. After the success of the first product, a 500-ml tetrahedron-shaped milk carton (Tetra Classic^{TM}) invented by engineer Erik Wallenberg, the company introduced different packaging shapes and sizes. Its greatest success, the Tetra Brik^{TM}, introduced in 1963, extended the company's market reach to include fruit juices and other beverages.

Tetra Pak revolutionized the milk packaging industry with the introduction, in 1961, of aseptic cartons that used UHT pasteurizing technology. Dairies could now offer milk that could be stored at an ambient temperature with a six-month shelf life. The technology was quickly extended to fruit juices, and Tetra Pak cartons became one of the most widely used forms of liquid food packaging in the world. By 2001, aseptic packaging accounted for over two thirds of Tetra Pak's business. (*See Exhibit 1 for Tetra Pak's product line.*)

Ruben Rausing's tenet "a package should save more than it costs" stimulated a constant emphasis on efficiency, in terms of package weight, cost, and filling line productivity. While it focused on maintaining quality, minimizing waste, and reducing distribution costs, Tetra Pak quickly built an environmentally friendly image. Much of Tetra Pak's success was linked to: (1) its proprietary "roll-fed, form-fill-seal" system technology; (2) the strength of its broad network of entrepreneurial market companies; and (3) a strong emphasis on customer and retailer efficiency. Tetra Pak firmly believed that its excellence in technology and product innovation was a direct result of carefully listening to its customers. (*See Exhibit 2 for technical terminology.*)

In 1991 Tetra Pak acquired Alfa Laval, mainly to access its liquid food processing technology.[1] By integrating it with its existing filling system business, Tetra Pak could offer turnkey processing plants to its dairy, juice, or beverage customers. This acquisition also provided the company with an opportunity to diversify into food packaging.

By 2001, Tetra Laval – as the new group was called – saw its annual revenues reach € 7.65 billion. Its main subsidiary, the Tetra Pak group of companies, operated in more than 150 countries, sold over 90 billion packages annually, and employed about 20,000 people. Tetra Pak sold three types of products: filler line equipment (systems), packaging materials (printed rolls), and processing equipment. Geographically, Tetra Pak was divided into two regions – Tetra Pak Europe and Africa, and Tetra Pak Asia and Americas – with similar sales volumes.

---

[1] This included sterilizers, separators, homogenizers, standardizers, heat exchangers, flow equipment, and other systems needed by customers to process their liquid food before packaging.

2 Starting Project Phoenix: Genesis (October 1993 – March 1996)

After the acquisition, Alfa Laval business units were integrated in Tetra Pak (liquid food processing equipment), spun off (industrial applications), or set up as separate divisions (dairy farming and milking equipment). Tetra Laval Food (TLF) was set up in 1993 as a business unit within Tetra Laval to provide packaging and processing solutions for convenience foods, prepared foods, ice cream, and oils.

2.1 Focusing on a Can Replacement

TLF's CEO, Nils Björkman, and his CTO, Stefan Andersson, were eagerly looking for growth opportunities in solid foods. They had long toyed with the idea of finding an alternative to the ubiquitous metal can, which represented a huge market of 160 billion packs a year. Canners had lines that were mostly or fully depreciated and had a singular emphasis on increasing the efficiency of the lines, which had typical speeds of 24,000 cans/hour. Unlike other segments of the packaged food market, little had changed on the canning side, the biggest ambient packaged food segment. Most importantly for the company, targeting the canning industry avoided any risk of cannibalization of Tetra Pak's existing liquid food businesses.

2.2 Experimenting with Aseptic Technology

Given Alfa Laval's long aseptic technology focus and Tetra Pak's aseptic carton orientation, Andersson was given the mission to develop an "aseptic carton can." Carton packs would have several advantages over cans including convenience and a modern, environmentally friendly image for consumers; shelf space optimization and exposure quality for retailers; and distribution efficiency (higher load per pallet) for canners. In addition, it was hoped that aseptic technology would confer on solid foods the benefits of freshness and taste it had provided to milk and juice.

Tetra Pak and its primary competitor, SIG Combibloc, had made several attempts earlier on to package soup, tomato products, and sauces in aseptic cartons, with widely variable results. Tetra Pak's filling technology, unlike Combibloc's, proved ill adapted for packing foods with high viscosity or large particles.[2] But

---

[2] Working from blanks that were first formed into boxes, then filled and sealed in sequential processes, Combibloc had fewer problems with particles going into the seals than Tetra Pak, which continuously formed pouches from a roll, then filled and sealed them in a single operation before folding them into boxes.

the most vexing problem for Andersson was the unsatisfactory cost competitiveness of aseptic systems compared to traditional canning lines. With a 6,000 to 10,000 pack/hour speed, Tetra Pak's aseptic machines could not compete on productivity or cost with canning lines and their 24,000 pack/hour capacity. Andersson was convinced that the company needed to shift away from the aseptic paradigm. Other approaches deserved to be pursued, and the "retorted carton" concept seemed the most obvious.

2.3 Setting up a Dedicated Research Project: Phoenix

There had been several unsuccessful attempts at Tetra Pak to retort cartons in the late 1980s. Boiling a cardboard package for long periods (up to three hours) at temperatures of 120°C to 130°C was a bold undertaking. Nevertheless, and with the full backing of Björkman, Andersson started a small R&D project on carton retorting in October 1993 while continuing his aseptic carton research (to avoid antagonizing the "aseptic believers" within Tetra Laval's management and board.)

The project, called "Phoenix," was entrusted to a team of two engineers and a secretary. The project leader, Ulf Ringdahl, reported directly to Andersson. The second engineer, Jan Lagerstedt, had considerable experience in filling machine development at Tetra Pak. A small steering group, chaired by Björkman, was set up to guide and supervise the project.

2.4 Defining a Target Application

The steering group and the team decided that the project would initially focus on a 400-gram carton, the largest segment in the 30-million-ton food canning business. The project team initially concentrated on making the solution suitable for packing four product categories chosen for their market volume and perceived compatibility with a carton package: vegetables, ready-made meals, soups, and pet foods. (*See Exhibit 3 for canning category volumes.*)

The carton would be printed on a four-color flexo machine and would have a laser perforated opening to allow consumers to open the package by hand in four easy steps – no scissors or openers should be needed. It would target a minimum shelf life of two years since, it was felt, the five-year shelf life of the can was seldom utilized, except by the army. The carton, which ultimately weighed 18 grams (as opposed to 60 grams for a typical can of similar size) would be easily recycled; the fiber would fit into existing waste streams.

2.5 Testing the Concept with Customers

TLF management felt it was important to gauge the interest of some of the largest canning customers in Europe prior to committing significant time and money to Phoenix. So, very early on, members from the project steering group, senior management, and project team started visiting food canners to explain their concept and present hand-made package prototypes. Four or five companies were approached at a very high level between 1993 and 1995, all of them leaders in their product categories.

Bonduelle in France, a family-owned company and Europe's largest vegetable canner, was one of the first prospective customers TLF visited. Nestlé was also invited to Lund because of its large canned products portfolio and the fact that it was an important customer for Tetra Pak. Rupert Gasser, Nestlé's head of manufacturing and R&D operations, and his senior packaging specialist were, thus, exposed to the retortable carton concept in 1995 and showed interest. They saw a greater opportunity for the carton in pet foods than in canned meals or soups, two relatively less important categories for Nestlé.

Despite initial interest in the concept, most of the other companies approached generally felt that the project outlook was too uncertain for them to commit resources at such an early stage. They wanted to see a working prototype and many of the technological obstacles resolved before making any commitments.

2.6 Developing a Demonstration Machine

Encouraged by its earlier customer contacts, TLF now had to begin developing a working prototype of its system. In the fall of 1994, it embarked on a design cooperation with a small, entrepreneurial Italian machine manufacturer, Galdi, which had designed one of Tetra Pak's earlier blank-fed machines. The project team recommended building a "transportable" prototype so customers could test the process in their own environment. Their proposal was accepted and they received the necessary funding to develop the first mobile test unit (MTU), dubbed "Junior." The machine was to use a batch retort provided by Getinge, a Swedish sterilization equipment manufacturer. The capacity of Junior was a mere 600 packs/hour; however, this was deemed enough for customers to try out the concept and develop samples for consumer research, storage trials, and recipe formulations.

2.7 *Experimenting with Retorted Cartons*

Simultaneously, the Phoenix engineers undertook all kinds of carton retorting tests. Initially, the tests were disastrous: water entered the package during the retort process and the package contorted as if it had spent hours in a clothes tumbler dryer. The team had yet to understand the interface between packaging materials and product during the retorting process.

These problems brought the project to a definite low point. At that time, only four or five people believed that there was still something to salvage from Phoenix – including the two originally assigned to the project and the key steering group members. What saved the day was the interest tentatively expressed by Nestlé's Rupert Gasser and his packaging technology specialist. This triggered the Phoenix team to work intensely with Tetra Pak's R&D group, which had strong competencies in materials development, and with its retort partner – Getinge – to solve the problem. Partnering with a retort specialist helped the team gain experience in sterilization and understand how to adjust temperature and pressure for optimum results. This cooperation bore fruit, and the team progressively managed to develop the right kinds of laminated materials and to control the retort parameters to obtain an impeccable retorted carton. (*See Exhibit 4 for a picture of Junior.*)

3 Restarting and Reorienting Phoenix (April 1996 – May 1997)

In January 1996 Tetra Laval decided to refocus on its core business and closed its TLF division, selling off parts of its portfolio. For a few months, the Phoenix project remained in limbo as its 35 full-time members wondered whether they would be needed in the new organization. Some senior managers in Tetra Pak were skeptical about the project that Andersson and his steering group colleagues had strongly supported.

Kjell Andersson, head of Tetra Pak R&D, was reluctant to continue the project in its current state with the large group of people. He believed in the concept of the retortable carton, which he had helped develop, but thought there were flaws in the system TLF had conceived. So he decided to take the project back one stage to the "prestudy" phase. He wanted the team to explore simpler alternatives than the slow "form-fill-seal" concept embodied in the machine platform.

3.1 Resizing the Team and Setting up a New Steering Group

So Phoenix was integrated into Tetra Pak's R&D center with a new project leader, Gunther Lanzinger, as Ringdahl left Tetra Pak. An engineer by background, Lanzinger had also been broadly exposed to the commercial side of the packaging business. The team, now down to just ten people, was reinforced with its first "commercial" member, an in-house marketing consultant, Erik Lindroth, who joined on a part-time basis before becoming full time in 1999. Lindroth initiated a series of market research studies to test the acceptance of the new carton by consumers and retailers, with encouraging results.

When Björkman was appointed head of Tetra Pak-Europe, a new steering group was set up. Stefan Andersson, who had joined Björkman in Region-Europe, stayed as steering group chairman, providing a vital link with the past. Other steering group members included Kjell Andersson, one of his senior engineers and the head of Tetra Pak-France. The latter was brought in to nurture the two most promising customer contacts developed thus far: Nestlé's pet food development center and Bonduelle's vegetable canning headquarters, both located in France. The steering group decided to meet every two to three months to revisit the commercial viability of the project and supervise its technical aspects.

3.2 Redesigning the System

The new Phoenix team was asked to reassess each aspect of the system. New filling and sealing processes needed to be developed and tested. Taking the filler offline quickly emerged as the best solution to "decomplexify" the system and increase its operating speed. The team decided to outsource the development of the filler to Zacmi, an Italian specialist in processing equipment for the food industry, known for developing some of the best rotary fillers in the world.

In the meantime, the team gathered fresh resources from within R&D to tackle the remaining issues. The biggest challenge was the development of a packaging material that could withstand a retorting process while preserving the food's freshness and qualities. Within a year, with the help of Tetra Pak's specialists on packaging material and good relationships with raw material suppliers, they had overcome all the major hurdles.

Once the individual technologies had been verified, the components of the system had to be assembled into a single rig and tested. The team needed a prototype – a new Junior MTU – that could be sent to the customers' facilities for testing with real products. The new Junior machine was ready by early 1997.

4 Clarifying the Value Proposition

From the very beginning, the project team had established a list of targeted product benefits for each player in the value chain: customers (the food packers), retailers, and consumers. They had also speculated about potential concerns to be addressed. Early contacts with customers (in 1993 and 1994) then with retailers and consumers (in 1994 and 1995) had, by and large, validated the team's assumptions. So the concept and value proposition were reconfirmed during the prestudy.

4.1 Customer Benefits and Concerns

Food canners would appreciate a number of the new package's features:

- Logistical efficiency: The package weighed less than one third of an equivalent-sized can and a single pallet of printed carton blanks contained the same number of packs as 15 trucks full of empty cans.
- Cost competitiveness with cans: Since the baseline cost of paperboard was lower than the price of steel or aluminum, Tetra Pak's retortable carton could conceivably cost less than cans.
- Its novelty and appealing, modern image: This could help revitalize highly commoditized canned product categories. Its larger printable front surface area would, undoubtedly, create more consumer impact.

However, customers would probably hesitate about investing in a completely new production line if they had spare capacity in their fully amortized canning plant. A new retortable carton line, it was calculated, would require an investment of several million euros (from €5 to 7 million, depending on the line configuration and capacity). The team was also aware that customers might object to a shorter shelf life (two years for cartons *vs.* five or more for a can). Finally, they suspected that customers would be reluctant to be tied to a single supplier of packages, as they were used to buying their cans on the spot market. (*See Exhibit 5 for details of Phoenix's efficiency and Exhibit 6 for a breakdown of Phoenix system costs.*)

4.2 Retailer Benefits and Concerns

Retailers had generally been very positive about the concept and liked the following features:

- The shelf-efficiency of square and stackable packages. This would increase their shelf space utilization by 30%, thus contributing to a higher DPP.[3]
- The cartons' resilience to light shock and the fact that – unlike for cans[4] – food remained safe in cartons even in case of a slight deformation.
- The product novelty and appeal, with its four-color flexo printing, which could contribute to a fast turnover and a premium image.

On the negative side, retailers were concerned about the vulnerability of a carton package to handling accidents or tampering. Cans could be stacked very high on pallets in their warehouses without any risk. Would it be the same with cartons, and what about the food safety risk if cartons were inadvertently punctured? But the biggest risk for retailers, of course, was linked to package acceptance by consumers. Would the retortable cartons move as fast as Tetra Pak suggested?

4.3 Consumer Benefits and Concerns

Tests had confirmed most of the team's hypotheses regarding consumer benefits:

- Opening ease and safety were the first benefits spontaneously mentioned by consumers. The carton could be torn open by hand without a can opener or scissors, removing any danger of cuts, a frequent occurrence with metal cans.
- Retortable cartons, due to their compactness and light weight, were also perceived as easier to carry home and more convenient to handle and store.
- Finally, the carton was perceived as an environmentally friendly material, much easier to dispose of than cans, and recyclable. They would also benefit from the positive image of Tetra Pak's milk cartons.

On the negative side, consumers were not easily convinced that food in retortable cartons would have the same shelf life as food in cans.

Overall, despite the fact that the concept allowed differentiation in the commoditized packaged food market, the team was aware of the potential difficulty in changing deep-rooted industry habits.

---

[3] DPP (direct product profitability) was an activity-based costing model many retailers used to calculate the profitability of products per thousand units.

[4] When a can has been even slightly dented during in-store handling operations, the layer of enamel inside the can may be broken, which puts the food in direct contact with the raw metal and potentially creates a food safety problem. This is why educated consumers do not accept crushed cans, obliging retailers to withdraw them at a loss.

5 Starting a Partnership with Nestlé (Project Arizona)

Initial discussions with Nestlé and its packaging technology group had been encouraging, and toward the end of 1996 the two companies started Project Arizona to identify and pursue specific applications for retortable cartons. The partnership was based on a precommercial agreement signed at a high level by both companies and specifying the following:

- Sharing of intellectual property (IP), whereby Tetra Pak would control the IP rights to the package and Nestlé would retain any IP associated with the food products jointly researched.
- Sharing of all costs associated with developing specific applications suited to the new package. There would be no monetary exchange between the two companies.
- Nestlé would have exclusive rights to the particular applications that would be pursued and specific geographies for a limited period of time.
- A time schedule within which both companies would do field tests.

In 1999, through Project Arizona, Tetra Pak made inroads into several different groups within Nestlé. The company was initially hesitant about committing to a highly uncertain experiment, so it suggested involving a European contract packer in the field test. However, this was not an attractive proposal for Tetra Pak, which wanted to see real proof of commitment from its partner. Nestlé's pet food division was the first to commit to the concept, a proposal that created some questions on the Tetra Pak side: If pet foods were to become the first application for such a new package, would it prevent the company from building a necessary premium image in human foods? However, the attractiveness of landing an "A-brand" as the first test customer dissolved those early fears. Besides, the potential volumes within this group were very attractive for Tetra Pak.

Product trials with the new Junior machine began in June 1997 in Amiens, France, where Nestlé had established its pet food R&D center. Initial results on cat food were positive, as cats showed a $3:2$ preference for food from Tetra Pak's retorted cartons over cans. A project team member, delighted with these results, remarked: "We can't fool cats!" However, Nestlé rapidly concluded that a 400-gram package was too big for cats. So when Nestlé Italy expressed interest in a new premium line of dog food for its Friskies[TM] brand, the trials moved to target dog food and concluded positively toward the end of the year.

Tetra Pak had made a conscious effort to reduce its overall system cost, a key driver in the pet food industry. Management could, therefore, convince Nestlé that the pricing of Phoenix would be very competitive with its canning lines. Sacrificing initial profitability to win a global brand owner like Nestlé's Friskies[TM] as pilot customer was seen as a good strategy.

6 Reinforcing the Project Organization and Visibility

In 1997 Tetra Pak's top management embarked on a major drive to improve its innovation process. The effort was led by a small team of senior managers, chaired by Bo Wirsén, former head of Tetra Pak-Europe and a member of group management. Wirsén, who came to Tetra Pak from Alfa Laval, had been involved in early discussions about the Phoenix concept and in the first precommercial agreement negotiations with Nestlé.

Wirsén was convinced that Tetra Pak should manage its important innovation projects more professionally. Borrowing one of 3M's management ideas, Tetra Pak's key projects (also known as Pace Plus projects) would have a fully empowered project leader and a high-level supervisory board. That board, composed of representatives from Tetra Pak's technical and business community, would be chaired by at least one member of group management. Phoenix was now a Pace Plus project and Wirsén accepted the chairmanship of its steering board. Tetra Pak's newly appointed CTO, Göran Harrysson, joined the Phoenix board shortly thereafter. The two senior managers were strongly instrumental in placing the project firmly under top management's spotlight.

The project would be subjected to the new innovation gate process that Tetra Pak management had devised. It consisted of four phases (see *Exhibit 7 for the project path along the four phases*).

Wirsén strongly believed in Phoenix and was intent on quickly turning it into a true business endeavor. This, he felt, would require putting a senior business leader at its head. In 1999 he hired Joakim Rosengren, managing director of Tetra Pak-Sweden, as general manager for the project. Lanzinger, who had previously been the project manager, now concentrated efforts on technical development. Rosengren, who kept his job in Tetra Pak-Sweden, was to spend 50% of his time on Phoenix. The project team liked having a real businessman at its head, and it started building its own identity within R&D. Around this time Wirsén started toying with the idea of spinning off the project into a separate business unit:

> I wanted our Phoenix people to dream about the business! They couldn't remain diluted within a big R&D unit. So, in 1999, I proposed to create a new company around it. It was important for the team to build their identity and for management to build pressure for commercialization. But people in R&D objected! They didn't like the idea! It was too early!

Wirsén was keen on increasing the visibility of the project with top management and the company board.

> When Joakim came on board, we started building a return map for the project and calculating a break-even. We forecast to spend hundreds of millions of Swedish Kronor up to launch. I wasn't surprised by that investment, but I saw the importance of showing to all who were involved how much we were spending. I also wanted other members of our Global Leadership Team to be aware of it. They were all supportive, and so was the family [Tetra Pak shareholders].

7 Starting a Second Partnership with Bonduelle (Project Edix)

By mid-1998 Bonduelle had expressed an interest in the Phoenix carton. The team was happy to bring a strong second potential customer on board for several reasons:

- It proved the legitimacy of retortable cartons as a viable alternative to the can, even for its most traditional applications – vegetables.
- It reinforced Tetra Pak's motivation to keep funding the project, now that it had a second potential customer lined up.
- It introduced Tetra Pak to a new product category and a new type of customer after Nestlé, *i.e.*, a large and private regional player.

So a new project – Edix – was set up and several product varieties were tested before beans were chosen. Bonduelle proved to be extremely demanding on visual and taste quality; so many adjustments had to be made to retorting times and temperatures before the team could meet their customers' quality specifications. These initial tests lasted a whole year. Then Bonduelle specified that the system it would use should be equipped with the filler of their preferred French supplier, Hema, instead of Zacmi. This slowed down the testing process further as the team had to adjust Junior to accommodate their customers' specific request.

The second relationship taught management an important lesson: each individual customer was bound to have its own requirements and processes. Tetra Pak would therefore have to conduct extensive retorting tests in each new product category, probably over a year or more, before meeting its customers' quality requirements. On the positive side, the Phoenix team was seeing its knowledge and confidence level grow with each new product category, as it added new test results to its database.

However, management was aware that it could not leverage its knowledge by pursuing customers in the same applications, be it pet food or beans, at least for a certain period of time. Its partnership agreement with trial customers included a period of protection of exclusivity. New customers would have to be found in new product categories.

8 Moving to Real Life Field Tests and the Creation of Tetra Recart™

By January 2000, Tetra Pak had signed a field test agreement with Nestlé's Friskies™ for a new line of premium dog food to be launched in Italy. By mid-2000, a similar agreement had been signed with Bonduelle, for a line of beans, also in Italy. Both customers had chosen Italy, by coincidence, because it was a market where they had a relatively low market position for the categories they had chosen. This ensured that they would not take a major risk with a hitherto untested concept.

The team started work immediately on a full-scale commercial system for Friskies™. It should have taken only about nine months to produce and install a new line on site, but a last-minute change in package dimensions necessitated a new range of tests. Consumer research had shown that the original 400-gram carton was too deep to allow pet owners to scoop out food with a spoon without touching the carton with their hand, something most people disliked.

The two field test contracts reinforced management's confidence in the commercial viability of the Phoenix concept. It was now time to move from "project" mode to "business" mode. The first step was to conceive a proprietary brand for the new product. "Tetra Recart™" (a contraction of "retortable carton") was chosen by mid-2000 and the Phoenix team, which was still part of the R&D group, was renamed accordingly. The second step was to refine the breakeven and return calculations that Rosengren had initiated when he joined the team. Breakeven would now occur when an annual sales volume of one billion packs could be reached. This volume would be possible by 2005, assuming four to five high-volume customers signed up.

Discussions about spinning off the project team into a new business unit intensified, and management decided to create a new Tetra Recart™ business unit by January 2001, in time for the start of the actual Friskies™ Italian market tests. Wirsén and his steering group members believed that the new unit should have a dedicated business manager. Rosengren had chosen to accept full-time responsibility for the newly created "Nordic market cluster," which had just been set up by Region-Europe, so a search started for a new managing director for Tetra Recart™. The steering group pondered whether the new job should go to an "outsider" familiar with the food industry. But Tetra Pak had relatively little experience with hiring "outsiders" in senior positions, and besides, management wanted the new unit to fully leverage the company's technical, production, and commercial capabilities and relationships. Wirsén commented:

> In the end, I wanted a Tetra Pak-cultured person. I knew Jan Juul Larsen who was reporting to me as managing director of Tetra Pak-Taiwan. He had proven leadership qualities, was very focused, and showed an appetite for the challenge. We could always bring in outsiders with category knowledge under him.

In May 2001, before Larsen had relocated in Lund to take over the new Tetra Recart™ business unit, the Friskies™ field tests started in Italy. The line "Friskies Freschezza" gained rapid acceptance with retailers and consumers, and initial consumer tests showed that Tetra Recart™ beat the can on most key attributes (*Exhibit 8*). Despite its supposedly easy-to-tear, laser-perforated opening, some consumers instinctively used scissors to cut the package open, while others tried unsuccessfully to tear it open before using scissors. This highlighted the need for greater attention to consumer education and opening instructions.

While field tests proceeded, the Tetra Recart™ team noticed that can prices actually fell dramatically, thereby negating any pricing advantages Tetra Recart™ had with its own package pricing. In addition, with the advent of pouches, there was strong competition brewing,[5] which led to the conclusion that the cost situation and competitive comparisons were customer dependent, not necessarily absolute. (*See Exhibit 9 for can price fluctuations.*)

9 Developing the Tetra Recart™ Organization and Business

When Larsen took over in July 2001, his first concern was to build a professional business organization and establish an appropriate governance process.

> When I arrived, there were 50 people in Tetra Recart. The steering group was very engaged in operational decisions; they felt ownership! But you cannot run a business when all decisions move from one steering group meeting to the next one. We needed to gain our management freedom and start operating as a normal business unit. However, the project was getting very big so we needed to maintain a high degree of top management visibility and support. Tetra Recart AB was already set up as a legal unit, so I proposed to create an embryo for a completely independent business unit.

Wirsén endorsed Larsen's proposal. The project steering group and its bimonthly operational meetings was dismantled. The new company, Tetra Recart AB, would have a board of four members, three of them members of Tetra Pak's Group Leadership Team. (S*ee Exhibit 10 for the evolution of the steering group membership and Tetra Recart's board membership.*) Bo Wirsén (head of Region Asia-America) remained as chairman of the new board, which held quarterly meetings. Larsen was given carte blanche to sign new contracts with customers. The creation of a board at the top management level was well accepted by everyone within Tetra Pak and welcomed by the two field test customers, who saw it as yet another sign of commitment by Tetra Pak's management.

Larsen was aware that he had to focus his team entirely on making a success of the two field tests and adding new customers to his backlog. The Friskies™ market test was very positive and the brand saw its market share increase regularly from a very slow starting point. But the Bonduelle field test had yet to start, as the customer was unhappy with the results of its protracted retorting tests. A new material had to be developed from scratch to meet Bonduelle's high quality standards.

While trying to solve his daily load of technical problems and maintaining his two field test customers happy, Larsen also had to try to build a professional organization. This, he felt, would require a mix of "insiders" with a strong experience of Tetra Pak and "outsiders" who would bring in-depth knowledge of the food industry. His board gave him free rein to recruit from both sides. Thus a nucleus of organization started to take form. (*See Exhibit 11 for the Tetra Recart initial organization.*)

---

[5] By 1998, the worldwide market for standup plastic pouches had grown to 2.3 billion packages, of which 700 million were retortable pouches, many of them for the pet food market. However, the bulk of the latter were 100-gram packs, a size more suited to cat food than dog food.

10 Addressing Tetra Recart Growth Challenges

Larsen was aware that top management, although fully supportive of the new business unit, would expect him to show results and to do so quickly. He saw four major challenges ahead.

10.1 Choosing Target Customer Segments

Tetra Recart's two field test customers represented two worlds with very different ways of conducting business. Nestlé was a multinational with global ambitions and brands. Bonduelle was a privately owned regional company, with very different attitudes and concerns.

Tetra Recart had focused its initial sales contacts primarily on big and global canning brand owners such as Heinz, Campbell, Green Giant, Del Monte, and the like. The world's top ten canners, indeed, represented 30% of the total canning industry volume. If any one of these giants decided to switch from cans to cartons, even progressively, Tetra Recart would see its orders surge and it would reach breakeven rapidly. However, initial contacts with these big players had received a fair amount of attention but no demonstrated commitment, except from Nestlé.

The largest players in the market might not easily be persuaded to put their winning market strategy at risk by introducing a radical packaging innovation that had the possibility (however remote) of hurting their respective market positions. Lindroth and Larsen realized that this would require a large number of tests, elaborate business justifications, and, hence, a long decision process.

Also, multinationals were apt to change strategies. In early 2001 Nestlé announced its intent to acquire the pet food business of Ralston Purina, a $10 billion company. With the majority of Purina's pet foods being of the dry variety, it remained to be seen how Nestlé would view its long-term relationship with Tetra Recart[TM] and its wet-pet-food retortable carton.

A second segment of smaller regional players, which were often privately owned, constituted another market. Their decision process, in the hands of a few family members, might be simpler and shorter. Being national or regional, they might be less demanding than multinationals on issues of geographic exclusivity. However, the Bonduelle experience had shown that these companies would not take any testing shortcuts. Consequently, the process from initial contact to full commercial operation could take as long as for the multinationals, but for a much smaller volume.

A third segment to consider could be "noncanners," *i.e.*, food packers without a canning line, yet interested in launching retorted products like ready-made meals.

Should Tetra Recart target the big multinational brand owners, like Nestlé, which controlled the volume, go after smaller regional canners, like Bonduelle, or pursue "noncanners" – or a combination?

10.2 Defining a Product Development Strategy

Larsen felt that Tetra Recart would soon have to decide whether and when to respond to the various demands for product variants expressed by some of the potential customers it had approached. Its system had barely been introduced in a field test with a 400-gram carton and a first-generation system before canners – including Bonduelle – started talking about different carton sizes. The fastest-growing segment in cat food was a 100-gram portion, and a large producer (and potential Tetra Recart customer) of cat food had bet on the success of this segment. Fruit and vegetable producers, on the other hand, were seeing strong growth in the 200-gram segment. Some potential customers had even mentioned their desire to see Tetra Recart develop microwaveable cartons, which would require a complete reformulation of the packaging materials.[6] Other manufacturers asked for different shapes and even "see-through" cartons.

In addition, some members of the Tetra Recart organization were wondering whether the original high-speed capacity targeted by the initial system to compete with canning lines (24,000 packs/hour) was the only speed the market needed. That speed would be a must for large-volume categories, like the ones produced by global brand owners (assuming that they would use the system to pack their core products). However, a lower capacity and (presumably) cheaper machine would be better adapted to those niche applications that several potential customers considered as ideal for testing the Tetra Recart™ system.

Rosengren had hired a very experienced technical director, Karl Håkansson, and Larsen had added package development specialists to the team to address these issues. Håkansson wondered whether it made sense to start developing new products before commercially releasing the existing carton and system on the market. If Tetra Recart decided to develop new products to broaden its range, would the news make potential customers postpone their decision in anticipation of a package size better suited to their needs, thus slowing down Tetra Recart's business growth? And how could Tetra Recart develop new package types or machines without enlisting test customers?

However, rumors were circulating that Combibloc – Tetra Pak's main competitor – was getting ready with its own retortable carton. Competition could become fierce, not just with cartons but also with standup pouches. Given competitive pressures, when should Tetra Recart start developing its product pipeline and what should be its priorities?

---

[6] The Tetra Recart carton was composed of fiberboard (carton), plastic film, and aluminum foil; the latter was used as an oxygen barrier. For a microwaveable carton, the aluminum foil would have to be substituted with another barrier material – a daunting challenge and significant R&D effort.

10.3 Choosing How to Go to Market

The fact that both Nestlé's Friskies™ and Bonduelle happened to have their development centers in France led Tetra Pak-France to be involved in customer contacts early on. Its market company head joined the initial Phoenix steering group and allocated two people within his organization to work with the two potential customers on technical and commercial issues. Both started with little or no knowledge of the industry, but they learned fast and proved useful in smoothing the interface between Swedish and French engineers. As the promising outlook of Tetra Recart began to be publicized internally, a number of market companies – not all, though – expressed an interest in taking an active part in selling the new product. However, none of them could claim much knowledge of, or contacts in, the canning industry and few had the resources to pioneer a new market.

So the Tetra Recart team had thus far made all the initial business development contacts on their own, often with the help of steering group members. The market companies were informed and even associated with some early sales calls, but the marketing and sales process had so far been entirely managed by Tetra Recart. Now that the new company was ready to move into higher gear, Larsen had to decide on the extent to which the market companies ought to be involved with his business. On one hand, he felt a strong pressure to build volume and hence needed help in selling to and managing customers. On the other hand, he realized that his business required a lot more front-end investments, including training and product testing.

For example, Larsen was conscious of the risks inherent in overextending his reach and losing control over the quality of product and service delivery. He felt the need to maintain a firm control over the type of customers that ought to be approached as a priority in order to build certain reference applications and attract customers. Market companies, by contrast, were used to running their business fairly independently. Would they agree to follow guidelines imposed by Tetra Recart for selecting their customers or preparing business cases? Would they also accept dedicating staff to Tetra Recart for business prospects that would not immediately bear fruit?

Larsen had to invent a market approach that would guarantee his control over the business while leveraging the strong Tetra Pak market organization.

10.4 Reducing "Time-to-Profit"

Finally, and most importantly, Larsen was aware that, after nurturing the project for eight years and several hundred million Swedish Kroner, Tetra Pak management was eager to reach breakeven. While there were 50 people dedicated to the company full time, there were as many as 90 additional people who spent at least 50% of their time on Tetra Recart. This represented a considerable "burn rate" for a startup that was still in its market trial phase. Tetra Recart was not anticipating positive operating cash flow for another five years or breakeven for at least another decade.

Corporate management had declared it imperative that Tetra Recart reach its objective of one billion packs by 2006. Larsen needed to reassess the business and devise a plan to attract customers faster, grow sales, and generate critical mass to fend off competition. "You have to be very honest that all you have is a crystal ball," said Larsen as he pondered the goal of 30 installations, 15 to 20 customers, and one billion packs in sales by 2006/2007. Market research had helped identify that the target segments for Tetra Pak constituted 36 billion of the 160 billion retortable food packs in the worldwide market. From here, the number had been further narrowed down to identify key players in these segments. Target setting had become more structured and, concurrently, time to contracts was anticipated to reduce[7] *(See Exhibit 12 for original sales estimates.)*

Nevertheless, Tetra Recart's original growth estimates had proved extremely optimistic. Despite all the benefits, potential customers hesitated over investing in the new system for fear of unsatisfactory returns and were taking much longer than originally anticipated to complete their trials. The time from initial identification of potential customers and tests with the MTU through contract development could take up to 24 months. Once a contract was signed, Tetra Pak would initiate the production of a customized line, which could take another 12 months to be ready for delivery and commercial production *(Exhibit 13)*.

Larsen wondered what strategy and sales approach would accelerate his customers' decision process. To satisfy the capital investment concerns of some of its smaller customers, Tetra Recart had set up a financing solution through its own finance company, Tetra Laval Credit AB. But what else could he do to speed up revenue growth without further burdening his payroll and business development costs? Larsen summed up his dilemma with the following concluding remark:

> As head of the team, my temptation could be to focus only on the short term, to sell and get revenues. But in our business, we need also to build an organization. We need to be able to deliver on time! We need to continue to develop new applications and new generations to support our customers!

---

[7] It took six years for Tetra Pak to turn the first contacts with Nestlé (1995) into a successful field test (2001), and seven years had already been spent with Bonduelle since the first contact, with a field test planned for 2003!

Exhibit 1
Tetra Pak's Product Line

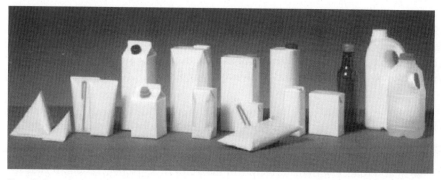

Tetra Pak packages, from left to right: Tetra Classic, Tetra Wedge, Tetra Rex, Tetra Prisma, Tetra Brik, Tetra Fino, Tetra Top, Tetra Recart, Glaskin bottles and EBM bottles.

Tetra Classic

In September 1952 the first Tetra Pak machine for tetrahedron-shaped cartons was delivered to the Lund Dairy in the south of Sweden. Packing of cream in 100-ml cartons began that November.

Tetra Classic Aseptic

This aseptic version of the tetrahedron-shaped package launched in 1961 permitted the packaging of perishable food products, which could be stored and distributed under nonrefrigerated conditions.

Tetra Brik

In 1963, the Tetra Brik package was launched. It has a modular shape specifically designed to comply with international standards for loading pallets.

Tetra Brik Aseptic

An aseptic version of the Tetra Brik package was introduced in 1969. In the Tetra Brik Aseptic system the packages are formed from sterilized packaging material and filled under sterile conditions.

Exhibit 1 (continued)

Tetra Top

Tetra Top is a reclosable, square package with rounded corners. It has a polyethylene lid, which is injection-molded and sealed to the package in a single process. The Tetra Top package addresses consumer demand for ease of opening, pouring, and reclosing. The package was introduced in 1989.

Tetra Prisma Aseptic

Tetra Prisma Aseptic, commercially introduced in 1997, is an octagonal package made in accordance with the principle of the Tetra Brik Aseptic system. It first appeared in the 330-ml volume.

Tetra Wedge Aseptic

The Tetra Wedge package was commercially introduced in 1997. The package is based on the same reliable technology as the Tetra Brik Aseptic system. Its new, innovative shape enables products to be easily distinguished on shop shelves and keeps packaging material consumption to a minimum.

Tetra Fino Aseptic

Tetra Fino is a carton-based pillow-shaped package that was introduced in 1997. It is a roll-fed packaging system based on the Tetra Brik Aseptic technology. The Tetra Fino Aseptic packaging system offers good economy for producers as well as for consumers.

Aseptic EBM Bottle

In 1995 Tetra Pak extended its product range to include molded plastic bottles, EBM (extrosion blow molded) bottles. In 1998 the company had developed an aseptic version of the EBM bottle, and the multilayer aseptic EBM bottle was introduced for UHT milk.

Source: Company literature

Exhibit 2
Technical Terminology

Aseptic Technology

An aseptic package is one that has been sterilized prior to filling with sterilized food, resulting in a product that is shelf stable for over six months. Aseptic technology has been called the "most significant food science innovation of the last 50 years" by the Institute of Food Technologists.

Tetra Pak aseptic cartons are made of three basic materials that together result in a very efficient, safe, and lightweight package.

- Each material provides a specific function:
    - Paper (75%): to provide strength and stiffness
    - Polyethylene (20%): to make packages liquid tight and to provide a barrier to microorganisms
    - Aluminum foil (5%): to keep out air, light, and off-flavors – things that can cause food to deteriorate

Combining each of these three materials enables the creation of a packaging material with optimal properties and excellent performance characteristics:

- Higher degree of safety, hygiene, and nutrient retention in foods
- Preserving taste and freshness
- Can be kept for months with no need for refrigeration or preservatives
- Efficient (a filled package is 97% product and only 3% packaging material), using a minimum quantity of materials necessary to achieve a given function; a good example of resource efficiency
- Lightweight (among the lightest packages available)

Canning (Retorting) Technology

Canning or retorting is the process of hermetically sealing cooked food for future use. It is a preservation method in which prepared food is put in glass jars or metal cans that are hermetically sealed to keep out air and then heated to a specific temperature for a specified time to destroy disease-causing microorganisms and prevent spoilage. Low-acid foods, such as meats, are heated to 240°-265°F (116°-129°C), while acidic foods, such as fruits, are heated to about 212°F (100°C).

Cans used for foods that react with metals, causing discoloration (usually harmless), may be coated with a lacquer film. Highly specialized machinery, knowledge of bacteriology and food chemistry, and more efficient processes of cooking have combined to make the commercial canning of food extremely successful. Still, canning leads to a loss of nutrient value in foods, particularly of the water-soluble vitamins.

Exhibit 2 (continued)

Filling System Technology

Unlike its competitors' blank-fed systems, which started with precut and pre-printed carton blanks that were formed into boxes, then filled traditionally and sealed, Tetra Pak machines started from rolls of printed materials that they formed into continuous tubes, cut and filled into pouches, sealed, and then folded into boxlike cartons. The Tetra Pak process made it difficult to handle products with a high particle concentration.

UHT Technology

UHT, or ultrahigh temperature, is a process by which liquid food is "flash-sterilized" (*i.e.*, subjected to high temperature for a very short duration) before being poured into aseptic containers in a sterile atmosphere. The UHT process eliminates all microorganisms that might cause spoilage and reduce shelf life with a minimum alteration of the product's organoleptic qualities, at least as compared with traditional sterilization.

Source: Company literature and public sources

Exhibit 3
Statistics from the Food Can Industry

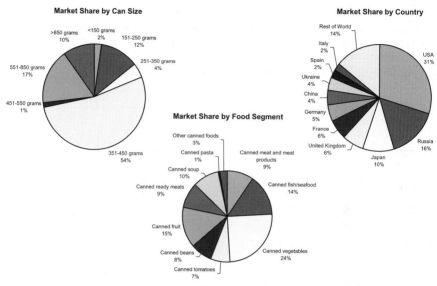

Source: Euromonitor

Exhibit 4
Junior

Source: Company literature

Exhibit 5
Efficiency vs. Traditional Food Can

	Can	Phoenix	Result
Filled packs per pallet	1,152	1,920	+67%
Filled packs per truck	49,536	57,600	+15%
12-tray footprint	23 × 31 cm	18 × 30 cm	-35%
12-tray weight	5.5 kg	5.1 kg*	-8%
Package weight	45–60 g	17 g*	-60–70%

*With 400 g content
Source: Company literature

Exhibit 6
Overall Cost Breakdown of Tetra Recart System

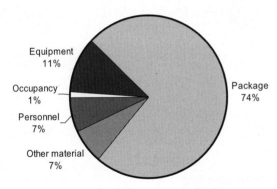

Equipment 11%
Occupancy 1%
Personnel 7%
Other material 7%
Package 74%

Source: Company literature

Exhibit 7
Tetra Recart Project Path – 1993 to 2001

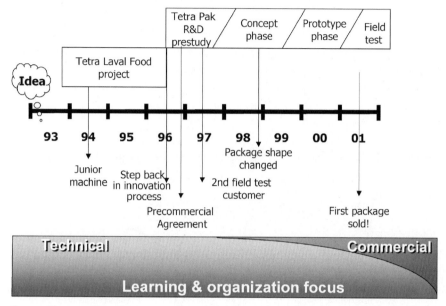

Source: Company literature

Exhibit 8
Results of an in Home Test (Germany): Can vs. Tetra Recart
(% Who Agree Completely with the Attributes Indicated)

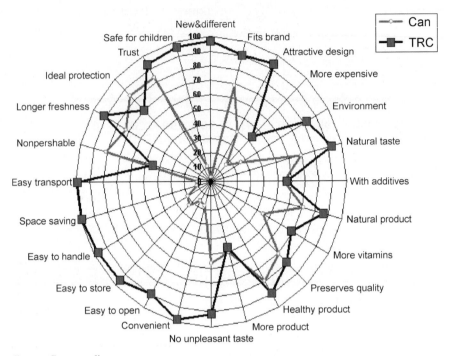

Source: Company literature

Exhibit 9
Can Price Fluctuations

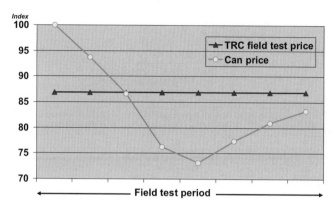

Source: Company literature

Exhibit 10

Steering Group Composition			Tetra Recart AB Board Members
1999	**2000**	**2001**	
Kjell Andersson, MD Tetra Pak R&D	Kjell Andersson	Stefan Andersson	Stefan Andersson
Stefan Andersson, CTO	Stefan Andersson	Mark Atkins, MD Tetra Pak R&D	Nils Björkman
Sven Andren, R&D	Sven Andren	JeanLouis Cheyrou	Göran Harrysson
JeanLouis Cheyrou, MD Tetra Pak France	JeanLouis Cheyrou	Göran Harrysson	Bo Wirsén, Chairman – Tetra Recart AB Board
Lars Danielsson, Director Tetra Pak Converting Europe	Lars Danielsson	Gunther Lanzinger	
Gunther Lanzinger, Phoenix Project Manager	Michael Kenyon, Legal Affairs	Jan Juul Larsen, MD Tetra Recart	
Bo Wirsén, Chairman – Steering Group	Gunther Lanzinger	Bo Wirsén, Chairman – Steering Group	
	Joakim Rosengren, General Manager- Phoenix		
	Bo Wirsén, Chairman – Steering Group		

Source: Company literature

Exhibit 11
Organization Chart – October 2001

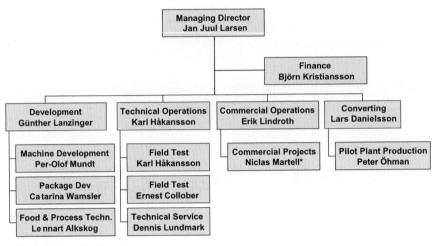

* Hired from outside Tetra Pak
Source: Company literature

Exhibit 12
Tetra Recart Sales Targets

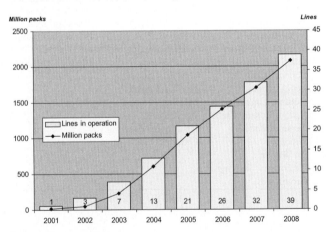

Source: Company literature

Exhibit 13
Tetra Recart Sales Cycle

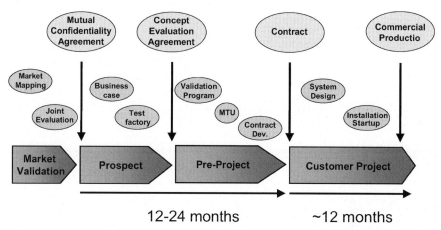

Source: Company literature

CASE 6-7
Innovation and Renovation: The Nespresso Story

Kamran Kashani and Joyce Miller

The business has two – and only these two – basic functions: marketing and innovation. Marketing and innovation produce results; all the rest are costs.

Peter Drucker
Management: Tasks, Responsibilities, Practices

To just keep pace in this industry, you need to change at least as fast as consumer expectations. That's renovation. To maintain a leadership position, you also need to leapfrog, to move faster and go beyond what consumers will tell you. That's innovation.

Peter Brabeck-Letmathe, CEO, Nestlé S.A.

Individualisation is a driving force in today's markets – in everything from the telephone to beer to tea to personal computers. The Nespresso System is an innovative concept that offers consumers individual portions of freshly-ground coffee in a range of tastes that result in an exceptional cup of espresso every time. We're convinced of the power of the idea and the technology behind it, and we're aiming to grow this business eightfold to CHF 1 billion in the next decade.

Willem Pronk, CEO, Nestlé Coffee Specialties

Copyright © 2000 by IMD – International Institute for Management Development, Lausanne, Switzerland. Not to be used or reproduced without written permission directly from IMD.

1 Introduction

Nespresso was the leading brand in portion coffee, a small but fast-growing category in the world coffee market. The product consisted of high-quality coffee packed in aluminum capsules for exclusive use in specially designed machines. Developed at Nestlé, the Nespresso System, as the capsule-machine combination was called, offered the consumer a refined quality and individualized cup of espresso coffee with speed and convenience at the push of a button. In the stagnating global market for coffee, Nespresso had created a new growth segment.

This case is about the history of Nespresso's development and the challenges for its future. *Part I* describes Nestlé's strategies for growth through innovation and renovation and traces the history of Nespresso from its uncertain beginnings to becoming one of the fastest growing businesses in the company. This section also describes the management's practices and views on promoting a climate for innovation. *Part II* explains the challenges facing management as the company aimed to grow the Nespresso business toward the CHF 1 billion level.[1]

PART I

2 Nestlé and the History of Nespresso

Established in 1867 in Vevey, Switzerland, by 1999 Nestlé was the world's leading food company, employed over 225,000 people, had sales of more than CHF 70 billion, and operated 495 factories in over 80 countries. Peter Brabeck, who took over the reins in June 1997 from long-time CEO Helmut Maucher, believed that continuous innovation and renovation would be the keys to meeting the company's ambitious 4% growth target in an industry where the norm was half that level. While changing demographics and lifestyles in industrialized countries, where Nestlé generated the bulk of its sales, meant opportunities for new business development, these markets were also increasingly characterized by discriminating consumers, powerful retailers, growth of private labels, and pressure on prices and margins.

In seeking new growth markets for its existing products, Nestlé had recently launched its strategic brand Nescafé, the pure soluble coffee that the company had invented in 1938, into the much wider area of ready-to-drink coffee beverages. Similarly, Nesquik, the company's global brand for chocolate powder, was extended into new liquid forms. In addition, a completely new product concept, a remineralized table water marketed under the brand Pure Life, was successfully tested in a number of markets in Asia and Latin America. For Brabeck, these were

---

[1] In 1999, US$ 1.00 = CHF 1.50 (one Swiss franc = 100 centimes).

examples of new concepts that combined technological and market innovation. Brabeck commented:

> The success of leading brands is proportional to their capacity for innovation and renovation. Innovation is taking something into the next category or inventing a whole new category. Renovation is keeping the product in the same category and improving it. These concepts are not only technical and production-related, but must be applied throughout our business: from organisational structures to finance, administration, marketing, communication, *etc.* To make a quantum jump into another direction, we need to look for new opportunities and we must be prepared when they arrive.

In Brabeck's views, certain management qualities were prerequisites to attaining the company's growth targets. These qualities were used as criteria by senior management in appointing individuals to important positions (*Exhibit 1*). Although Brabeck was patient, he was not complacent about his objective of long-term growth through innovation and renovation. He remarked:

> Nestlé has the lowest risk profile that a company can have. While we get pressure from the financial analysts to do things differently, the market is finally beginning to understand. We're not interested in compromising long-term shareholder value for short-term share price maximisation. We have a long-term vision for growth.

In generating growth through highly differentiated new products, Brabeck wondered what policies and practices might be needed at Nestlé to promote a climate for change and innovation. In the same context, he wondered what lessons the Swiss giant could glean from its long learning curve in bringing the Nespresso System to market.

3 Nespresso System: Product Innovation Against the Odds

The Nespresso System consisted of individually portioned aluminum capsules containing 5 grams of roast and ground (R&G) coffee made for exclusive use in specially designed coffee machines (*Exhibit 2*). Hermetically sealed to preserve the freshness of the coffee for six months after production, the Nespresso capsules also had the advantage of being clean and easy-to-use compared to traditional, hand-measured espresso coffee. Each capsule corresponded to a single cup of coffee and worked only in Nespresso machines. Consumers could choose from eight available coffee varieties (*Exhibit 3*).

The outward simplicity of the Nespresso technology masked its complexity. The coffee-making process involved three stages: *prewetting* (where water was sprayed on the coffee to expand it), *aeration* (where air was passed through the coffee to create small irrigation channels), and *extraction* (where water flowed through the coffee at optimal pressure and heat). The extensively researched Nespresso machine contained a chrome-plated capsule holder equipped with a built-in opening and filtering system, a microchip-monitored pump that delivered varying levels of pressure depending on the coffee blend, and a high-precision thermoblock system to continuously heat water in the ideal range of 86° to 91° centigrade. This extraction process was seen to be highly innovative, incorporating years of research and learning that would be difficult for a competitor to replicate without patent infringement. The Nespresso concept currently outperformed all competitors in quality and convenience.

Nestlé held the patents for the Nespresso machines but licensed their production to manufacturers who distributed and sold them through selected household appliance retailers. The manufacturers also offered after-sales service. Nestlé did not make any money on the sale of the machines. It was only the sale of the capsules that determined the profitability of the business. Industry analysts estimated that profit margins on the capsules could be as high as 50%.

With its distinctive brand, luxury image, and focus on the small portion of the espresso coffee segment, Nespresso represented a major departure from most Nestlé lines of business reputed for large-scale production and mass marketing. Outsiders sometimes remarked that it was a company for "lifers" – managers who stayed with the company their entire careers.

3.1 The R&D Push

The technology behind the Nespresso System actually originated in Geneva's Battelle research institute. Nestlé acquired the rights to commercialize the idea in 1974. At the time, Nestlé dominated the instant-coffee market, which accounted for 30% of coffee consumption worldwide. Instant coffee generated over 80% of the company's coffee sales and commanded large promotional budgets and managerial attention. On the other hand, Nestlé had no significant presence in the much larger R&G segment, which represented 70% of total world consumption. Furthermore, while the total coffee market was stagnant, even declining in some countries, management realized that a growth opportunity in the "gourmet" segment of the coffee market was rapidly developing in the USA and elsewhere. In the same way that Grand Cru appellations distinguished top-quality wine from everyday wine, "gourmet" referred to premium-end coffee (*e. g.,* espresso), where special care and attention were devoted to the coffee variety, roasting process, and brewing phase.

The aim of the Nespresso product development was to combine Nestle's R&D strength with its deep knowledge of the coffee business to bring a high-quality coffee product to the market. But a number of technical problems had to be overcome first. For example, the projected production costs were too high, the freshness of the enclosed coffee could not be maintained, and the quality of espresso[2]

---

[2] A "good" cup of espresso was distinguished by its fine taste and its "crema," the visually distinctive creamy foam on the surface, which coffee connoisseurs considered an essential part of the espresso experience.

was inconsistent from cup to cup. Nevertheless, a few people in Nestlé's R&D organization firmly believed in the project and pushed it forward. They improved both the coffee capsule and an espresso machine later produced by Turmix, a Swiss manufacturer. Other problems were resolved in time.

The project gained early support from Nestlé's food service division, which saw Nespresso as an entry into the restaurant market, thereby positioning Nestlé more strongly in this segment. But in 1982, following a market test with eight automatic machines in Swiss restaurants, that strategy was abandoned in favor of the office coffee sector. Sobal – a Swiss company that distributed appliances and was already present in office coffee – was approached to distribute the Nespresso System (the machine and the capsules) in offices and other institutions.

3.2 Nespresso: A "Satellite" Outside of Nestlé's Main Coffee Business

To further develop, produce, and market the Nespresso System, a separate company was created in June 1986 as a 100%-owned Nestlé affiliate, located across the street from the company's headquarters. It was thought that the new unit, later called Nestlé Coffee Specialties (NCS), would be able to move faster in seizing the market opportunity in the newly created individual-portion-coffee category. Exclusively responsible for promoting products under the Nespresso brand name, the company could develop its own commercial, distribution, and personnel policies separate from Nestlé. A special production line was opened to make the coffee capsules at the Nescafé plant in Orbe, Switzerland.

Camillo Pagano, at the time senior executive vice president in charge of several worldwide strategic product divisions and business units, commented:

> I saw this contraption in Orbe that somebody said was going to be fantastic. I discovered that the R&D team had developed this system without even really talking to the marketing side. There is no doubt that technical development can bring innovation. But internally, there was a lot of scepticism about the possibility to commercialise Nespresso. The business was physically moved out of Nestlé so that it could establish credibility and so that it didn't have to fight against all the company's rules.

> People thought I was a 'nut' to spend so much time on this small thing and to support the idea. Nespresso is so different from what the company does in its day-to-day business. I felt that developing this business would take time and patience. You need champions at the top for a new idea. You need to give an idea support against criticism. Any innovation immediately hits resistance in an organisation. Small satellites like this can help people gain insight into how a business could be developed differently. These offshoots provide an opportunity to train and test people. If they make a mistake there, it's not so costly.

The Nespresso System was first launched in Italy, the world's largest espresso-drinking nation, and in Switzerland, Nestlé's home market. Japan was also later included as it was one of the world's fastest-growing coffee markets. Sales initially focused on the office-coffee sector. The Nespresso machines were designed internally by Nestlé and a Swiss design company and manufactured under license by Turmix for European markets. Sobal purchased the machines from Turmix and the coffee capsules from NCS and then sold the two together as the Nespresso System. Additionally, several companies that distributed vending machines in the office sector added Nespresso machines to their product lineup. It was thought that, compared with households, the office segment would be less sensitive to the relatively high unit price of the machine which, at the time, retailed for close to CHF 800. A box of ten coffee capsules sold to offices for CHF 4.

3.3 Worrying Signals from the Marketplace

By the end of 1987 only half of the machines that had been manufactured were actually sold. The company was behind target on sales for both the machines and the capsules. And without the machines, further coffee capsules would not sell. There were other worrying signs as well. Machine defects were absorbing a large part of the maintenance and service budgets, which Nestlé was covering in total. Slow sales in nontraditional espresso markets like Japan were also deepening the losses. Arguments were made to bring in someone from the outside with new ideas to save Nespresso from an early death. In 1988, Yannick Lang, a former Philip Morris executive, was recruited to the venture.

4 Rethinking the Strategy of Nespresso

Swiss-born and US-educated, 33-year-old Lang came into the Nespresso team with a reputation for flair and creativity and a solid commercial and marketing track record, having catapulted the Marlboro Classics clothing line from CHF 20 to CHF 100 million within a few years. Pagano recounted:

> The Nespresso business was at the point where we needed an entrepreneur to take it further. We needed to find somebody who wouldn't react like a Nestlé manager. People in our organisation are good, but at the time, everyone was asking, 'How could we sell this thing in supermarkets?' Nestlé doesn't bring people in from the outside as a common practice, but we needed someone who understood what it meant to sell a premium system – something between Louis Vuitton fashion accessories and Maggi bouillon. This needs a special mentality. What really convinced me was Lang's background, coming from Philip Morris where going into the clothing business was also a stab in the dark; it was not the usual thing to do.

Rupert Gasser, executive vice president in charge of technical, production, environment, and R&D, to whom Lang subsequently reported, remarked:

> Lang was purposely hired externally rather than internally. He was ambitious and strong-headed. He had a strong drive for recognition. He wanted to do something outstanding. Lang had personality; he was a force. And importantly, he did not carry all the trappings of the company history. He was outside of Nestlé's main coffee structure.

4.1 A Strategic Shift to the Household Market

The idea of spearheading the development of Nespresso – a small operation in a separate company with a separate product – was very appealing for Lang. But the team of eight people he inherited were acting in a "skunkworks," and the business was losing money in the office segment with flat sales. Lang soon concluded that the prospects for the Nespresso System in the office sector were limited, but that there was potential, albeit unproven, in the household market. Lang thought that households headed by well-educated, affluent 35- to 45-year-old men and women who enjoyed drinking restaurant-quality espresso at home could constitute a profitable segment for the Nespresso System. Market trends evident in the late 1980s supported this idea. Cafés and bistros were opening up across Europe, and the success of gourmet and specialty coffee chains in the USA (*e.g.*, Starbucks, The Coffee Beanery, and Caribou) had led espresso to be perceived as a trendy, socially elite drink. The vast majority of espresso drinkers tended to be city dwellers with discerning tastes in food. The growth in popularity of Italian lifestyle, cuisine, and fashion had also fuelled interest in this method of extracting what was said to be the best aromas and positive components of coffee [1].

Lang believed that to build long-term business with such discerning consumers, Nespresso had to be in the household market. Lang presented his revised strategy to Nestlé's general management in early 1989. Despite substantial internal skepticism about the prospects for the Nespresso System, he received the "green light" to launch the household strategy, but only in Switzerland. The rationale was that it was a high-risk strategy that still needed to be proven, and that a single market could be shut down more easily than multiple markets. If Lang could deliver on his target to triple the sales volume within a year, the top management agreed, then the business could continue.

Very little market research on the household market existed at the time. What research had been done indicated low potential for the Nespresso concept. Some data showed that the perceived consumer value of the coffee capsule could not exceed the low figure of 25 centimes (as opposed to the target consumer price of 40 centimes). A market test with five upscale household appliance stores in Switzerland was similarly disappointing. The objective was to sell 100 Nespresso machines through intensive retail merchandising and demonstration. The actual sales, however, were less than half of the target. To avoid being shut down, Lang interpreted the market test results he presented to corporate management rather liberally, *i.e.*, his presentation showed a more favorable picture than the evidence suggested.

Although most people in Nestlé considered the commercialization of Nespresso to be impossible at the time, the Nespresso team was completely convinced of its potential, despite the grim market research findings. Pointing to the failure of the first market tests for fax machines and mobile phones, their attitude was that consumer research tends to reflect past experience – which does not necessarily lead to innovation. The continuing conviction of Lang and his team was based on intuition and a strong personal belief in the Nespresso System.

4.2 Positioning Nespresso at the Top

Lang's new strategy involved positioning Nespresso away from the more utilitarian office coffee and targeting consumers at the top of the household market. Nespresso machines would be produced in Switzerland under exclusive agreement with a leading manufacturer of espresso machines. Nespresso machines were sold through selected household appliance retailers, department stores, electrical shops, and kitchenware outlets. Nespresso machines retailed in the range of CHF 350 to CHF 900. Retailers earned a 25% to 35% margin on the sale of the machines. Machine makers earned a 30% to 35% margin, typical for a product of this type. Again, Nestlé did not profit from the sale of the machines. To accompany his global ambitions, Lang subsequently developed "machine partnerships" with several international producers and distributors of household appliances like Matsushita, Krups, Philips, Alessi, Jura, and Magimix.

During this period, Lang and his team introduced modifications to the design of the coffee capsules that cut material costs while making them recyclable.

Gasser commented on the new strategy:

> There were not many people in the company who believed in Nespresso, but Lang did. He was totally convinced of the opportunity. Nespresso was purposely run at arm's length and not built into Nestlé's main coffee structure. Our CEO challenged Lang a lot. He found the challenge motivating; he liked it.

> Individually-portioned coffee was an idea mainly pushed forward by R&D. There was a lot of internal criticism at the time, but the project got support from Nestlé's CEO, who dared to do something different. When something is new, it will always meet with resistance. There will always be a lot of ifs and buts. Most people didn't think Nespresso would be a revolutionary idea. There was a concern that it would distract us from our core business in instant coffee. It was seen as competition to instant coffee.

4.3 Debut of the Nespresso Club

The shift to the household market soon led to a reevaluation of the strategy for distribution. The idea of channelling capsule sales through supermarkets was explored but rejected as Lang felt that this would simply transfer the profitability of the business away from Nespresso. An earlier attempt to sell the coffee capsules in US food outlets had actually failed because the consumer base had not been broad enough and retailers had been left with a considerable stock of stale coffee capsules. Experience showed that it could take up to three months for the Nespresso capsules to arrive on store shelves, cutting in half the time remaining until the best-before expiry date. With such a short shelf-life, quality could not be assured.

As there were still some nagging problems with the reliability of the Nespresso machines, and quick service turnaround was considered critical, Lang seized on the idea of using a direct marketing channel to stay close to the consumer. Turning a technical constraint into an elegant marketing solution, Lang conceived and launched the Nespresso Club, which, in addition to handling service calls, offered consumers:

- *Around-the-clock order taking*: Consumers could telephone a national call center 24 hours a day, 7 days a week. Orders for Nespresso capsules could be placed by toll-free telephone, fax, mail, and, eventually, over the Internet.
- *Prompt delivery of fresh coffee*: Within two business days.
- *Personalized advice*: Trained coffee specialists were on hand to advise consumers about the different flavors and provide technical assistance on the machine. Club members could also benefit from recipe suggestions, Nespresso accessories, and special offers and get information about new coffee blends.

Lang also put in place some stringent operating rules. For example, in the Nespresso Clubs, telephones had to be answered within three rings, and he insisted that staff and managers dress in a way that reflected well on the luxury image of the system they were selling.

The Club was an immediate success. Purchasers of the Nespresso machines, sold through household appliance stores, automatically became members of the Nespresso Club. The Club concept was the first direct marketing experiment within the larger Nestlé organization. In 1990 there were 2,700 Club members in Switzerland, France, Japan, and the USA. In that year, the stretch sales targets set a year earlier had been surpassed. Nespresso Clubs were later established in Germany and the Benelux countries (1992) and in Spain, Austria, and the UK (1996). The Nespresso System was also available through agents in the Middle East, southeast Asia, and Australia.

4.4 Spreading the Word

As part of the strategy to further internationalize and position the Nespresso System as a premium product, the company sought the patronage of British Airways, Cathay Pacific, and Swissair, among others, which began serving Nespresso coffee on board their long-haul first-class flights. Top restaurants, mostly in France and Belgium, were also solicited. In countries where Nespresso Clubs existed, heavy investments were made in training sales clerks in retail stores where Nespresso machines were sold. Experience showed that sales clerks needed to give a high level of attention to consumers in order to sell the Nespresso machines. The training and financial incentives were seen as keys to supporting and motivating retail clerks to demonstrate and sell the machines.

Despite the step-by-step international growth, no major advertising or public relations campaigns were undertaken. The Nespresso Club was the chief means of communication. The company relied mainly on word of mouth on the part of its extremely loyal consumer base to promote the Nespresso System. Alfred Yoakim, who had been part of the original Nespresso product development team, said:

> Advertising is good for the image. It can reassure consumers and communicate an image of quality. But I'm not convinced it will make a consumer change to espresso and buy the Nespresso System. For this, we can only rely on word-of-mouth.

Rupert Gasser concurred:

> We don't need mass media or television. We're targeting the crème de la crème. We need simpler means, unconventional ways of reaching new consumers.

4.5 Achieving Growth and Profits

Under Lang's leadership, NCS achieved breakeven in 1995 and subsequently became one of the fastest-growing business units in the Nestlé organization. By 1997 Nespresso Club members numbered 220,000 (reflecting a 30% average yearly increase), the company's ristretto variety had won special recognition from the International Institute of Coffee Tasters, and the company had installed a second production line for the capsules and was manufacturing them around the clock.

Lang was known as a tough manager by the team of "young tigers" he had built, some of whom he had hired away from first-class consumer-goods companies. Lang was also known as a strong-willed person who did not give up easily. People recalled that every win gave him a boost. He resisted all efforts to apply methods that might lead to risk avoidance as he felt the Nespresso business had to go on taking risks. However, some who worked closely with

Lang found his strong-headedness a barrier. In the words of one of his associates:

> He was passionate about his own ideas but not those of his colleagues. He was impatient, wanting to make things happen in no time. He was a tough boss, a difficult person to work with and manage.

5 Creating a Climate for Innovation

Different opinions were expressed regarding what the experience of bringing the Nespresso System to market meant for the larger Nestlé organization. Lang believed that he had created an environment where his team could take intelligent risks and where everyone could be an innovator. Lang felt that if he had stuck to an "armylike structure," Nespresso would have failed, as he believed that anything militarylike would stamp out innovation. For him, having an infantry that marches when it's told to march and assumes that the chain of command knows best would not lead to innovation. Lang was known to accept structure where it served his purpose and "helped people to do a good job." He rejected it in instances where "it stopped people from innovating."

Lang went to great lengths to keep the Nespresso concept alive, sometimes doing things at the edges of the Nestlé organization. He also established privileged contacts with high-level members of the top management. Former executive vice president Pagano, who before his retirement at the end of 1991 held private discussions with Lang, remarked:

> Nespresso was developed as a totally innovative system. Lang crafted a strategic image that was consistently carried through. At one time, he was so convinced of the concept's potential that he even tried to buy Nespresso.

Pagano felt that an individual's personality was tremendously important with respect to innovation:

> It's not necessarily the inventor of an idea who has the capability to take it forward. This needs courage, leadership, temperament, and charisma. An organisation also needs to build in the right to make mistakes. This message is not always going down through the 'bunker' of our middle ranks. It's stuck or not being properly explained. Nestlé needs people who can take an idea, believe in it, and bring it to fruition against the odds.

Rupert Gasser, who had overseen the entire Nespresso development and believed in giving broad responsibility to risk takers, the "youngsters" as he called them, offered his insights into the innovation process at Nestlé:

> Our problem is not a lack of good ideas. We have too many of them. The key is to be able to extract an idea, carry it forward in the organisation, and transform it onto the supermarket shelf. The real barrier to innovation is perceived risk. People ask themselves, 'Why should I engage in a process with an uncertain outcome?' People reject risk-taking because they feel it can endanger their current status, their jobs. At Nestlé, we

need to create a risk-friendly environment where people don't feel unduly exposed. We must accept that people make errors. But failure due to trying something new shouldn't be a career risk. Sometimes you have to shelter the real innovators, the risk takers.

At the same time, Gasser believed that innovators needed to be occasionally challenged to make sure they had not ignored important details. He said:

Sometimes people who push for innovative ideas forget some basic things. But by being challenged in a dialectic process, they improve their solutions. It's a positive process.

Gasser recognized that innovation in a large corporation like Nestlé was a complex process:

Growth through acquiring brands and companies is relatively low risk. Building new business is different; it's more complex and carries higher risks. In our company, we must be able to get R&D, manufacturing, and the local Nestlé market companies to work in close proximity. This needs to be a constructive process; otherwise, innovation will not be successful.

He agreed that innovation and renovation were both necessary for Nestlé[3]:

To stay viable in the long term, we need both innovation and renovation. Innovation is a quantum leap that is not necessarily only related to products. It can also be related to manufacturing process improvement or trade channels. Renovation, on the other hand, is more incremental; it allows you to sharpen your claws every day.

At Nestlé, identifying high-potential individuals who would foster change and innovation was a senior management responsibility. Brabeck personally followed the careers of several individuals with the aim of picking the right people for tough assignments and helping them to grow in their jobs. He believed that Nestlé needed managers to keep the core businesses running who would assure order and discipline and provide security and longevity. But he also saw the need to provide space for what he called "change drivers." But this change was complex. Brabeck explained:

You can't impose change from the top. You can only create an environment that stimulates change. Many managers in large institutions have been trained to keep things running as they have been. They have learned to comply with an enormous number of detailed procedures and systems. They were taught by experience that they are better off following the expected and accepted tracks of routine rather than venturing out into the new and unexpected.

We need to create a climate where there is a certain freedom to fail, and where those people who are promoted have made decisions and carried them out, even if they were not always 100% successful. We don't want to advance the careers of those who have never made a mistake because they've never done anything except apply the rules. We have to identify, foster, and mentor people who have proven that they're willing to stick their neck out, who made a mistake, learned from the mistake, and are willing to continue taking risks.

---

[3] Innovation and renovation referred to as radical and incremental changes, respectively.

On the other hand, Brabeck observed:

> If you're only being innovative and creative, but not producing reliable products, you'll soon be out of business. For Nestlé, the quality and security of our products is fundamental.

For Alfred Yoakim, the head of the Nespresso development team from its early days, there were lessons on innovation from Nespresso's experience:

> By staying in an ivory tower, you can improve an existing product, but you can't innovate. To innovate, you need the freedom to make your own decisions, to not follow the old rules. Marketing people can innovate because they are free of technical restrictions. Technical people can innovate because they are free of commercial considerations. What we have to do is put the two together. There's still room to innovate in the espresso market. The only limit to innovation is our imagination.

6 Nespresso Under New Leadership

Lang's tenure at NCS was not without controversy. He was thought by some to be a maverick, an "out-of-the box" manager who forcefully pushed for his ideas, and who found being challenged by his superiors motivating. Gasser called him a real force behind the Nespresso phenomenon and a "rare bird." Lang was subsequently nominated to a select list of managers whom Nestlé considered as "corporate assets." Late in 1997 an opportunity came along for Lang to move to the USA to work on Nestlé's large but underperforming ice-cream business. The appointment was considered by everyone, including Lang himself, a timely career move. However, less than six months later, Lang announced that he was leaving Nestlé, having been headhunted by another food company.

In the eyes of Nestlé's top management, NCS had now reached a critical stage in its evolution needing to shift gears for sustained future growth. In the words of Gasser:

> The challenge for any small, innovative organisation is to make such a transformation. Nespresso needed structure, better operational efficiency, control of costs, quality, and inventory, in addition to modern personnel policies. As the company grew, these issues became increasingly important.

Willem Pronk, 44, a career-long Nestle manager, was appointed to spearhead the transformation at NCS. He was considered to be a very structured manager and a good marketer.

For Nestlé's CEO, Nespresso's development reflected solid technical know-how combined with the marketing spark of an individual embedded in a supporting organization. Brabeck commented:

> There was a long learning curve to transform the Nespresso innovation into an acceptable, perceivable end-product for the consumer. With Pronk, we've agreed to take the business from the CHF 150 million where it stands now to CHF 1 billion within the next 10 years. Then it becomes an interesting business for a company the size of Nestlé.

At the start, we realised that we had to separate this business so that it could develop its own life. At a certain point, a different management style was needed. Not everyone is good for all phases of a business. As a group, Nestlé has the resources to cover the whole spectrum. In a crisis moment, we may bring someone in from the outside, but normally we try to avoid this. We're not looking for saviours. We look for people who will become long-term collaborators at the service of the company.

PART II

7 Towards the One-Billion Target

7.1 Pronk Takes the Helm

In his own words, Willem Pronk was "born in Nestlé." Over the course of his 20-odd-year career at Nestlé, Pronk had developed a reputation for being an innovator and a risk taker. While managing Nescafé in the Netherlands, for example, he had introduced a new "*café à la carte*" concept to rejuvenate the brand's stagnating business in that market. Despite the skepticism of his superiors, he had allocated an entire marketing budget to promote the new concept: a selection of different Nescafé blends in a stylish wooden box. The concept worked with spectacular results, pushing sales up by more than 200%. This and other moves brought Pronk to the attention of headquarters. He quickly began to climb the Nestlé ranks. Reflecting on his early days in Nestlé Netherlands, Pronk said:

> As Peter Brabeck says, 'If we had continued to do what we'd always done, we would have continued to have what we always had – an unsatisfactory performance.' I pushed for trying something new. If it didn't work, the worst case scenario was that I would be fired. But so what? People ask me if it's frustrating to work in a multinational like Nestlé. My answer is: No, I'm probably more frustrating to other managers than they are to me. In a big company like Nestlé, you have access to resources and all the sources of innovation. You can either accept the system or challenge it. There are no hard-and-fast rules. I'd rather ask for forgiveness than permission.

Soon after arriving at NCS, Pronk began working to enhance the consumer service side of the business. The Nespresso Club database was segmented according to consumption habits and the length of membership, and targeted communications were launched. Fifteen to twenty percent of Nespresso capsule orders currently came from small offices. "New" members (less than a year) were tracked closely as it was felt that their long-term consumption habits were formed by usage patterns established in the first few months. A Nespresso Club staff member personally followed up on "heavy" capsule users (more than 100 capsules per month) who did not place an order by the computer-anticipated date to ensure that their machine was functioning, take orders, and answer questions. When machines needed service, the Club arranged for free home pickup and return of the repaired appliance. The machine manufacturer did the repair. A Nespresso machine was loaned free to members during the period.

Pronk believed that the above renovations, among others, were needed to create customer intimacy, build long-term loyalty, and put Nespresso on the road to becoming a meaningful concept in the competitive world of coffee. (*See Appendix for background information on the world coffee market.*)

7.2 Issues for Growth

With its ten existing Nespresso Clubs in 1999, NCS was receiving 7,000 orders worldwide for coffee capsules each day and had the capacity to produce 350 million capsules annually (*Exhibit 4*). Close to 100,000 Nespresso machines had been sold in the past year alone, bringing the installed base up to about 350,000. New consumers were also coming on board (*Exhibit 5*).

While the Nespresso concept had created and dominated an entirely new gourmet coffee category – the portion-coffee market – several issues still had to be resolved if NCS was to accelerate its growth. Pronk was very concerned that NCS had, to date, only achieved an average awareness rate of less than 5% in its international markets and penetrated less than 1% of households. Some asserted that these ratios were significantly higher among upscale households, the current target segment for Nespresso. Nevertheless, the fact remained that very few coffee consumers actually knew about the Nespresso System.

Pronk saw the individualized-portion concept on which the Nespresso System was based as a vital means for rejuvenating the mature coffee market. But little market research had been undertaken to date. Pronk remarked:

> The quality of the future depends on the quality of the plans you put into action today, and the information on which they are based. That's why market research is key. For me, this isn't about getting reports; it's about being present during the interviewing and being totally immersed with the consumer. We need to know three things: Why do people buy our product? Why do they buy a competing brand? And why don't they buy any product in this category? Then we can look at the three ways to grow the business: sell more to current consumers, sell to consumers of the competition, and sell to noncategory users.

Looking to the future, Pronk was concerned about five issues: how to attract new consumers, whether to introduce a lower-cost system, how to serve a wider market, whether to diversify into noncoffee products, and how to preempt competitors.

7.2.1 Attracting New Consumers

Most people drank espresso "away from home" in restaurants, hotels, and bars and from vending machines in offices and public areas. Home consumption was limited, in part because of the perceived high price of espresso-making machines. In Italy, espresso was inexpensive at 1,000 lire (CHF 0.75) a cup and widely available at the neighborhood bar. Only one in five Italian households consumed espresso at home. In France, the world's second-largest espresso-drinking nation, 66% of adults regularly drank espresso but only 12% drank espresso at home. In northern countries, like the UK and Germany, espresso drinking was in its infancy, but growing. Most US households still used electric drip coffee makers.

Less than 1% of US consumers had an espresso coffee machine at home [2]. Pronk reflected:

> The Nespresso System guarantees high quality and offers consumers individual choice, consistency, and convenience. These are the values of the future. One type of coffee for all is an old-fashioned concept. But despite the success of the Nespresso Club and our comprehensive knowledge of consumer preferences as a result of it, our awareness remains low in priority markets. Word-of-mouth is a slow process. We have to find other means of making the Nespresso concept known.

One idea that had been implemented a few years ago was to give free machines and coffee capsules to opinion leaders, like politicians, journalists, *etc.* in key markets. This was thought to be less expensive than a media campaign.

Until now, little advertising had been used to raise consumer awareness about Nespresso. Old-timers like Alfred Yoakim were strongly against the idea. He said:

> Advertising won't do us any good. It's just a waste of money.

Camillo Pagano was also cautious about the new growth strategies being discussed. He said:

> I'm worried about the mass marketing mentality that is entering the picture. In this business, you can't talk about household penetration and market awareness.

To promote the Nespresso concept and sales of the machines, NCS continued to invest significant resources (upwards of CHF 200 per machine sold) in training retailers and doing merchandising and in-store promotions. Recent studies had shown that retail demonstration accounted for 40% of new converts to the Nespresso System. The remaining 60% of new customers came through word of mouth. In spring 1998, 2,100 Nespresso machines had actually been given away free to Belgian retailers as part of an incentive program initiated by General Electric Belgium. Upon seeing that 55% of these new machine owners never called the Club to order coffee capsules, and a year later only 26% of the target group were active customers, Pronk noted:

> The experience confirmed what we already knew: that people to whom the Nespresso System is targeted must have a desire to drink espresso coffee. If we promote the machines too much, we'll succeed in getting people to buy the machines, but not the complete concept.

7.2.2 Going Downmarket?

By most measures, the Nespresso System was more expensive than alternative methods of making coffee. Compared with the cost of traditional filter coffee machines, the Nespresso machine was five times more expensive. Compared to regular espresso coffee, Nespresso coffee was three times more costly. Yoakim said:

> Like all espresso machines, the Nespresso System is targeted to an affluent group of consumers, the top 10% of households. If they can afford the espresso machine, they can surely afford the coffee capsules. We're not selling a machine and then a capsule. We're selling a result: a cup of espresso. And I want the consumers to get the best result in their cups.

Pronk was adamantly against offering a lower-priced system and the idea of marketing the system under a different brand name. He believed there was still enormous growth potential in Nespresso's current strategy.

7.2.3 Distribution to Serve a Wider Market

In markets like Switzerland, where a critical mass of consumers had been created, retailers had approached NCS about the possibility of selling the high-margin capsules. Going through traditional food retailers would be a departure from the past, a move Pronk was not entirely convinced of:

> If we give away 25% margins to retailers, there won't be much left here to show for profits. Besides, the Club gives us a fantastic knowledge of consumers that would be lost if we went through retailers. On the other hand, if we continue to sell exclusively through the Nespresso Clubs, non-users won't see the product.

Another idea was to showcase the Nespresso concept through a chain of exclusive boutiques established in high street locations in major European cities. It was estimated that each boutique would involve a one-time investment of CHF 250,000.

7.2.4 Diversification into Noncoffee Products

New uses for the Nespresso System based on the "portion" concept had yet to be explored. For instance, individually portioned soups or teas could broaden usage among current consumers and attract new consumers. The idea of noncoffee products was not very popular, however. Pronk and Gasser were both concerned that such a product extension would dilute the original concept.

7.2.5 Blocking Competitors

NCS currently held a 90% share worldwide in the household segment of the portion-coffee market, with Switzerland, France, and the Benelux countries being its most important markets. The remaining 10% share mainly belonged to Belgium's

Malongo, with its 1,2,3 Spresso system, and Italy's Illycaffè, which had yearly sales of US$130 million. Having pioneered the ESE Standard (*Easy Serving Espresso*), Illycaffè had begun supplying "L'Espresso" portions and its own espresso machine to hotels, restaurants, and bars. The company dominated this sector in Italy but was less established in the mass consumer market. Since 1994, Lavazza, a family-run Italian company with US$700 million in annual turnover, had marketed four individual portions, containing 6.5 grams of ground espresso coffee, custom-made to fit its own espresso machines. Close to 80% of Lavazza's sales were in the office coffee channel [3]. These companies had entered the portion coffee market with lower-quality products and espresso-making machines retailing in the range of CHF 300 to CHF 350.

Although the technical complexity of the Nespresso System and related patents were major barriers, the NCS team was convinced that international food companies with their large resources could challenge Nespresso's dominance by extending their long-established R&G brands into the portion coffee segment. In fact, NCS had learned that Sara Lee had recently registered a patent for a portion-coffee concept with a different extraction process, which was as yet commercially unproven. In 1997 Kraft Jacobs Suchard had expanded its *Carte Noire* line in France to include "Espresso Dosettes Filtres." Additional developments were also under way that catered to the single-person households in NCS's target group for espresso drinkers.

Pronk expected a serious entry by one or more players into the high-quality, portion-coffee market where Nespresso was currently positioned. Nevertheless, he believed that NCS had several advantages: advanced technical knowledge, the Club, and a critical mass of consumers in some of its international markets. It was estimated that it would take a newcomer four years to establish a foothold in any national market, and a minimum of 15,000 to 20,000 installed machines to break even.

8 Conclusion

More than 13 years after Nespresso's debut, NCS had seemingly reached a satisfactory performance but still fell far short of Nestlé's high ambition of CHF 1 billion sales. At a time when Nestlé was seeking new growth opportunities, a couple of questions preoccupied those who were familiar with Nespresso:

1. What lessons could be learned about innovation or renovation from the Nespresso story?
2. What must NCS do now to reach its ambitious growth targets?

Exhibit 1
Qualities and Characteristics of a Nestlé Manager

The higher the level of the position and the responsibility of a Nestlé manager, the more he/she should be selected on the basis of the following criteria (in addition to professional education, skills, and practical experience):

- Courage, solid nerves, and composure; capacity to handle stress
- Ability to learn, open-mindedness, and perceptiveness
- Ability to communicate, to motivate, and to develop people
- Ability to create a climate of innovation
- Thinking in context
- Credibility: in other words "practice what you preach"
- Willingness to accept change and ability to manage change
- International experience and understanding of other cultures

In addition: broad interests, a good general education, responsible attitude and behavior, and sound health.

Source: The Basic Nestlé Management and Leadership Principles

Exhibit 2
The Nespresso System

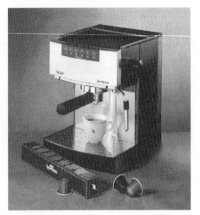

Source: Company literature

Exhibit 3
Available Varieties of Nespresso Coffee Capsules

Ristretto	Composed of pure Arabica beans from Latin America for its finesse and a touch of Robusta for its intensity, it is savoured very short in a demitasse filled halfway.
Arpeggio	The Grand Crus of Central America give strength and fullness to its aroma while a touch of Brazilian Santos harmoniously balances its flavour.
Roma	A blend of Latin American Arabica for delicacy, this gives the strong, intense espresso of Italian "baristas" with a thick, rich cream. It is the ideal base for making a *cappuccino.*
Capriccio	A satisfyingly smooth espresso made from early Latin American Arabicas for a rich and straightforward taste, Brazilian Santos for full and consistent body, and a touch of Central African Robusta for intensity. Top with whipped cream for a true *Viennois.*
Livanto	Made from the best Arabicas of South and Central America, this espresso is rich, intense, and smooth. Savoured in a demitasse or cappuccino cup, with or without milk, it is the morning espresso par excellence.
Cosi	A pure Arabica with all the fullness and flavour of the Grand Crus of Central America, the density of mountain grown coffees from Eastern Africa, and a touch of mildness from Brazilian Santos. With milk, this espresso can be savoured as a *con latte.*
Volluto	A delicate yet full-bodied pure Arabica espresso. The mellow richness of early Latin American beans gives this blend a distinct, elegant, subtle bouquet. Characterised by its mildness, it is ideal any time of the day.
Decaffeinato	Naturally decaffeinated, it has a delicate, balanced, and enticing flavour. Its freshness and intensity come from the great early Arabicas of Central America reinforced by a touch of Central African Robusta.

Source: Promotional material provided to Nespresso Club members

Exhibit 4
Growth of Nestlé Coffee Specialties
Turnover, Employees, and Nespresso Club Members 1989–1999

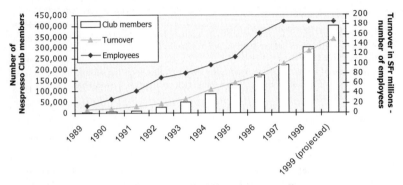

Source: "Nespresso Prodige." *Bilan*, April 1998, and company literature

Exhibit 5
Development of NCS Capsule Sales 1990–1999

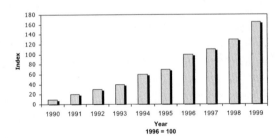

Source: Company literature

Appendix
The World Coffee Market: Mature and Highly Competitive

Globally, the coffee market was in a mature stage. Growth had been achieved only through increased retail prices and growing consumer demand for high-quality, gourmet coffee (*Figure 1*). European countries led the world in per-capita coffee consumption, with Scandinavian nations heading the list (*Figure 2*). The coffee market was split between R&G coffee and instant coffee, and secondarily between the "away-from-home" and "in-home" sectors (*Figures 3 and 4*). Worldwide, R&G accounted for almost 70% of coffee sales or about US$198 billion at average consumer retail price, whereas instant coffee accounted for 30% of sales or US$85 billion (*Figure 5*).

The coffee industry was dominated by food multinationals renowned for their strong brands, aggressive pricing, and high marketing expenditures. Nestlé was actually the world's largest coffee company with a 23% share of the total market (*Figure 6*). The company dominated the instant-coffee segment with a 56% share from its flagship brand, Nescafé, and its numerous product varieties, but the company was ranked fourth in the R&G segment with a 7% share. Philip Morris (Kraft Jacobs Suchard, Maxwell House), whose food business generated over US$27 billion, was the second largest coffee company with a 14% share of the total market. The company was the leader in the R&G segment with a 13% share and second in instant coffee. Sara Lee (Douwe Egberts), with an annual turnover of US$20 billion, was the third largest coffee company with a 7% share of the total market. It was the second largest player in the R&G segment. Procter & Gamble (Folgers Coffee), with an annual turnover of US$37 billion, was the fourth largest coffee company with a 6% share of the total coffee market.

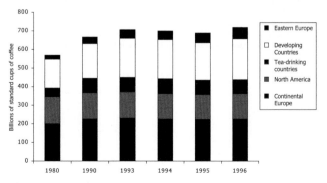

Figure 1 Total Worldwide Coffee Consumption in Billions of Cups: 1980–1996 (all coffee, all channels)
Source: Company literature

In recent years, profit margins for coffee producers had been narrowing as a result of growing competition in the R&G segment. This trend had encouraged companies to enter the smaller, but higher-margin, high-growth gourmet-coffee segment. In 1999, this segment was estimated to be worth US$10 billion worldwide and was growing at a rate of 20%.

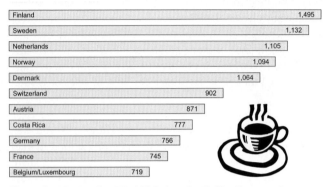

Figure 2   The Leading World Markets for Coffee Consumption
Per-capita Consumption in Total Cups, 1996
Source: Company literature

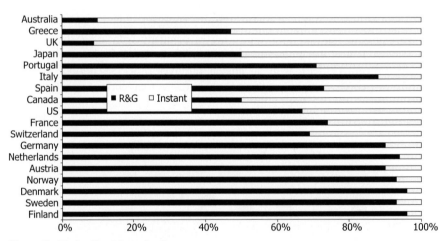

Figure 3   Market Breakdown by Country
R&G *vs.* Instant Coffee, 1995
Source: International Coffee Organisation

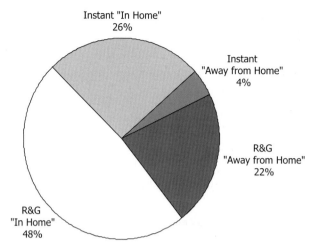

Figure 4 "Away-from-Home" *vs.* "In-Home" – R&G and Instant Coffee (worldwide consumption in Billions of Cups: 1996)
Source: Company literature

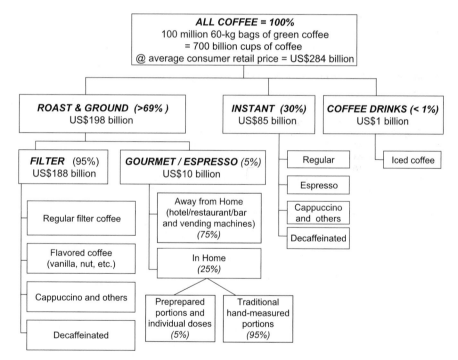

Figure 5 Partial Structure of the Global Coffee Market, 1999
Source: International Coffee Organisation and company literature

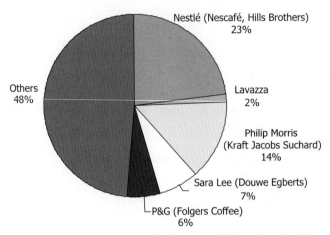

Figure 6 World's Leading Coffee Roasters – Market Share by Volume, 1995
Source: International Coffee Organisation, Encyclopaedia of Global Industries, (Food, Drink, and Tobacco Products), and company literature

References

1   "Coffee International File, 1998 to 2002." *Market Tracking International*: 216–17.
2   Wheeler, Michael. "Coffee to 2000: A Market Untamed." Economist Intelligence Unit, 1995.
3   "Espresso, Where the Leaders Stand." *Tea & Coffee Trade Journal*, November 1998, and "Coffee International File 1998 to 2002." *Market Tracking International*: 219–22.

Index